Seals and Sealing Handbook

Seals and Sealing Handbook

Fifth Edition

Robert Flitney

AMSTERDAM • BOSTON • HEIDELBERG • LONDON • NEW YORK • OXFORD
PARIS • SAN DIEGO • SAN FRANCISCO • SINGAPORE • SYDNEY • TOKYO
Butterworth-Heinemann is an imprint of Elsevier

Butterworth-Heinemann is an imprint of Elsevier
Linacre House, Jordan Hill, Oxford OX2 8DP
30 Corporate Drive, Suite 400, Burlington, MA 01803

Fourth edition 1995
Fifth edition 2007

Copyright © 2007 Elsevier Ltd. All rights reserved.

No part of this publication may be reproduced, stored in a retrieval system, or transmitted in any form or by any means electronic, mechanical, photocopying, recording or otherwise without the prior written permission of the publisher

Permissions may be sought directly from Elsevier's Science & Technology Rights Department in Oxford, UK: phone (+44) (0) 1865 843830; fax (+44) (0) 1865 853333; email: permission@elsevier.com. Alternatively you can submit your request online by visiting the Elsevier web site at http://elsevier.com/locate/permissions, and selecting *Obtaining permission to use Elsevier material*

Notice
No responsibility is assumed by the publisher for any injury and/or damage to persons or property as a matter of products liability, negligence or otherwise, or from any use or operation of any methods, products, instructions or ideas contained in the material herein. Because of rapid advances in the medical sciences, in particular, independent verification of diagnoses and drug dosages should be made

British Library Cataloguing in Publication Data
A catalogue record for this book is available from the British Library

Library of Congress Cataloging-in-Publication Data
A catalog record for this book is available from the Library of Congress

ISBN: 978 1 85617 461 9

For information on all Butterworth-Heinemann publications
visit our web site at http://books.elsevier.com

Typeset by Charon Tec Ltd (A Macmillan Company)
www.charontec.com
Printed and bound in the UK

07 08 09 10 10 9 8 7 6 5 4 3 2 1

Working together to grow
libraries in developing countries

www.elsevier.com | www.bookaid.org | www.sabre.org

ELSEVIER   BOOK AID International   Sabre Foundation

CONTENTS

Preface xi

Chapter 1   **Introduction**   1
- 1.1   Purpose of this book   1
- 1.2   What does it cover?   2
- 1.3   What applications does it cover?   4
- 1.4   What disciplines are involved?   4
- 1.5   The future   6

Chapter 2   **Static seals**   7
- 2.1   Introduction   7
- 2.2   O-rings   8
  - 2.2.1   Elastomer O-rings   8
  - 2.2.2   Plastic O-rings   19
  - 2.2.3   Metal O-rings   20
- 2.3   Alternative elastomer sections   25
  - 2.3.1   Rectangular section rings   26
  - 2.3.2   X section rings   27
  - 2.3.3   T-seals   28
  - 2.3.4   L section seals   30
  - 2.3.5   U-rings   30
  - 2.3.6   Integral backup flange seals   30
  - 2.3.7   Bonded seals   31
  - 2.3.8   Spring seals   34
  - 2.3.9   Hygienic seal arrangements   36
  - 2.3.10   Window seals   39
- 2.4   Alternative plastic sections   43
- 2.5   Alternative metal seal designs   45
  - 2.5.1   C-rings   45
  - 2.5.2   Low load C-rings   46
  - 2.5.3   Spring energized metal seals   46
  - 2.5.4   E-rings   47
  - 2.5.5   Sigma seals   48

2.6	Cured in place seals		48
	2.6.1	Moulded in place	49
	2.6.2	Cured in place	50
2.7	Formed in place seals and gaskets		51
	2.7.1	Sealing rigid flanges with anaerobic sealants	52
	2.7.2	Sealing flexible flanges	56
	2.7.3	Sealant selection	60
2.8	Bolted joints and gaskets		61
	2.8.1	Introduction	61
	2.8.2	Forces acting in a bolted joint assembly	62
	2.8.3	Gasket behaviour	62
	2.8.4	Flange types and standards	65
	2.8.5	Calculation methods	67
	2.8.6	Gasket installation	72
	2.8.7	Bolting	73
	2.8.8	Gasket types and applications	78
	2.8.9	Compact flanges and connectors	92
2.9	Selection of the optimum static seal design and material		97
	2.9.1	Temperature	98
	2.9.2	Pressure	98
	2.9.3	Industry design codes	99
	2.9.4	Manufacturing method	100
	2.9.5	Sealed fluid	100
	2.9.6	Life expectancy	100
	2.9.7	Leakage integrity, emissions	101
	2.9.8	Manufacturing volume	101
	2.9.9	Maintenance and accessibility	102
2.10	References		103

Chapter 3	**Rotary seals**		**105**
3.1	Introduction		105
3.2	Lip seals		105
	3.2.1	Basic lip seal design	106
	3.2.2	Dynamic sealing mechanism	109
	3.2.3	PTFE seals	120
	3.2.4	PTFE lined elastomer seals	124
	3.2.5	Excluders	127
	3.2.6	Cassettes	133
	3.2.7	V-ring seals	134
	3.2.8	Bearing seals	137
	3.2.9	Lip seals for pressure	140
3.3	Alternative elastomer and plastic seals		148
	3.3.1	O-rings	148
	3.3.2	Elastomer fabric seals	149

		3.3.3	Elastomer energized plastic seals	150
		3.3.4	Spring energized PTFE seals	152
		3.3.5	Positive lubrication rotary seals	155
	3.4	Mechanical seals		159
		3.4.1	Introduction	159
		3.4.2	Basic design	160
		3.4.3	Method of operation	161
		3.4.4	Basic seal types	165
		3.4.5	Seal housing design	183
		3.4.6	Typical applications and seal arrangements	186
	3.5	Compression packing for rotary shafts and valves		219
		3.5.1	Introduction	219
		3.5.2	Method of operation	221
		3.5.3	Packed gland arrangements	222
		3.5.4	Fitting and using packing	225
		3.5.5	Packing types	228
	3.6	Clearance seals		237
		3.6.1	Introduction	237
		3.6.2	Labyrinth seals	238
		3.6.3	Honeycomb and hole slot seals	249
		3.6.4	Brush seals	252
		3.6.5	Leaf seals	256
		3.6.6	Viscoseals	258
		3.6.7	Circumferential seals	262
		3.6.8	Centrifugal or liquid ring seals	264
	3.7	Magnetic fluid seals		268
		3.7.1	The particular attributes of a magnetic seal are:	268
		3.7.2	Ferrofluid types	270
		3.7.3	Typical applications	271
	3.8	Rotary seal selection		273
		3.8.1	Liquid sealing	273
		3.8.2	Gas sealing	275
	3.9	References		277
Chapter 4	**Reciprocating seals**			**282**
	4.1	Introduction		282
	4.2	Elastomer and plastic seals for hydraulic applications		282
		4.2.1	Background	283
		4.2.2	Seal stability	285
		4.2.3	Sealing mechanism	286
		4.2.4	Tandem seal arrangements	299
		4.2.5	Rod seal arrangements	304
		4.2.6	Piston seal arrangements	304
		4.2.7	Excluders	310

4.3	Pneumatic cylinder seals		315
	4.3.1	Pneumatic rod seals	316
	4.3.2	Excluders	318
	4.3.3	Piston seals	318
	4.3.4	Material considerations	320
4.4	Piston rings		321
	4.4.1	Introduction	321
	4.4.2	Internal combustion engines	321
	4.4.3	Compressor piston rings	332
	4.4.4	Piston rings for hydraulic cylinders	340
4.5	Compression packing		341
	4.5.1	Packing types and application areas for reciprocating duties	342
4.6	Clearance seals		345
4.7	Diaphragms and bellows		347
	4.7.1	Flat and dished diaphragms	348
	4.7.2	Rolling diaphragms	350
	4.7.3	Diaphragm materials	351
	4.7.4	Diaphragm applications	353
	4.7.5	Polymer bellows	353
	4.7.6	Metal bellows	354
4.8	References		357

Chapter 5 Materials — 358

5.1	Elastomers		358
	5.1.1	Why do we use elastomers?	359
	5.1.2	Elastomer material basics	362
	5.1.3	Factors to consider when selecting an elastomer	367
	5.1.4	Elastomer types and their applications	381
5.2	Plastics		393
	5.2.1	Benefits of plastics	394
	5.2.2	Potential disadvantages of plastics	395
	5.2.3	Mechanical properties of plastics	395
	5.2.4	Plastic material types and their applications	397
5.3	Carbon		403
	5.3.1	Manufacture of carbon	403
	5.3.2	Tribology of carbon	404
5.4	Silicon carbide		405
	5.4.1	Reaction bonded	405
	5.4.2	Self-sintered silicon carbide	408
	5.4.3	Graphite loaded sintered silicon carbide	408
5.5	Tungsten carbide		408
5.6	Silicon nitride		409
5.7	Alumina ceramic		409

	5.8	Hard/hard mechanical seal face combinations		410
	5.9	Metals		410
		5.9.1	Stainless steel	411
		5.9.2	Nickel alloys	412
		5.9.3	Low thermal expansion alloys	414
	5.10	Soft metal overlay		414
	5.11	References		415
Chapter 6	**Failure guide**			**417**
	6.1	Introduction		417
	6.2	Static seal failure guide		417
		6.2.1	Elastomer seals	417
		6.2.2	Plastic seals	423
		6.2.3	Metal seals	425
		6.2.4	Gaskets	426
	6.3	Rotary seal failure		428
		6.3.1	Lip seals	428
		6.3.2	Other elastomer and plastic seal designs	436
		6.3.3	Mechanical seals	437
		6.3.4	Dry gas seals	443
		6.3.5	Compression packing	445
		6.3.6	Clearance seals	446
	6.4	Reciprocating seals		447
		6.4 1	Polymer reciprocating seals	447
	6.5	References		452
Chapter 7	**General information**			**454**
	7.1	Glossary of sealing terms		454
		7.1.1	Elastomer and plastic seals, compression packing	454
		7.1.2	Expansion joints and flange gaskets terminology	458
		7.1.3	Mechanical seals terminology	463
	7.2	Standards		466
		7.2.1	General	466
		7.2.2	O-rings	467
		7.2.3	Hygienic seals	468
		7.2.4	Gaskets	469
		7.2.5	Rotary shaft lip seals	471
		7.2.6	Mechanical seals	471
		7.2.7	Compression packing	471
		7.2.8	Reciprocating seals	471
		7.2.9	Material properties	472
		7.2.10	Surface texture measurement	474
		7.2.11	Standards organizations	474

7.3	Surface texture measurement		476
	7.3.1	Introduction	476
	7.3.2	Ra arithmetic mean	477
	7.3.3	Root mean square deviation of the profile from the mean line	478
	7.3.4	Rt – total height of the profile	480
	7.3.5	Rz (ISO) – maximum height of profile	481
	7.3.6	History of peak parameters	481
	7.3.7	Rsk skewness	482
	7.3.8	Material ratio curve	484
	7.3.9	Conclusion	485
7.4	Organizations with a direct interest in sealing technology		486
	7.4.1	Trade and industry organizations	486
	7.4.2	Research organizations	488

Appendix 1 Sealing Technology – BAT guidance notes   490

1	Executive summary	496
2	Preface	501
3	General introduction	504
4	Generic BAT for sealing technologies	509
5	BAT for bolted flange connections	520
6	BAT for rotodynamic equipment	541
7	BAT for reciprocating shafts	557
8	BAT for valves	559
9	Conversion factors	573
10	Further reading	578
11	References	580

Appendix 2 The application of European ATEX legislation to mechanical seals   582

2.1	Ignition Hazard Assessment	592
2.2	Maximum Surface Temperature	595

Index   597

PREFACE

Seals are everywhere. My career has provided an ever-expanding experience of new applications, seal designs and material developments which show no sign of diminishing. They are a key component of our houses, domestic appliances and vehicles while transport and virtually all industries are totally dependent on them for safe and reliable operation. They are often treated as a trivial commodity product, but the costs and consequences of a failure can be substantial. One failed fitting in the water system of a house can cause a lot of damage. In our vehicles they can be costly to replace, and, in a hydraulic braking system, fundamental to our safety. In many industries reliable seals are equally fundamental to reliable operation of the plant and containment of fluids. The costs of seal replacement on process plant can be very high and the implications of sudden failure can be substantial in terms of lost product, clean-up and pollution. For this reason nearly all engineers require some appreciation of seal selection and application and this book sets out to provide an unbiased overview of the potential options.

The progress in seal development over the last two decades has been enormous covering both seal design and materials. During this time our understanding of how seals operate and the ability to optimize the design has progressed substantially. This is one reason why this book has been completely re-written, rather than trying to update the earlier editions. Another reason was to ensure that all the similar application information is close together. There are individual chapters for rotary, reciprocating and static seals, so that it is relatively easy to explore the potential variants available for an application. It has been another aim when preparing this book to provide a balance of suitable information as a basis for initial seal selection across all the potential seal types available in each application area. This is something that has been sadly lacking in most of the general sealing books that have been produced.

Almost every sealing application is some sort of challenge. At high temperature pressure or speed there is the technical challenge of reducing leakage and friction or increasing life. With apparently simple applications the challenge can be ever present demands to reduce cost and improve reliability. Designs that will operate at wider tolerance bands or are easier to assemble may then be required. You will find that design for assembly occurs frequently throughout this book.

Working with seals may seem extremely specialist, and indeed it does require specialist knowledge of the design criteria, materials and the limitations of an individual arrangement. But it is also important for the seal specialist to have a very broad understanding of engineering including manufacturing techniques and machinery operation. I have seen many 'good ideas' fail because at least one of two vital criteria have been overlooked: the requirement for assembly in a real working environment, and appreciation of the fact that machinery does not run at constant conditions forever; in the real world it stops and restarts.

The preparation of this book has been supported by many of the friends that I have been fortunate to make across the sealing industry and whose organizations are acknowledged where relevant. I have also made many new contacts who have been equally supportive. My thanks go to all who have contributed to making this book as up to date and relevant as possible. My special thanks are also due to my wife, Christine, without whose unstinting support it would never have happened, and son Jonathan who spent several days proofreading the entire text.

Robert Flitney
May 2007

CHAPTER 1

Introduction

1.1 Purpose of this book

This book sets out to introduce the subject of sealing technology and provide the reader with an understanding of the issues involved and sufficient information, or sources of information, to assist with achieving reliable and effective sealing both when designing new equipment or as a solution to a maintenance or performance problem.

It is intended for design and development engineers from the many industries that use seals and the operations and maintenance engineers involved with the plant and machinery once it is in use. The reader should find information that will help to select an appropriate seal arrangement and optimize a current design, and if necessary explore alternative sealing methods in the search for a more reliable or cost-effective solution.

Even a relatively simple looking seal, such as the humble O-ring, depends on some quite unique properties of the elastomer material from which it is manufactured to enable it to function reliably. Failure to understand the properties of the material and adhere to some well-developed design rules can rapidly cause problems. New materials have dramatically extended the potential application areas with respect to temperature range and fluid compatibility. But, these materials often necessitate specific design rules to accommodate their characteristics.

Dynamic seals have to be optimized to provide both minimum leakage and minimum friction and wear, often mutually exclusive tribological objectives. A better appreciation of the factors involved should permit a more informed dialogue with potential seal suppliers. Practically all seals, either static or dynamic, depend on the properties of the materials from which they are constructed and the interaction of the mating surfaces to achieve a reliable seal with minimum leakage. Again, developments in materials and design of the seals have progressively extended the performance of dynamic seals in most application areas.

User demand and legislation restricting emissions are creating requirements for tighter control of leakage, longer seal life and reduced life cycle costs. By introducing a wide variety of seal designs and methods together with details of the materials it is the intention of this book to provide information on methods used in many different industries that may then promote healthy debate on the potential options leading to an informed selection of a cost-effective and appropriate seal arrangement.

1.2 What does it cover?

The following chapters cover seals for static, rotating and reciprocating applications in turn. Each of these chapters also includes guidance on the selection of an appropriate seal for individual applications. The specific properties of materials used for seals or sealing systems are covered in a separate chapter. The Failure Guide sets out typical symptoms that may be observed with failed seals and the factors to consider, many of which involve attention to the fluid system rather than the seal. Each industry and technology has some specific terminology, standards and key organizations. Information on these areas, together with an explanation of surface texture terminologies is included in Chapter 7. Compliance with emission regulations is a fact of life for industrial plant. The European Sealing Association has made a major contribution to this area and the BAT Guidance Note prepared by the association members is provided as an appendix together with information on compliance with the European ATEX regulations for operation of mechanical seals in explosive atmospheres.

Static seals are discussed in Chapter 2 and this starts by explaining the use of elastomer O-rings and then extends to alternative materials and seal sections. The relative merits of the many potential elastomer, plastic and metal seals are discussed together with key design criteria. The basics of gasketed joints and the various material options are discussed together with the fundamental differences between gasketed and sealed flanges. A number of the specialized flange sealing techniques are also covered.

Static seals are far more numerous than dynamic seals. They have therefore been the focus of considerable attention with respect to production engineering for volume components. A variety of techniques are presented for consideration that integrate the seal or allow it to be formed or moulded in place.

Chapter 3 discusses rotary seals. The rotary seal produced in the highest volume is undoubtedly the humble lip seal. Although they have been manufactured for many years a thorough understanding of their operation is relatively recent and still developing. The basic operation, key design features and the many potential variants that may be considered are discussed together with developments of plastic seals that can be used to extend the performance envelope. Exclusion of contaminant is often just as important as the sealing of lubricant, so this aspect is also discussed. Alternative designs of elastomer and plastic seals, some of which can be used to high pressures in specialized applications, are covered.

Mechanical seals cover a very wide market and range of applications from domestic white goods to turbo machinery. The basic designs, key design features and areas of application are discussed.

Compression packing is still a relevant contender in many applications, so the key features and modern material options are presented.

A wide range of other seal types are also used in rotating machinery, particularly in high speed turbo machinery. As efforts are made to increase machine efficiency and reduce emissions both seal designs and new materials are continually being investigated. Current developments for a number of seal types, labyrinth, honeycomb, leaf seals, brush seals and viscoseals, are summarized.

O-rings are used as reciprocating seals in some areas. They do not provide the optimum seal and present certain hazards. Chapter 4 discusses these problems and introduces many specialist designs of elastomer or plastic seal that may be used to improve performance. The reasons for this are explained, together with summary charts to guide readers to the most appropriate style for a given application area.

Piston rings are a very commonly used form of reciprocating seal. They fall into quite distinct application areas covering internal combustion engines and compressors. Although dry compressors have been in use for many years there is an increasing requirement for 'bone dry' gas compression which has further stretched material development. Clearance seals must again be considered as are compression packings.

In sealing technology extensive use is made of specialized materials and these are discussed in Chapter 5. Elastomers have quite unique properties that make them especially useful as seals. Much of the development of high temperature and chemically resistant elastomers has been prompted by the demand of the seal industry customers. The plastics used are also developed specifically for seal applications as are certain carbons. The important considerations relevant to material selection are discussed, especially the importance of considering the entire range of fluid constituents.

The Failure Guide in Chapter 6 presents key areas to investigate and a range of specific examples across the seal types discussed in the earlier chapters. It will normally have to be used in conjunction with the relevant technical section to ensure that the important criteria for the individual seal type are considered.

Chapter 7 includes a range of general information. A comprehensive glossary for much of the terminology used across the sealing industry is provided. There is a summary of some of the key standards across the seal types and numerous industry sectors, together with contact details for standards organizations. Appendix 3 provides additional details for UK readers of locations where they may access reference copies of British Standards. An introduction to surface texture measurement is also provided and the most frequently used texture parameters that the reader may come across.

The first two appendices provide valuable additional information, especially for seal applications in the process industries. The Best Available Technology Guidance Note prepared by the European Sealing Association is a very comprehensive guide to the performance and cost effectiveness of mechanical seals, compression packing and gaskets in the process industry. With the cooperation of ESA it is reproduced in full in Appendix 1.

The European Union ATEX Regulations have created much discussion and raised a number of questions. The current requirements of ATEX with respect to mechanical seals are provided in Appendix 2.

1.3 What applications does it cover?

The use of seals in our modern world is very wide ranging. In fact it can be a challenge to think of anything in our daily lives that does not involve some form of sealing. We are surrounded by them in our homes from food containers, body care products and pharmaceutical containers to water system and glazing seals. All forms of mechanized transport and any manufacturing process involve a bewildering array of seals. The various utilities are totally reliant on seals to either contain water or gas within the distribution network or exclude water and contamination from electrical and electronic services.

As with any technical area, life does not stand still. Advances in analytical and measurement techniques have provided a much improved understanding of the behaviour of seals and the critical parameters to provide improved performance. When combined with advances in material technology this has led to quite dramatic improvements in seal performance over the last two decades.

The impetus for this change has been created by a number of factors including demands for higher reliability, emissions legislation together with a realization of the costs associated with product loss, demands to reduce manufacturing costs while still improving reliability and all combined with a change of corporate culture which has pushed responsibility further down the supply chain. An indication of the variety and speed of change is evidenced by the number of new patents issued, approximately 200 are reviewed in *Sealing Technology Newsletter* each year.

1.4 What disciplines are involved?

The manufacture and application of seals involves a number of disciplines:

- Materials, covering the entire range of polymers, ceramics and metals. Many materials, especially polymers and ceramics are developed specifically for the manufacture of seals. Many seal designs involve high strains, and often plastic and metal static seals depend on plastic flow to provide an effective seal. The stress/strain properties, fluid compatibility and corrosion effects can be critical to success. Very often either minor constituents of the sealed fluid or impurities in a material can be a source of severe problems.
- Tribology, all dynamic seals involve friction, lubrication and wear. There is seldom the luxury of selecting the lubricant, it is most often the fluid to be sealed. For many seals instead of optimizing the lubricant and lubricant film for reliability the amount of lubricant is minimized to meet leakage requirements. Dry or boundary lubrication is involved in an increasing number of seal applications. Even static seals can be subject to small repeated movements with pressure and temperature cycles that can cause fretting damage and wear.
- Manufacturing technology impacts on both the seal manufacture and design. Considerable effort can be involved in the development of manufacturing techniques for new elastomers and plastics. Seal designs may be

limited by factors such as the moulding of elastomers, geometry and characteristics of ceramics, and production facilities for specialized plastics.
- Production engineering of the equipment to be sealed is equally important. Seals can require specially prepared surfaces machined to close tolerances and installation of seals without damage can be crucial to equipment reliability. There is an increasing trend to design the seal into equipment at an earlier stage in the development process and even incorporate the seal into a component to facilitate rapid and reliable assembly. The use of cartridge or cassette seals for rotary applications removes much of the uncertainty concerned with correct assembly of the seal.
- Fluid engineering is important to understand the flow in the sealing interface for a study of the lubrication and around the seal for factors such as heat dissipation and avoiding deposition of solids.
- Both chemistry and physics are important to provide an understanding of the chemical and physical interactions between the seal, sealed fluid and both static and dynamic counter-faces. The subtle changes in behaviour of the seal and fluid are due to changes in even apparently minor ingredients such as fillers in the seal and additives or contaminants in the sealed fluid. In many industries an appreciation of the environmental aspects and complexities of emission measurements are now important.
- An overall appreciation of engineering systems is also important to appreciate the application conditions in which a seal will operate. On occasions seal designs and materials have been developed which can provide impressive performance under steady state conditions in a laboratory, only to find that in the real world of stops and starts, pressure cycles and varying temperatures they are quite inadequate. The potential for disturbances such as pressure surges, lubrication starvation, temperature excursions, sudden changes to the fluid state or even a different fluid can all affect the design and selection of a sealing system.
- The specific requirements of individual applications or industries. This can include factors such as safety and backup systems, hygiene, reliability and life expectancy, purchase cost, life cycle cost, leakage/emissions, obligatory standard procedures, customer preference and the working conditions under which assembly and maintenance may occur.

Many problems that occur with seals are either because the supplier has not been made aware of the full operating spectrum, the user has not been made aware of potential limitations of a particular design or material, or there has been a failure by all parties to understand the implications of some apparently insignificant factor in the application. This can be during assembly, testing, normal operation, off-design operation or maintenance. It can range from storage conditions or the fluid used in assembly, some aspect of the sealed fluid that is not considered, testing with an inappropriate fluid, spikes in temperature or pressure either up or down, assembly in an unfamiliar environment or without realizing the implications of a specific design feature to a multitude of factors concerned with standby and spare equipment and how it should be maintained or exercised.

Hopefully the reader will be better informed to address these potential problem areas.

1.5 The future

The author has experienced an ever-evolving career, which has continually been exposed to new applications, seal and material developments, which have extended application areas for existing designs, and fresh approaches to the solution of traditional problem areas while in many areas leakage has reduced from drops per minute to being measured in ppm. It is intended that this book will assist with the development of more reliable components, provide potential solutions to existing sealing problems and perhaps even inspire a fresh approach to well-established areas leading to further developments in the application of technology. And, as we become ever-more conscious of the environment in which we live, any leakage that is avoided and power that is saved will be a welcome contribution to the world we share.

CHAPTER 2

Static seals

2.1 Introduction

The designs and types of sealing methods for static duties cover a vast range. To discuss the various types it is useful to divide them into two basic categories, seals and gaskets. Seals are typified by an elastomeric O-ring in a seal groove and gaskets by fibrous materials clamped between a pair of flanges.

Seals, within this context are designed such that they are expected to be self-energizing whereas it is normal for gaskets to be clamped with sufficient force that the pressure is resisted by the stored energy within the gasket. Hence the relative assembly load for a seal may be much lower than a gasket for the same duty.

There are many variations on these two individual themes and several that cover a grey area between them. The selection will depend on a number of factors. These include:

- Temperature range.
- Pressure range.
- Fluids to be sealed.
- Environment.
- Integrity of sealing required.
- Material of counter-faces.
- Life requirement.
- Maintenance requirements.
- Volume to be manufactured.
- Assembly methods.
- Testing and inspection criteria.
- Historical experience of designers.
- Degree of flexibility and relative movement of counter-faces.
- Requirement to withstand pressure, thermal or mechanical shock.
- Industry standards and practice.
- Sterility and hygiene requirements.
- Established custom and practice within a particular industry.

Hence the type of static seals chosen for an automotive gearbox casing, a pipe joint in a food plant, a vacuum chamber working in electronics manufacture, or a high pressure steam pipe in a process plant may all be very different. The reasons will be driven not only by the operating conditions but also by the individual requirements of safety, life expectancy, maintenance and very different volumes of manufacture for these applications. Also, individual material classes achieve a satisfactory seal by quite different mechanisms, and it is necessary to take this into account at the selection and design stage. This is covered in the individual sections in this chapter.

2.2 O-rings

2.2.1 Elastomer O-rings

The elastomeric O-ring is probably the most common form of general purpose static seal. A typical arrangement is shown in Figure 2.1. The design of the groove is relatively simple, but compliance with some well-developed design rules is necessary to achieve a reliable seal. These rules are often ignored either because of lack of knowledge or for reasons of economy and this can lead to problems.

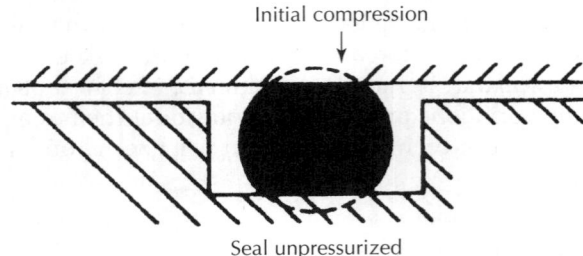

Figure 2.1 General arrangement of an O-ring in a groove.

The O-ring groove is designed to provide an initial compression on the seal across one axis of approximately 15 to 20%. The actual range depending on the design standard used and cross-section of seal may be between 7 and 30%. This compression, which is also variously described as interference, squeeze or nip, is usually perpendicular to the axis of action of pressure in the sealing area. Selection of the correct amount of interference is important. There has been a recent trend to increase the amount of interference to a nominal value in the range 20 to 25%, particularly for flange seal arrangements. It is important to consider both the application and also seal material before finalizing the groove design. Most design standards are for the regular materials in general purpose applications. Some materials, such as perfluoroelastomers, require special consideration as the high temperatures of operation and low strength at high temperature mean that care is required to avoid overstressing the material and interference is normally recommended to be limited to 13–15% maximum.

It will be noticed that the O-ring is only compressed on one axis and there is free volume in the groove on the other axis. This is most important and several reasons for this free volume will be discussed.

2.2.1.1 How does an elastomer O-ring work?

When the seal is installed in the groove and compressed, as described above, it creates some initial interference force on the groove counter-faces. This initial assembly force is quite small. For a relatively common O-ring section of 2.65 mm an average hardness seal will require 20 or 30 N/cm of circumference.

When pressure is applied the seal will usually be compressed against the downstream wall of the groove, Figure 2.2. The sealing action is provided by the unique properties of an elastomer. These materials at their normal working temperature

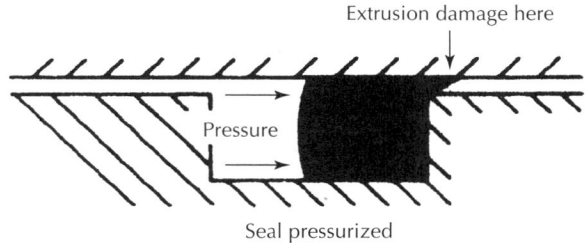

Figure 2.2 O-ring subjected to pressure will be energized against the wall of the groove.

are both virtually incompressible and also have very low elastic modulus. They are thus highly deformable, but maintain a constant volume as the Poisson's ratio is very close to 0.5. When initially installed the installation stress is reacted on the opposing axis by tension within the material, Figure 2.3(a). As pressure is applied to the seal the inherent flexibility of the material, and resistance to change of volume, transfer the pressure to the axis of compression, Figure 2.3(b). Due to the properties of the elastomer material it is behaving as if it is a liquid. Hence changing the applied pressure creates an interference stress between the seal and counter-face that is always equal to the applied pressure plus the initial interference stress. While this situation remains the O-ring will continue to function and a standard O-ring arrangement can reliably seal to several hundred bar. Assuming the seal arrangement is correctly designed there are four main reasons for failure:

- Movement of the counter-faces reducing the amount of squeeze on the O-ring.
- Extrusion of the seal out of the groove.
- Ageing of the seal causing loss of elastomeric properties.
- Low temperature causing loss of either squeeze or the elastic properties.

Designing the equipment to be sealed to avoid excessive movement of the sealing counter-faces is obviously extremely important, but outside the scope of this book. It is, however, a key factor for the designer to consider when attempting to

10 *Seals and Sealing Handbook*

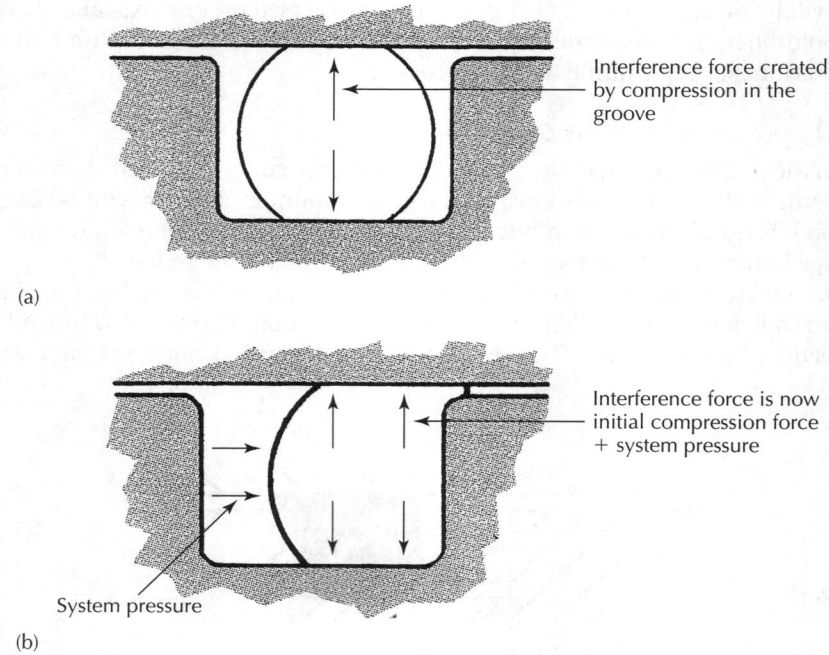

Figure 2.3 *The sealing action of an elastomeric O-ring: (a) with no applied pressure; (b) when pressure is applied.*

achieve reliable sealing. Often, moving the seal groove to an alternative location or reconsideration of the bolting arrangements can provide significant benefits. A standard design of O-ring groove can be used to seal 100 bar and above. With suitable precautions to avoid the problems highlighted above it is possible to seal 1000 bar or more quite reliably. Extrusion, ageing and low temperature are discussed in sections 2.2.1.5, 5.1.3 and 5.1.3.5 respectively.

The rectangular section of the O-ring groove is important for several reasons:

- The elastomer material is incompressible, so when it is compressed on the axis of the seal interference it is necessary to let it expand on the perpendicular axis.
- Both the O-ring and the groove will have a manufacturing tolerance, sufficient free space is necessary in the basic design to permit a variation in the lateral expansion due to variations in amount of squeeze and O-ring cross-section. These tolerances lead to quite a wide variation in compression of the O-ring. Figure 2.4 shows the variation that can be expected using standard design criteria for BS 1806 O-rings.
- The elastomer material will have a thermal expansion coefficient that is approximately an order of magnitude higher than any surrounding metalwork. At elevated temperature space for this expansion is required.

Static seals 11

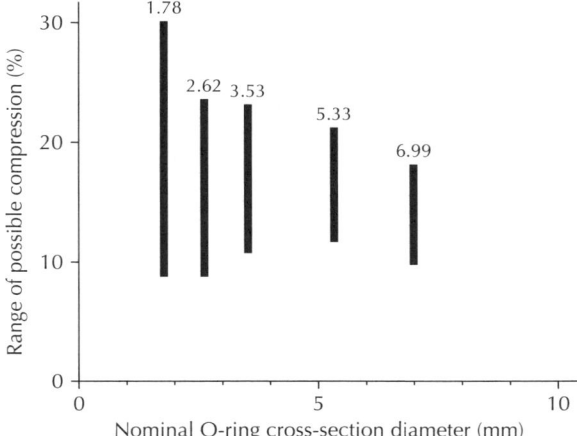

Figure 2.4 Variation in the compression of the standard inch cross-section series of O-rings using the seal's cross-section tolerances and groove manufacturing recommendations in British Standard 1806.

Table 2.1 gives some typical values of relative expansion of an elastomer; it can be seen that at 150°C a volume expansion of up to 9% may occur compared with room temperature installation.
- The sealed fluid may swell the elastomer. It is quite typical for this swell to be in the range of 5 to 10%. It is necessary to allow for this swell in the design of the groove.

Failure to take account of these factors can cause damage when fitting or subsequently due to excessive extrusion forces if the seal overfills the groove due to thermal expansion or swell.

The normal design criterion for the majority of standards and manufacturers' catalogues is that the O-ring occupies 70% of the groove volume. This is often known as 'groove fill'. There are special applications where this value may be varied but this should only be where necessary, and with full consideration of the operating conditions and design tolerances.

Table 2.1 Typical thermal expansion of an elastomeric material

Operating temperature °C	% Expansion Linear	Volumetric
38	0.4	1.1
93	1.5	4.5
149	3.0	9.0
204	4.3	13.0
260	5.5	17.0
316	7.0	22.0

2.2.1.2 Selection of an O-ring and design of the groove

Once the overall size requirements are known the starting point should be a data list of standard O-ring sizes. There are well-established standard size ranges covering both inch and metric sizes. The application, industry involved and freedom of design will dictate selection of the appropriate standard.

A common inch series of sizes is available from BS 1806[1] in the UK, SAE AS 568[2] in the USA and more recently as metric dimensions of the same basic O-rings as ISO 16032.[3] O-rings manufactured to the sizes in these documents are very widely available.

Due to the delays in production of a viable ISO standard covering pure metric sizes the availability is not as widespread as may be expected. The most widely used pure metric seals are those based around the size ranges of the British Standard BS 4518[4] and the similar Swedish document SMS 1588.[5] Seals to these metric standards are readily available from the major European manufacturers.

Future reference in these pages to inch O-rings will mean the BS 1806/AS 568 O-rings and metric will refer to the BS 4518/SMS 1588 series. There are also a number of alternative series that have been devised for specialized applications, and a number of countries also have national standards. In general the aerospace industry has a series of standards for higher precision and quality seals but the size ranges are the inch sizes.

Having decided on an appropriate standard the cross-section to be used must be selected. Each of the standards has a range of cross-sections. The inch standard has sections of: 1.78, 2.62, 3.53, 5.33 and 6.99 mm (0.070, 0.103, 0.139, 0.210 and 0.275 inches). The metric series has comparable sections of: 1.6, 2.4, 3.0, 5.7 and 8.4 mm. The typical range of O-rings available to these standards is shown in Table 2.2. Consultation of the size charts will show that at most average component sizes there will be a choice of two or three cross-sections. For instance if a 2 inch OD seal is required there is the choice of 0.070, 0.103,

Table 2.2 The O-rings available in the main inch and metric standards

O-ring cord section (mm)	Smallest inside diameter (mm)	Largest inside diameter (mm)
Metric series		
1.6	3.5	37.5
2.4	4.0	70.0
3.0	20.0	250.0
5.7	45.0	500.0
8.4	145.0	250.0
Inch series		
1.78	2.0	132.0
2.62	1.5	245.0
3.53	4.5	455.0
5.33	11.0	655.0
6.99	114.0	655.0

0.139 and 0.210 inch sections. The equivalent in metric series at 50 mm has the option of 2.4 and 3.0 mm. The choice of cross-section will be a compromise of technical and commercial considerations. Figure 2.4 provides an initial clue to one technical factor. With a larger cross-section it is easier to control the O-ring squeeze as the tolerances are a lower proportion of the seal cross-section. Larger cross-sections also provide improved resistance to extrusion and some work indicates that resistance to ageing is improved.[6]

However, a larger cross-section requires more surrounding metalwork and so space and weight limitations may dictate a smaller cross-section. The cost of the elastomer material may also be a factor, especially in high volume applications or where a specialist polymer is required, so pricing considerations may dictate a smaller cross-section.

One exception to this rule is in high pressure gas applications where a minimum cross-section possible should be used. This is discussed in section 6.2.1.8.

Catalogues and standards provide the basic information necessary to design seal grooves, but a number of important factors must be considered to achieve reliability. With flange type grooves, Figure 2.5, it is necessary to select the dimensions relevant to internal or external pressure according to the application. If this is not done then the O-ring will move excessively in the groove which can cause wear and potential rolling. With internal pressurization it will also stretch the ring causing some loss of section. As the flanges will normally be tightened down metal to metal the groove depth, H, is a known value, within the tolerances, and extrusion gap will be dependent on machining quality and retention of the flange under pressure.

When considering a radial arrangement, Figure 2.6, some additional factors must be taken into account. The dimension governing the squeeze of the seal is now dimension F, and this depends initially on two diameters: C and the groove outside diameter for Figure 2.6(a) and D minus the groove diameter for Figure 2.6(b). The groove dimension therefore depends on the tolerance range for the two diameters. It is also important to consider any potential eccentricity. The F min and max values must be maintained when the two components are fully eccentric. The clearance, C, is therefore also important and the design rules

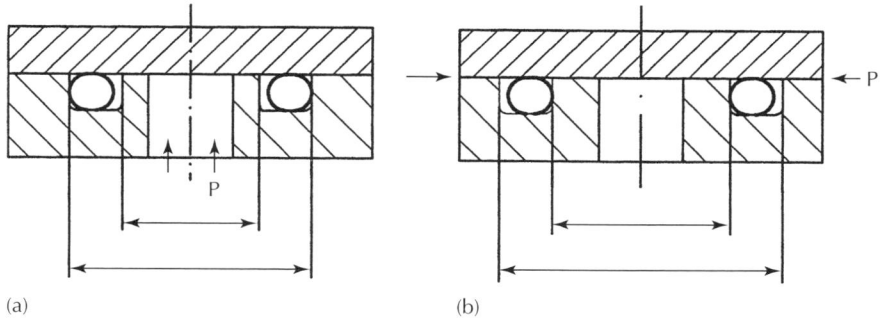

Figure 2.5 O-ring grooves for a flange assembly: (a) for internal pressure, O-ring o.d. is normally equal to groove o.d.; (b) for external pressure, O-ring i.d. is normally equal to groove i.d.

14 *Seals and Sealing Handbook*

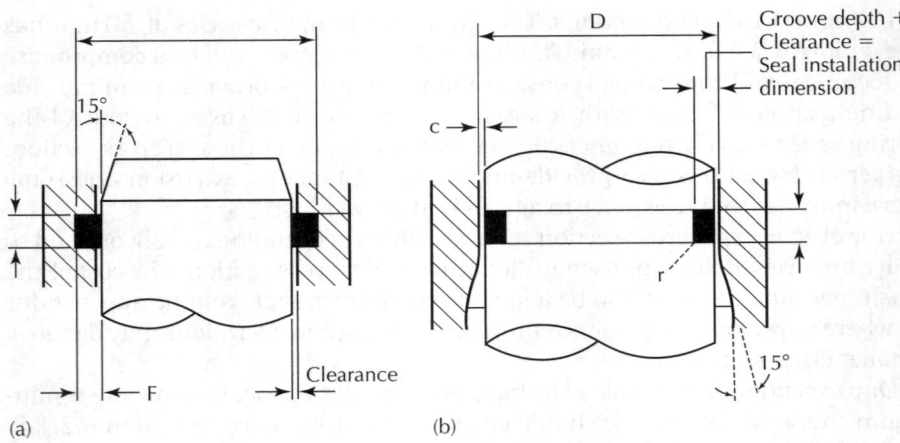

Figure 2.6 O-ring grooves for radial sealing: (a) seal in the cylinder bore; (b) seal in the piston.

should not be exceeded. For critical duties, and where tolerances are relatively wide, the tolerance of the O-ring must also be taken into account, Table 2.3. A further important requirement for radial configurations is the assembly taper as shown on the diagrams in Figure 2.6. Failure to provide these will cause damage to the seals on assembly. Any internal cross-drillings and steps must also be treated to avoid damage to the seals.

Table 2.3 O-ring cross-section tolerances

Inch series seals	Cross-section (mm)	Tolerance ± (mm)
	1.78	0.08
	2.62	0.08
	3.53	0.1
	5.33	0.13
	6.99	0.15
Metric series seals	Cross-section (mm)	Tolerance ± (mm)
	1.6	0.08
	2.4	0.08
	3	0.1
	5.7	0.12
	8.4	0.15

The groove radii given in design documents are also important. The groove base radius, r_1, must not be too small or it will become a stress raiser. If it is too large it will take up some of the groove volume and groove fill will be compromised. The corner radius r_2 is necessary to avoid damage during fitting, to both the seal and the fitter. However, if it is more than the values given it can cause premature extrusion damage.

An outward taper of up to 5° can be used. This can facilitate production of the groove by a form tool.

The 'dovetail' groove configuration, Figure 2.7, is popular in some applications as it can help retain the O-ring in the groove during assembly, inspection or maintenance. It is therefore often used in the lids of equipment such as chemical vessels. Care must be taken with the taper of the groove, a maximum angle of 60° is permissible. If it is too steep the seal may be distorted away from contact with the counter-face and leak.[9]

Triangular grooves, Figure 2.8, are useful for saving space and can make assembly more convenient. However, the dimensions of the 45° chamfer and the radius on the spigot fitting are critical. Performance of this design can be limited due to high installation stresses.

A number of suppliers and other organizations have software available in spreadsheet form to assist with groove design. This can be very useful as it enables consideration of tolerance extremes, eccentricity, temperature variations, stretch and other factors very rapidly.

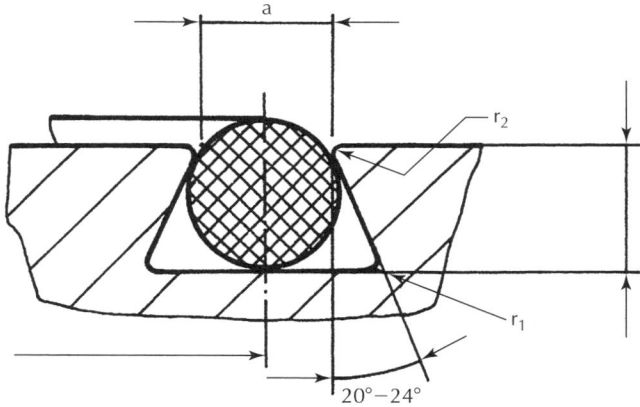

Dimension a = Approx 85% of nominal O-ring cord section.

Figure 2.7 A dovetail groove to aid retention of the seal in the groove during assembly.

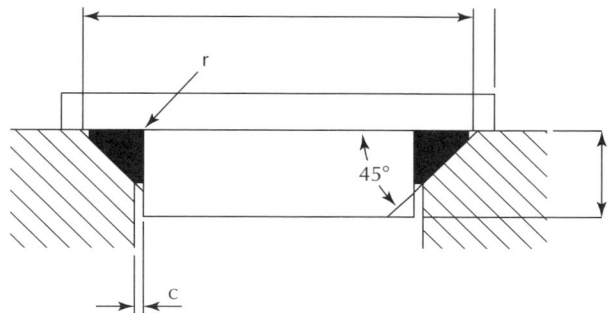

Figure 2.8 A triangular groove configuration.

2.2.1.3 Tandem grooves

In the quest for added reliability and sealing security for critical sealing situations there is a temptation to think that two seals are better than one and design the equipment to have a pair of seals in tandem. This approach is often taken with the thought that if one seal fails there is a further seal as a backup.

Extreme care in the design and assembly is required if a pair of seals are used. Space constraints will usually dictate that the volume sealed between the seals is relatively small. If this space becomes filled with liquid then very high pressures can be generated. The sealed liquid and the seal, elastomer or plastic, will have a coefficient of thermal expansion at least an order of magnitude higher than the surrounding metalwork. Once the space is filled with liquid the pressure rise will be of the order of 5 to 15 bar/°C temperature rise. On small components it is possible for the space to be filled with the lubricant used for assembly, so this situation can occur quite easily.

2.2.1.4 Surface texture

For static applications the surface texture is often considered to be less significant. However, it is important to prevent wear during pressure oscillations and as a source of interstitial leakage past the seal. The recommended maximum for sealing surfaces of 0.8 μm Ra should be used. The overall texture is also important, this is discussed in more detail in section 7.3. For critical applications requiring very low leakage benefits have been shown by using improved surface texture.

2.2.1.5 Extrusion

As elastomers are relatively soft materials they can readily extrude under pressure into the clearance gap between the metal components. Table 2.4 provides data from a manufacturer on the recommended clearance gaps that can be used for pressures up to 100 bar. It should be noted that the acceptable clearance increases with cross-section as discussed in section 2.2.2. However, other suppliers do not make this distinction, as shown in Figure 2.9, but do allow for harder materials which are more extrusion resistant. In reality any of this data must only be taken as a general guide as the material properties of the O-ring can vary significantly and for some materials much closer clearances are required, especially at elevated temperature. See section 5.1.

Table 2.4 Permitted extrusion gap for use of O-rings up to 100 bar

O-ring cross-section (mm)	Maximum diametral clearance (mm)
1.6	0.12
1.78	0.13
2.4	0.14
2.62	0.13
3	0.15
3.53	0.15
5.33	0.18
5.7	0.18
6.99	0.2
8.4	0.2

(Source: James Walker)

Static seals 17

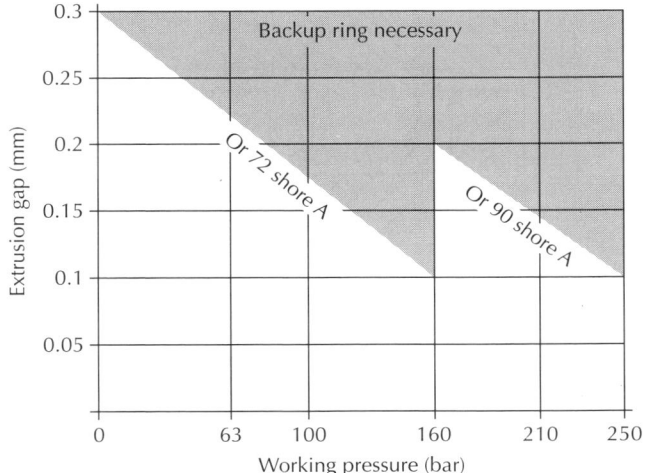

Figure 2.9 Extrusion gap limits for two different hardness materials. (Source: Freudenberg Simrit)

If there is any doubt concerning the ability of the seals to withstand extrusion then anti-extrusion precautions must be taken. This involves either:

- Reducing the metalwork clearances.
- Selecting a harder or more extrusion resistant material.
- Providing anti-extrusion or backup rings.

Reducing the metalwork clearances is generally expensive but may be justified if there are also tolerance and alignment issues.

A harder or otherwise more extrusion resistant material may not be possible as it may provide less resilience and also higher assembly forces. Premature ageing may also be an issue.

The usual course of action is to use backup rings, Figure 2.10. Choice of backup ring material is a compromise. The most common material in Europe is glass filled PTFE. In the USA the use of a hard rubber is also popular. The material has to be sufficiently hard to resist extrusion, but resilient enough to conform to the housing and fill the clearance gap. For more arduous duties involving high pressures and temperatures more temperature resistant plastics such as PEEK may be used. These require careful sizing as they will not flex to absorb tolerance variations. At very high pressures a softer metal, such as bronze, may be used but these generally require a tapered configuration to provide some energization, and may well have a plastic intermediate backup, Figure 2.11.

The fitting of backups can be a problem. They may have a simple split, but this can itself be a source of damage. A solid ring can be difficult or even impossible to fit depending on the configuration. With PTFE rings a spiral cut ring is popular,

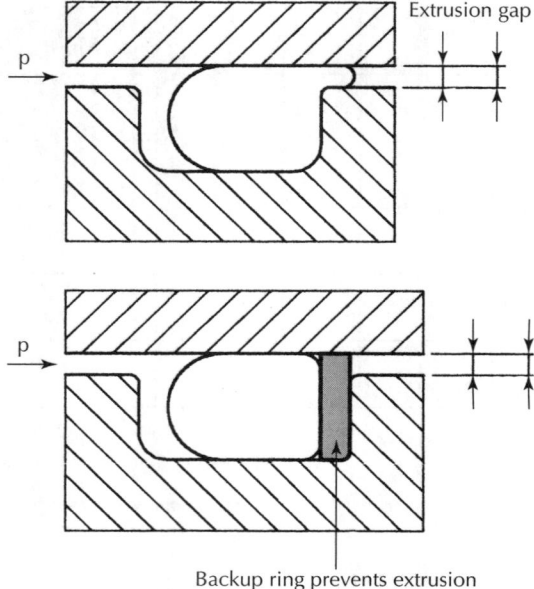

Figure 2.10 A typical backup ring configuration.

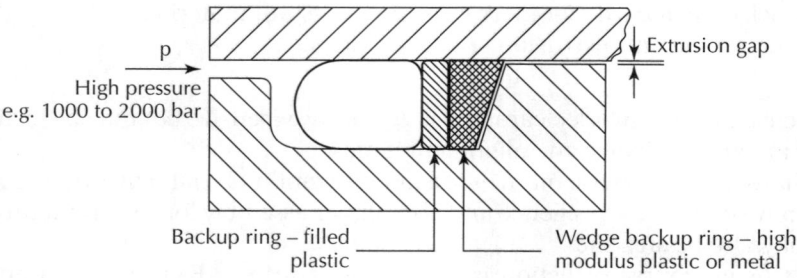

Figure 2.11 Backup ring configuration for high pressure.

Figure 2.12. This facilitates fitting and partially eliminates the extrusion gap in the ring. They are, however, as with all split rings, susceptible to damage during fitting, especially with small sizes and narrow cross-sections.

It is important to remember that the length of the O-ring groove must be increased to accommodate the backup ring. They should not be fitted in a standard groove intended only for an O-ring. Some O-ring standards provide guidance on the design of grooves to accommodate backup rings, for example BS 5108[7] and in the USA, AS 5782 and 5860.[8] Backup rings to suit these standards should be readily available from major suppliers.

Static seals 19

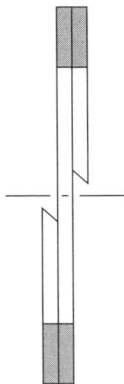

Figure 2.12 Spiral backup ring.

2.2.2 Plastic O-rings

O-rings made entirely from plastic, such as PTFE, are available but their use is restricted. The material has little energization properties and plastic flow will lead to rapid loss of sealing force. They can only be contemplated where a rigid flange arrangement is available so that no relative movement of the metalwork occurs. If a plastic O-ring is required the most common method is to use a plastic sheath surrounding an elastomer ring, Figure 2.13. These are normally used where extreme chemical resistance is required and alternative cross-sections

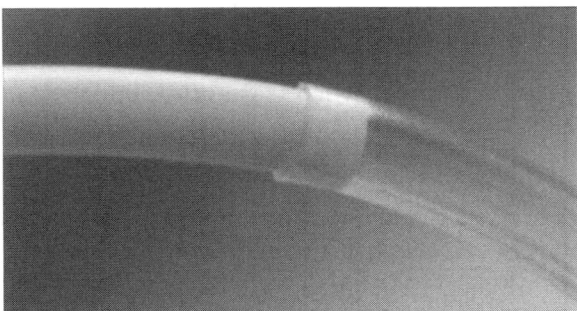

Figure 2.13 PTFE sheathed O-ring for extreme chemical resistance. (Source: Sealing Technology)

are not suitable, often because of sterility considerations. The integrity of the plastic sheath is important to protect the elastomer from attack. The thickness of the plastic then becomes a compromise between having sufficient thickness to provide a coherent plastic sheath and being thin enough for the elastomer to provide the necessary energization force. Because the plastic does not conform so readily to the texture of the metal counter-face a higher interference force is required, and as the elastomer has a comparatively reduced cross-section this can be a design constraint. The interference of the seal must also take into account the creep of the plastic after installation.

Because of these problems, these seals have a relatively restricted range of applications, and can usually be replaced either by a chemically resistant elastomer, see section 5.1, or an alternative cross-section of plastic seal, as discussed in section 2.4.

Plastic coated O-rings are generally manufactured to be a direct replacement for elastomer seals and therefore are available to fit the standard O-ring groove sizes. It may be found that they have a higher interference to provide the necessary sealing force. Additional care is required when fitting as the plastic is easily damaged and will not recover rapidly. Groove surface finish will also require an improved finish for optimum sealing. Even so, they are unlikely to provide the integrity of sealing obtainable from a correctly applied elastomer seal.

2.2.3 Metal O-rings

Metal O-rings are designed for extreme conditions that exceed the capabilities of elastomer or plastic seals. They can be used from cryogenic temperatures up to 850°C. They have the benefits of being radiation tolerant and not outgassing under high vacuum. They are therefore used in applications such as: nuclear power plant, high vacuum systems, gas turbines, oil and gas plants and as cylinder head and exhaust seals. They are only suitable for flange applications. The usual materials are stainless steel, Inconel or copper, depending on the application. The rings are manufactured from tube, but can be solid. The tube may be gas filled or can be of a vented design, Figure 2.14. Some designs are close to those of a C-ring, which are covered in section 2.5.1. Different thicknesses of tube are available which will provide a range of sealing capabilities, with a trade-off of resilience against sealing force.

Metal O-rings depend on elastic deflection to create the sealing force. They do exhibit some elastic recovery when dismantled, but this is not considered sufficient to permit reuse. The metal thickness of the tubing can be varied depending on the application and the sealing force required. Typically they will be between 0.25 and 0.4 mm thickness for small cross-sections and 0.5 and 1.25 mm for seals of 6 mm cross-section. They are also often coated in a softer material that will yield and flow into the surface texture of the groove on assembly to provide improved sealing contact. The coating can be a soft metal such as silver, nickel or plastic, usually PTFE. A typical coating thickness is 0.030 mm, but double or triple coatings can be applied to give a coating thickness up to 0.080 mm. A very thin coating of 0.01 mm may also be used. The coating provides improved sealing, particularly in gas and vacuum applications. Care must be taken to ensure that the coating is compatible with the fluid and operating temperatures. Metal coating materials are discussed further in section 5.10. The coating does not form part of the structure of the seal and is not taken into account in the groove interference calculations. The coating will flow when the seal is assembled and not contribute directly to the sealing stress.

Static seals 21

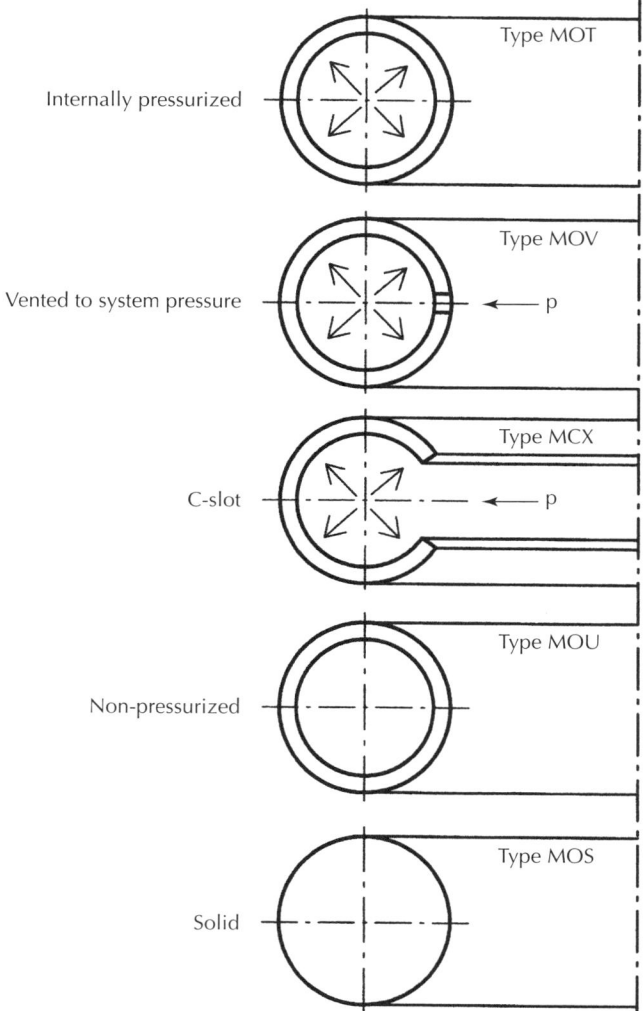

Figure 2.14 Metal O-ring types. (Source: Trelleborg Sealing Solutions)

The differing designs can be selected depending on the application. An internally pressurized seal will have improved resilience which will help to maintain sealing force in applications where there is relative movement of the mating flanges, or vibration. Expansion of the gas at elevated temperature will help to increase sealing force as the metal properties reduce. For very high pressures a vented seal design is used which allows the system pressure to energize the O-ring. Table 2.5 provides details of the application range of the different types of metal O-rings, note that the pressure limits vary widely depending on seal type. The temperature limits recommended for various metals are shown in Table 2.6.

Table 2.5 Application of different metal O-ring types

Seal Code	Description	Page	Extreme conditions (Note 1)	Seating loads (Note 2)	Springback (Note 3)	Vacuum sealing (Note 4)	Pressure (MPa)	Max. working temp. (°C)	Standard material Seal (Note 5)	Coating
Type MOT	Pressurized	8	A	C	C	A	40 MPa	850°C		
Type MOV	Vented internal	8	B	B	C	C	1000 MPa	600°C	Mild steel	PTFE
Type MOW	Vented external	8						Cryogenic to	Stainless steel 316L 321	Silver
Type MOU	Non-pressurized	9	C	B	C	C	4 MPa	400°C	Inconel® 600	Nickel

Type MOS	9	Solid	C	D	D	B	4 MPa	500°C	Copper	
Type MCX	11	Internal pressure	B	A	A	C	200 MPa	750°C	Inconel® 718	PTFE
Type MCY	11	External pressure							Inconel® 750	Silver

Properties: A = Excellent, B = Good, C = Satisfactory, D = Poor.
Inconel® is a trade mark of INCO Alloys International, Inc.

Notes:

1) Extreme conditions could be radiation, searching gases, long life.
2) Thin wall material should be used to give low seating loads. This must be specified as seals are made in standard wall unless otherwise requested.
3) The elastic recovery of the seal is known as the 'Springback'. Springback depends upon wall thickness, and also heat treatment for Wills Rings® C. Higher seal resilience gives higher springback, and higher seating loads.
4) Ability to seal a hard vacuum to meet a leakage rate of $Q < 1 \times 10^{-9}$ mbar.l.s^{-1}.
5) Other material options are available. Not all materials are available in all sizes. Not all coating available for all materials. See temperature limitations (Table 2.6).
(*Source: Trelleborg Sealing Solutions*)

Table 2.6 Temperature limits of metal O-ring materials

Standard material	Temperature (°C)
Copper	400
Mild steel	550
Stainless steel AISI 316L (1.4435)	800
Stainless steel AISI 321 (1.4541)	800
Inconel® 600	850

(Source: Trelleborg Sealing Solutions)

It is important to note that metal O-rings cannot be directly substituted for elastomeric seals. The details of the groove design are quite different. Although the typical metal O-ring compression is approximately 20%, similar to that for an elastomer O-ring, the tolerances and surface texture required are significantly more stringent. The differences between groove machining requirements for sample 50 mm diameter seals are summarized in Table 2.7. Note that the seal compression forces are also very much higher. The compression force to install an elastomeric ring of 2.5 to 3.0 mm cross-section to 20% is in the range of 2–5 N/mm of

Table 2.7 Comparison of groove design criteria for elastomer and metal O-rings

	Metal	Elastomer
Tolerance on groove depth (h)	0.08 mm	0.25 mm
Tolerance on groove outside diameter (D)	0.05 mm	0.5 mm
Flange load to compress seal	10 000–50 000 N	150–800 N
Surface finish	0.4 µm Ra	0.8 µm Ra

For 50 mm o.d. seal of 2.5 mm section

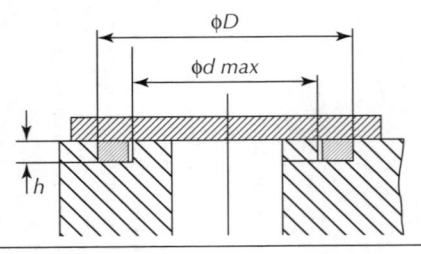

(Source: BHR Group)

circumference. The equivalent figure for a metal ring is 100–250 N/mm. The surface texture requirements for the groove surface depend on both the application and any coating on the O-ring. A surface texture of 0.1 µm Ra may be necessary for uncoated seals in gas or vacuum applications. Where a soft coating such as PTFE is used then a texture of 0.4–0.8 µm Ra may be used. This is still more stringent than that required for elastomers.

A metal O-ring does not introduce the factors of high thermal expansion, or O-ring swell introduced by elastomers, so it is not necessary to provide the additional groove volume discussed in section 2.2.1 above. For applications that experience extreme cyclic pressures, such as high performance internal combustion engines, a groove design that securely locates the O-ring may be used, Figure 2.15. This groove design also avoids trapping pressure in the groove during pressure cycles.

Some of the major variations available for metal O-rings include:

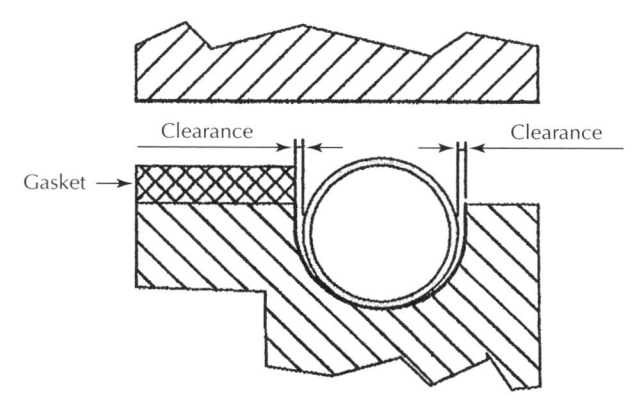

Figure 2.15 Metal O-ring groove design for automotive cylinder head applications, the seal is held securely, and there is no spare groove volume to trap pressure. (Source: Trelleborg Sealing Solutions)

- Metal type.
- Metal treatment, such as age hardening.
- Coating type.
- Tube thickness.
- Pressurized.
- Non-pressurized.
- Venting.

They cannot therefore be considered as a stock item and a detailed consideration of the application is required to make the optimum selection. They are also available in non-circular configurations where there is a requirement to seal complex shaped components.

2.3 Alternative elastomer sections

O-rings are by far the most common form of elastomer seal. They are relatively easy to manufacture, readily available in standard sizes and there are many standards and guides to facilitate design. However, they do have a number of limitations, and hence an alternative style of seal should often be considered to achieve improved performance. Problems that may be encountered with O-rings include:

- During pressure cycles the O-ring will flex and change shape, this can cause wear and a minute pumping action that will lead to a minute leak.

- Action of pressure and any movement can cause rolling of the ring.
- They may not provide sufficient flexibility to accommodate wide tolerances, particularly at large sizes.
- Separate anti-extrusion rings may be required.
- They may be twisted during fitting, particularly small cross-sections or large components with difficult access.
- Retention in the groove may be an issue.

A variety of alternative seal geometries have been developed to overcome some of these problems.

2.3.1 Rectangular section rings

Rectangular section rings are widely used. They can generally be considered to be interchangeable with elastomer O-rings. They may be manufactured by moulding for high volumes but are also lathe cut from extruded tube. A particular benefit of rectangular rings is that they will not flex under cyclic pressure. They will therefore be less prone to leakage due to flexing and wear, Figure 2.16. For this reason they are used as an alternative to O-rings on some high duty hydraulic equipment and similar applications. Another familiar application is oil filters on automotive engines, where improved stability during fitting is an important consideration. The rings

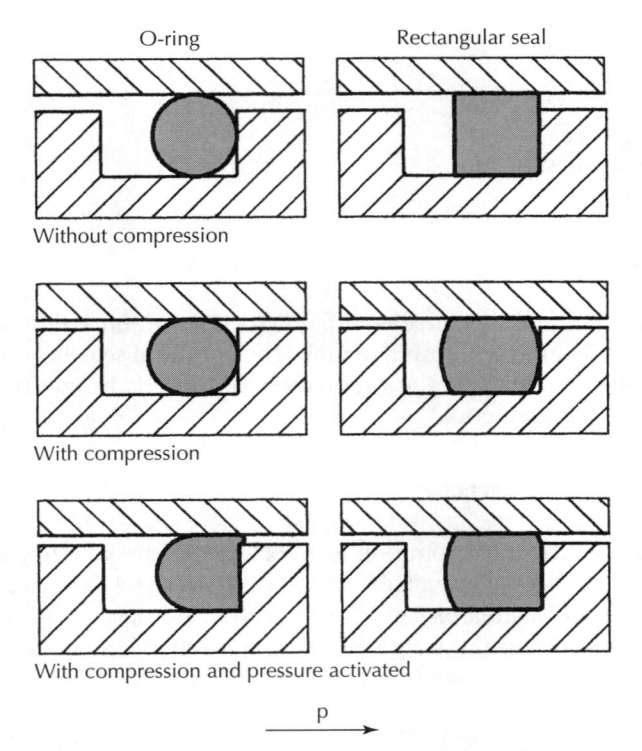

Figure 2.16 Comparison of the action of a rectangular section seal and O-ring during pressure cycles.

may be of rectangular or square cross-section, depending on the application and manufacturing methods.

2.3.2 X section rings

Seals that approximate to an X section are manufactured as an alternative to O-rings. The cross-section may vary from something approaching a square section with lobes, Figure 2.17, to a genuine X section, Figure 2.18.

X-rings offer some potential benefits for certain applications. There are two individual sealing surfaces, so minor damage is less likely to lead to leakage. The cross-section is more resistant to spiral twisting with cyclic pressure. The mould parting line is not on the sealing surface of radial sealing applications, as it is with an O-ring. Of particular interest is that the elastomer interference is achieved by a combination of squeeze and deflection of the seal section. The style of seal in Figure 2.18 is particularly suited to situations where the tolerances are wider than

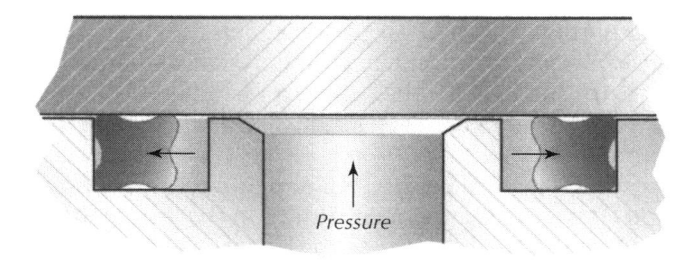

Figure 2.17 The Quad ring is a popular substitute for O-rings. (Source: Minnesota Rubber)

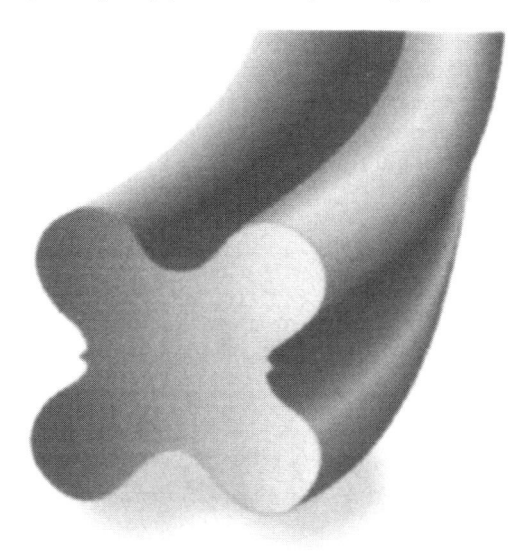

Figure 2.18 Modified Quad ring with a more flexible cross-section. (Source: Minnesota Rubber)

would be acceptable for O-rings. However, because of the more compliant cross-section the pressure range is more limited. Depending on the actual cross-sectional geometry it may be limited to pressures in the range of 8–10 bar.

Although they are more resistant to rolling than O-rings, the standard designs of X-rings are not immune to this form of failure. Although they are considered to be a direct replacement for O-rings, to achieve improved stability a shorter groove is recommended. As there is less need to allow for the lateral displacement of the seal as it is compressed in the groove, less groove width is required. This is normally some 15% less than the standard O-ring grooves.

2.3.3 T-seals

The T-seal is a well-established design that has been in use for many years. It is widely used as an O-ring substitute in reciprocating applications, section 4.2, but is equally suitable for static applications. The cross-section, as the name implies, resembles a T, Figure 2.19. The basic T section is elastomer and performs the sealing function. The overall rectangular section is completed by a pair of backup rings. These will normally be made of plastic and in the standard configuration are in a split form. T-seals are normally used in radial configurations where the additional anti-extrusion benefits they offer can be utilized. Due to the T cross-section separate designs are necessary for outside diameter and inside diameter sealing configurations.

The sealing action is performed in a similar manner to an O-ring. There is initial compression across the vertical axis of the T in the groove. Under the action of pressure the T section will be forced to the low pressure end of the groove. As the pressure continues to rise the fluid-like characteristics of the elastomer will continue to energize the low pressure side backup ring radially outwards, thus closing off the extrusion gap, Figure 2.20. The use of an energized backup ring has several benefits. It does not have to be manufactured to a close fit in the bore as it is expanded by the elastomer as the pressure rises. This also means that it can be provided in harder, more extrusion resistant materials than a simple O-ring backup which is not energized effectively by the seal.

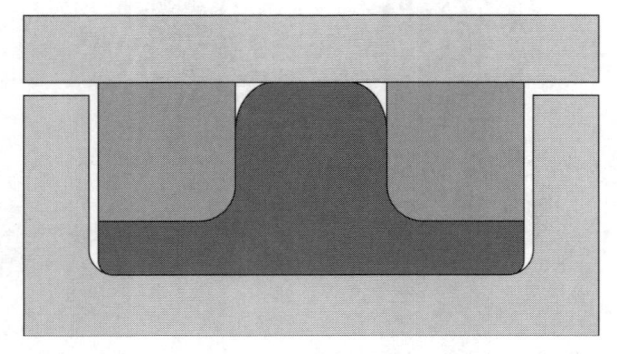

Figure 2.19 Basic T-ring configuration. (Source: Greene Tweed)

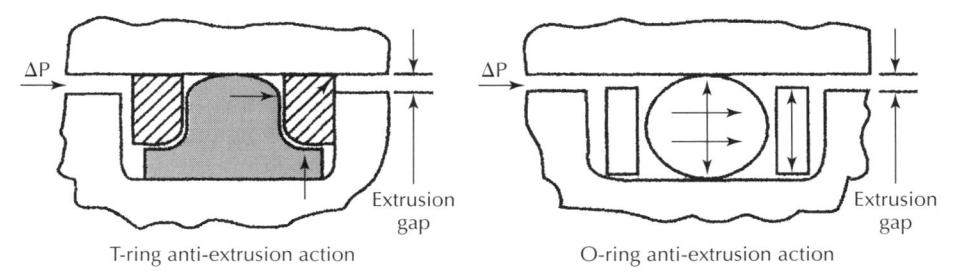

Figure 2.20 The action of T-seal under pressure to energize the backup ring into close contact with the sealing counter-face. (Source: Greene Tweed)

The other benefit of this design is again resistance to spiralling. The basic rectangular overall configuration provides effective resistance to any tendency to twist in the groove.

A number of basic configurations are available from simple T, Figure 2.19, a version with radiused elastomer and backup rings, Figure 2.21 and also those with tapered flanks. The purpose of the radius is to reduce the stress concentrations in the elastomer at the change of section and hence prolong the life of the seal. The tapered flanks provide additional energization of the backup ring for high pressure or other arduous applications.

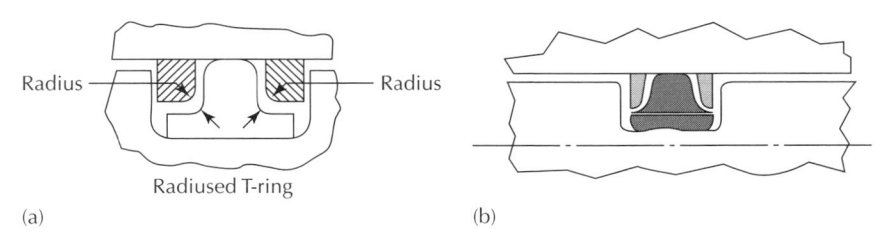

Figure 2.21 T-seal with radius at the transition between sealing section and flanges (a) and tapered flanks to enhance pressure energization (b). (Source: Greene Tweed)

The standard ranges of T-rings are usually available in three widths. The basic seal size will fit within a standard O-ring groove. An intermediate width is suitable for a groove designed for an O-ring plus a standard width backup ring and the widest section, a heavy duty ring is available for a double backup ring groove. As an example, for a 5.33 mm cross-section seal the groove widths for the three options would be 7.1 mm, 8.5 mm and 10.7 mm respectively.

The backup ring material can be selected according to the application. Typical options are nylon, virgin PTFE, filled PTFE and for higher temperature and pressure applications PEEK may be used.

The normal configuration for the seals is to use a split backup ring. This permits easy assembly and readily allows expansion under pressure from the elastomer beneath. However, with small cross-sections and in difficult fitting situations it is possible for the backup ring to become trapped on assembly. To overcome this problem a non-split backup may be specified. This may need careful treatment and resizing during fitting.

2.3.4 L section seals

These are a variation of the T-seal. They are particularly intended for situations where the pressure only acts from one direction. The base of the L will still energize the backup ring into position providing the same benefits as the T section. An example is shown in Figure 2.22. Assuming that the seal only has to act in one direction this design can offer a more robust elastomer and backup in the same space as a T-seal. For critical applications or where access is difficult the possibility of fitting the wrong way round must be considered.

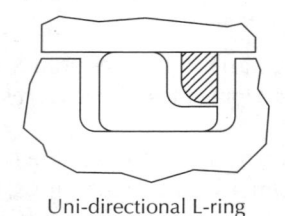

Uni-directional L-ring

Figure 2.22 An L section seal that can be used as a single acting T-seal. (Source: Greene Tweed)

2.3.5 U-rings

U-rings, such as those used for reciprocating cylinder seals, may be used as static seals. These are generally of a larger cross-section than O-rings, and so will require larger grooves and hence more space. The major benefit of such seals is the ability to operate effectively with a much wider range of tolerances. The flexibility of the lips will provide some sealing force over a wider range of groove dimensions. However, a relatively robust U section should be used to ensure adequate sealing force under all conditions. Although the geometry of these seals means that they can be pressure energized it is possible for very rapid pressure transients to buckle the U inwards if an insufficiently robust design is used.

2.3.6 Integral backup flange seals

Seals that are a composite construction of elastomer and a backup ring are produced as flange seals. The backup ring may be designed to provide extrusion resistance, but also for other purposes such as permeation reduction.

Figure 2.23 is a seal that has an elastomer sealing component, with a plastic backing that serves the purpose of reducing vapour permeation from an automotive fuel system. Alternative designs that cater for more arduous duties provide extrusion resistance in applications where the pressure and relative metal work movements are such that good extrusion resistance is required.

An alternative to this design is a complete polyurethane flange sealing ring, Figure 2.24. These seals are available to fit standard hydraulic system flanges

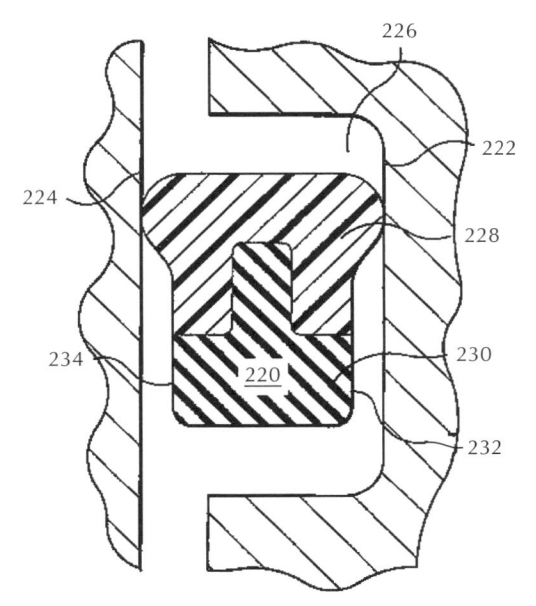

Figure 2.23 Elastomeric flange seal with integral plastic backup ring to reduce fuel permeation. (Source: Patent No.: WO 2003/016756)

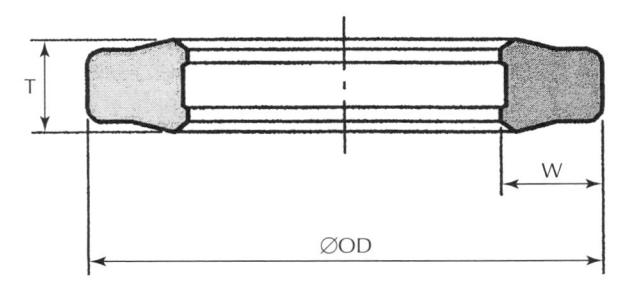

Figure 2.24 Polyurethane flange seal for hydraulic system pipe flanges. (Source: Trelleborg Sealing Solutions)

and offer good abrasion resistance, the possibility of use on inferior surface finish and high extrusion resistance.

2.3.7 Bonded seals

The bonded seal is a combination of an elastomer sealing lip and a metal washer bonded together during the moulding process of the elastomer. The sealing lip is a pre-described amount thicker than the washer, Figure 2.25. They are primarily designed for the sealing of hydraulic fittings, other pipe joints, or under the head of bolts. As the connection is tightened the elastomer is compressed until

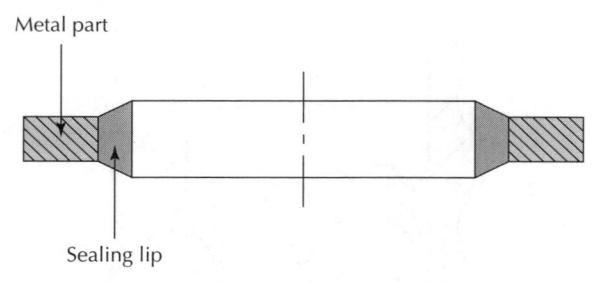

Figure 2.25 Basic design of a bonded washer. (Source: Trelleborg Sealing Solutions)

metal-to-metal contact is achieved. The connection can then be tightened to the desired torque for secure retention. The metal washer effectively replaces the groove wall around other designs of elastomer seal and provides a controlled compression. Once in place the elastomer is retained and will function as a pressure energized seal much as the O-ring and other elastomer seals. This type of washer therefore makes an effective and economic seal for applications such as pipe joints. It is necessary to ensure that the washer is of the correct size to seal on plane parallel surfaces and does not lose contact because of the thread start. This can be ensured by providing some means of retaining the washer concentric to the bolt or fitting, Figure 2.26. For very high pressures a close tolerance recess is necessary to prevent 'panting' of the metal washer which can lead to wear. An alternative is to use a self-centring design, Figure 2.27. This type of washer ensures concentricity and as it is an interference fit to the thread of the fitting it can be pre-assembled. This is useful for applications where fitting or access is difficult as it avoids the problem of potential loss of the washer. A further alternative design provides for sealing actually on the thread of a bolt if this is necessary.

Bonded washers of these various designs are widely available to fit the standard thread sizes of bolts and hydraulic fittings.

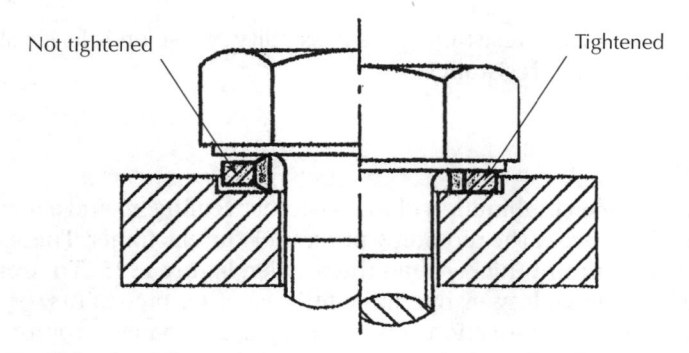

Figure 2.26 Bonded seal installation showing a retaining recess. (Source: Trelleborg Sealing Solutions)

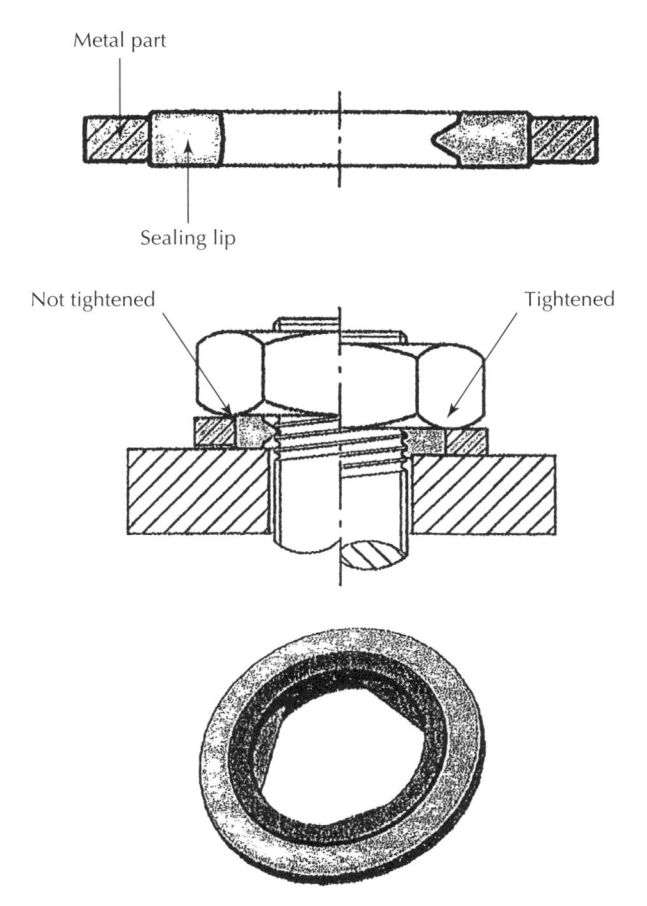

Figure 2.27 Self-centring bonded seal design that can be retained on a bolt by the thread. (Source: Trelleborg Sealing Solutions)

The principle of the bonded washer has been extended to produce what are variously called integral seals, sealing plates or composite seals. The elastomer sealing element is moulded integrally to a metal, or possibly plastic, retainer in the same way as a bonded seal with the thickness of each controlled to provide the required compression of the elastomer. In these seals the metal component may be a complex shape to fit directly to a component such as a valve body, engine accessory or pump housing, etc. The metal plate may have sealing around apertures of a complex shape or a number of individual seals around multiple fluid ports in for instance a valve block flange seal, Figure 2.28. With a seal of this type there is no need to machine individual seal grooves around ports and there is one easily handled seal to fit instead of numerous O-rings. It can also be considered as a replacement for a relatively fragile gasket of complicated geometry. As they are individually designed and manufactured for a specific application this is only an economic proposition for relatively large volume production.

Figure 2.28 Purpose designed integral sealing plates that extend the bonded seal principle to a complete component. (Source: Parker)

2.3.8 Spring seals

The spring seal, or 'S'-seal, is a derivation of the T-seal which is used particularly in the oil and gas exploration industry. Instead of a pair of separate backup rings the spring seal has a pair of toroidal wound springs moulded into the body of the elastomer at the outer edges of the sealing surface, an example is shown in Figure 2.29. These springs will normally be of corrosion resistant material such as 316 stainless steel or Inconel. This design of seal has two potential advantages over the T-seal. As the backup rings are moulded into the elastomer they will not become

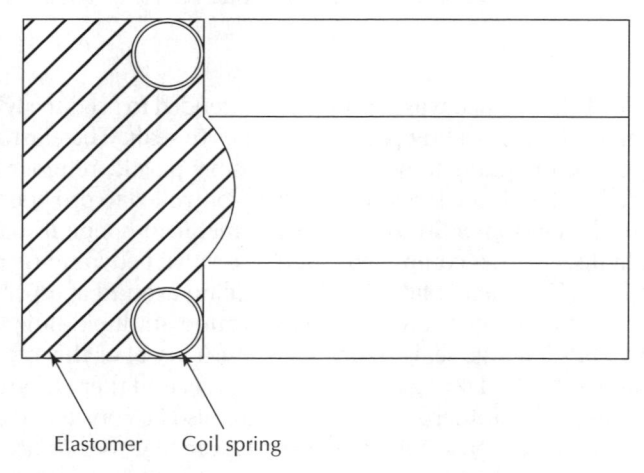

Figure 2.29 Cross-section of an elastomer spring seal, which contains a pair of toroidal springs moulded into the corners of the seal to provide extrusion resistance. (Source: BHR Group)

dislodged during assembly of the components. This is an important factor in large complex assemblies or if access is difficult. The second advantage is that the metal coil spring can provide extrusion resistance at larger gaps than the T-seal design as the backup ring, apart from having inherently higher extrusion resistance, will not dislodge into the gap. This type of seal is very specifically a static seal. Should relative motion occur between the sealed components the metal spring can cause wear or damage to the sealing counter-face.

The manufacture of this design of seal requires particular attention to the moulding process as integration of the spring into the elastomer is crucial to providing a reliable seal. The seals are still capable of being stretched during assembly so can be assembled in typical O-ring grooves. They are typically available, like T-seals for standard O-ring grooves and also the single and double backup ring designs. Spring seals are not available for the very smallest O-ring groove sizes.

The amount of interference designed into these seals can be higher than that of O-rings. This is to ensure sealing contact during extremes of temperature and with relative movement of the sealed components. A further factor with these seals is that the relatively rigid spring can inhibit movement of the elastomer in some circumstances so they may not always be as effective in transient situations as some other seal designs.

To cater for particularly wide clearance gaps and increased relative movement of the surrounding metalwork variations of the basic spring seal design are available. An example is shown in Figure 2.30. This seal has a more flexible profile to maintain sealing contact over a wider range of tolerances and metal movement.

Figure 2.30 A variation of the spring seal to work with wider clearances and metalwork tolerances. (Source: James Walker)

2.3.9 Hygienic seal arrangements

In industries concerned with processing food, drink, pharmaceuticals and personal care products plus the biotech industry the plant must be designed to permit hygienic operation. Pipe work may be designed for rapid regular dismantling to permit cleaning or the plant may be designed for cleaning or sterilizing in place (CIP) or (SIP). In the latter case a cleaning or high temperature sterilizing operation will take place without dismantling the plant. In such equipment it is important to avoid crevices where product residues can accumulate and potentially begin to decompose. With CIP or SIP it is also important to avoid steps, of more than 0.2 mm, or depressions in the pipe work where the flow of cleaning fluid may not thoroughly wash the surface.

These requirements mean that it is not feasible to use conventional groove designs, such as O-ring grooves, which leave a considerable free space in the groove. A further aspect to be considered is that the seal material must be compatible with both the system product and also the cleaning fluids which may be at a much higher temperature, a topic that is covered further in Chapter 5.

A number of specific pipe couplings and also seal arrangements have been developed for hygienic applications. Some types are covered by national, international or internal company standards, but many of these have been in use for some considerable time and are not considered to be compatible with current requirements in some areas of the food and drink industry.[23]

2.3.9.1 Regular hygienic fittings

The well-established couplings have been assessed for applications in the biotech industry.[22] This assessment concluded:

- Ring joint type (RJT): This type of coupling is recommended for use where piping systems are frequently dismantled (BS 4825: Part 5), Figure 2.31(a). This type is not suitable for CIP as there is a crevice between the liner and the male part and would result in process fluid becoming trapped between the two metal components. This coupling has the advantage that it does not need to be accurately aligned when the connection is made and can withstand a degree of 'rough handling'. It is used in the brewing and dairy industry in applications where pipework is manually cleaned.
- International Dairy Federation (IDF): This coupling is recommended for applications where CIP is normally practised. The seals are flanged and are compressed between the specially shaped ends of the male part and liner to give a crevice-free joint, Figure 2.31(b). These fittings are reported to be used successfully in some biotech applications.[22] A variation of this type of fitting uses a T gasket that is an elastomer seal bonded onto a metal T section support, Figure 2.31. This design is covered in ISO 2853.
- Clamp type: These, Figure 2.31(c), may also be used in CIP applications. They are also suitable for frequent dismantling if that is necessary, BS 4825: Part 5. A flanged seal is held between two grooved liners by a clamp which is ring shaped with a taper sided channel section. The clamp can be

Static seals 37

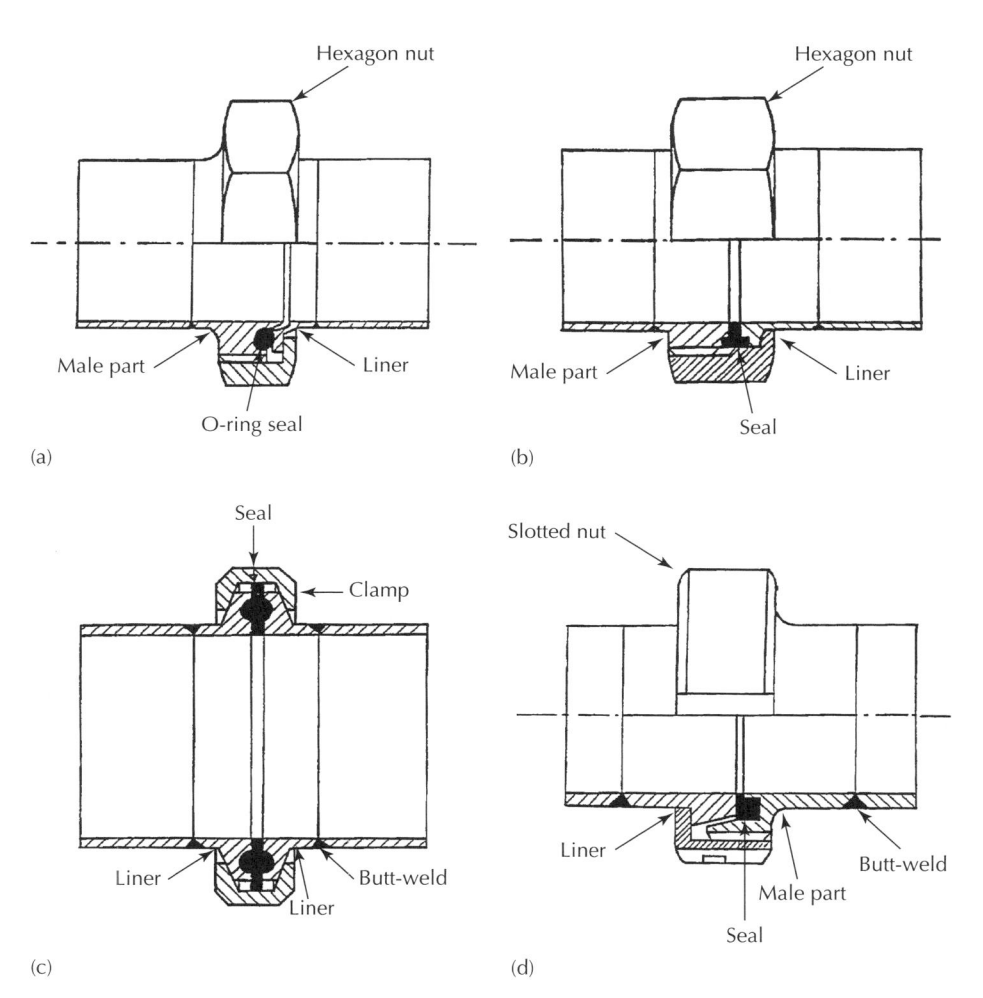

Figure 2.31 Some standard food industry and hygienic pipe fittings.

secured by a locking bolt or a quick-release mechanism. The seals supplied for clamp type couplings are flanged to permit easy assembly and to prevent any displacement by vacuum or pressure. The seal is considered to form a smooth crevice-free joint between the liners, which makes clamp type couplings suitable for CIP duties.

Care is needed with certain designs and sizes of these fittings to ensure that they are not overtightened which could cause the seal to distort. The pipework must be accurately aligned to prevent mechanical distortion of the seal when making the connection. In addition to the food industry these fittings are used in the biotechnology industry. Some user companies have indicated a preference for clamp fittings rather than screw type couplings because in the event of a spill, screw threads cannot be decontaminated effectively. Clamp type couplings are perceived to have the

38 Seals and Sealing Handbook

advantage that in the event of a product spillage at the fitting there is no thread to become filled with product that may be difficult to clean.
- DIN 11851: This fitting, Figure 2.31(d), is similar in concept to the IDF fitting. It has the disadvantage in the standard fitting that there is a crevice and a number of authorities do not consider this fitting to be suitable for CIP.[23, 24] A further potential problem with this design of fitting is that it has a clearance on the cone fitting which means that the two pipes are not automatically aligned, leading to a potential step in the pipe joint.

2.3.9.2 High reliability hygienic fittings

A detailed study has been carried out by a group covering suppliers, users and research organizations in the food and drink industry. This group has provided considerable detail on the requirements for hygienic sealing and seal designs to provide acceptable performance in CIP applications.[23] The design objective is to eliminate the free space in the groove but still allow for seal thermal expansion and swell, etc. Problems can occur with the regular fittings described in section 2.3.9.1 especially with repeated CIP operations. This can include excessive stress on the seals due to thermal expansion during CIP and sterilization leading to surface failures and cracks in the elastomer. The level of interference stress to ensure that product cannot gain access to the interstices between the seal and metal surface texture have also been investigated.[23] This work has made a number of specific recommendations and developed two seal configurations specifically for CIP, with the aim of achieving a minimum of one year life. The seal designs, Figure 2.32, have been developed so that the seal will provide adequate seal interference at the product side of the groove to prevent product access around the seal where it cannot be cleaned. Some space is then allowed in the rear of the housing for thermal expansion to avoid excessive protrusion into the product flow or high stress concentrations in the elastomer. A particular concern has been to provide a seal that is sufficiently constrained at the product side

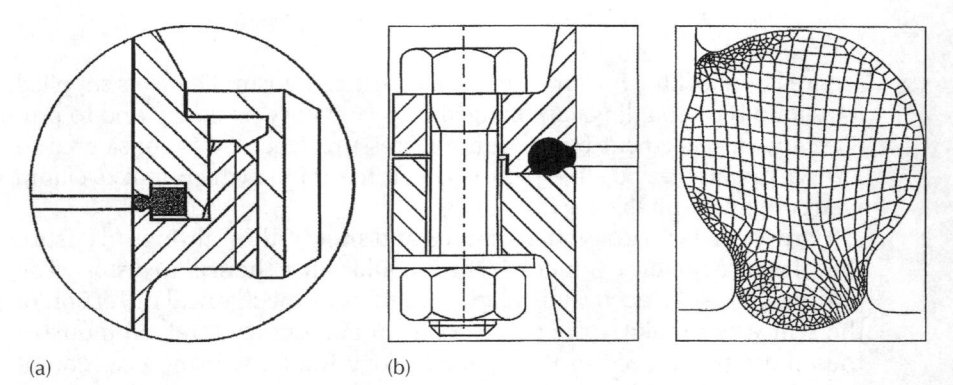

(a)   (b)

Figure 2.32 (a) The diamond section seal profile for hygienic fittings with thermal expansion space in the rear of the groove. (b) An O-ring groove design with sealing at the product end of the groove to prevent trapping of product around the seal.

of the groove, to prevent bacterial growth in the seal housing, and also permitting thermal expansion of the seal during high temperature sterilization and CIP without either excessive seal stress or protrusion into the pipe bore.

The profiled diamond-shaped seal, Figure 2.32(a), is designed to seal on the front faces of the diamond. The block of elastomer behind the seal will accommodate the thermal expansion, relieve stress buildup on the sealing faces and limit expansion into the product stream to a minimum.

Figure 2.32(b) is a design to permit the use of O-rings within a hygienic fitting. To prevent any trapping of product residue the groove is designed to be an interference on the O-ring at the inside diameter edge of the groove. The housing is designed to accommodate thermal expansion and swell to minimize protrusion of the O-ring into the pipe bore.

It is necessary to observe a number of design criteria to achieve satisfactory long-term CIP reliability. Factors to be considered include:[23]

- Elastomer to be 70 Shore A. (Harder gaskets are not sufficiently resilient to fill the surface texture of the metal surfaces.)
- Metal surfaces to be better than 0.8 μm Ra (4 μm Rz).
- Minimum interference stress of the elastomer to be 1.5 MPa.
- Pores on the elastomer surface to be less than 1 μm. (High quality seals are required.)
- Maximum recess of seal within groove to be 0.2 mm.
- Maximum protrusion of seal from groove to be 0.2 mm.
- Avoid tensile stress in the elastomer as it will cause surface pores and liability to cracking.
- Limit compression to 25% to minimize surface tension on seal.

The two seal arrangements in Figure 2.32 are included in hygienic fittings manufactured to DIN 11861-1 and 11861-2.

2.3.10 Window seals

A major application of static elastomer seals is in glazing gaskets for buildings and vehicles. The introduction of the large-scale use of glazing for buildings was facilitated by the development of a sealing system based on that used for vehicle windscreens. This led to the use of the glass curtain wall for high rise buildings.

The seal was based on the two piece windscreen seal design in use at the time. Current large building window seals continue with this theme. They have been developed to attach panes of glass to a supporting framework of metal, concrete or other construction material. To attain the sealing stress required to secure and seal panels to frames the seals are a two part design with the window seal and a separate locking strip. The locking strip is manufactured from a harder material, typically 10 degrees harder elastomer than the actual seal. This extra hardness of the locking strip means that more pressure can be transmitted to the sealing lips.

The locking strip is inserted into a groove in the gasket to put the body of the seal under compression to produce the sealing force. An example of a design for a single pane of glass is shown in Figure 2.33(a). A more complex design for a pair of double glazed panels is shown in Figure 2.33(b). A seal of the type shown in 2.33(b) also seals and protects the building structure and provides some insulation as no steel or concrete framework is exposed to the external environment.

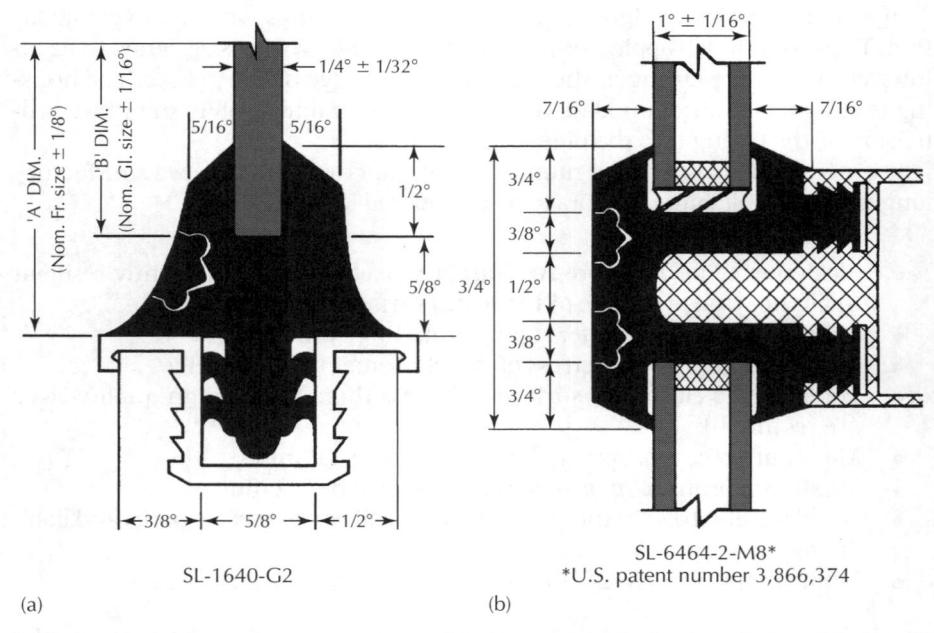

Figure 2.33 Glazing seals for high rise glass wall buildings: (a) single glazed panel; (b) seal for a neighbouring pair of double glazed panels. (Source: Stanlock)

A wide variety of sealing profiles are also used on more conventional building windows, with double glazed panels having several sealing strips per window. An example of a typical seal section is shown in Figure 2.34. There is a seat area for the glass and sealing lips in two directions. It will also be noted that there is a hollow section to provide additional compressibility.

An important factor with all of the building glazing seals is the requirement to maintain a seal with considerable relative movements of the glass and surrounding structure. Not only is there the difference in thermal expansion coefficient between the glass, plastic and brick, concrete or steel but the windows will change in temperature relatively rapidly compared to the building structure, especially for instance when exposed to sunlight.

There appears to be some difference between the USA and Europe on the sealing material. The USA continues to promote the use of Neoprene, used when the seals were first developed, whereas EPDM appears increasingly popular in Europe.

Vehicle window seals have to be designed for similar conditions to buildings in that there will again be relatively large thermal movements, but the metal structure of a vehicle will behave very differently to a building. In such a high

Figure 2.34 Typical seal profile for a domestic double glazing system.

volume production-oriented environment the fitting of the seals is a major design consideration. Facilitating opening windows which still provide an adequate seal is also an area in which there is a lot of interest. In this area development is focused on facilitating a low load while the glass is moving and then increasing the sealing force when it is in position.

The design shown in Figure 2.35 has features on both the seal and the door to facilitate retention of the seal. The weather strip (22) sealing against the window

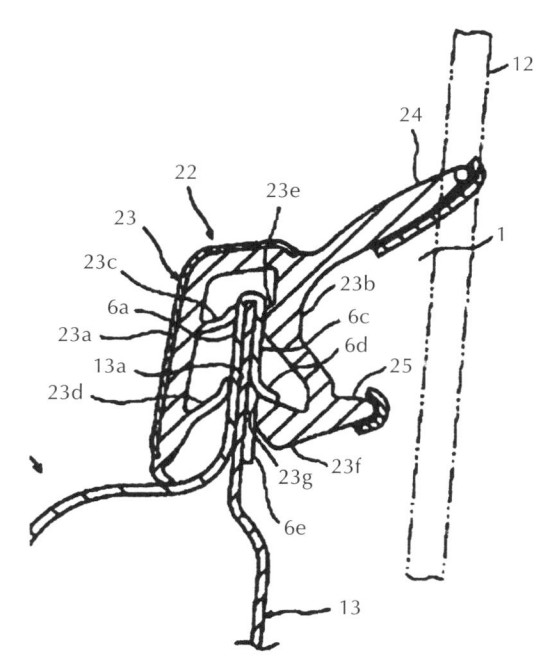

Figure 2.35 A method of retention of a door window seal on vehicle bodywork. (Source: Patent number WO 2005/042286)

42 Seals and Sealing Handbook

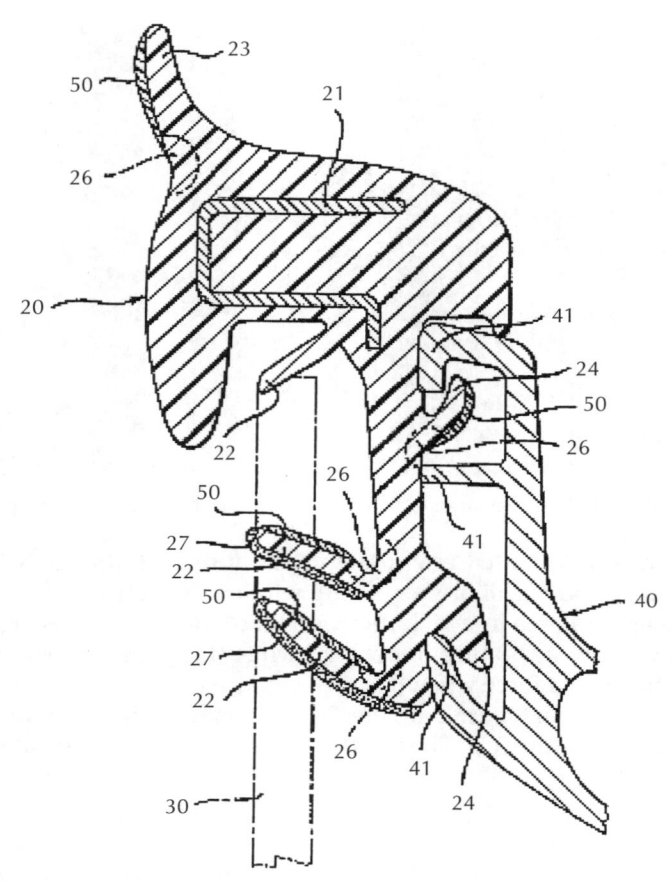

Figure 2.36 A vehicle door window seal that contains piezo-electric actuators to increase the sealing force when the window is closed. (Source: Patent number WO 2003/104008)

(12) is installed on the upper side part (6a) of the door outer panel (6). The claws (6d) engage with the lower end of the hook (23f) of the weather strip to prevent it from being extracted. The additional door sections (6a, d, e) abut the weather strip to prevent lateral movement.

There is considerable development in the area of activated seals to provide sealing in the closed position. An example is shown in Figure 2.36. The elastomer seal (20) is fitted to a frame (40) and closed against a window (30). It may have one or several sealing lips (22) that rest against the window. To improve the sealing system and obtain a variable sealing effect, which is adapted to environmental factors, the sealing lips are provided with a piezo-ceramic actuator (50) and can be pressed against the window using a voltage applied to the actuator. An alternative approach to achieving the same benefit is to use a seal activated by memory metal.[25] This type of approach is particularly necessary with coupé type cars where there is a large glass area and no support structure from the door frame. This creates the potential for considerable relative movement between the glass and surrounding structure.

2.4 Alternative plastic sections

Plastic materials are both harder than elastomers and have a limited elastic limit and resilience. They cannot therefore apply sufficient sealing force within their own elastic limits and retain inherent energization properties as discussed for elastomers. It is therefore generally necessary to provide separate energization of a plastic material to ensure an adequate seal. The most popular method for using plastics, most often PTFE, is a U section seal with a metal spring energizer, Figure 2.37. The metal coil spring provides the initial interference and ensures a seal at low pressure. As the pressure is increased the U section ensures that the system pressure energizes the seal in addition to the spring load, ensuring a continued seal.

Figure 2.37 Spring energized U section PTFE seal for static applications. (Source: Balseal)

For static sealing applications a coil spring manufactured from metallic strip is used. This provides a high interference load to create adequate sealing contact between the seal and counter-face. It should be noted that the spring design is different to that used for dynamic seals of this type discussed in sections 3.3.4 and 4.2.

These seals are again available to suit standard O-ring groove sizes. However, as they are individually machined they are also readily available in special sizes. The design will also be adjusted to suit the application. As an example, for cryogenic duties a higher spring load and initial interference will be used to counteract the additional thermal contraction of the seal, and compensate for the fact that the plastic will be harder at very low temperatures.

This design of seal is unidirectional so specific designs are required for flanges and radial sealing applications. With flange applications it is also necessary to have separate types for internal and external pressure, Figure 2.38. The strip coil spring will also limit the amount of stretch that is possible when fitting a radial seal so application in blind grooves may be limited.

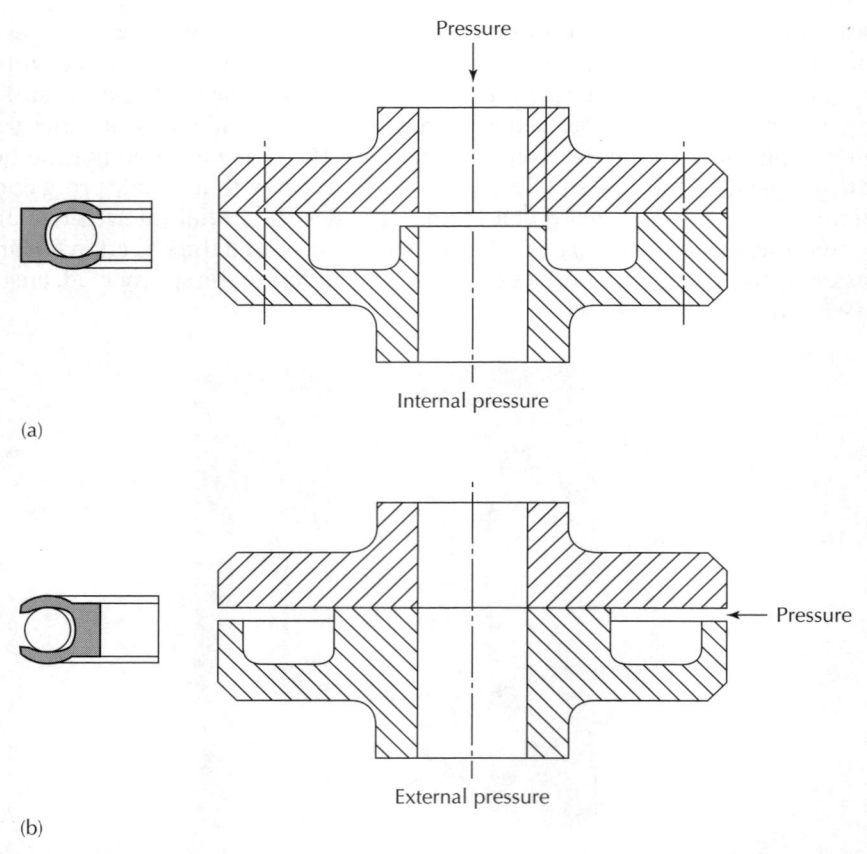

(a) Internal pressure

(b) External pressure

Figure 2.38 PTFE spring energized seals for internal and external pressure. (Source: Trelleborg Sealing Solutions)

These seals will be more expensive than most O-rings, and more difficult to fit, with greater care required in handling and with surface finish, etc. They provide a wider range of temperature capability and practically unlimited fluid resistance. For aggressive chemical environments the spring material can be selected from a high nickel alloy. They can also be used in cryogenic environments where elastomers are unsuitable.

The materials will be chosen to suit the application. For many applications an unfilled PTFE will be used to provide the optimum sealing integrity and widest fluid resistance. A filled PTFE may be used if improved extrusion resistance is required. In some applications, such as food, water and cryogenics, ultrahigh molecular weight polyethylene (UHMWPE) may be used.

2.5 Alternative metal seal designs

A variety of metal seal designs are used as an alternative to O-rings. They are intended to provide either reduced flange loading, improved flexibility for flange movements or a combination of the two.

2.5.1 C-rings

C-rings are used in similar application areas to metal O-rings but offer improved flexibility and lower flange loading requirements compared with the equivalent O-rings. The linear compression on assembly is similar at approximately 20% for flange seals but the load required to achieve this is approximately 50% of the equivalent metal O-ring. As the C is open to the pressure it must be correctly specified and fitted for external and internal pressure applications. The improved flexibility also means that C-rings can be considered for axial sealing configurations, where they are compressed radially. Examples of each configuration are shown in Figure 2.39. The compression applied to the axial seals is much lower, approximately 5%. They will require considerable care when fitting to avoid damage and a shallow taper of 5–6° is required. Although an axial configuration can be considered this is still for static seals only.

C-rings are available in a similar range of metals and coatings to metal O-rings.

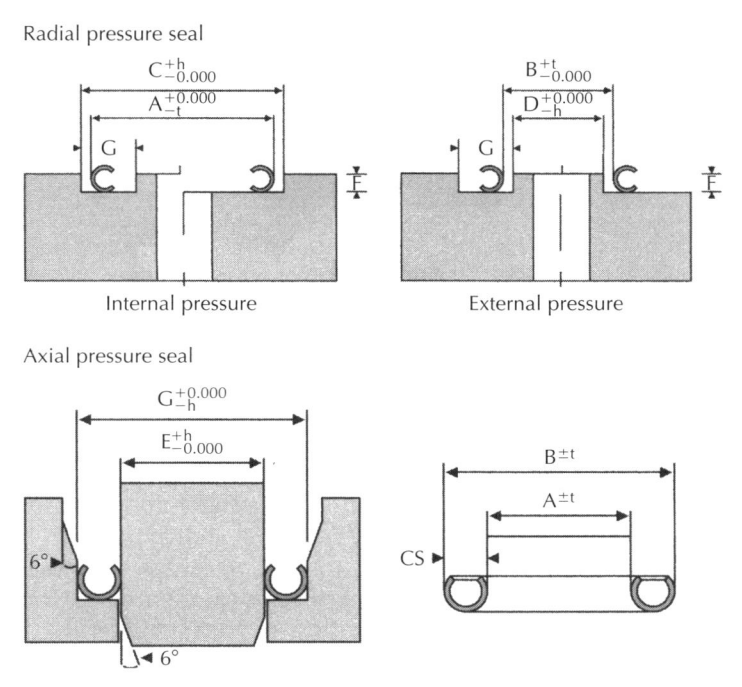

Figure 2.39 Metal C-ring configurations. (Source: Garlock)

2.5.2 Low load C-rings

To address the high load limitations of conventional metal seals one company has introduced a 'C' section metal seal called Ultra-Flex, primarily intended for high vacuum applications. This is designed to offer the user improvements in the areas of high and low temperature resistance, improved tightness, lower permeability, zero outgassing, reduced contamination and improved chemical resistance. It has offset contact ridges in order to have them positioned at the main reaction axis when optimum deflection is achieved. A simple forming technique is used to keep the soft sealing material of aluminium, silver or copper, attached to the elastic metal substrate. This avoids problems that occur when trying to bond two metals. This seal, Figure 2.40, has a face loading requirement that is up to an order of magnitude lower than standard metal O- and C-rings at approximately 20–30 N/mm of circumference. The lower overall compression loading is achieved by creating the area of high stress at the contact ridges which will cause plastic flow in the area of these ridges.

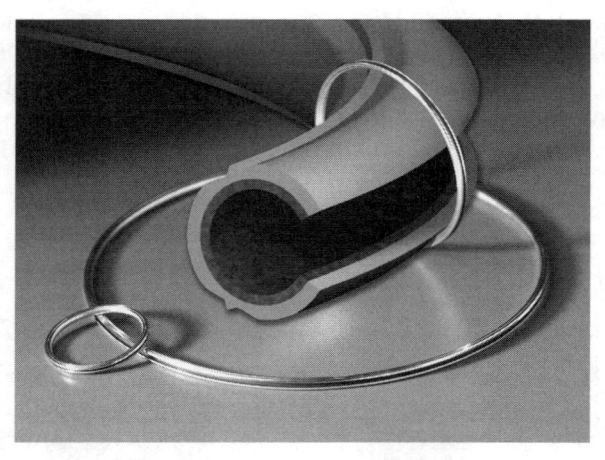

Figure 2.40 Low load metal C-ring, Ultraflex. (Source: Garlock)

2.5.3 Spring energized metal seals

These are a special version of metal O-rings intended to provide high integrity sealing in the nuclear industry and other applications where a minimal leak is required. A closely wound helical spring is surrounded by two metal jackets. The inner metal ring provides the energization from the spring to the outer metal layer. The outer layer is designed to be more resilient than the flange groove surfaces and is manufactured from a range of coating materials to suit the application including high purity silver. The construction is shown in Figure 2.41.

The application areas cover vacuums to 350 MPa and from cryogenic to 980°C. The spring provides additional resilience to cater for flange movements. The compression of the spring for energization involves a higher flange loading than metal O- and C-rings. Compression to the design criteria of 16% requires

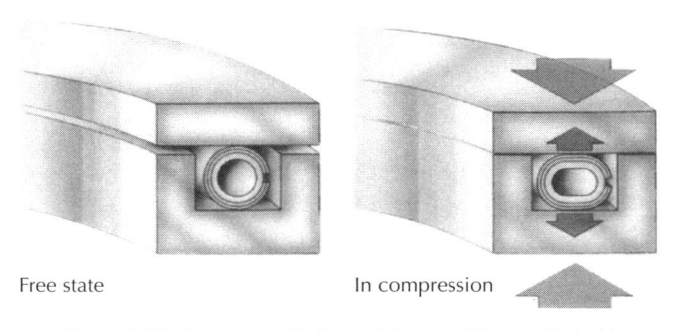

Figure 2.41 Spring energized metal O-rings. (Source: Garlock)

450–700 N/mm. The leakage rates are as low as 10^{-10} atm cc of He at 1 bar, some two to three orders of magnitude lower than standard pressurized metal O-rings or the permeation losses through an elastomer seal.

2.5.4 E-rings

E-seals, such as those shown in Figure 2.42, provide a lower flange loading and more flexibility than O- or C-rings. The improved resilience or springback makes them more suitable for applications where some flange separation may occur in service. They are used on aerospace flange and ducting applications plus other

EIS series Internal pressure short leg	
EES series External pressure short leg	
EIL series Internal pressure long leg	
EEL series Internal pressure long leg	

Figure 2.42 Metal E-seal configurations for internal and external pressure. Long leg designs provide lower flange loading. (Source: Nicholsons)

high temperature duties. They are available in cross-sections from 2.3 to 5 mm. Two basic designs are available, short leg and long leg. The long leg design will allow lower flange loadings and more resilience. The flange loadings are in the ranges 15–25 N/mm for short leg and 5–15 N/mm for long leg seals. The standard material for the seals is a high nickel alloy N07718 (Alloy 718) as discussed in section 5.9.2. A range of soft metal coatings are provided depending on the application and temperature, Table 2.8. As these seals have a low compressive loading for a relatively wide sealing face area the sealing contact stress will be comparatively low compared with most other metal seal designs. This will make the sealing integrity very dependent on the surface finish of the flanges and selection of the most appropriate coating. If high sealing integrity is required they are unlikely to be the most appropriate choice of metal seal.

Table 2.8 Soft metal coatings for metal E-seals

Plating or coating	Comments
None	
Silver	Suitable for most applications
Nickel	
Copper	
Gold	Very expensive
Lead	Soft, max temp 150°C/300°F
PTFE	Max temp 220°C/340°F
Indium	Very soft, low temp only

Plating may be incomplete away from sealing surfaces.
(Source: Nicholsons)

2.5.5 Sigma seals

The sigma seal, Figure 2.43, is a variant of the E-seal which provides greater flexibility for flange movement. It is particularly intended for applications such as high temperature ducting on gas turbine engines where considerable relative flange movement may occur. They are available in larger cross-sections, 3.0 to 6.0 mm. The flange loading requirements will vary depending on the seal section and material thickness, but are broadly comparable to E-seals. The seals may be compressed between 5 and 20% of their free height. The material may be one of several high temperature nickel-based alloys or stainless steel for lower duty applications. Again a range of soft coatings may be used.

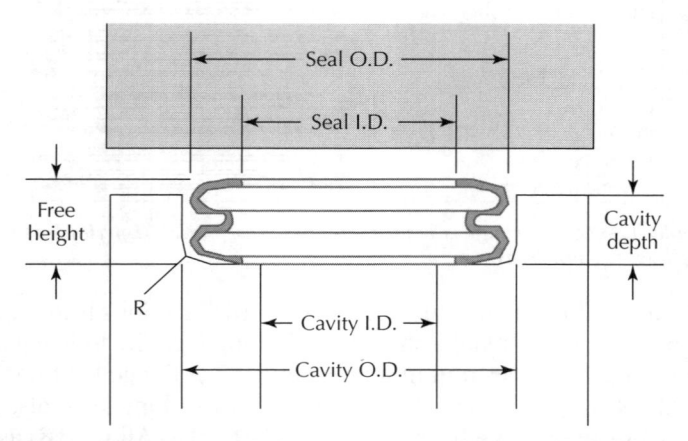

Figure 2.43 Sigma seals permit more relative flange movements than most other metal seals.

2.6 Cured in place seals

Where a high volume of units are manufactured it is feasible to consider providing an elastomer seal that is cured in place on the component. There are two basic methods which for convenience are termed mould in place and cure in place.

2.6.1 Moulded in place

With a moulded in place seal the component to be sealed forms one face of the mould and a seal is moulded directly onto the substrate. This provides several advantages over a separate seal:

- Complex seal shapes can be considered around a component that is optimized for function and not compromised to facilitate the incorporation of a standard seal geometry.
- The seal is retained on one face and is resistant to being omitted or dislodged during assembly.
- A wide range of elastomers may be considered. The choice may be limited by the temperature resistance of the substrate to which the seal is moulded.
- Seal will be retained during occasions such as service access in the field.
- Good control over seal material and spares usage.

Cured in place seals can still be used with a conventional groove design and subjected to controlled compression as if it were a conventional O-ring. Some consideration may be required of the tolerance implications but these will still be close to those of other moulded elastomer seals.

If this type of seal is to be used it obviously requires the involvement of a seal manufacturer at an early stage in the product design. This can lead to potential simplification of the component design as shown by the valve plunger in Figure 2.44. The supply chain to the production line can also be simplified.

One version of the cured in place seal is the integral seal, or sealing plate, discussed in section 2.3.7. In a sealing plate the elastomer is generally supported by

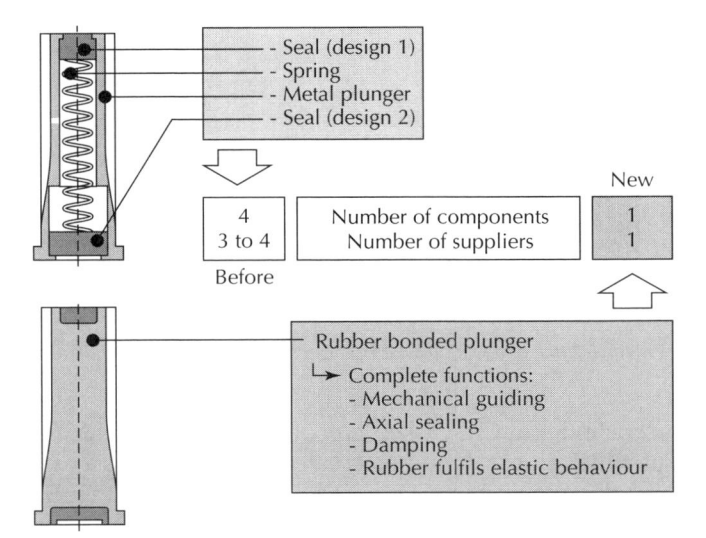

Figure 2.44 Moulded in place seals can make the component simpler and avoid loss of seals during assembly. (Source: Freudenberg Simrit)

the plate and compressed between two mating seal surfaces. A moulded in place seal is typically moulded directly to one of the surfaces to be sealed and compressed against the other face. This saves the additional seal component compared with a sealing plate, but the seal cannot be replaced in isolation. The expected life of the component, maintenance requirements and general serviceability must therefore be considered before selecting between these options.

2.6.2 Cured in place

A system that will require less investment and provide more flexibility than moulding in place is termed cured in place (CIP), not to be confused with clean-in-place, section 2.3.9. With this system the seal is applied as a fluid by tracing machines in precise beads to one of the flanges, Figure 2.45. The bead is then cured, to form an elastomeric material that adheres to the flange surface. Sealing is accomplished through compression of the cured seal during assembly of the flanges, again just as previously discussed for O-rings or other elastomer seals. Seals in this category can be readily used for sealing joints that have to be broken for service.

A major benefit over the moulded in place seals is that a change in component only requires the reprogramming of the dispensing head, rather than manufacture of a new mould. It is therefore applicable to relatively low volume production.

Figure 2.45 Application of a sealing bead for a cured in place seal. (Source: DuPonT Vertex)

Manufacturers find that CIP gaskets can be a considerable advantage because of the speed at which the seals can be created. It is, however, a technology that requires specific designs and precision and repeatability during dispensing and curing. The beads must be of consistent shape and position. As a result, decisions regarding foamed or solid material, cure mechanism and one- or two-component materials need careful consideration. CIP seals offer the advantages previously discussed over die-cut rubber, die-cut foam rubber or moulded seals. Among them are automatic application, serviceability, labour savings, improved quality, reduced inventory and flexible manufacturing.

CIP seals must be cured prior to assembly and several systems are available:

- Fast ultraviolet light.
- Heat cure.
- Two part mix.
- Slow moisture curing which can take as long as 7–14 days.

Low production rates can use the slower curing methods and avoid investment in costly equipment. However, there is risk to parts being contaminated or damaged while curing, so generally the faster cure methods are preferred.

CIP is currently rather limited in the sealing materials available because the elastomer must be compatible with the curing system. All four curing methods mentioned above may be used with silicone elastomer and this is the most common material used. Polyurethanes are also widely available. For high production rates the UV curing method is popular as it is rapid and does not involve heating the material and component. This saves considerable time and also energy costs and may be used with components manufactured from materials that may not be heated or could distort at moulding temperatures.

A wider range of elastomer compounds are becoming available for CIP[10] and currently ethylene acrylic, EPDM and FKM elastomers are available with UV curing systems via the patent DuPont elastomers system, Vertex.

An important factor in the development of a CIP process is achieving an uncured elastomer mix of the correct consistency for application via the applicator nozzle that will also cure satisfactorily in a free state. The finished and cured sealing bead requires both a controlled aspect ratio and acceptable tolerance on the bead height to provide an acceptable alternative to a conventional moulded seal. The current methods claim an aspect ratio of 0.8 to 0.9, but this will vary with elastomer compound and application method.

2.7 Formed in place seals and gaskets

Formed in place (FIP) seals and gaskets are applied as a fluid sealant to one of the flange surfaces before the parts are assembled. When the parts are assembled the sealant spreads between the flanges, filling gaps, voids, scratches and surface irregularities. After assembly the gasket cures and forms a durable seal.

Many factors influence gasket choice but formed in place gaskets are available for a wide range of applications. To achieve the required sealing performance on such a wide range of surfaces, two types of FIP materials are available:

- Anaerobic sealants for rigid flanges.
- Special silicone products for lightly clamped and flexible flanges.

2.7.1 Sealing rigid flanges with anaerobic sealants

Whether or not a flange is classified as rigid or flexible depends on its design and function within the whole assembly. Rigid flanges are designed to achieve the optimum stiffness between two mating parts and to minimize movement between those components. In addition, they may be used to transmit forces from one part to another. To fulfil the requirements of rigid flanges the clamp load of the bolts has to be distributed as uniformly as possible over the mating surfaces.

Anaerobic gasketing materials can provide a number of potential benefits compared with a conventional gasket:

- They can add structural strength to the finished component.
- There is no gasket relaxation because anaerobic gaskets allow flanges to come together with metal-to-metal contact, and ensure the correct clamp load is maintained throughout the life of the assembly.
- No retightening of the bolts is required to compensate for gasket relaxation.
- The metal-to-metal contact means that no allowance is needed for gasket thickness – resulting in tolerances being more accurately maintained. This can be important if bearings are preloaded by the assembly of two halves of a housing.
- High shear strength which can be used to stop movement due to side loading. This reduces any tendency to bolt loosening and fretting between flanges and increases the structural strength of the whole assembly.
- They also allow relaxation of surface finish and flatness tolerances and enable scratches and scored surfaces to be sealed without rectification.
- Sealant can be used for flanges of varying sizes and shape; ready moulded or cut seals are not required.

Anaerobic sealants cure on metal surfaces in the absence of air. This allows them to offer extensive on-part life when exposed to the atmosphere. This makes multiple application methods possible and reduces the problems associated with the use of volatile and/or moisture-cured materials.

As anaerobic gaskets cure only between metal parts, excess material can be wiped away from exterior surfaces or flushed away from interior faces. Liquid anaerobics are miscible in many fluids which can avoid problems of internal passages and channels becoming blocked. However, the compatibility of the sealant with product within the assembly, and the consequences of possible contamination, must be considered.

Cured anaerobic gaskets demonstrate excellent solvent resistance to petroleum-based fuels, lubricating oils, water/glycol mixtures and most other industrial chemicals.

In addition to manual application from a tube or dispenser, Figure 2.46, they can be applied by several fully automated robotic dispensing, screen or stencil printing systems.

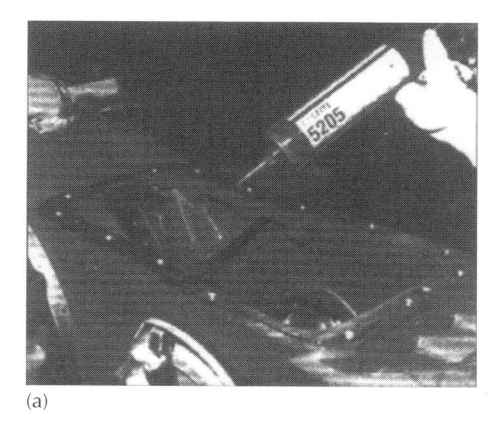

Figure 2.46 Two systems that can be used for manual application of an anaerobic sealant: (a) manual application using a dispenser tube; (b) using a Trax roller to provide even dispersion of the sealant on a flange. (Source: Henkel Loctite Adhesives)

2.7.1.1 Flange design considerations for anaerobic sealants

A number of design considerations should be taken into account to achieve the optimum performance, as a sealant cannot be used to overcome a basically inadequate design.

The size and number of bolts are important as the bolt load is usually the only force that holds the flanges together. As a result, the size, location and the distance required between the bolts will depend on the forces that are acting on the flange. In many cases forces are not acting equally over the entire flange so bolts of a larger diameter or higher grade should be used in places where higher loads are expected. In addition bolt spacing and positioning should be considered. The best clamping pattern is invariably a combination of the maximum practical number of bolts and optimum positioning. When flanges are bolted together, the force applied through the bolts should result in a pressure distribution equal to at least the minimum required over the entire face. Bowing of flanges can be minimized if the distance between bolts is 4 to 10 times the bolt diameter.

One of the industry standard models for bolt spacing reveals that the force under the bolt is distributed in a 45° cone. As a result, the effective bolt length and flange thickness are important parameters in optimizing spacing. Applying this rule, the bolt spacing is also dependent on the stiffness of the flange.

Taking this a step further, bolt positioning is at least as important as the bolt spacing. When the bolts are located in the wrong places, the forces applied through those fixings will not act to create the necessary contact on the flange. Indeed, they will cause bending of the flange or separation of the faces. Straight lines drawn from bolt to bolt should follow the centre line of the contacting surfaces of the flanges as closely as possible. However, the inherent properties of an anaerobic sealant and the reduced flange load compared to a gasket can have important design and production advantages.

For example, changing from paper gaskets to an anaerobic has helped to produce more efficient joints for valve fittings. The need for an improved seal became apparent when a gunmetal valve was redesigned. The smaller, lighter and more user-friendly unit meant that the bolt pattern had been changed. In particular, the distance from the central valve to the perimeter holes was extended. At the same time, the flange thickness was reduced to meet a design target of 50% less weight. However, the flanges bowed very slightly when the four M5 stainless steel clamping screws were tightened. In fact, to seal the flanges by compressing a paper gasket required a torque greater than that of the threads tapped into the valve body and as a consequence, the threads stripped. By applying an anaerobic flange sealant to the smooth-milled gunmetal flanges, the gap of 0.5 mm was filled and provided an instant low pressure seal. It also cured fast enough to permit an on-line 5.5 bar air-pressure pre-despatch test of all the valves.

- Flange rigidity: Rigidity has a great influence on the compressive stress distribution and the amount of bowing of the flanges. By increasing the flange thickness, the compressive shear distribution improves. If thicker flanges are not possible, stiffening ribs, ideally positioned mid-span between the bolts, can help to avoid bowing of the flanges. Theoretical pressure at a given point of the flange surface may be almost impossible to calculate because of the complex design of the component. However, a more practical and realistic approach is the use of pressure sensitive film to evaluate stress distribution in a joint.[11]
- Surface requirements and minimum dimensions: The following parameters are important with anaerobic gasketing products:
    - Surface roughness should be between 0.8 and 3.2 µm Ra.
    - Surface flatness should not exceed 0.1 mm over a 400 mm length.
    - Maximum gap at surface imperfections such as scratches or machining marks must be within the maximum cure-through range for the selected anaerobic product.
    - Flange overlap width must be generally 5 mm minimum and bolt holes 3 mm minimum, to ensure reliable curing of the anaerobic product.
    - Dowel and bolt holes should be chamfered to eliminate raised metal and shimming.

- The use of alignment dowels is recommended for the assembly of large metal parts in order to prevent smearing of the FIP gasket material.
- Regardless of the application method, a continuous pattern of FIP gasketing material should be applied inside or around bolt holes to eliminate secondary leak paths.

In addition to the correct design, effective sealing of an FIP gasket depends on adhesion to the flange surface. In order to achieve a reliable cure and maximum adhesion, it is generally necessary to clean the flange surface.

Although anaerobic FIP materials exhibit indefinite 'open' time once applied, it is recommended that assembly takes place within one hour in order to reduce the potential for particle contamination.

Similarly, silk screen/stencil application devices should incorporate dust hoods to prevent contamination.

Anaerobic compounds cure rapidly between metal surfaces, but this can be further accelerated by heat and/or activators. To ensure sealing success, all fasteners must be tightened to the specified torque immediately after assembly. Subassembled parts may require slave fasteners to assure consistent clamp load during the curing process. The maximum possible fixture time should be allowed prior to pressure testing, while the minimum possible air pressure and duration should then be used to assure a satisfactory assembly. In the short-term, before sealant cure, pressure resistance is dependent on flange width, FIP gasket material, viscosity of the product and induced gap.

A very wide range of anaerobic sealants is available to suit many different operating conditions, design criteria, application methods, cure time and type of joint, flange or thread sealing, etc.

A representative selection of examples is presented in Table 2.9.

- Application techniques: Pneumatic or manually powered caulking guns can be used to apply a bead of anaerobic sealant from either 300 or 850 ml cartridges. This is a simple and inexpensive method of applying these products.

It is essential that the proper sequence of assembly be followed for the particular equipment. Anaerobics can be left open, but a part should never be rested on its mating flange without applying the correct torque to the fasteners. This is because a partial cure can result where the cover comes into contact with the anaerobic material.

The sealant should be applied to only one of the sealing surfaces. For the sealing operation to be completely successful, the product must be in the correct location over the whole flange surface. It is important to remember to circle dowel and bolt holes when manually applying sealant.

When applying a bead, the cartridge or tube should be held slightly above the sealing surface in order to lay the bead onto the joint. The nozzle tip should not be held directly on the surface, as bead shape will be harder to control.

Once the application is complete, the bead should be inspected for uniform thickness, air pocket voids and continuity. Any imperfections should be repaired as soon as possible.

Table 2.9 Examples of the range of FIP sealants available

Gasket type	Formed in place							
Flange type	Rigid			Flexible		Flexible		
Cure method	Anaerobic			Moisture		Moisture		
Cure speed	Moderate	Fast	Slow	Slow		Slow		
Tack-free-time	N.A.			30 min	10 min	10 min	5 min	20 min
Temp. resistance	200°C		150°C	250°C	200°C	200°C		
Instant seal	Limited		Excellent	Limited	Limited	Excellent		Limited
Oil resistance	Excellent			Limited	Excellent	Excellent		
Water/glycol resistance	Good			Limited	Excellent	Excellent		Moderate
Colour	Pink		Red	Copper	Grey	Grey		Black
Product	**510**	**5205**	**5206**	**5920**	**5699**	**5999**	**5900**	**5910**
Dispensing systems								
Handheld	Yes					Yes		
Automatic applicator	Yes					Yes		
Special capability	Screen printable			Bead application		Bead application		

It is important to use the recommended bead size for each project and not apply more material than is necessary to seal a joint. Too much sealant will result in the material being squeezed out on either side of the joint. Wiping the exterior is acceptable for anaerobic sealants.

If leak testing is required, wait at least 45 minutes and use the lowest air pressure possible for the shortest time duration.

Anaerobic products are also applied using a number of other successful application methods.
- A PTFE coated steel stencil or silk screen can be mounted on either a wood or aluminium frame which is supplied with a 'squeegee'. Dowel pins mounted on the frame allow it to be indexed for repeatable bead placement. The quantity and pattern are dependent on stencil or screen thickness and sealant bead width.
- Repeatable bead applications can be achieved by using robots or X–Y machines. Anaerobic materials can be dispensed from 300 ml or 850 ml cartridges or 5 litre packages for high volume production lines.

2.7.2 Sealing flexible flanges

Unlike rigid flanges, their flexible counterparts are not used for applications where optimum stiffness of the component is necessary. These flanges do not usually support the function of the component, and are normally used to seal and keep a liquid inside a component, to protect from external contamination, to cover moving parts for safety reasons or to encapsulate components to reduce noise. Examples of flexible flange designs include sump to engine crankcase, engine timing chain covers, gearbox cover, various stamped sheet steel parts, housings and covers made of plastic and thin-walled metal castings. Some movement between the flanges can therefore be tolerated and an optimum clamp load distribution is not necessary provided that unnecessary distortion is avoided during assembly.

In addition, there are other types of flange designs that have the same need for flexible gaskets. These include parts where the compressive load distribution required for anaerobic sealants or fibrous gaskets cannot be achieved. Examples are assemblies with varying flange materials and large differences in their thermal expansion coefficients which can result in bowing, or flanges where more than two parts are mounted together forming T-joints.

O-rings and other elastomer configurations can also be used on flexible flanges. The use of an FIP sealant can again provide more design freedom on flange geometry and avoid seal stockholding requirements as discussed above for anaerobic sealants.

Flexible FIP materials are generally based around special silicones, formulated to meet all of the requirements of a gasket for flexible joints. Such products provide high per cent elongation to compensate for any micro-movement as well as providing excellent long-term adhesion to most substrates. They also possess a curing mechanism which is independent of the substrate meaning they can be used on metal, painted metal and plastics. Most single part compounds are offered as curing, or vulcanizing at room temperature (RTV). The curing systems are based on exposure to atmospheric moisture. Other benefits include good cure-through volume which means that large gaps can be sealed, up to several millimetres, and wide operating temperature range, -70 to $315°C$, with a potential for intermittent exposure up to $350°C$.

Some budget alternatives are available in the market, but these do not offer the flexibility, fluid resistance and temperature range of the silicone compounds, so should be treated with extreme caution.

2.7.2.1 Design considerations for flexible flanges

The method of sealing is through the adhesion of the cured gasket to all of the contact faces in the joint. To allow the gasket to compensate for micro-movement without overextending the elongation limit of the adhesive, a certain minimum thickness of gasket material is required. This means that a defined gap should be achieved between the flange surfaces. Assuming that the two flanges are to be clamped together to ensure integrity of the bolting, this requires a flange design specifically arranged for the sealant.

Reservoir groove is one method used to achieve a defined gap, which creates a design that incorporates a reservoir for cured gasket material by having a semi-circular groove stamped or cast into one of the flange faces. The surface finish should be $0.8–6\,\mu m$ Ra. The dimensions of the retention groove are optimized according to the total size of the flanges and the expected micro-movement.

The larger the extent of any expected micro-movement, the bigger the retention groove in order to allow for elongation of the gasketing material.

Chamfer or radius design is another proven configuration and works more effectively than the retention groove.

The Global Engineering Centre of Loctite in Munich, Germany, conducted extensive testing of RTV silicones to determine the optimum flange design. It was found that, to achieve the necessary thickness of cured gasketing product, a chamfer at the in-board surface of the joint can be cast or machined. If one part is made of stamped steel, a defined bending radius can be used to get a similar fillet comparable with the chamfer design.

There are basically six benefits of the chamfer design:

- It creates an instant seal capability due to the metal-to-metal contact.
- There is fast curing of the exposed 'wedge' of product.
- An ideal product thickness is produced.
- Squeeze-out of the silicone is constrained by the chamfer.
- Usage of the sealant remains low, with resultant low cost.
- The surface of the flange can be 'as cast'.

A similar design utilizing the radius on pressed steel components such as oil pans and engine timing case front covers has proved to be effective in production.

Comparison testing

Fatigue testing was carried out in the laboratory to compare the chamfer design with both groove and flat face metal-to-metal designs. The tests were based on standard pieces that had been exposed to predetermined levels of flange movement at a frequency of 20 Hz. The duration at which any leaks occurred was then recorded. A fatigue diagram was produced showing the 10% probability of failure characteristics for the different joint designs.

As a reference point the actual joint movement was measured on an engine bed plate under load conditions. The maximum movement was found to be in the order of 25 µm. The results revealed that with this level of movement:

- A traditional flat face metal-to-metal joint could fail after 10 000 cycles.
- The groove flange design could fail after 400 000 cycles.
- The chamfer design remained leak free even after 10 000 000 cycles.

This is a convincing demonstration of the effectiveness of this approach to flange design for silicone sealants.

RTV silicone dispensing

To achieve satisfactory sealing as described, accurate product dispensing is critical. As these products cure in contact with atmospheric moisture they are not compatible with screen-printing systems because they would cure on the mesh.

For volume applications a seven axis robotic dispensing system with product fed by drum pump from either 20 litre or 200 litre drums with a volumetric control is the most viable delivery method. Flow monitors can be integrated to enable generation of SPC data, while automatic vision systems can be used to detect dispensed bead positioning.

Low volume and service applications are achieved using manual application direct from cartridges. A service procedure with guidance on bead position should be provided for the service departments to ensure correct application of the product.

Correct handling
RTV silicone gaskets seal by adhesion, not compression, so it is necessary to assemble joints before the sealant skins over. And because adhesion must be maintained for the life of the joint, the RTV silicone sealant must elongate as the components move. This causes the stress at the bond line to increase and if cohesive strength of the sealant is exceeded, the seal will fail. It is therefore necessary to ensure selection of the correct silicone to meet the demands placed upon the joint.

Surface cleanliness is also critical for successful sealing. Contaminated surfaces lessen the joint movement capability by reducing the force that would be required for adhesive failure. Thermal cycling and/or severe loading may shear fully cured material from contaminated surfaces.

It is essential that a continuous bead of silicone sealant be applied inside or around dowel and bolt holes to eliminate secondary leak paths.

As silicone sealant begins curing immediately when exposed to atmospheric moisture, parts should be assembled quickly after applying the sealant. Maximum 'open' time will be between 10 and 25 minutes depending on product grade used.

RTV silicone cure speed is humidity dependent which means that it will be slower in low humidity environments.

Application techniques
Pneumatic or manually powered caulking guns can be used to apply beads of silicones from either 300 or 850 ml cartridges. This is a simple and inexpensive method of applying these products.

It is essential that the proper sequence of assembly be followed for the equipment and sealant involved. RTV silicone sealants applied to a joint must be assembled immediately.

The sealant should be applied to only one of the sealing surfaces. For the sealing operation to be completely successful, the product must be in the correct location over the whole flange surface. It is important to remember to circle dowel and bolt holes when manually applying sealant.

When applying a bead, the cartridge or tube should be held slightly above the sealing surface in order to lay the bead onto the joint. The nozzle tip should not be held directly on the surface, as bead shape will be harder to control. When applying RTV silicone materials the work should be carried out as quickly as possible.

Once the application is complete, the bead should be inspected for uniform thickness, air pocket voids and continuity. Any imperfections should be repaired as soon as possible.

It is important to use the recommended bead size for each project and not apply more material than is necessary to seal a joint. Too much sealant will result in the material being squeezed out on either side of the joint. Wiping the exterior is not recommended for silicone sealant applications as there is a risk of leakage due to excessive removal of material.

Excessive silicone material on the inside may break away in the assembly and could plug passageways and/or filters. The excess silicone on the inside of a flange is minimized by the use of a chamfer design described above. If leak testing is required, wait at least 45 minutes and use the lowest air pressure possible for the shortest time duration.

Repeatable bead applications of RTV silicones can be achieved by using robots or X–Y machines. The materials are normally dispensed from 20 kg pails or 200 kg drums with pressure pump systems using, ideally, positive displacement bead control pumps.

Service and repair

It is sometimes necessary to remove covers and disassemble components that have been sealed with FIP gasketing materials. Both RTV silicone and anaerobic gaskets have good shear resistance and in order to separate components it is recommended that a peel or cleavage load be applied by lifting the end of the cover or part. Pry slots or jacking screws are recommended where possible.

Removal of old sealant and cleaning the surface is the only way to promote successful resealing. Chemical gasket removers are available that will soften or dissolve sealants. Plastic scrapers should be used to prevent gouging of the sealing surface. Abrasives or wire wheels should not be used on aluminium or plastic parts.

Cleaning should not be carried out using general petroleum cleaners or mineral spirits, as these can leave a residue that prevents adhesion. Cleaning fluids that are specified by the adhesives/sealant manufacturer should be used.

2.7.3 Sealant selection

The type of flanged component, as described in sections 2.7.1 and 2.7.2, will decide whether an anaerobic or silicone sealant is required.

Thereafter the primary factors that influence product selection are:

- Operating conditions, pressure, fluid and temperature.
- Flange design or detail of other components to be sealed such as screw threads.
- Other specific product process and performance features may also need to be considered.

Examples of the range of products available are shown in Table 2.9. However, there is such a wide range of products available that for any critical and especially for high volume applications the assistance of a reputable supplier will be invaluable.

Section 2.7 has been prepared primarily from material supplied by Henkel Loctite Adhesives, whose contribution is gratefully acknowledged.

2.8 Bolted joints and gaskets

2.8.1 Introduction

2.8.1.1 What is a gasket?

What is the difference between a seal and a gasket? There is no clear definition of these terms and different industries often use seal and gasket in a very loose manner. In the *Seals and Sealing Handbook* the following definitions, as discussed in the introduction to this chapter, are being used to discriminate between the two:

- Seals are designed to be pressure energizing such as the method described for an O-ring in section 2.2.1. The compression load on the seal is sufficient to provide an initial low pressure seal, but when pressurized further activation comes from the properties or design of the seal.
- Gaskets are clamped with sufficient force such that the pressure is resisted by the stored energy within the gasket. A gasket has been described as 'a device for sealing two surfaces, by storing energy between them'.[12] Thus the gasket has to react to the forces generated by the bolts, and therefore the work and energy imparted to the bolted joint becomes 'stored' within the gasket itself.

The division between the two is far from clear cut, and a particularly grey area is the range of proprietary connectors that use metal-to-metal seal rings, which are partly pressure energized. These are included as a separate section of metal gaskets.

Gaskets can be classified into three main categories:

- Non-metallic (elastomers, cork, compressed fibre, graphite, PTFE, etc.).
- Semi-metallic (spiral wound, kammprofile, metal cased, etc.).
- Metallic (API ring joints, lens rings, etc.).

Each has is own advantages, and these will be described further in this section.

2.8.1.2 Bolted joint assemblies

Unlike a typical O-ring assembly where the two flanges to be sealed are bolted metal to metal forming a relatively rigid joint, gasket assembly can typically be assumed to be a 'floating' assembly where there is no metal-to-metal contact after bolting the flanges.

For example, the common raised face flange having a fibre-based material gasket, Figure 2.47. Here, for an increase in system pressure a higher gasket stress would be required. However, with increasing bolt force the gasket undergoes greater compression. Additionally the bending moment on the flanges is increased, so that the bolts, flanges and gaskets could all be considered to be spring elements making up the gasket system. It is important to realize that the gasket is often highly non-linear in load recovery behaviour.

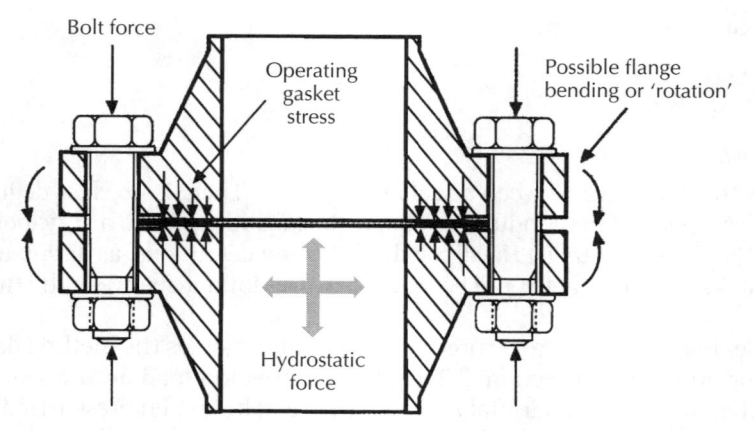

Figure 2.47 A typical raised face bolted joint assembly. (Source: James Walker)

2.8.2 Forces acting in a bolted joint assembly

The initial bolt load created on assembly is transferred to the gasket via the flanges. This initial seating stress compresses the gasket and will compact any voids within the fibres, etc. When the system is pressurized the hydrostatic force will tend to increase the tensile force on the bolts and unload the stress on the gasket.

The stress remaining on the gasket when it is at the system operating pressure is termed to be the 'operating' or 'residual' stress. This is the stress at which the gasket will be expected to operate and it will determine the degree of 'tightness' achieved by the joint. The term tightness as used in gasket terminology requires some explanation. It does not refer to the tightness or torque of the bolts or indeed the load on the gasket. It is an abbreviation for leak tightness, and describes the performance of the gasket with respect to measurement of emissions.

It should be noted that on a raised face assembly such as the one shown in Figure 2.47, there will be some deflection of the flanges themselves, termed flange rotation. This will vary depending on the load applied, the flange material and the geometry of the flanges. The effect of this rotation will be to create an operating stress at the outside edge of the gasket that is greater than on the inside.

Flat faced flanges, see Figure 2.51, will behave quite differently. The overall contact area of a full-face joint can be typically twice that of an inside bolt circle (IBC) joint.

2.8.3 Gasket behaviour

Gasket material behaviour is covered by a range of standards. Typical standards are BS 7531 or DIN 3535 and ASME F36. Material properties such as stress relaxation, compression, recovery and gas permeability are covered.

2.8.3.1 Stress relaxation

Retention of the operating stress within the gasket is important to maintaining the level of energy stored in the joint and hence the ability to continue to seal.

Relaxation can occur at the flange/gasket interface as well as in the gasket material itself. This characteristic is particularly relevant in relation to non-metallic sheet jointing materials. Materials having high elastomer content can be expected to relax significantly as the elastomer decays at temperature. It should also be noted that with fibre sheet jointing materials, thick materials will relax more than a thinner sheet of the same material. This is why a manufacturer will recommend that the thinnest gasket possible should be used. It should be as thick as is required to accommodate flange distortion and misalignment.

2.8.3.2 Tensile strength

Tensile strength is not usually considered as a prime property for gasket materials. However, if the joint is relatively thick and inadequately compressed, then the internal pressure forces on the inside edge have to be resisted by the tensile strength of the gasket. This again demonstrates the benefit of a thinner gasket. Less relaxation will occur internally and less area will be exposed to the system pressure.

2.8.3.3 Flange surface finish

Pipe flanges manufactured to one of the usual standards will be supplied with a light gramophone-finish groove across the gasket seating face to the values given in Table 2.10. These are suitable for non-metallic and semi-metallic gaskets. This finish prevents creep of the gasket across the flange faces. It is also important that the surface finish is not so rough that the gasket is unable to deform effectively to fill the gramophone groove. The use of pastes or lubricant may reduce the sealing performance. They can fill in the surface finish and accelerate stress relaxation.

Table 2.10 Machining of flange faces for gasketed joints

Gasket type	Machining method	Approx. serration depth	Approx. tool nose radius	Approx. pitch of serrations	Ra (μm) min	Ra (μm) max	Rz (μm) min	Rz (μm) max
	Turning	0.05 mm	1.6 mm	0.8 mm	3.2	12.5	12.5	50
	Other machining				3.2	6.3	12.5	25

Note: both faces of a joint flange should be machined to the same specification.
(Source: James Walker)

Although the flange finish values given in Table 2.10 will give acceptable performance, improved results may be achieved if the range is restricted further[12] to the following values:

- Spiral wound gaskets: 3.2–6.3 μm Ra.
- Compressed fibre materials: 6.3–12.5 μm Ra.

It is preferable for the two flanges to be of the same material and machined identically and to the same specification. Particular attention is required for applications involving gases with high permeability, such as helium or hydrogen and also for high vacuum sealing.

The surface finish for metallic gaskets is specified within the individual standards to which they apply or the manufacturer's catalogue for specific gasket types.

2.8.3.4 Load tightness

In this section tightness refers to the sealing ability of the gasket. The extensive work on emissions since the 1980s has shown that all gaskets leak to varying degrees. This may not be apparent from visual observation but detailed monitoring of gas tests will demonstrate low levels of leakage. Testing of gaskets is carried out with helium test gas and leakage monitored by a mass spectrometer.

As a fibrous gasket is compressed it will be compacted and any porosity between the fibres will be reduced. The material will also become effectively denser. A load-leakage test on a gasket will tend to produce an exponential decay curve as shown in Figure 2.48.

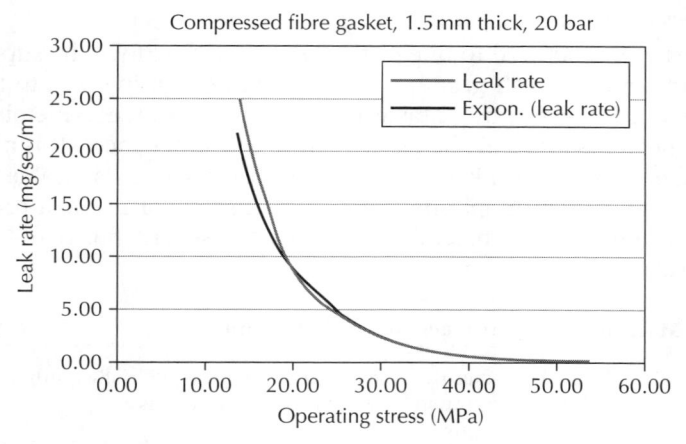

Figure 2.48 A graph showing how leak rate reduces with increased operating stress for compressed fibre gasket. (Source: James Walker)

The normal procedure for a gasket tightness test is to carry out a loading/unloading cycle, Figure 2.49. This will measure the leakage as the gasket stress is reduced, after being loaded to a higher value, as will occur in practice due to the hydrostatic force when the system is pressurized and as the gasket material relaxes. It can be seen in Figure 2.49 that this test produces a reasonably straight line.

It can also be seen on this graph that as the gasket becomes denser at higher loads the slope of the unloading-leakage data lines change. This helps to demonstrate that the sealing properties are quite complex, depending upon the initial and operational stress levels that are likely to occur in service. Other factors such as flange rotation make calculation of actual gasket stress extremely complicated.

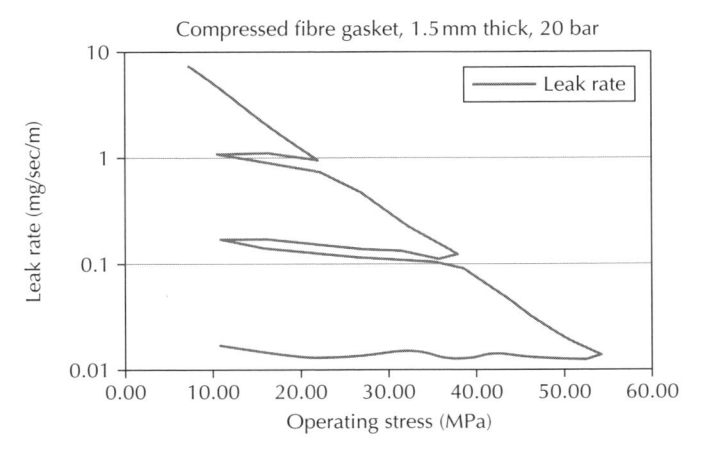

Figure 2.49 Typical fibre gasket performance during a loading–unloading test. (Source: James Walker)

An important lesson from this work is that for a given operating stress on a specific gasket, the leak rate will increase with increasing system pressure, as demonstrated in Figure 2.50.

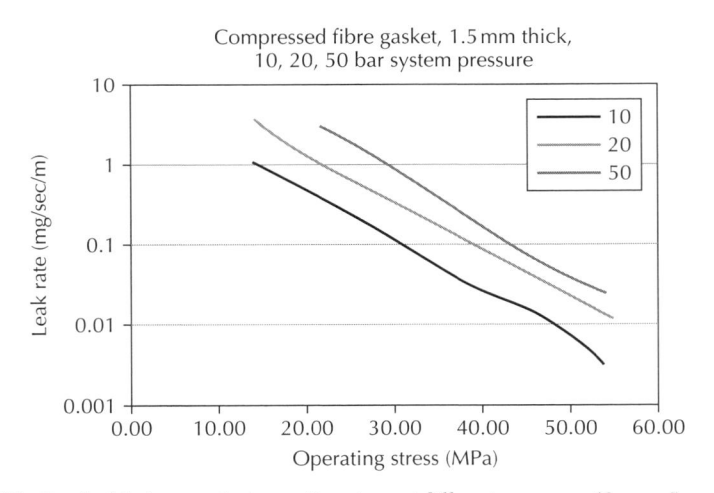

Figure 2.50 Graph of leakage against operating stress at different pressures. (Source: James Walker)

Also, for a sheet gasket material, the leak rate increases approximately in proportion to the material thickness, so a gasket of double the thickness will give twice the leak rate.

2.8.4 Flange types and standards

There are many common flange standards available, though perhaps the most widely used are ASME, DIN and BS. There are also a large number of other national standards but many of these have their origins in the ASME or DIN series flanges.

There are a wide variety of flange styles, and configurations, as shown in Figure 2.51, but the raised face flange is probably the most common type regularly employed.

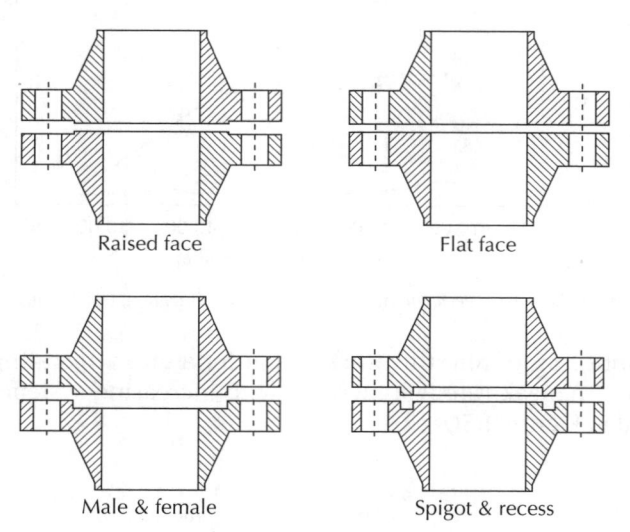

Figure 2.51 Examples of typical flange face configurations for gasketed joints. (Source: James Walker)

2.8.4.1 ASME

ASME B16.5 flanges are used globally in power stations, refineries, chemical plants and most other major industrial facilities. This standard covers flanges from 12.5 mm (0.5 inch) to 610 mm (24 inches) nominal bore. They are classified with pressure ratings in pounds per square inch (p.s.i.), in classes such as 150, 300, 600, 900, 1500 and 2500. It is important to note that these pressure ratings are at elevated temperatures and so, for example, a class 150 flange is rated to 290 p.s.i. (20 bar) at ambient temperature.

Large diameter ASME flanges above 610 mm and up to 1500 mm (60 inches) nominal bore are covered by the ASME B16.47 standard. This has two main categories, Series A and Series B. The Series A covers flanges formerly known as MSS-SP44 (Manufacturers Standardization Society) whereas Series B covers those from API 605 (American Petroleum Institute) which tend to be more compact. There are other large diameter flange standards such as the Taylor Forge classes 175 and 350, as well as the American Water Works Association (AWWA) flanges.

Heat exchangers are commonly produced having male and female flanges of class 150, 300 or 600, etc., but manufactured to TEMA (Tubular Equipment Manufacturers Association) dimensions.

2.8.4.2 DIN

By comparison to the ASME flanges, the DIN series are rated with PN numbers which indicate the nominal pressure rating in bar. These, for example, are PN25

Static seals 67

and PN40. A major difference between this standard and ASME is that the pressure ratings relate to ambient temperature. This metric series of flanges are now covered by European standards such as EN 1092, for these PN rated flanges.

2.8.4.3 British Standard

The BS 10 flange standard is now rarely used. Many of these flanges still remain in service at a large number of industrial sites as it was a widely used standard for power and industrial plant in the 1960s and 1970s. These flanges are classified by a letter system, for example Table E, Table H, Table J, etc. in increasing order of service pressure rating.

2.8.4.4 Relevant gasket standards

ASME flanges, gaskets from non-metallic sheet materials, are covered by ASME 16.21 or EN 12560 Part 1. Spiral wound joints for these are made in accordance with ASME B16.20 (formerly to API 601), or to BS 3381 or EN 12560 Part 2. The ASME B16.20 standard also covers metal-jacketed gaskets and API ring joints. For DIN series flanges, gaskets are cut from sheet materials in accordance with EN 1514 Part 1, with spiral wound joints being made to EN 1514 Part 2. The EN 12560 and EN 1514 have additional sections covering other gasket types such as PTFE envelopes and corrugated metallic gaskets in Parts 3 and 4 respectively. As for BS 10 flanges, the cut gasket dimensions are given in BS 3063.

2.8.5 Calculation methods

As discussed in section 2.8.3, the load-sealability characteristics of a gasket are quite complex. Incorporating these effects into a reliable flange design method has been the objective of designers for many years, and a number of flange design codes are now well recognized.

The ASME VIII and DIN 2505 codes are well established and widely used. However, certain limitations in these codes have led to the research and development of alternatives such as the PVRC and CEN methods.

There are advantages and disadvantages to each of these methods and these are summarized in sections 2.8.5.1 to 2.8.5.4. Individual design parameters will dictate the most appropriate method and for further advice on gasket selection and relevant load sealability requirements please contact technical specialists.

2.8.5.1 ASME VIII

The ASME code has been used for many years to design flanges, though it has a number of recognized flaws when it comes to determining a suitable bolt load with regard to gasket sealing. Calculations are performed to determine the greater of either operating or initial forces using the following formulae.

Initial load requirement:

$$W_{m1} = \pi b G y$$

Operating load requirement:

$$W_{m2} = \frac{\pi G^2 P}{4} + 2b\pi GmP$$

The factors m and y are the gasket factor and initial seating stress values respectively. One problem is that the code does not use the whole gasket contact area in the calculation. In the formulae above the 'effective width' is b and the effective diameter is G. The system pressure is P.

For a common raised face flange the contact width of the gasket element, from the raised face outside diameter to the gasket inside diameter, is designated N. The basic width is then calculated as being half this value and called b_0. The effective width will depend on the value of b_0 being greater or less than 6.3 mm (1/4 inch), though in the majority of cases this is likely. If b_0 is greater than 6.3 mm then the effective width is calculated as $b = \sqrt{b_0/2}$.

Or in cases where b_0 is equal to or less than 6.3 mm then $b = b_0$.

The effective diameter G is simply the outside contact diameter less $2 \times b_0$.

Therefore the actual gasket contact area is usually far greater in reality than the figure calculated by this method.

Note also that the gasket factor m is effectively a multiplier of the system pressure as an operating stress. However, as shown in section 2.8.5.4 the actual sealing performance of a gasket is more realistically a three-dimensional exponential decay curve, rather than a single number.

This section also shows that the operational stress defines the gasket sealability. There seems to be no reason to try to link initial gasket seating loads (i.e. for compression purposes to take up flange flatness, etc.) with the m factor which is related to the operating stress requirement for a given system pressure. However, when originally devised in the 1930s and 1940s the relationship existed between the factors of $(2m - 1)^2 \times 180 = y$ (using units for y of p.s.i. and rounding of the m factor to the nearest 0.25).

Thus for compressed fibre gaskets, a lower value of y was determined for the thicker materials presumably because they would deform more readily to make a crude seal against whatever flange distortion existed. Thus, from the relationship above, the thicker materials also gained a lower m factor, suggesting that they would give better sealing performance. However, it has been shown that thicker materials not only have a greater tendency to stress relax, but also have a greater number of micro-porosity channels where leakages can occur. In gas sealability tests 3 mm thick compressed fibre jointing material tends to leak at approximately twice the rate of a 1.5 mm sample of the same material at a given operating stress. Hence this method has been superseded by sections 2.8.5.2 to 2.8.5.4.

2.8.5.2 DIN 2505

This method also employs gasket factors that are used to determine the bolt load requirement. These are the maximum and minimum stress levels for installation at ambient temperature, as well as a maximum stress at elevated temperature:

(σvu σvo σBu σBo)

The maximum initial gasket stress allowable is a function of the width to thickness ratio to avoid crushing effects on the soft materials. This method also has an m factor, though uses the actual contact width of the gasket in the calculations. Thus, this method is less likely to produce insufficient gasket stress than by ASME VIII, especially as the minimum initial gasket stress values given in the code are fairly conservative.

Another popular European design method used to determine bolt load requirements is the AD-Merkblatt B7, where again the initial and operating bolt load requirements are determined using gasket factors. These are k1, k0 and KD, which relate to the effective widths and deformation resistance of the gaskets concerned.

Note that both of the above methods are now generally tending to be replaced by the new EN 1591 design method.

2.8.5.3 PVRC

For a number of years the Pressure Vessel Research Council has recognized some of the inadequacies of the ASME code and sought to provide a better means of calculating the required gasket loading. The proposed calculation requires gasket data to be developed from a series of loading and unloading leak tightness tests in order to obtain coefficients to describe the sealing performance of the gasket.

This method uses the concept of 'tightness parameter' where the gasket sealability is related to a dimensionless number thus:

$$Tp = \frac{P}{P'} \frac{\{L'\}^{0.5}}{\{L\}}$$

This formula essentially relates the test pressure to atmospheric pressure and the measured leak rate against a unitary leak rate measured in mg/sec/mm diameter of gasket. The index of 0.5 suggests that doubling the system pressure will in fact produce a fourfold increase in leak rate, though this is perhaps a worst case scenario for most gaskets.

The coefficients used in the calculation are G_b, α and G_s, and are derived from the sealability test as shown in the typical PVRC chart, Figure 2.49.

There is some debate about the reliability and repeatability of the coefficients as they are derived from the log/log plot of an assumed square law relationship, so relatively minor changes in the leak rate during the test could affect the final data reduction.

However, the coefficients are at least derived from actual tightness testing, so the calculation becomes realistic in terms of typical in-service performance of the gasket.

2.8.5.4 CEN

Like the PVRC method, this uses loading/unloading gasket sealing tests to determine the gasket performance characteristics. They are used in a flange calculation method, EN 1591, which is quite iterative and complex in nature. The data required for these calculations is derived using test procedures specified in EN

13555.[20] As a gasket is loaded, some flange rotation may occur, changing the effective stress on the gasket from the inside to outside contact position, as described in section 2.8.2.

The gasket seals differently in the unloading phase of the test compared to the loading cycle. If we consider that a gasket is loaded to point Q_A initially and unloaded by the hydrostatic to the operating stress Q_{SminL} as shown in the diagram, Figure 2.52, then we can see that the sealability decreases slightly between these two points. However, the sealability is generally better than when this stress level was applied during the loading phase.

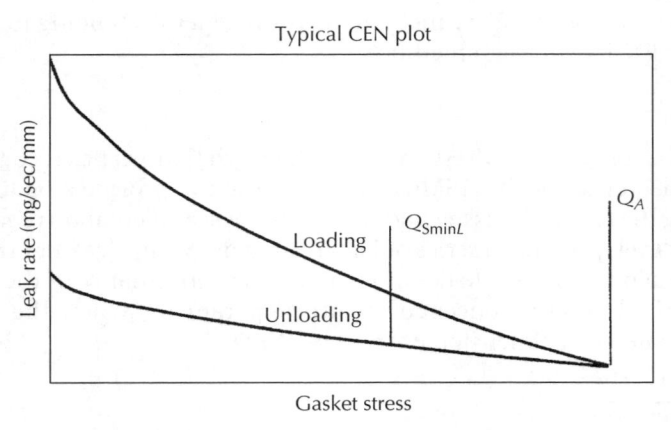

Figure 2.52 Example of a result from a CEN test. (Source: James Walker)

The unloading modulus of the gasket changes with increased initial stress. Therefore the change in sealability between Q_A and Q_{SminL} will change depending upon the actual value of Q_A applied. This standard uses the unloading modulus of the gasket in the calculations to examine the stiffness of the gasket as part of the overall joint stiffness.

The EN 1591 method looks at the required sealability from loading/unloading tests. By knowing the degree of unloading on the joint from the hydrostatic forces and the tightness level desired, it is possible to calculate the flange deflection and effective gasket stress for a given bolt load. Therefore changes to the flange geometry and applied bolt load will determine the amount of flange rotation, the change in gasket stress across the sealing element, in both the initial and operating conditions, and therefore the degree of tightness achieved.

The problem with this method is that it can be quite iterative and complex, though computer programs are becoming available which will make the calculation method easier to perform.

The degree of gasket testing required can also be considerable, and work is under way to look at combining the methodologies of both CEN and PVRC testing to harmonize the test protocols.

Static seals 71

(a) E-1585A Floating head final torque, clad gasket, used studs, wire brushed and lubricated.

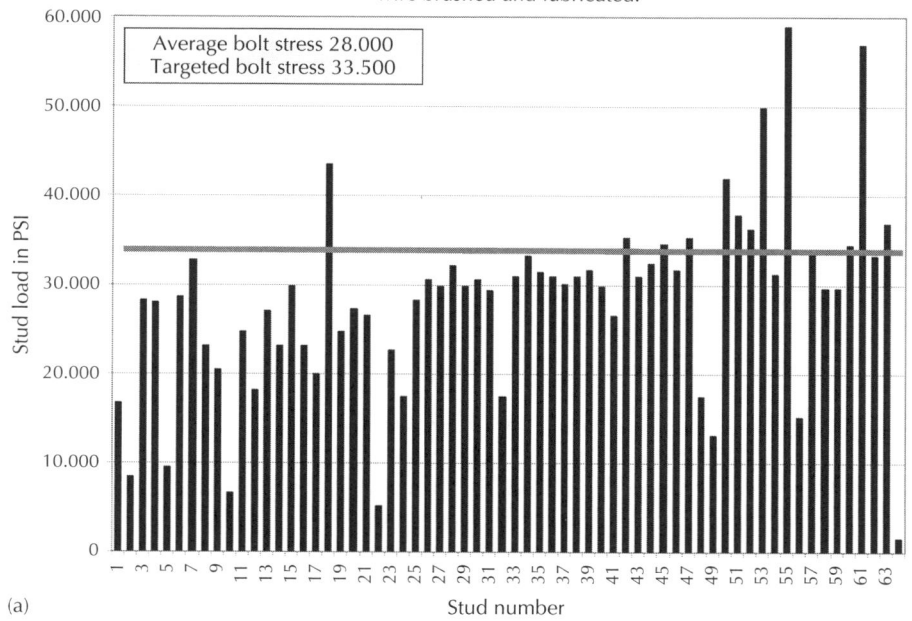

Average bolt stress 28.000
Targeted bolt stress 33.500

(b) E-1585B Floating head final torque, clad gaskets, new studs, no washer.

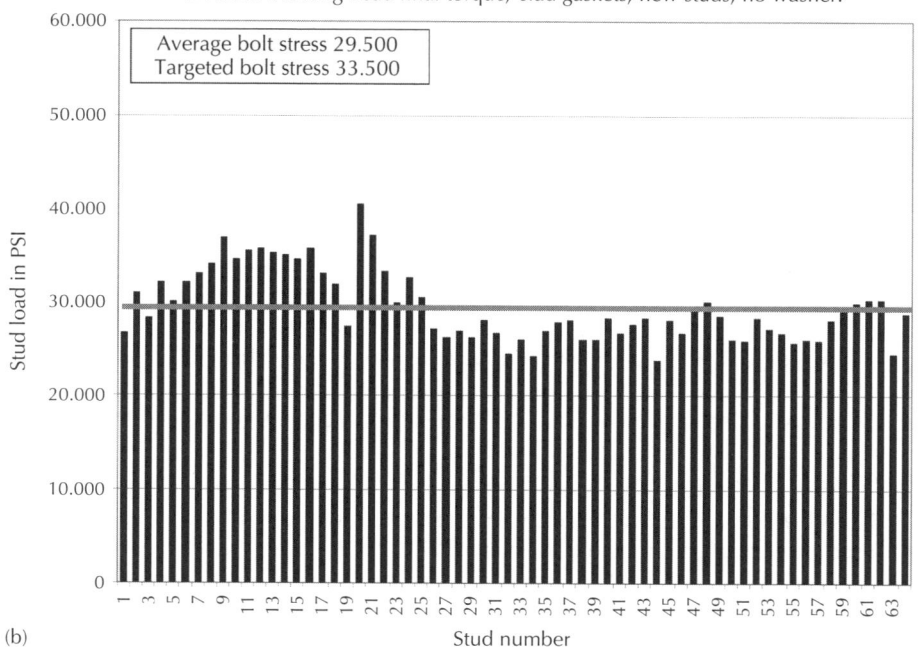

Average bolt stress 29.500
Targeted bolt stress 33.500

Figure 2.53 The potential variation in bolt stress depending on the friction between the bolt and nut and the nut spot face.

2.8.6 Gasket installation

2.8.6.1 Best practice

- Mating flanges should be of the same type and correctly aligned.
- Fasteners should be selected to ensure that they do not exceed their elastic limit at the required tension.
- Do not retorque elastomer bound compressed non-asbestos gaskets after exposure to elevated temperature. They may well have hardened and are at risk of cracking.
- Ensure that fasteners show no signs of corrosion which might affect their load-bearing capacity.
- Nuts should have a specified proof load 20% greater than the UTS of the fastener.
- Hardened steel washers of the same material as the nuts should always be used.
- A thread lubricant or anti-galling compound should be used on bolting as appropriate, but only a thin uniform coating should be applied. Where stainless steel is used, it should be ensured that such coatings are suitable for use.
- Fasteners and/or gaskets should never be reused.
- Good quality gaskets should always be procured from reputable suppliers only.
- Gaskets should be as thin as possible.
- Gaskets should never be hammered out against the flange. Not only can this cause damage to the flange, it will also damage the gasket material and thereby reduce gasket performance.
- When cutting full face gaskets, the holes should be cut first, followed by the gasket outer and inner diameters. Note that if the bolt holes are fairly close to the gasket o.d. then punching out the holes last may produce enough stress to crack the gasket at this point.
- Gaskets should be stored in a cool dry place, away from heat, humidity, direct sunlight, ozone sources, water, oil and chemicals. They should also be stored flat, i.e. not hung on hooks.
- Avoid the use of jointing compounds and pastes, these can lubricate the flange/gasket interface and encourage stress relaxation effects.

Appendix 1 also contains useful information on gasket good practice.

2.8.6.2 Bolt tightening

Gaskets should be tightened evenly in at least three or even four stages using an opposed pattern as illustrated in Figure 2.54.

Be aware that 'cross talk' exists between bolts during the tightening process so that as one tightens and the gasket compresses, another bolt may loosen. Therefore, a final pass around all of the bolts at the end is suggested to ensure that all remain tight.

This subject is also covered in some detail in Appendix 1.

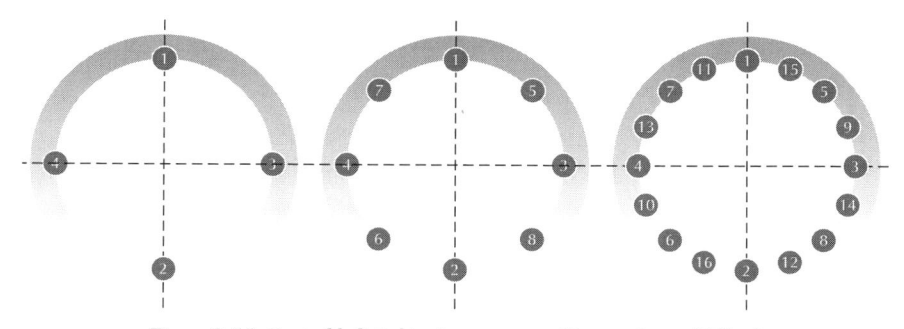

Figure 2.54 Typical bolt tightening sequence. (Source: James Walker)

2.8.7 Bolting

Having determined the gasket loading requirement, consideration must be given to the best materials and tightening methods to achieve this loading. As mentioned previously the overall joint integrity is essentially a function of three main criteria:

- Correct gasket selection to suit the operating conditions and the overall bolted joint strength/stiffness.
- Quality of the joint components, the gasket manufacture, flange and bolt materials, etc.
- Installation competence, ensuring that the gasket is fitted correctly, with the design seating stress applied both accurately and evenly.

2.8.7.1 Materials

Commonly used bolting materials and standards include BS 4882 and ASTM A193.

It should be noted that stainless steel fasteners, e.g. B8, have significantly lower strength than alloy steel materials, B7. Care should be taken to select a bolt material having sufficient yield strength so as to be able to apply adequate gasket stress while retaining a margin of safety, bearing in mind the variance of torque/tension scatter that may be possible during tightening.

Note also that especially with some exotic bolt materials and at elevated temperatures, the true onset of yield may be below the theoretical value. Bolt material standards such as those mentioned above should be consulted for details of maximum recommended stresses and operating temperatures. Some initial guidance is provided by Table 2.11 and Appendix 1.

2.8.7.2 Tightening methods

Torque vs tension

While torque is often recommended as a method for loading bolts in order to achieve a reasonable gasket stress, it should be noted that because of the variance

Table 2.11 Bolt materials

Material	Temperature (°C) Minimum	Maximum
Carbon steel	−20	300
B7, L7	−100	400
B6	0	500
B8	−250	575
B16	0	520
B17B	−250	650
B80A	−250	750

in nut and thread friction which is particularly difficult to control, the theoretical relationship is not particularly accurate.

$$T = \frac{F}{4}\left(\mu_h(A + D) + \mu_t(2d_e \sec \theta) + \frac{2p}{\pi}\right)$$

In this formula:
F = the axial load requirement
A = the across the flats dimension of the nut (i.e. the outside diameter of the nut and washer contact)
D = the washer inside diameter
μ_h = friction coefficient under the head of the nut
μ_t = friction coefficient of the thread
d_e = effective bolt diameter
p = thread pitch
θ = half the thread form angle

The effective bolt diameter, thread pitch and angle θ are all readily available in engineering manuals for common thread forms.

However, the friction coefficients under the head of the nut and on the thread will almost certainly be difficult to determine.

Simplified formula
In simplified form, for lubricated fasteners, washers, nuts, etc., the approximate relationship between torque and fastener may be represented as:

Torque = 0.2 × Load × Diameter

The units must be consistent with the system being used. For example, if the torque is required in ft-lb then for bolts in inches the value must be divided by 12.

Compared to the more complex formula given above the simplified formula is often deemed sufficient for the purpose, as accurately gauging the coefficient of friction can be extremely difficult. However, it should be noted that the accuracy of applied torque vs actual fastener tension achieved can typically be ±60%.

A nominal 0.15 has been used as the friction on both the thread and the nut washer interface on a conventional UNC threaded fastener. Using the more complete formula given above a change of only ±0.05 on the friction coefficient can vary the torque required by typically 30%. The potential extent of the problem is demonstrated by the results in Fig. 2.53 (p. 71).

Hydraulic tensioning
Hydraulic tensioners have a number of advantages particularly when tightening large bolts where they can provide a very high tensile force if required.

They offer the ability to be linked together, so that a number of bolts may be tightened in unison. However, some compliance exists in the tensioner system (embedding of the nut and washer, thread deflection, etc.), so that some overload is required to compensate for the relaxation once the tensioner has been depressurized. This requires careful analysis to avoid crushing the gasket or overloading fasteners in a highly stressed gasket assembly. The degree of torque imparted to the nut collar when the nut is 'run down' the fastener also has an effect on the amount of load transfer relaxation that occurs when the tensioner is depressurized.

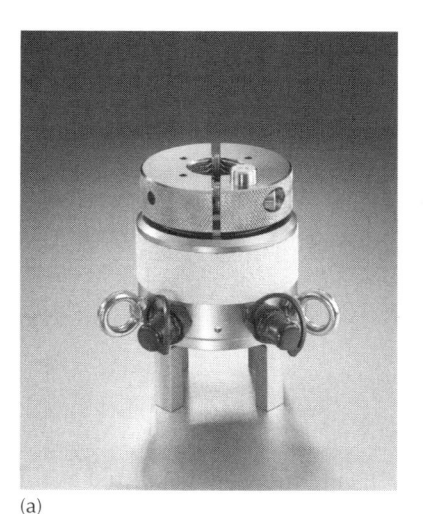

(a) (b)

Figure 2.55 Hydraulic nut tensioners. (Source: Hydratight)

Ultrasound
This can be a relatively accurate method for measuring the extension of the fastener, but it should be noted that the fasteners should ideally be spot faced for good contact with the signal probe. The bolts or studs also require calibration to the ultrasound equipment, as load extension characteristics of fasteners may vary between batches. This method requires a degree of operator skill and training can be relatively time consuming.

Load disc
The load disc is a bolt-thread connected washer of same size as the standard hexagonal nut. It allows a torque-based tightening system to be used with

increased accuracy. The load disc is prevented from turning by the tool that turns the nut and this creates a counter-nut effect. The bolt itself is not subjected to the torque, the thread segment inside the disc moves axially according to the bolt elongation. Once the desired bolt load is obtained it is retained by the nut. The bolt load accuracy of the disc is thought to be the result of several factors:

- Bearing face friction between the nut and the disc is more consistent.
- Side load is eliminated.
- Bolts are stretched only to the required load.
- Inadvertent gasket crushing by high, initial pulling forces are avoided.

Figure 2.56 Torque discs designed to give consistent friction for bolt tightening. (Source: Sealing Technology Hytorc)

2.8.7.3 Load monitored fasteners

RotaBolt

This is a system where a modified bolt or stud has an indicator cap fitted at the end which locks when the required tensile load is achieved.

There is a small gap between the end of the bolt and the cap when the fastener is in the unstressed state. As the fastener is tightened it stretches and the gap closes, Figure 2.57. The gap setting depends on the desired tension to be achieved, and once the cap locks up then this value has been attained. This method ensures rapid and easy installation as well as simple in-service condition monitoring of the fastener tension.

The RotaBolt2 system has dual indicators offering max and min settings to show a loading band for greater accuracy in situations where some overall joint relaxation may be anticipated.

Note that each set of fasteners is factory preset to a load requested by the user. Thus they should only be used for the flange or system that they are intended for as a different system or gasket type may require a different load setting. This system is not suitable for fasteners of less than 16 mm diameter.

RotaBolt converts existing or purpose-made studs.

- Control cap
- Grease packed
- Normally 3 threads clear unless otherwise specified.
- Rota load indicators are made in stainless steel, but can be supplied in other materials.
- Stand off
- Gauge pin
  Made in compatible material to parent stud to match the thermal coefficient of expansion.
- Gauge pin
  Positively anchored here and tested in hostile conditions to prove reliability of the anchoring system.

(a)

(b)

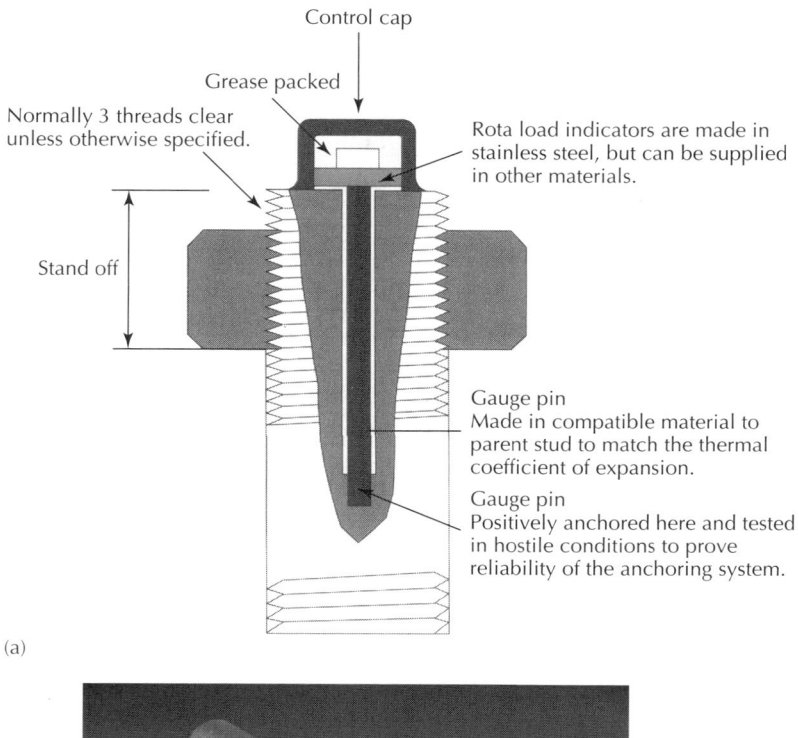

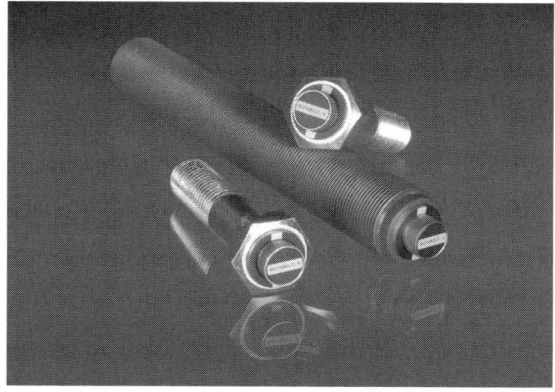

Figure 2.57 Bolt with a built-in strain measurement system. (Source: Rotabolt)

This system is used for critical flange jointing duties in the offshore, petrochemicals, mining, defence and power generation industries.

Truload
This system uses the same overall principle as the RotaBolt of directly measuring the bolt. The bolt load is measured by a transducer inserted into a pre-machined

hole in the fastener, Figure 2.58. As the fastener is tightened or alters during service the transducer provides an output of the load. This permits correct initial tightening and subsequent monitoring of the bolt. A bolt load variation after a second tightening of less than 3% is said to be achievable. Two systems of continued monitoring are available. A handheld meter can be used in mass production or for periodic checking of fastener performance on plant. The alternative for critical equipment is a permanent monitoring system.

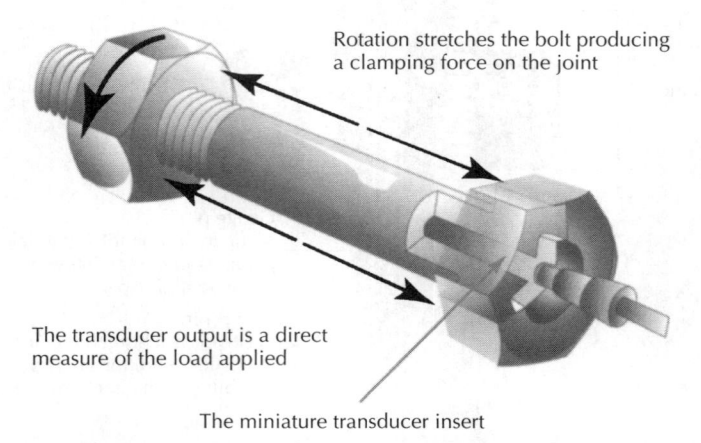

Figure 2.58 Instrumented strain measurement of bolts. (Source: Truload)

Sections 2.8.1 to 2.8.7 have been prepared with the assistance of James Walker & Co. whose contribution is gratefully acknowledged.

2.8.8 Gasket types and applications

Gaskets fall into three broad category designations: non-metallic, semi-metallic and metallic. Within each category there is a wide range of both materials and constructions of gasket to meet varying application requirements and historical codes of practice for individual industries.

2.8.8.1 Non-metallic

Elastomers
Elastomer gaskets are only used for relatively low pressure applications. As described in section 2.2.1 the elastomer will extrude as it is tightened due to the incompressibility and low shear modulus of the material. Elastomer gaskets are found as a typical flanged gasket for low pressure water and oil applications up to 10 bar. They are readily applicable to custom shapes as they can be moulded to suit the flange design, such as engine and compressor sumps and engine covers. Due to the characteristics of the elastomer, the bolt load must be kept to a low value, which can cause problems with achieving a satisfactory bolt tension to avoid slackening in service. To achieve satisfactory control of compression

and sufficient bolt stress it is necessary to design in compression limiters to either the flange, or within the gasket itself, Figure 2.59.

Fluids: Wide range depending on choice of elastomer
Temperatures: −40 to 200°C depending on elastomer
Pressure: 10 bar

Note: redesigning the assembly to use the elastomer as a seal rather than a gasket can often improve performance. In a sealing configuration elastomers can successfully seal over 1000 bar.

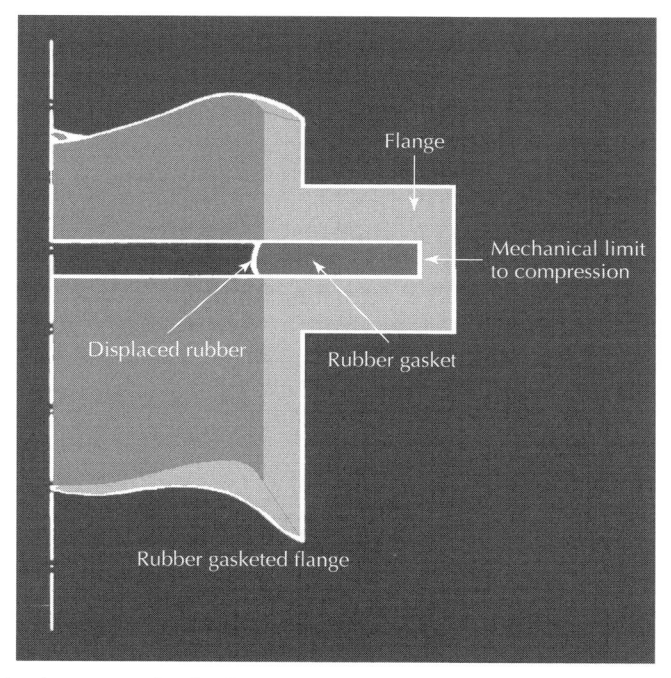

Figure 2.59 An elastomer gasket fitted to a stepped flange to provide retention of the gasket. (Source: BHR Group)

Cork
Cork jointing is usually bonded with elastomer. It has traditionally been used for a wide range of low pressure oil duties such as engine and transmission covers. Again the bolt loading is low which can cause problems of retention of bolt load with temperature cycling and vibration. Cork has been superseded by sealants in very many applications, see section 2.7.

Fluids: Oil, water, mild chemicals
Temperatures: −20 to 150°C
Pressure: 10 bar

Compressed fibre

Compressed fibre materials are based around the production techniques originally used for compressed asbestos fibre (CAF), which is no longer available. There is no direct equivalent to asbestos fibre which possessed quite unique fibrous and chemical properties. The compressed fibre materials available are alternatives but the application range can be much more restricted than that achieved by asbestos and very dependent on the fibres used and the construction.

Compressed fibre gaskets are bonded during manufacture using elastomer, usually nitrile. This provides the integrity and flexibility for cutting, handling and fitting. During use, especially over 100°C the elastomer will harden and degrade. While this may not affect the integrity of the gasket in situ, it can affect the ability to provide a reliable performance during temperature and pressure cycles and they should not be retightened after a period of use. Generally higher quality gaskets will have a lower quantity of binder.

Undefined compressed fibre

These may use a mixture of polyamide, aromatic polyamide (Aramid) and other fibres bonded with elastomer. They are manufactured for economy and are only suitable for use at low temperatures and pressures with oil, water and some mild chemicals. There are a wide variety of materials available and the individual properties should be carefully evaluated if anything other than low temperature and mild fluids are involved.

Fluids:	Water, oil, mild chemicals
Temperature:	−100 to 150°C maximum
Pressure:	40 bar

Aramid fibre

Aramid fibres are a derivative of polyamide and are exceptionally strong. However, the raw fibres are very brittle and as they originate from the polyamide family the water resistance is not as good as some other materials. These materials have been progressively developed to become the general purpose material for the low to medium pressure and temperature range of gasket materials.

Fluids:	Water, steam, oil, many chemicals
Temperature:	−200 to 400°C maximum (200°C max in steam)
Pressure:	70 bar

Premium mixed fibre

A number of materials are available based on a mixture of Aramid fibre with glass fibre. These can provide improved properties over a straight aramid material. Properties will vary depending on the fibre mixture and manufacture of the material.

Fluids:	Water, steam, oil, solvents, wide range of chemicals
Temperature:	−200 to 440°C (200°C max in steam)
Pressure:	100 bar

Carbon fibre
This is a compressed fibre material designed to provide improved steam and chemical resistance performance compared to Aramid fibre.

Fluids: Water, steam, oil, solvents, chemicals
Temperature: −200 to 450°C
Pressure: 110 bar

Multilayer compressed fibre
This material is designed to extend the performance of fibre-based gaskets by using a combination of elastomers, Figure 2.60, for binding the fibres to optimize the gasket performance.[13] Currently one material is available using this technology and a longer life is claimed compared with equivalent materials.

Fluids: Water, steam, moderate chemicals, oil, solvents, refrigerants
Temperature: −200 to 400°C (200°C in steam)
Pressure: 70 bar

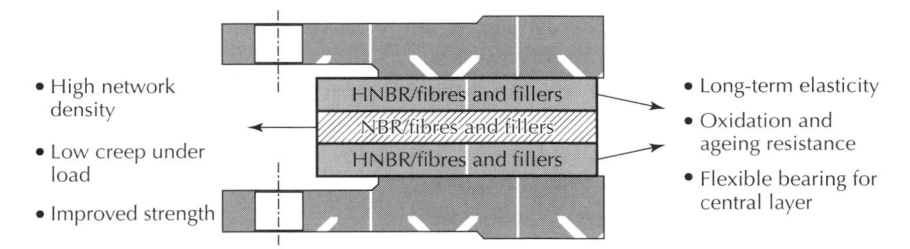

- High network density
- Low creep under load
- Improved strength

HNBR/fibres and fillers
NBR/fibres and fillers
HNBR/fibres and fillers

- Long-term elasticity
- Oxidation and ageing resistance
- Flexible bearing for central layer

Figure 2.60 Aramid fibre gasket. (Source: Sealing Technology, Klinger)

PTFE
PTFE is generally used because of its outstanding chemical resistance and for food and pharmaceutical applications. The widespread use of this material is limited as it can be prone to relaxation and creep.

For most gasket applications it is used as filled or as expanded PTFE. The majority of filled PTFE is now usually supplied as what is known as bi-axial or stabilized PTFE. This material is mechanically treated to provide additional strength to the material to resist creep, and hence relaxation and extrusion. It is also possible to incorporate a higher percentage of filler loading without adversely compromising the material properties compared with traditional skived sheet. A variety of fillers are used to provide additional strength, but these need to be assessed individually to ensure that fluid resistance and hygiene requirements are not compromised. Typical fillers are glass, silica, barytes and calcium fluoride.

- Silica is resistant to mineral acids but attacked by hydrofluoric acid and by concentrated alkalis.

- Barytes is resistant to concentrated alkalis; suitable for most mineral acids but restricted to sulphuric acid concentrations of 75%. It is suitable for use with aqueous hydrofluoric acid but not with the anhydrous form.
- Calcium fluoride has good resistance to alkalis but is attacked by highly concentrated sulphuric acid and has limited resistance to hydrofluoric acid.

Expanded PTFE, as typified by the Gore material, provides a softer and more flexible gasket. It is therefore useful where additional conformity may be required or where flange loadings must be low, such as for glass lined vessels. However, it should be noted that once it has been compressed this material is still subject to the creep experienced by other PTFE materials. Expanded material may be supplied in sheet gasket form, but also as cord or tape that can be placed on a joint surface. Note that this may be difficult unless the joint already has studs fitted. A further problem with materials such as expanded PTFE is that their flexibility can make them difficult to use particularly with larger flanges or if assembly in a constrained location or vertical flanges are involved. A material designed to overcome these difficulties has been developed[16] which comprises a bi-axial PTFE core with porous outer layers. This provides a material with a relatively stiff core for ease of handling together with a soft conformable layer to mate with flanges that may be uneven and where low loading is required.

For higher duty applications PTFE can be used as the filler material for semi-metallic gaskets such as spiral wound and kammprofile, see section 2.8.8.2. It can also be used as a sheath to protect fibrous gasket material to provide most of the fluid resistance of PTFE with the characteristics closer to those of a fibre gasket. These are normally called envelope gaskets.

Fluids: Virtually all fluids
Temperature: -200 to $250°C$
Pressure: 60 bar (filled PTFE), 30 bar (expanded PTFE sheet)

Exfoliated materials

Exfoliated graphite Exfoliated graphite is used as a high temperature material with wide chemical resistance. The exfoliated form consists of a series of 'worms' of exfoliated material, Figure 2.61, which interlock on compression to form the gasket material. It can be manufactured without a binder which permits the good fluid resistance and lack of stress relaxation and provides good long-term retention of gasket stress. It is softer and more compressible than sheet fibre jointing, as well as providing a higher level of sealing tightness at a given gasket stress. However, graphite is readily damaged, so requires more care in handling and storage. It is also supplied with reinforcing layers to increase its strength and rigidity. The highest performance grades generally have a tanged stainless steel mesh layer to provide reinforcement. This can present problems with cutting and cause sharp edges. Abrasive water-jet cutting is preferable whenever possible as this avoids creating the sharp edges.

(a)   (b)

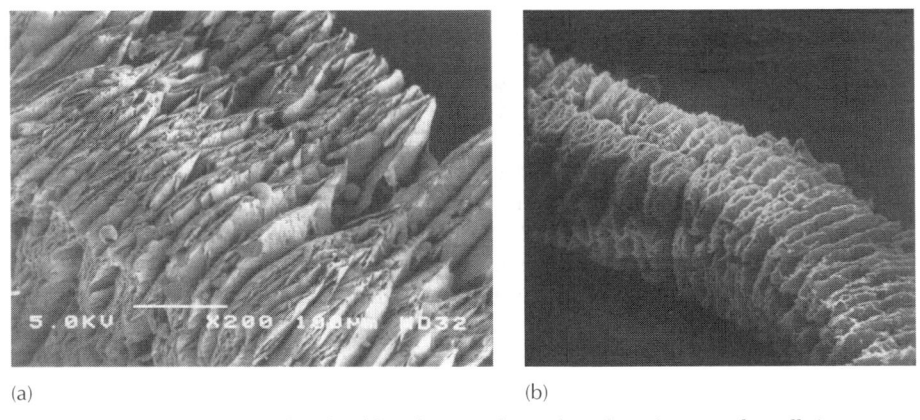

Figure 2.61 Examples of exfoliated vermiculite and graphite. (Source: Flexitallic)

The material has been widely quoted as being able to operate at up to 1000°C in inert atmospheres. A problem with most gaskets is access of air to the external side of the gasket. It has been found that graphite gaskets at high temperature can eventually fail due to oxidation from the outside progressively working inwards across the gasket. This has led to temperature restrictions on the use of graphite by many users.[17,18,19] The service life can be extended by the use of semi-metallic gaskets, section 2.8.8.2, as the metallic spirals or serrated teeth create a partial metal-to-metal seal which restricts access of air to the graphite. The temperature ratings listed below are those considered viable to provide a reliable seal for five years' operation.[19]

Fluids: Water, steam, chemicals, except strongly oxidizing media such as concentrated acids
Temperature: −200 to 350°C flat sheet gasket
                     450°C spiral wound gasket
                     550°C kammprofile gasket
                     1000°C inert environment with no exterior access of oxygen
Pressure: 180 bar

Mica   The use of mica fibres provides a high temperature and chemically resistant material. Unfortunately the nature of standard mica fibres requires the use of a binder, usually silicone. The material is therefore relatively weak and cannot be used at high pressure. An alternative is to use a tanged stainless steel core. It is used for high temperature, low pressure applications such as exhaust pipe gaskets and furnace or flue sealing. Alternative forms of mica fibres are under development to make them more suitable for high temperature gaskets.

Fluids: Hot air, oil, chemicals
Temperature: −50 to 900°C
Pressure: 5 bar

Exfoliated vermiculite

This material is manufactured from exfoliated flakes of the mineral vermiculite, a derivative of mica. It may be created by either chemically or thermally exfoliating the material. The thermally exfoliated material provides similar particles to exfoliated graphite, Figure 2.61. This is bound with some elastomer binder to form a gasket material that can be more flexible and easier to handle than non-reinforced graphite material. Chemically exfoliated material provides a different structure that has helped to provide a flexible gasket sheeting.[14] For higher strength this material is available in reinforced sheet such as the tanged material. It is also used as a filler for semi-metallic, spiral wound and Kammprofile gaskets. Some forms of the material without binder are available for high reliability applications where loss of sealing stress is critical.

Fluids: Water, steam, chemicals, solvents
Temperature: 980°C
Pressure: 100 bar

2.8.8.2 Semi-metallic

For higher duty applications involving increased pressure or temperature, higher flange loadings, high process flows, etc. then semi-metallic gaskets may be used. The combination of non-metallic filler for compressibility and metal for strength, resilience and chemical resistance provides a more reliable gasket in many conditions.

Spiral wound

The most familiar semi-metallic gasket is the spiral-wound type which is widely used on high pressure joints. A profiled strip of metal is wound into a spiral with a layer of gasket material interleaved between the metallic spirals, Figure 2.62. The gasket material will be selected to suit the application and will normally be either PTFE or graphite depending on the application. An alternative offered is exfoliated vermiculite. A variety of configurations are used depending on the type of flange and application. On raised face flanges, the gaskets have an outer support ring which locates inside the bolt PCD, and they can also be supplied with an inner ring. These are usually for the higher pressure systems, or processes having high flow rates or abrasive media, as the inner ring reduces turbulence at the pipe bore. On spigot/recess flanges a simple seating element gasket is usually used with no additional support rings, and the flanges are dimensioned to achieve the correct

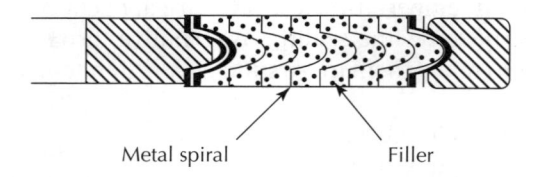

Figure 2.62 Spiral wound gasket construction.

gasket compression when metal to metal contact is reached. In these cases the gaskets should be designed having the correct inner and outer clearances for the recess used.

The standard pipeline gaskets are nominally 4.5 mm thick and compressed to a working thickness of 3.2/3.45 mm, although a thicker gasket will be used on large flanges in excess of one metre diameter.

With this design of gasket the metallic strip resists the pressure forces and because it is loaded onto the flange faces will provide some sealing. The softer gasket material provides a gap filling function to create the high integrity sealing. The flexibility of the spiral winding provides improved sealing force retention during cyclic operation compared with a flat face gasket.

It has been found that under adverse conditions such as uneven flange loading there is the possibility of inward buckling of the gasket spirals, Figure 2.63. This has proved to be a particular problem with graphite filled gaskets, due to the incompressibility of the graphite.[21] The recommended arrangement is to use an inner support ring with the gasket to prevent buckling.

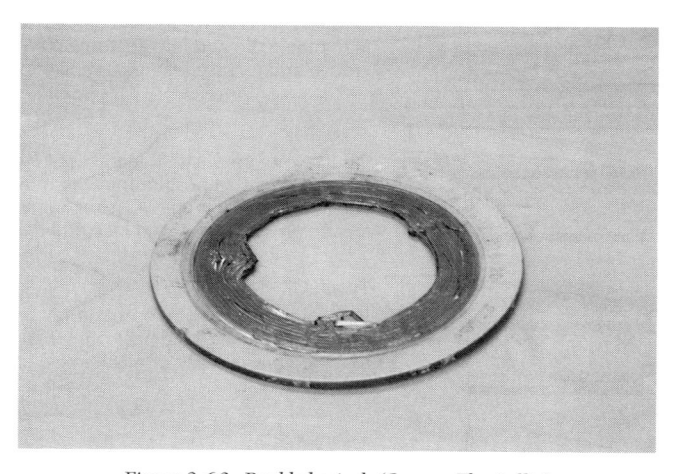

Figure 2.63 Buckled spiral. (Source: Flexitallic)

Conventional spiral wound gaskets are specified for high pressure and high flange load applications. To provide leak tight sealing at lower pressures and flange loadings spiral wound designs with the filler, usually PTFE, standing proud of the spirals have been developed to meet the current emission specifications for bolted joints.

Fluids: Dependent on filler, see sheet gasket materials
Temperature: Maximum of 260 to 1000°C depending on filler
Pressure: 300 bar

Kammprofile
Kammprofile gaskets are a solid metal ring having a concentric serrated tooth form profile on each side. A covering layer of graphite, PTFE or vermiculite is

applied, which becomes compressed into the serrated surface when the gasket is loaded. These gaskets have a good level of sealing tightness combined with the potential benefit of low seating stress and are frequently used to replace metal-jacketed joints on heat exchangers. A number of metal core profiles are available, Figure 2.64. This style of gasket is preferred in many applications where there are wide temperature and pressure fluctuations.

Fluids: Dependent on filler, see sheet gasket materials
Temperature: Maximum of 260 to 1000°C depending on filler
Pressure: 200–300 bar depending on filler material

Spiral-wound and Kammprofile gaskets can be used on many similar duties depending on supplier and user preference. However, the relative potential merits of each can be summarized as:[19]

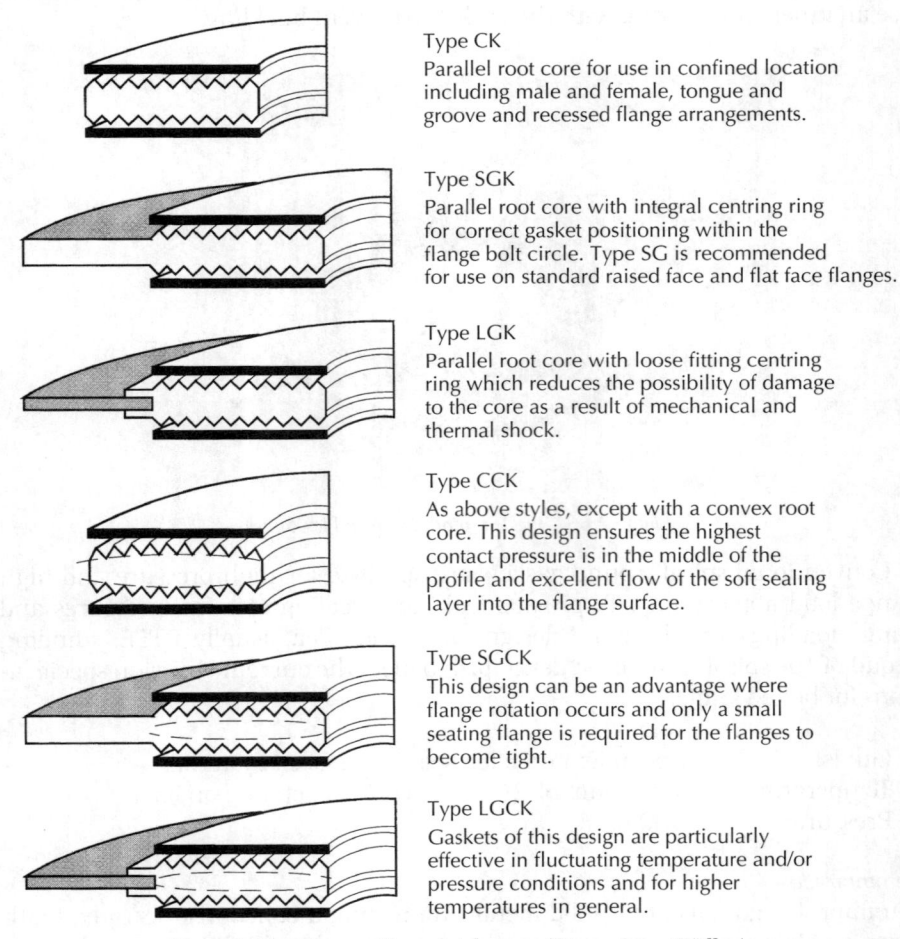

Type CK
Parallel root core for use in confined location including male and female, tongue and groove and recessed flange arrangements.

Type SGK
Parallel root core with integral centring ring for correct gasket positioning within the flange bolt circle. Type SG is recommended for use on standard raised face and flat face flanges.

Type LGK
Parallel root core with loose fitting centring ring which reduces the possibility of damage to the core as a result of mechanical and thermal shock.

Type CCK
As above styles, except with a convex root core. This design ensures the highest contact pressure is in the middle of the profile and excellent flow of the soft sealing layer into the flange surface.

Type SGCK
This design can be an advantage where flange rotation occurs and only a small seating flange is required for the flanges to become tight.

Type LGCK
Gaskets of this design are particularly effective in fluctuating temperature and/or pressure conditions and for higher temperatures in general.

Figure 2.64 Kammprofile gasket designs. (Source: James Walker)

Spiral-wound applications:

- Flanges are bowed.
- Flange rotation occurs.
- Flange faces are out of parallel.
- Thermal movement of flanges is likely.
- Radial shear of gasket expected.
- Most convenient for cleanup of flanges during replacement.

Kammprofile applications:

- Low sealing stress.
- Assembly stress on gasket is very high.
- Grooving or damage on the flange causes localized high stresses.
- Gasket dimensions, such as narrow cross-section, would make a spiral wound unstable.
- Use of compression limiter not possible due to dimensional constraints.
- Specially tailored thickness required.
- Highest available levels of leak tightness are required.

Corrugated metal gaskets

Plain corrugated metal gaskets have been available for many years. A development of these is to incorporate a filler of gasket material, usually graphite or PTFE, into the corrugations. A number of proprietary profiles are available, as shown by the examples in Figure 2.65. The recess traps the material, preventing it from being extruded under load.

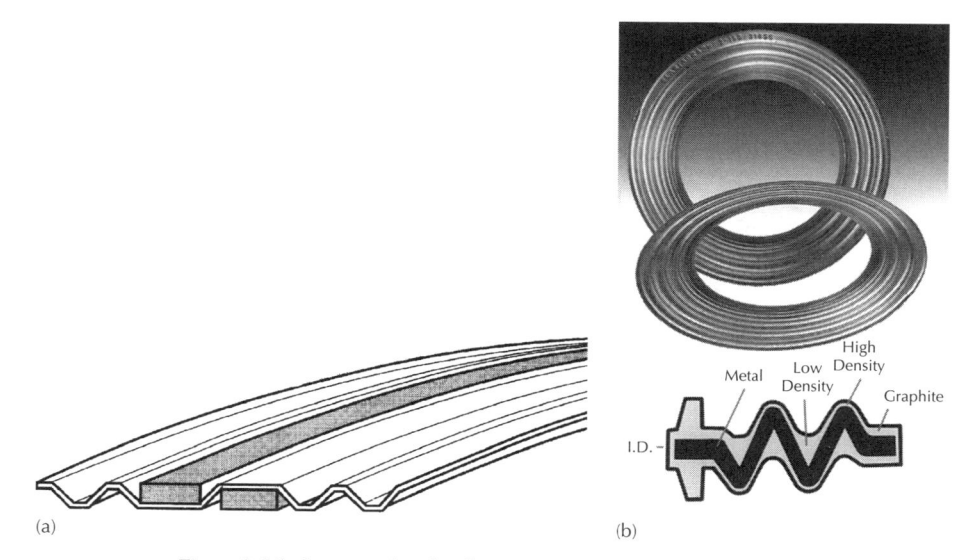

Figure 2.65 Corrugated gasket designs. (Source: James Walker, Novus)

Fluids: Dependent on filler, see sheet gasket materials
Temperature: −200 to a maximum of 260 to 650°C in steam depending on filler
Pressure: 200 bar depending on filler material

Metal jacketed
In these gaskets a metal jacket encloses a soft filler material. There are a wide range of jacket designs, Figure 2.66, normally using metals such as soft iron,

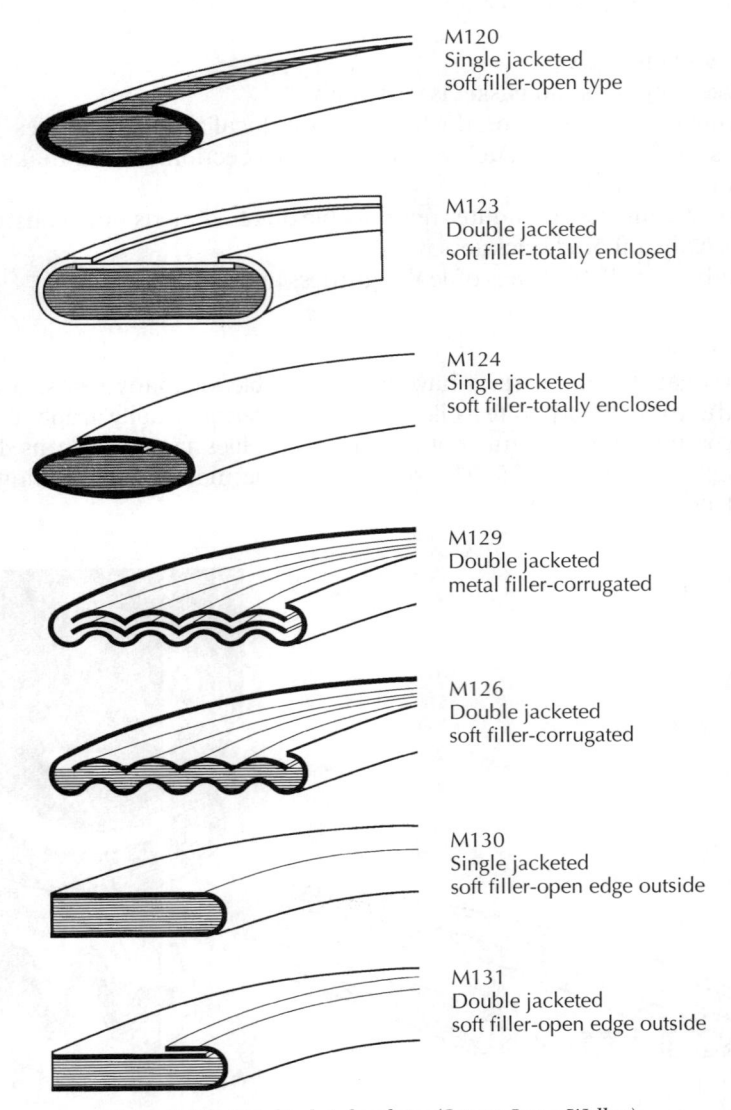

M120
Single jacketed
soft filler-open type

M123
Double jacketed
soft filler-totally enclosed

M124
Single jacketed
soft filler-totally enclosed

M129
Double jacketed
metal filler-corrugated

M126
Double jacketed
soft filler-corrugated

M130
Single jacketed
soft filler-open edge outside

M131
Double jacketed
soft filler-open edge outside

Figure 2.66 Metal jacketed gaskets. (Source: James Walker)

carbon steel and stainless steel. The filler may be a non-asbestos millboard but alternatives such as exfoliated graphite are also used for high temperatures and resilience. The selection of jacket design and orientation is very dependent on the application. It should be noted that the orientation of many jacketed designs is important, especially as some flanges are designed with stress raisers to increase the sealing force, Figure 2.67.

A recent metal cased gasket has been designed to combine low seating load, high sealing performance, and ultrahigh cleanliness in a metal seal.[15] The seal construction is based around a graphite core that provides good springback and deflection properties, as well as an ability to withstand temperatures beyond 1000°C. Two metal cups moving axially relative to one another during seal compression control the die-formed graphite. The graphite ring and metal cup assembly is encapsulated by hermetically laser welding, to prevent any risk of high temperature oxidation, Figure 2.68. This process is carried out because the

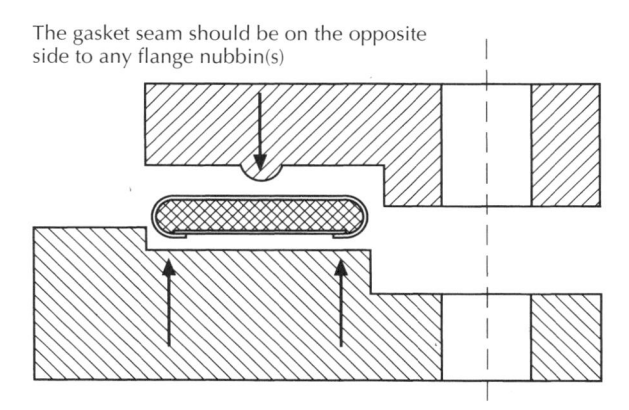

Figure 2.67 Flange designed with stress raisers for metal gaskets. (Source: James Walker)

Figure 2.68 Jacketed metal gasket with a graphite core. (Source: James Walker)

upper temperature limit of exposed graphite is substantially reduced. The whole assembly is then sealed with an envelope of the actual metal sealing material, depending on the application.

The sealing function of this design relies on the ridges machined on the external face of the spring metal cups acting through the sealing envelope, to concentrate the seating load over a limited contact area, thus maximizing the seating stress.

Fluids: Dependent on metal
Temperature: 1050°C
Pressure: 100 bar depending on gasket and flange design

2.8.8.3 Metallic

Metal-to-metal joints are generally used in arduous sealing applications involving high temperature, high pressure or both, particularly where reliability is critical. The division between metal-to-metal seals and gaskets is even less distinct than for other classes of seals and gaskets. As far as this book is concerned the difference is that a metal seal has some inherent resilience and ability to be energized by the pressure, such as a hollow O-ring or a C-ring. By contrast a metal gasket depends on the initial sealing force to provide sufficient interference stress at the seal interface to resist the system pressure. Some of the metal joint rings discussed below depend to some extent on pressure energization and the flanges are clamped metal to metal, so could arguably be called seals.

Metal joint rings are manufactured from one metal or a combination of metals in a variety of shapes and sizes. There are a number of specific standards which provide the geometry of the joint ring and the mating groove. In addition a number of proprietary joint rings and systems are available, often originally devised to solve a particular sealing problem. All have one thing in common, high contact stress at the sealing interface is required to cause plastic flow of the metal to create a seal. Depending on the joint design this may be the gasket ring, the flange or both. This can create problems with respect to reusing the assembly as it may be necessary to remachine flanges before a new joint ring can be fitted. Hence in general the gasket material will be softer than the flange to reduce damage to the seating area. It is also possible to use a softer ductile layer on the metal ring to ensure plastic flow in the contact zone.

Oval and octagonal

These are commonly used on oilfield applications and details of these joints are given in standards ASME B16.20 and API 6A. The gasket sits in a recess in the flange face which has 23° angled walls, Figure 2.69. This requires additional machining of the flanges compared with a standard flange. The oval section ring creates an area of contact stress concentration, but this is not well defined. The octagonal rings concentrate the stress in the differential taper and will perform reliably to much higher pressures.

Oval

Octagonal

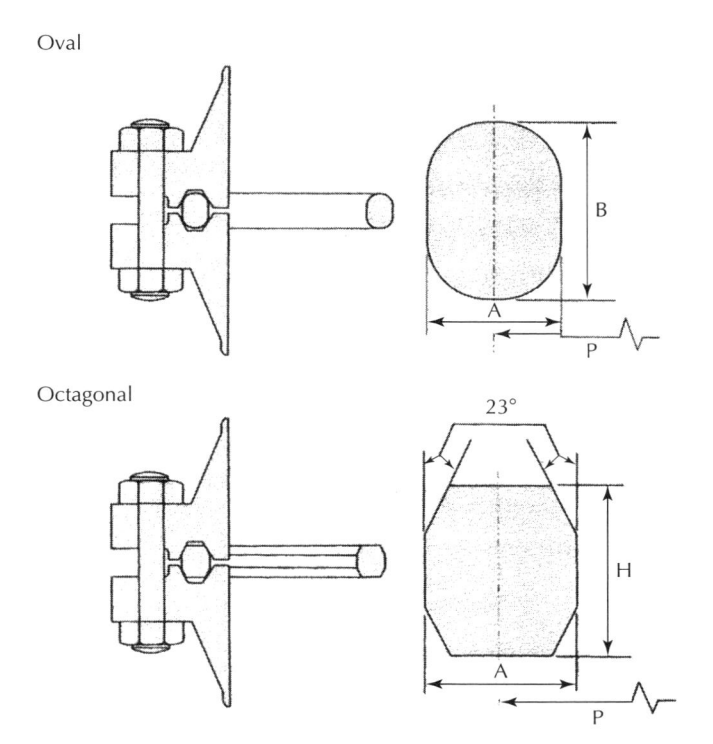

Figure 2.69 Oval and octagonal metal gaskets. (Source: Flexitallic)

RX and BX
The RX joints are an unequal bevel octagonal ring and are considered to be a pressure energized or pressure assisted seal. The BX is also octagonal though shorter in profile and designed to go into a recess that becomes metal to metal when the flanges are tightened. These are used on very high pressure flanges up to 1380 bar (20 000 p.s.i.) rating.

Some standoff exists between the flange faces though PTFE inner spacers can be supplied to reduce the effects of erosion and the buildup of dirt on the flange face inside the gasket. Similarly sponge rubber protectors are sometimes used to keep the area clean outboard of the gaskets and around the bolts. For hydro testing purposes rubber covered ring joint gaskets are available to avoid damaging the flange recesses.

Lens ring
This is again a taper metal sealing ring, which mates with a tapered flange. In this case the taper is radial at the inside diameter of the flange, Figure 2.70. This has the advantage of reducing the effective pressure area compared with the API rings such as the RX discussed above. They are covered by German standard, DIN 2696.

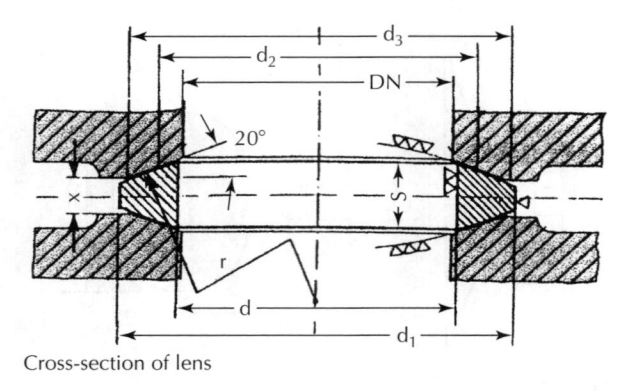

Figure 2.70 Lens ring. (Source: Flexitallic)

Convex metal gaskets

The purpose of this geometry, Figure 2.71, is to provide a high stress concentration and reduce flange loading when compared to a flat gasket. The material must be chosen to ensure plastic flow of the metal gasket.

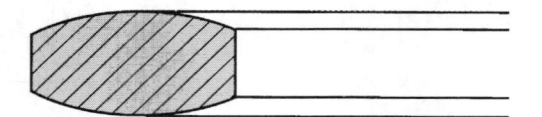

Figure 2.71 Convex metal gasket. (Source: James Walker)

2.8.9 Compact flanges and connectors

2.8.9.1 Compact flanges

The Compact Flange System (CFS) is an alternative flange design that is not based around the conventional bolted flange designs. As the name implies the flange assembly is considerably more compact than a standard flange assembly. The method of operation is also different with a key characteristic being the flange face geometry. It includes a slightly convex bevel with the highest point called the heel, adjacent to the bore and a small outer wedge around the outer diameter of the flange face. During makeup of the connection, the bevel is closed and flange face-to-face contact is achieved. The compact flange has one seal which is activated by the bolt pre-stress, i.e. the heel seal and one self-energizing and pressure energized seal, which is the seal ring. This seal ring is an elastic seal energized by radial deformation caused by the wedging action of the seal groove. It is pressure energized when subjected to internal pressure. It also has an external pressure capacity of approximately 100 bar, or 1000 metre water depth, which is important for subsea applications. The flange diameter is considerably reduced and it uses a larger number of smaller bolts to achieve the bolt load required. A comparison of equivalent flanges for a similar pipe, in this case a 6 inch CL2500, are shown in Figure 2.72.

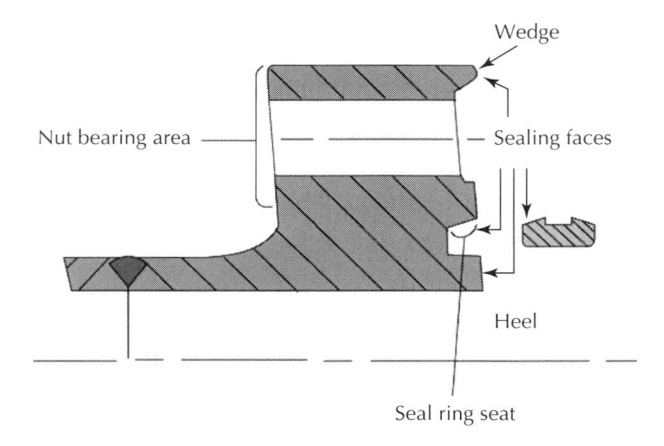

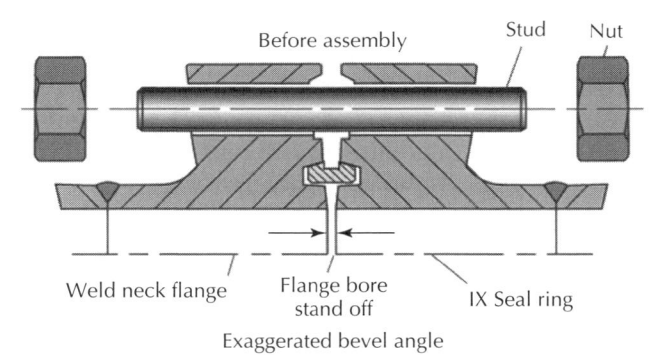

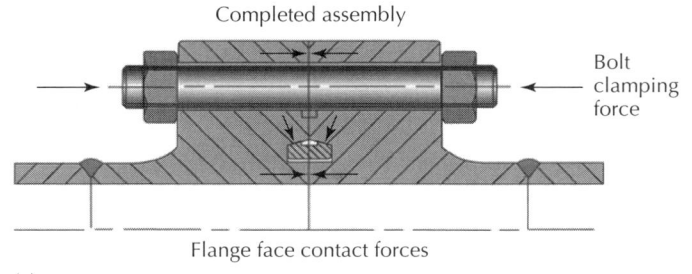

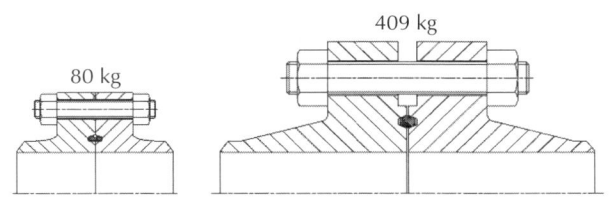

Static seals 93

(a)

(b) Comparison with a conventional flanged joint.

Figure 2.72 Compact flange design. (Source: Vector International)

Among the benefits claimed for this style of joint are:

- Significant reduction in weight and volume.
- Reduced hydrostatic force as the effective seal diameter is reduced.
- Reduced moment arm.
- Increased stiffness.
- Load transfer through the flange rather than the gasket.
- No relative flange movement.

These flanges are recommended for the following application limits:

Fluids:	Water, steam, oil, chemicals, depending on metal selection
Temperature:	250°C
Pressure:	1700 bar

2.8.9.2 Clamp connectors

These are manufactured as proprietary couplings by a number of manufacturers, for example Destec, Grayloc, Techlok (Vector International). An example is shown in Figure 2.73. All these clamp connectors work on a similar principle, although there may be detail differences in the construction and they cannot necessarily be considered as interchangeable.

The seal ring is a T in cross-section. The leg of the T forms a rib that is held by the hub faces as the connection is made. The two arms form lip seals that create an area of sealing surface with the inner surface of the hub. The clamp fits over the two hubs and forces them against the seal ring rib. As the hubs are drawn together by the clamp assembly, the seal ring lips deflect against the inner sealing surfaces of the hubs. This deflection elastically loads the lips of the seal ring against the inner sealing surface of the hub, forming a self-energized seal. Internal pressure reinforces this seal, so that the sealing action of the connector is both self-energized and pressure energized. These fittings offer a number of advantages compared with a conventional flanged joint including:

- Low weight when compared to conventional flanges.
- Reduced bolting requirements with only four bolts.
- Radial bolting allows 360° orientation around the pipe.
- Quicker and easier assembly and disassembly, no bolt holes to align.

Fluids:	Water, steam, oil, chemicals, depending on materials of construction
Temperature:	200°C
Pressure:	1500 bar

2.8.9.3 Taper lok

This is another proprietary connector design. It is supplied as a bolted flange design, or as a four bolt connector similar to the T-ring clamp connectors,

Static seals 95

(a) Prior to assembly — Self-energized seal

Fully assembled — Pressure energized seal

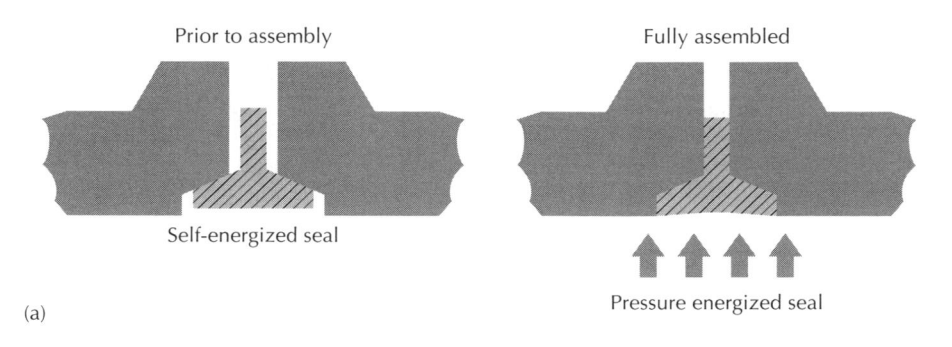

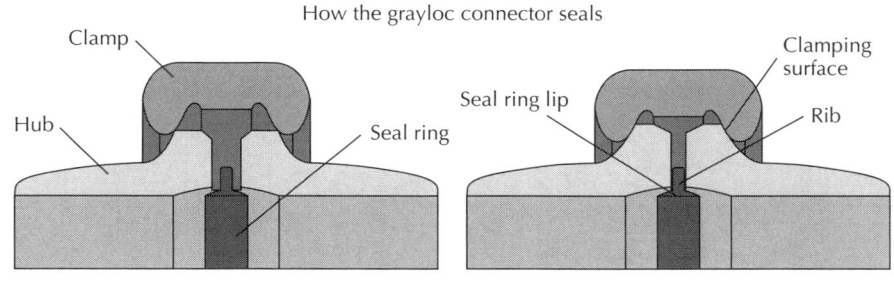

(b) How the grayloc connector seals

Clamp, Hub, Seal ring, Seal ring lip, Clamping surface, Rib

Rib of the seal ring is clamped between hub faces. Lips of the seal ring engage inner hub surface in an interference which detects the lips to achieve a seal.

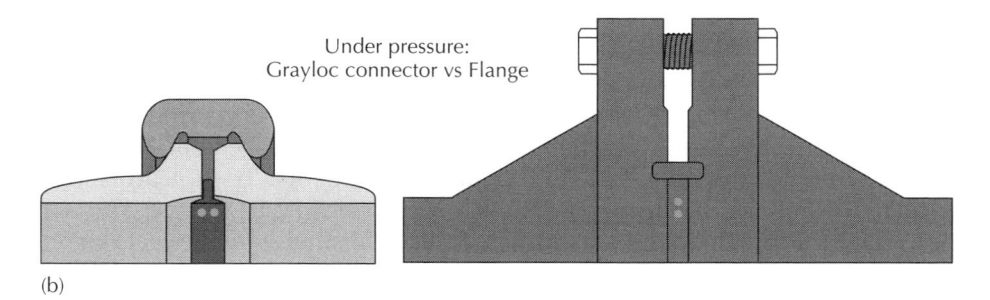

Under pressure: Grayloc connector vs Flange

Figure 2.73 Examples of clamp connectors. (Source: Techlok, Graylock)

section 2.8.9.2. The pipe flanges are machined with matching male and female tapers with a taper metal seal that is trapped between the two matching tapers on assembly, Figure 2.74. The joint is self-aligning and allows up to 2° axial pipe misalignment. Special models can be designed for up to 10–20° misalignment. Unless otherwise specified, the seal ring is typically made of the same material as the mating flanges which has advantages in avoiding thermal expansion/contraction concerns and galvanic corrosion between dissimilar metal components. A feature of this fitting is the significant length of the seal surfaces contacts on both the male and female components. The seal is therefore achieved by exposure of an extended contact area of metal surface, rather than

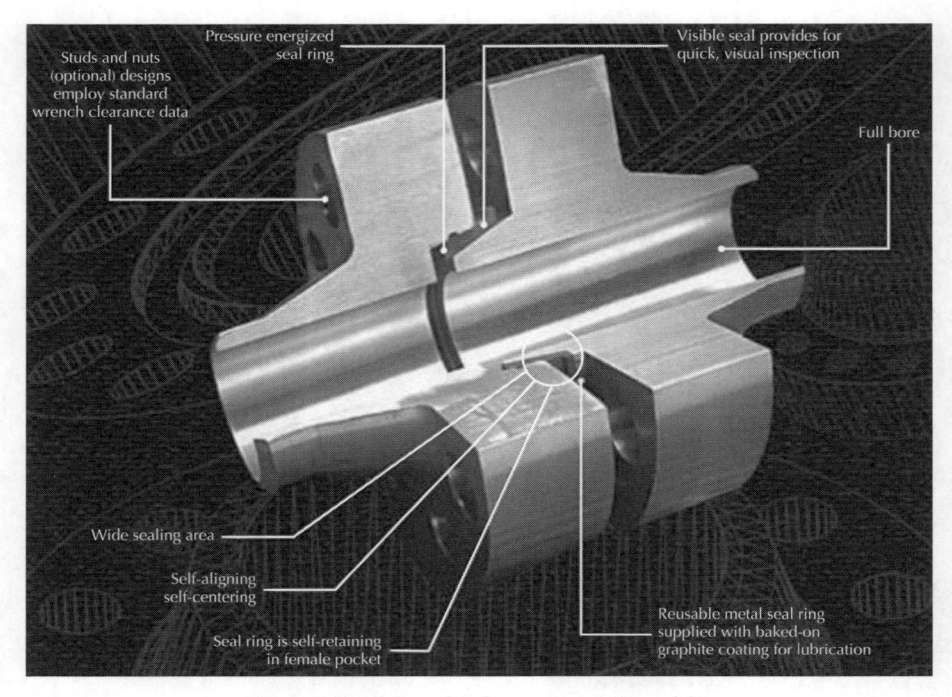

Figure 2.74 A Taper-lok flange. (Source: Taper lok)

the line-or-point contact characteristic of conventional metal joints such as the octagonal designs. The tapered seal ring is also pressure energized. As the system is pressurized, the fully confined seal ring is further wedged into the annulus formed by the mating male and female components. These features are considered to make this fitting particularly suitable for cyclic service. The seal rings have a small lip on the top of them. In the unbolted state the seal ring lip will stand off the face of the flange. When the flange is assembled the seal ring is forced into the female pocket until the lip rests against the flange face. This lip resting against the face is visual proof that the seal is seated correctly. As the metal sealing ring is not subjected to plastic flow it is claimed to be reusable, up to eight times being recommended.

If this type of joint is to be used then all pipe fittings have to be manufactured with the necessary tapers. Arrangements must therefore be made to have all components such as valves, etc. prepared with the correct flange taper. Care is also required to protect the exposed male taper from damage during installation.

Fluids: Water, steam, oil, chemicals, hazardous fluids
Temperature: 800°C
Pressure: 1400 bar

2.9 Selection of the optimum static seal design and material

A very wide range of static seal types are available for selection. Several important criteria must be considered to achieve an optimum selection. In addition there is the factor of individual or industry experience or traditional practice. While this is obviously very valuable it can also often become a 'tradition' that inhibits consideration of alternatives that may be either a technical improvement or provide similar performance more economically. Occasionally an industry tradition or practice should be tested to consider whether it still provides the optimum solution. The primary factors to be considered are:

- Temperature is a starting point that defines which materials may be considered for the application. The whole life cycle temperature range must be considered carefully.
- Pressure dictates the type of construction and seal designs that may be considered.
- Industry standards can define basic criteria of a design. Examples are pressure vessel codes, aerospace standards and hygienic design codes.
- Manufacturing method and equipment type. The mating surface material, surface finish, tolerances and flatness together with the assembly method will directly influence the types of seal and materials that may be considered.
- Sealed fluid and external environment will directly influence the selection of materials and in turn influence whether a particular seal type is feasible, or economic.
- Leakage integrity and emissions requirements. If a high sealing integrity is required this will influence the design of the assembly. A requirement to minimize vapour emissions can influence both the detail design and selection of materials.
- Life expectancy is important. If the life required is only a few minutes or hours then materials such as elastomers may be considered for considerably higher temperatures than if they are expected to survive several years.
- Manufacturing volume will influence selection. For one off or low volume manufacture it is usually most sensible to use standard available seal designs or methods wherever possible. As volumes increase it becomes viable to consider specific designs to suit the application. This can involve both achieving the required geometry for design reasons and facilitating product assembly.
- Maintenance and accessibility are important. If the seal is likely to be separated in normal service then a design and material that retains the seal in place, and does not adhere to the opposing surface, will be required. Examples are inspection and access panels, hygienic fittings, doors, windows and filler caps.

2.9.1 Temperature

Temperature directly influences the materials that may be feasible for an application and in turn the potential seal designs that may be considered. The whole life cycle temperature range must be considered. Excursion outside the working temperature range for a material may cause temporary leakage, damage leading to a significant reduction in seal life or in some cases gross failure.

- −50 to 200°C: This is the overall operating range for elastomer materials. Considering the selection of elastomer as the first option will provide the lowest assembly forces and a self-energizing seal. It will potentially provide the most economic construction and often the lowest leakage solution.

    In this temperature range the major reason for investigating alternative options to an elastomer would be achieving a material selection where the temperature range of the elastomer and compatibility with the sealed fluid meet the application requirements. Elastomers may also be unsuitable if a very high resistance to permeation losses is required.
- 0°C to 300°C: Perfluoroelastomers may be considered in this temperature range and will also provide a very wide fluid compatibility where resistance to aggressive chemicals and solvents is required.
- Cryogenic to 250°C: Plastic materials, most often PTFE or UHMPE, can be considered in this range. With materials such as PTFE the limiting factor in compatibility is usually the energizing element which, outside the range of elastomer capabilities, as above, will usually be a metal spring.
- Cryogenic to 850°C: Metal seals are available to cover this range. Actual temperature limits will depend on the metal and also seal section. The highest temperatures require materials with good high temperature strength such as high nickel alloys. N06600 is used for the highest temperature metal O-rings.

    The use of metal seals will involve closer machining tolerances and improved surface texture compared with either elastomer or plastic seals.
- Cryogenic to 1000°C: Gaskets and clamp connectors can be used over this entire temperature range. This will still require consideration of individual material capabilities for the temperatures involved in the application. Note that metals can often be used at higher temperatures in a gasket configuration than in a metal seal as there is less reliance on the elasticity of the metal to retain sealing force.

2.9.2 Pressure

If static seals are considered, sections 2.2 to 2.5, then the limit of pressure capability is a function of a combination of seal design and hardware arrangement.

It is possible to use an elastomer seal up to 2000 bar, provided the hardware is designed to avoid extrusion gaps at all pressures, excessive groove movement is avoided and adequate anti-extrusion precautions are included in the design.

Assuming that the component hardware is adequately designed to prevent separation of the sealed components due to pressure, the important considerations are seal energization and control of extrusion. Generally it is easier to seal flange arrangements as the flanges can be bolted metal to metal, minimizing the extrusion gap. With cylindrical arrangements there will always be some relative metal movement at very high pressures which create additional problems.

With elastomer seals the energization will continue, as described in section 2.2.1, provided that the elastomer is within the correct working temperature range. There can be an increase of material glass transition temperature with increasing pressure, and so for very high pressures the minimum temperature of the material may be increased. This can be a particular problem with highly fluorinated elastomers.

The general limit for plastic seals such as metal energized U-rings is up to 500 bar, but special designs with special provision of springs and anti-extrusion rings can be used to 2000 bar.

Metal seal can be used for very high pressures, up to 3000 bar, but it is necessary to select the correct geometry. Vented O-rings, or spring energized metal rings, which are also open to the pressure, are preferred. These designs will continue to be energized by the system pressure as it increases.

Pressure capability of gasketed joints is dependent on the gasket material and the design and bolting of the joint. Pressure ratings of standard flange designs are provided in the individual standards. The rating will depend on the materials, temperature and grade of bolts used. The gasket material will dictate the amount of pre-stress that can be applied to the joint to withstand the internal pressure. Assuming an adequate flange specification is used most sheet gaskets are limited to between 50 and 200 bar depending on material, gasket thickness and temperature. Higher pressures are possible with semi-metallic and metal gaskets as they are more resistant to blowout. However, they are still dependent on the ability of the bolt preload to withstand the pressure and maintain the required load on the gasket. Any assessment of pressure capability of a gasketed joint must allow for relaxation of the bolt and gasket stress.

2.9.3 Industry design codes

The starting point for seal selection in many industries will be any relevant established design codes. These may cover basic operational requirements, such as clean in place, which place specific design constraints on the designs that can be used. In other industries it may be necessary to comply with established design codes, such as those for pressure vessels, in order to obtain regulatory approval for the plant. These are often based on established practice and may not be the optimum design for individual installations however; considerable design justification may be involved if alternatives are considered.

Sections 7.2 and 7.4 provide information on many relevant documents and sources of further information that impact on individual industries.

2.9.4 Manufacturing method

The component surface texture, flatness and finishing method will directly influence the type of seal that may be used. Surface texture and tolerance requirements are provided for individual seal and gasket types in the relevant sections.

- Pressed, stamped, as cast or as moulded surfaces: If it is not feasible to maintain tight tolerances or a sufficiently fine surface texture it may be necessary to use a more conformable seal, such as using an elastomer instead of a metal. If surfaces are not flat, something that can provide satisfactory gap filling will be required. Soft elastomers and gasket materials are available but the pressure will be limited.
- Machining method and surface texture: Typical requirements for seals and gaskets are discussed in the appropriate sections. If it is not feasible to meet the requirements for a particular seal type the options may be either an alternative seal, or degraded performance. Other factors to be considered include assembly method and size.
- Very small seals: Small seals that are orientation sensitive, such as U-rings, require care in assembly procedures. If seals have to be stretched for assembly this can be a problem even for elastomers at small sizes as the elongation to break limit may be approached or exceeded.
- Large seals: Storage procedures and methods of retention during assembly must be considered.

2.9.5 Sealed fluid

This will directly affect the material selection. It is important to remember that a seal separates two fluids, oxidation from high temperature air can degrade a number of elastomer and gasket materials and may be the limiting factor rather than the system fluid.

All potential fluids should be considered, such as cleaning and flushing fluids, minor constituents of a process flow, etc.

2.9.6 Life expectancy

The life expectancy of elastomers, plastics and most gasket materials is generally directly related to operating temperature. They can be used above the normal recommendations if a very short life of a few minutes or hours is required. Conversely for good reliability and long life the temperature should be well below the recommended limit.

Predicting the life expectancy of polymeric materials has proved to be difficult but a number of methods are in use.[6, 9]

Specific details of graphite material limit recommendations are provided in section 2.8.8.

2.9.7 Leakage integrity, emissions

- Very high integrity, minimum vapour transmission: It will be necessary to consider:
  - Permeation through the sealing material.
  - Interstitial leakage around the seal through the surface texture interface between sealing material and the counter-face.

  All elastomers, many plastics and fibrous or plastic gasket materials will have some level of quantifiable diffusion. This can be minimized by choice of materials with lowest permeability, use of a diffusion barrier material or detail design of geometry to minimize surface area exposure for gas transmission.

  To eliminate leakage around the seal requires consideration of both material and surface texture to ensure that the seal has 'filled' the surface texture of the counter-face. Minimum contact stress figures for different materials are provided in sections 2.2 to 2.5. It is important to note that for both plastic and metal seals some plastic flow of the seal surface is required to provide high integrity sealing. Many metal-to-metal seals will not pass a stringent leak test with gas. Specialist designs of spring energized metal seals probably provide the highest leakage integrity.
- High standard of leakage control, low emission flanges, high reliability liquid seals: It is necessary to consider the standard of surface finish recommendations and machining methods as a maximum value and use optimum seal contact stress to achieve long-term reliable sealing. Suitable inspection procedures are also required to ensure that seals and other components are to specification.
- No special requirements: Use standard design recommendations, but beware that some long-term seepage may occur.
- Hygienic applications: Consult the specifications discussed in section 2.3.9 and standards in section 7.2.

2.9.8 Manufacturing volume

- One-off or relatively low volume production: Use standard catalogue seal or gasket profiles wherever possible to ensure economic and timely supply and proven design data. FIP sealing with manual application can also be considered.
- Regular production of repeat components, several hundred per week: Catalogue seal and gasket profiles will still be economic if they do not

compromise the design. Some automation may be considered such as robot applied FIP or cure in place may be considered. Specialist manufacturers may be considered to provide non-standard profiles.
- High volume production, several thousand per week: Specific seal designs may be considered to optimize the component design and assembly. Seal designs and profiles may be considered specifically for the component. Standard items such as O-rings may be colour coded or surface treated to specification to aid identification and assembly. Moulded in place seals may be considered or fully automated FIP or CIP.

2.9.9 Maintenance and accessibility

- Component is considered to be sealed for life: Manufacturing considerations are the priority and potential maintenance is not considered.
- Maintenance expected when seal will be replaced: Reasonable access to seal to be considered so that the replacement can be installed without damage and counter-faces cleaned and inspected reliably without damage.

    Potential sharp edges, working environment, ability to ensure seal remains in place during assembly, requirements for special tools for assembly all need to be considered. If precision tightening of bolts is required, is this practical once the equipment is installed in the working environment?

    Seals that may damage the counter-face, such as due to corrosion, or metal seals causing a witness mark on the counter-face should be avoided.

    Seals that are expected to leave considerable residue or adhere to the surface are not recommended.

    Retention of the seal if it is large, assembly is 'blind' or in an inconvenient position should be considered.
- Disassembly without seal replacement: This may occur with inspection covers, access panels for adjustment of instruments or machine settings, hygienic fittings, tank lids and filler caps, etc.

In these situations a seal that is retained in position is highly desirable. Options can include:

- O-ring retained in a dovetail groove. The design of the groove needs care to ensure that the seal is retained effectively but that the groove does not have excessive angles which can reduce seal contact stress.
- U or similar section seal with the heel clipped into a retaining recess or lugs.
- Specially made profile with retaining flange, lugs or other fittings.
- Moulded in place or cured in place seal. The selection of one of these techniques will depend on a significant volume of components. It will also only be feasible if the seal is expected to last for the entire life of the component to which it is moulded.

2.10 References

1. BS 1806, 'Specification for dimensions of toroidal sealing rings ("O"-rings) and their housings (inch series)', British Standards Institution.
2. AS 568 B, 'Aerospace size standard for O-rings', SAE.
3. ISO 16032-1, 'Aerospace fluid systems – O-rings, inch series: inside diameters and cross sections, tolerances and size-identification codes. Part 1: Close tolerances for hydraulic systems'.
   ISO 16032-2, 'Aerospace fluid systems – O-rings, inch series: inside diameters and cross sections, tolerances and size-identification codes. Part 2: Standard tolerances for non-hydraulic systems'.
4. BS 4518, 'Specification for metric dimensions of toroidal sealing rings ("O"-rings) and their housings', British Standards Institution.
5. SMS 1588, 'Metric O-rings, installation dimensions', Swedish Standards Institute.
6. D. Peacock, 'The effects of geometry on the compression set of elastomers', PhD Thesis, Cranfield University, 2002.
7. BS 5106, 'Back-up-rings specification for dimensions of spiral anti-extrusion back-up rings and their housings', British Standards Institution.
8. AS 5782 Aerospace Standard, 'Retainer, backup ring, hydraulic, pneumatic, polytetrafluoroethylene, resin, uncut', SAE.
9. E. Ho, 'Increased confidence in sealing system design using FEA simulations', 16th Int. Conf. on Fluid Sealing, BHR Group, 2000.
10. 'New technology simplifies seal production', *Sealing Technology*, September 2003.
11. 'Pressure-indicating films now available as dots', *Sealing Technology*, July 2003.
12. *Gasket Technology, Understanding Gaskets and Dimensional Guidebook*, James Walker Moorflex, 2005.
13. S. Piringer and U. Rustemeyer, 'Improving performance of fibre-based gaskets at elevated temperatures – the sandwich approach', *Sealing Technology*, March 2004.
14. J. Hoyes, 'The development of an advanced, coreless and flexible, sheet sealing material by the use of an inorganic binder', 18th International Conference on Fluid Sealing, BHR Group Ltd, 2005.
15. M. LeFrancois, SETE April 2004, 'Metal-to-metal seals: the innovative route in static sealing', *Sealing Technology*, April 2004.
16. D.A. Thomas and S. Woolfenden, 'Biaxially orientated PTFE sheet sealing materials – further developments', 18th Int. Conf. on Fluid Sealing, BHR Group, 2005 and *Sealing Technology*, November 2005.
17. W.F. Jones and B.B. Seth, 'Evaluation of asbestos free gasket materials', ASME/IEEE Power Generation Conference, Boston, October 1990.
18. W.F. Jones and B.B. Seth, 'Replacing asbestos steam path gaskets', *Power Engineering*, March 1992.
19. J. Hoyes, 'Gasket selection and developments in gasket technology', Design Maintenance and Care of Gasketed Bolted Joints Seminar, I.Mech.E., March 2006.
20. H. Kockelmann, 'Harmonised gasket testing procedures – EN 13555:2004', Design Maintenance and Care of Gasketed Bolted Joints Seminar, I.Mech.E., March 2006.
21. R.T. Mueller, 'Inward buckling of flexible graphite filled spiral wound gaskets for piping flanges', Sixth Annual Technical Symposium, Fluid Sealing Association, 1996.
22. G. Leaver, I.W. Stewart and J.S. Deans, 'Containment aspects of couplings for biotechnology plant', *Bio Safety Journal*, Vol. 1, Paper 2, April 1995.

23. 'Hygienic pipe couplings', European Hygienic Equipment Design Group, Doc. 16, Belgium, 1997.
24. EN 1672-2:2005, 'Food processing machinery – basic concepts – Part 2: Hygiene requirements'.
25. 'Active material based seal assemblies and methods for varying seal force', Patent Number WO 2005/089186, General Motors, 2005.
26. 'Guidelines for safe seal usage – flanges and gaskets', ESA 009/98, European Sealing Association, 1998.

CHAPTER 3

Rotary seals

3.1 Introduction

Sealing of rotary shafts covers a very wide range of equipment and applications including domestic equipment, all types of automotive and power plant machinery, industrial pumps, aircraft gas turbines, power generation turbines, and large industrial and pipeline compressors.

Each application presents an individual set of demands based around the space, weight, price and reliability demands of the user. A number of individual types of rotary seal have evolved and each has an individual place based around a range of applications. Within each class of seal the range of applications can vary enormously. Hence mechanical seals can be found on equipment as diverse as a domestic washing machine and aircraft gas turbines. Within any one application area it is quite usual for one type of seal to predominate, but at the boundaries in operating conditions between seal types then different seal types may be selected depending on the space envelope, friction, leakage and reliability objectives. For example, at low pressures there is a choice between using either a mechanical seal or lip seal. Examples include the stern shafts on ships and hydraulic pump and motor shafts. The selection will depend on individual manufacturer or user preference, and can also change as developments move the relative benefits of one seal type to another. A pertinent example are the recent developments in lip seals to work at four to seven bar as described in section 3.2.9.

If the reader is contemplating seal selection for a new piece of equipment it is therefore sensible to consider the full range of alternatives before finalizing selection of the seal type.

3.2 Lip seals

The lip seals currently used for sealing rotary shafts have evolved from the lip seals made from leather used during the 1930s. The development was largely empirical for some 50 years or more and it was not until the 1990s that detailed mathematical analysis of the seals gave a more established basis for understanding the detail of how these seals operate.

The types of seals in common use can be divided into three broad categories:

- Elastomer lip seals, which are used for sealing of shafts against exit or ingress of liquids.
- Plastic, primarily filled PTFE, which are used as an alternative to elastomer seals where fluid resistance or lack of lubrication may be a problem.
- Bearing seals, which are really a derivative of lip seals that usually operate at lower lip loads and have exclusion of contaminants from the bearing as a prime function.

This section covers the basic operation of lip seals and then discusses the detail of the different options available.

In common with other elastomer and plastic seal types it is important to remember that this component is only part of the seal. The housing to which it is fitted and the shaft that rotates within it are equally important. With lip seals the shaft is a particularly important component of the seal as in addition to the surface texture, the machining method is also an important factor affecting seal performance. The thermal conductivity of the shaft is the primary factor affecting heat dispersion from the seal so shaft material is also an important consideration.

Standard lip seals are designed to operate in a splash or nominally flooded environment but with little or zero applied pressure. The maximum pressure rating quoted by major manufacturers is 0.3 to 0.5 bar. Increasing pressure causes distortion of the lip, which increases the contact area, and a rise in the contact pressure of the lip. The combination of the two creates increased temperatures and wear. Special designs of lip seal are available for elevated pressure and these are discussed in section 3.2.9.

3.2.1 Basic lip seal design

A cross-section of a typical lip seal is shown in Figure 3.1. A number of different methods are used to describe the components of a lip seal, those used here are the same as those used by Johnston.[1] The seal design looks deceptively simple, but the detail has been developed by practical experience, and more recently numerical analysis, to derive the current designs.

The precise detail of individual seal designs will vary, depending upon the intended application and the preferences of a particular manufacturer, but the overall design principle will be the same. The function of the seal is derived from the asymmetric geometry. The seal is manufactured to provide a sharp seal lip that contacts the shaft. The geometry of the seal is such that the angle of the elastomer on the oil side of the seal is steeper than that on the air side. Typical design angles are in the region of 40–45° for the oil side angle β, and 25–30° for the air side angle α when the seal is in the free state. Once installed on the shaft these angles will change by perhaps 10°, giving angles in the working condition of 50° and 20°. A further contribution to the asymmetric stress distribution is governed by the

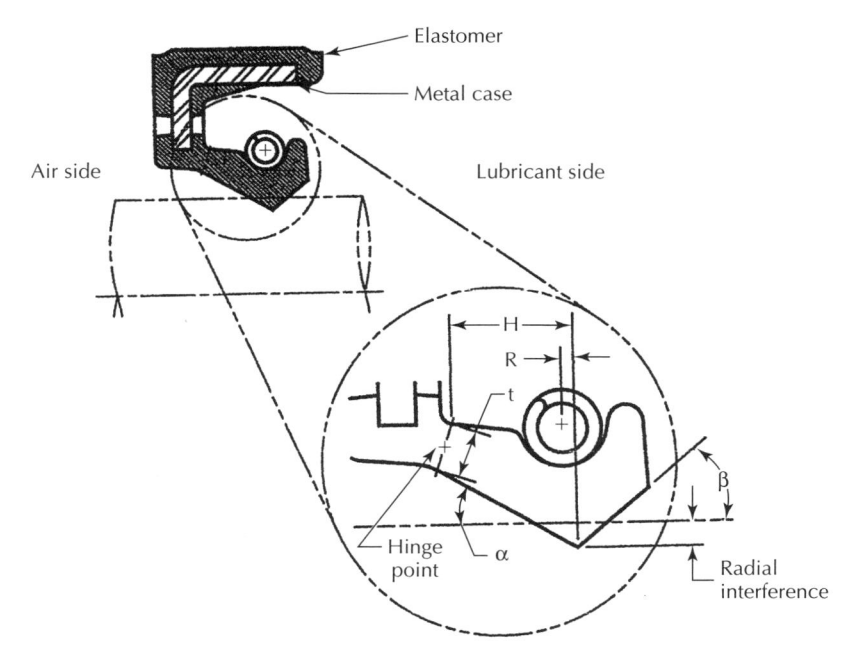

Figure 3.1 Cross-section of a basic lip seal design.

position of the spring. The spring is offset from the contact line of the lip, R, by typically 10% of the lip length, H. This offset is towards the air side of the seal. The spring position is important to provide the correct combination of direct loading to the lip, and a contribution to the asymmetric geometry. The offset described for R provides a definite value towards the air side without the potential for excessive lip distortion. If the spring is positioned such that a combination of manufacturing and installation tolerances move it to the oil side of the lip, this can cause tilting of the lip and potential leakage. The predominant choice for the spring is a garter spring. This provides a localized circumferential spring force in the location required. Some seals, generally those manufactured in low volumes, are fitted with a finger spring, Figure 3.2(f). This type of spring is also used to provide some reinforcement to the elastomer-only construction of this seal, it has no metal casing.

The overall geometry of the seal will vary depending on the application and size of seal. The flexibility of the seal will be governed by the lip length, H, and thickness, t, at the hinge point. This is normally close to the metal casing. For a standard seal the thickness will be something less than 50% of the lip length H. For increased flexibility the lip length may be increased, for instance to cope with larger than usual shaft radial deflections. Alternatively for seals designed to operate at pressure the thickness may be increased and the lip length shortened. This will have the effect of reducing the runout capability of the seal.

The seal lip is usually moulded to an L section metal casing. This supports the lip and forms the static location to mount in the seal housing. The metal case may be entirely or partially covered in elastomer. The outside diameter of the

108 Seals and Sealing Handbook

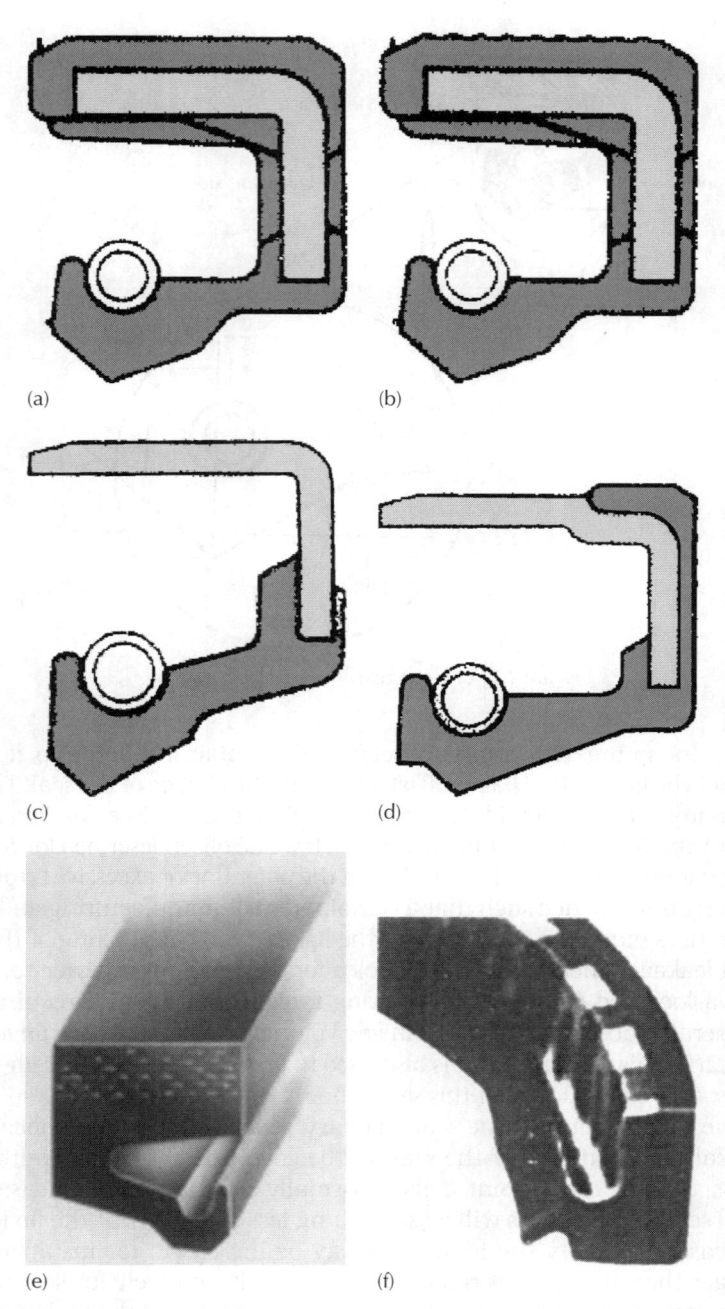

Figure 3.2 Examples of lip seal casing designs: (a) elastomer covered casing; (b) ribbed elastomer coating; (c) metal casing; (d) metal casing with partial elastomer sleeve on the o.d. to provide improved retention; (e) elastomer fabric casing that is used instead of metal on some large diameter seals; (f) complete elastomer seal with U section finger spring. (Source: (a)–(d) Freudenberg Simrit, (e) James Walker, (f) Garlock Klozure)

casing is an interference in the stationary housing to form the static seal. Using the standard lip profile a number of variations are used for the sealing of the casing on the outside diameter, and some of these are shown in Figure 3.2.

The ribbed elastomer coating, Figure 3.2(b), can be used where the housing tolerances are not controlled as closely as standard housings. This design can require care when assembling to make certain that it is[2] properly seated before the assembly press tool is removed.

The metal casing, Figure 3.2(c), without elastomer coating can be more economic as there is less elastomer material. It is used for industrial applications, and provides a more positive location, but closer tolerances and better finish for the housing diameter are required. It also requires careful consideration if there are temperature extremes and can only be safely used in steel housings. It should not be used in non-ferrous or plastic housings as the interference fit will be lost at high temperature. It is also favoured where a controlled fitting of the seal, such as field replacements, is involved. It is more difficult to damage the o.d. sealing surface, and ensuring that the seal is fitted square to the shaft is more assured.

The combination of both types in Figure 3.2(d) provides both positive retention and the additional sealing from the elastomer sleeve on the o.d. Variations are available without the rear coating or with the elastomer at the front edge of the casing. The variant shown in Figure 3.2(d) offers the benefit of corrosion protection to the back face of the casing and facilitates the inclusion of axial dust lips. This type of seal is preferred for heavy commercial vehicles and some agricultural applications.

The elastomer fabric seal casing, Figure 3.2(e), is offered for sizes in excess of 250 mm diameter. It avoids the necessity to manufacture the metal casing for large sizes, usually low volumes, and provides a more conformable seal to facilitate fitting on large diameters where it might be difficult to maintain adequate tolerances with the conventional seal designs. Versions of this seal design are also offered with the spring moulded into the seal to avoid accidental misplacement when fitting large seals in a difficult environment.

An alternative offered for some large diameter seals is the design with finger springs, Figure 3.2(f). Both (e) and (f) are also available as split seals which facilitates fitting at large diameters on equipment such as rolling mills and water turbines. The manufacturers of this type of seal also provide specialist on-site joining techniques.

A large range of geometries are available to cater for different applications; many of these are concerned with the exclusion of contaminant and are covered in more detail in section 3.2.5.

3.2.2 Dynamic sealing mechanism

The dynamic sealing mechanism of elastomer lip seals has been the subject of extensive research for over 50 years. The existence of a continuous oil film between the lip and shaft was first demonstrated in 1957.[2] During the intervening period it has required the development of modern experimental techniques and numerical analysis capabilities to confirm some of the hypothesis put forward in this pioneering work.

There are two mechanisms considered to contribute to the sealing action of lip seals which serve to continuously pump oil that enters the 'contact zone' between the seal and shaft back towards the oil side of the seal, usually known as inward pumping. These have been described as a micro mechanism, which provides both load support and inward pumping, and a macro mechanism, which primarily assists the pumping mechanism.

The micro mechanism has been the subject of considerable research and analysis over the past decade and much of this is summarized by Salant.[3] The basis of the pumping mechanism depends on two important characteristics of the seal:

- The asymmetric geometry as described in section 3.2.1.
- The surface texture of the elastomer in the contact zone after the seal has bedded in.

The performance of the seal is critically dependent on both of these characteristics.

A new lip seal has a sharp trimmed or moulded lip. During the first hour or so of running this lip beds in on the shaft to create a sealing or contact zone that is typically 0.2–0.3 mm wide. This may increase with continuing seal wear to be something in excess of 0.5 mm after some hundreds or thousands of hours' operation.

The seal contact stress profile is generated by the seal geometry and will have a peak at the lip and decay to zero at the edges of the seal contact, Figure 3.3. It can be seen that the contact band on the air side of the lip is more extensive than that

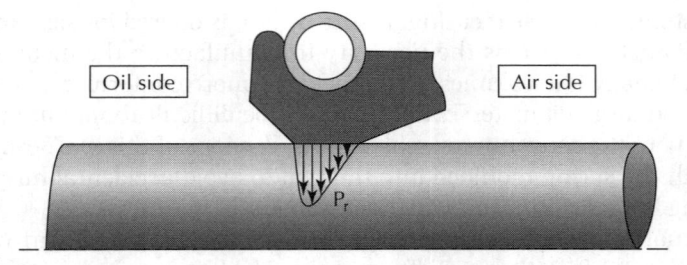

Figure 3.3 Seal contact stress profile. (Source: Freudenberg Simrit)

on the oil side, which is to be expected from the geometry and the position of the spring. When the shaft is rotated the lip will distort as the friction force is resisted by the elastomer material, with the maximum distortion at the lip. As the seal beds in the moulded surface will be abraded and the subsurface texture of the seal exposed. Lip seal materials expose a surface with pronounced asperities, or micro undulations. These asperities are deformed by the rotation so that they are oriented at an angle to the shaft axis, Figure 3.4. These angled asperities become in effect micro hydrodynamic wedges and serve to both lift the elastomer surface and at the same time pump the oil towards the lip. This action occurs on both sides of the lip, hence providing a lubrication film for the lip. However, as the asperity field on the air side is more extensive than that on the air side the inward pumping effect is stronger than the outward pumping. Hence the seal will tend to return the lubricating film to the oil side and will not leak.

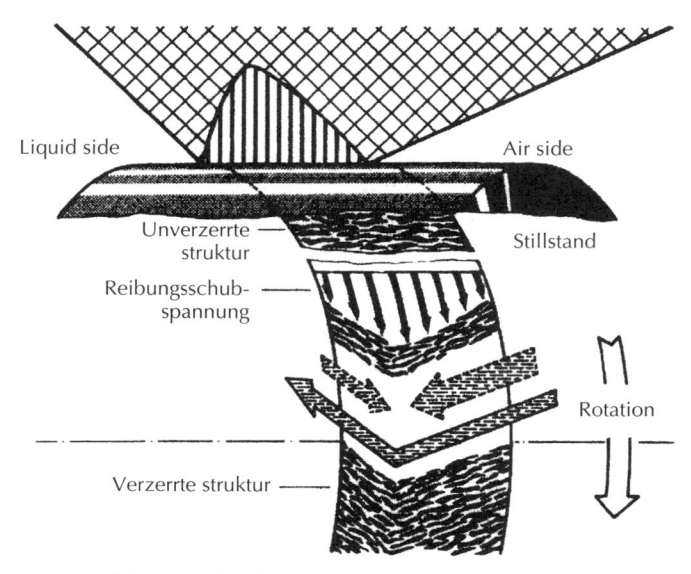

Figure 3.4 Distortion of the contact band creates asperities at an angle to the shaft axis. (Source: Salant)

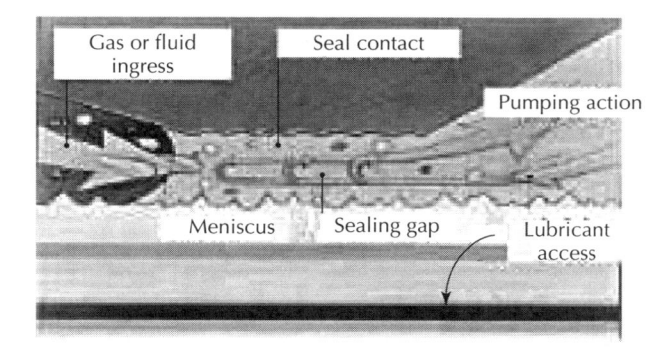

Figure 3.5 Diagrammatic representation of the fluid film transfer under a seal lip. (Source: Freudenberg Simrit)

With a new seal the asperity distribution will be a random pattern and the seal will provide a similar sealing function in either direction. However, after a period of time stress relaxation and set in the elastomer material will reduce the ability of the asperities to return to the neutral position when the shaft stops. They will therefore tend to become directionally oriented. The effect of this will be to make the seal more effective in the normal direction of rotation than in reverse rotation. This has been shown experimentally by carrying out reversing tests on a new seal and then repeating them when it has bedded in for a few hundred hours' running.

This pumping action is created by micron scale asperities. The shaft surface texture is therefore also a critical parameter in the operation of this sealing mechanism. In addition to the scale of the texture the 'lay' of the machining will influence the pumping action. Conventional cylindrical grinding and other finishing operations create what is effectively a micron scale screw thread on the shaft. Underneath the seal this will act as a micro screw pump and effectively pump oil past the seal. This action is dependent on the direction of rotation and so may be beneficial or create leakage. The normal recommendation for lip seals is that the shaft should be plunge ground in the area of seal contact, so that there is no shaft lead. The shaft will then not contribute to the oil transfer either positively or negatively. Under controlled conditions, such as an engine where the direction of rotation is assured then it is possible to provide a beneficial lead within the shaft texture. Some of the numerical analysis has shown that where the shaft lead is a low value, less than 0.1° then the shaft lead does not have an effect.[4] Production engineers are increasingly interested in applying modern machining techniques such as hard turning to shafts in high volume manufacturing. The concern with these is the effect of the lead on the shaft. Again, if the lead is sufficiently small the indications are that the seals will work successfully as the shaft pumping rate is less than that of the seal.[5] However, there are considerable variations in the measured pumping rates for different shaft finishing techniques. Figure 3.6 shows the pumping attributable directly to the shaft finish and Figure 3.7 summarizes the total pumping rate for the tests carried out by Kunstfeld.[5]

The micro pumping mechanism is accepted as providing the main load support function for lubrication of the seal. It is evident that this also depends on the seal texture after bedding in against the shaft surface. It is crucial therefore that the elastomer material provides the correct texture. Figure 3.8 is an illustration of the surface texture of successful and unsuccessful seal material.[8]

To ensure successful bedding in of the seal the shaft texture has been shown to be quite critical. The normal specification for plunge ground surfaces is 0.25

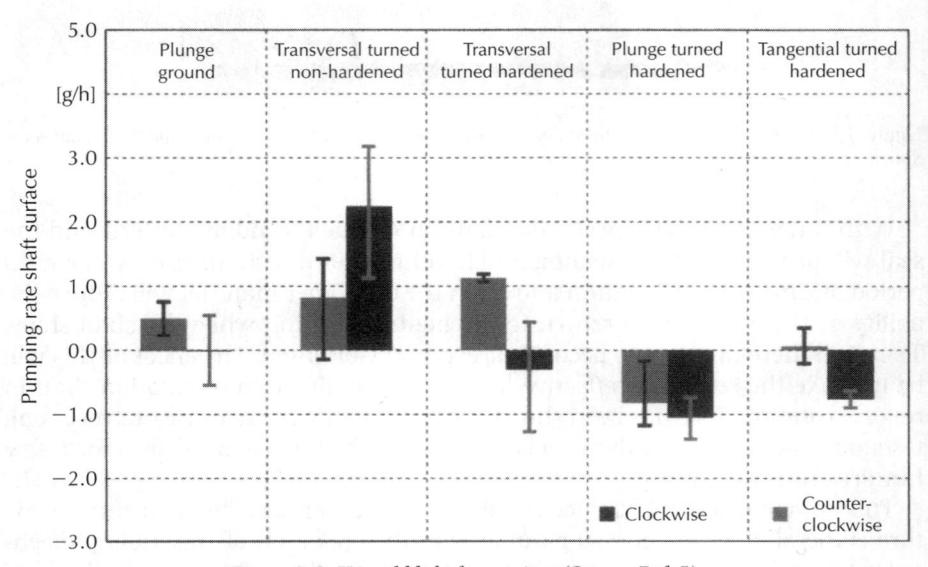

Figure 3.6 Kunstfeld shaft pumping. (Source: Ref. 5)

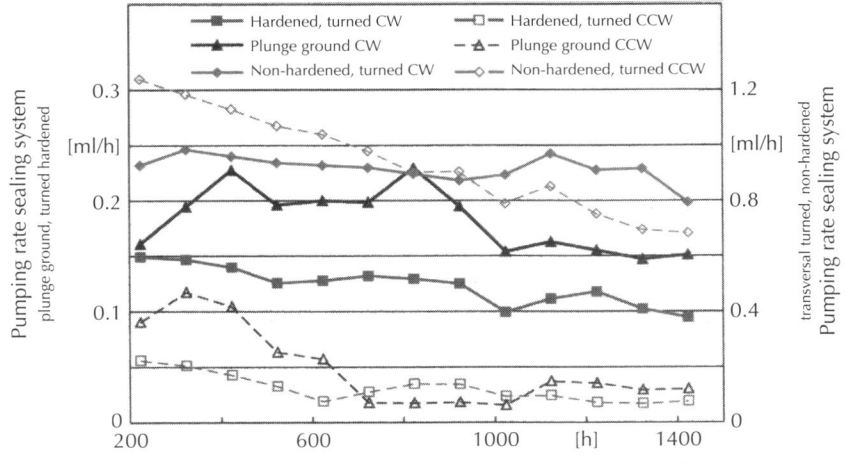

Figure 3.7 Kunstfeld seal system pumping. (Source: Ref. 5)

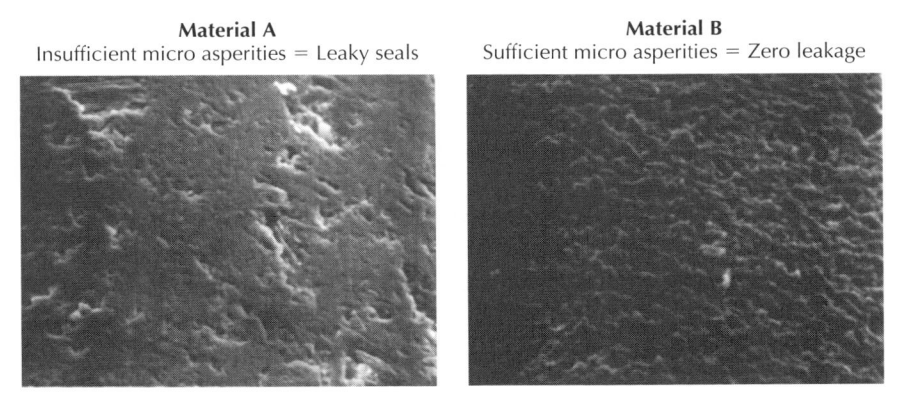

Figure 3.8 Examples of lip seal texture at the lip after bedding in showing an unsuccessful material (a) and successful material (b). (Source: SKF)

to 0.8 μm Ra. If the shaft is too smooth, or has a polished surface, the seal will not bed in correctly, especially in the presence of a viscous lubricant.

The second inward pumping mechanism is quite separate from the micro pumping and is directly attributable to the geometry and installation of the seal. As the seal rotates it will be subject to small disturbances caused by out of roundness of the shaft, dynamic runout, etc. These will cause small radial displacements of the lip as the shaft rotates. The radial displacement will translate into small axial movements of the lip. The geometry of the lip will act as a pump, encouraging liquid under the lip as it is displaced to the air side and wiping as it moves towards the oil. An additional contribution to this effect is the installation of the seal. Due to a combination of variables including machining of seal housing bore, installation of the seal square in the housing and manufacture of the seal to create a lip that is perpendicular to the shaft, the seal lip will be slightly out of square with the shaft. As the shaft rotates, this will create a wiping action on the area of shaft swept by the lip. This is considered to be potentially beneficial within limits as the heat dispersion from the lip is spread over a wider area. This wiping action will pump in a similar manner to the radial displacements caused by the shaft. All of these factors are potentially quite variable within the very small-scale displacements that are relevant to seal function. One seal manufacturer has a design that is intended to take advantage of this pumping mechanism in a controlled manner. This is the Chicago Rawhide (SKF Sealing Solutions) Waveseal. This seal has a wave profile imposed circumferentially around the lip, Figure 3.9, and has been described in more detail by Horve.[6] This macro pumping is due to the lip seal replicating a micro displacement reciprocating seal as the shaft rotates. If the geometry of a lip seal is compared with that of a reciprocating seal, described in section 4.2.3, it can be seen that they are very similar. This may seem surprising at first, but is to be expected as they are both in effect providing an inward pumping mechanism.

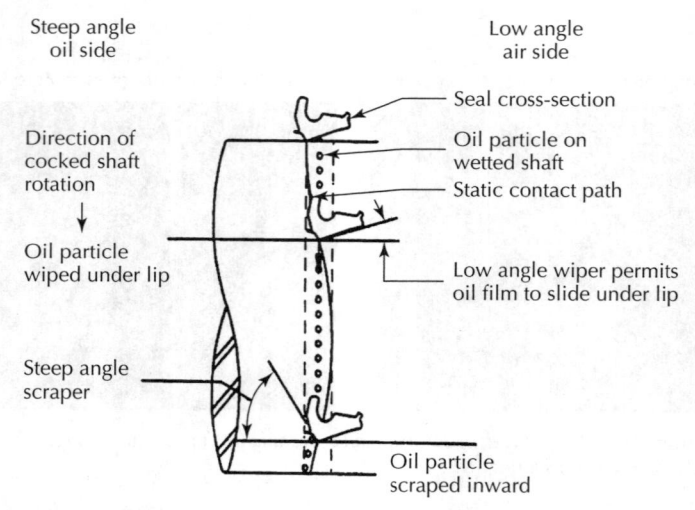

Figure 3.9 Chicago Rawhide Waveseal that is designed to promote a wiping action axially on the shaft. (Source: SFK)

3.2.2.1 Pumping aids

A large proportion of lip seals function effectively with the lip design and mechanism described in section 3.2.2. A number of seals are produced with what are variously termed pumping vanes, hydrodynamic aids or sealing aids. These comprise small vanes or webs on the air side of the lip.

Why should such sealing aids be necessary if the sealing mechanism has been shown to be effective? They produce a more effective pumping mechanism that overcomes potential leakage during critical phases of the seal operation. During the bedding-in process seals may be prone to leak as the lip is run in to an effective profile. They will also help to extend the performance parameters of a seal, it may continue sealing when it would otherwise be considered to have worn out, or enable it to operate with higher speed or runout conditions.

The majority of aids on modern seals are small raised ribs moulded on the air side of the seal lip, Figure 3.10. The ribs are in contact with the shaft and merge into the lip. If the ribs do not contact the shaft they will not be effective. These aids function

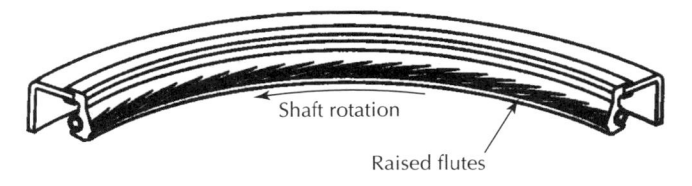

Seal with pumping features

Figure 3.10 Pumping aids on seal lip. (Source: BHR Group)

by collecting small amounts of leakage that pass the seal lip, and entrain it such that it is fed back into the seal lip area. The detail design of the various flutes and triangular shapes have been the subject of extensive patent coverage but most appear these days as flutes as shown in Figure 3.10. The relative benefits of different designs of aid and the effects of changing the angles are discussed by Johnston.[1] While different angles and shapes may be more or less effective as pumps there are two overriding considerations:

- The design must touch the shaft to be effective.
- The interference of the flute should not be sufficient to disturb the lip contact.

Many pumping aids are unidirectional which is quite adequate for applications which are entirely or predominantly for shafts that only have one direction of rotation. In situations that have intermittent reverse rotation, such as vehicle transmissions, the seal will still function. They do not pump leakage out when used in reverse, as they only function on fluid that has already leaked.

Where there is a true bidirectional rotation, or a seal may be used for a variety of applications, then bidirectional seal designs are available, Figure 3.11. It may appear at first that such seals would not work as the two opposing sets of pumping aids would work against each other. This is not so as the flutes for the 'wrong' direction present a diverging wedge to any leakage and therefore have little influence, as discussed for reverse rotation above. The only potential disadvantage is that the proportion of pumping in one direction is less than that available with a unidirectional design. For shafts that have occasional reverse rotation this style of seal will help overcome the potential small leakage that may occur due to the tendency for a plain lip seal to become unidirectional, as discussed in section 3.2.2, when the microasperity texture sets into a unidirectional orientation.

There are some concerns with the use of pumping aids. The most significant is that the very action of providing a pumping aid externally to the seal will help entrain contaminating liquid and pump it into the sealed system. This places a more exacting role on the exclusion of contaminant. For this reason the use of seals with pumping aids is normally restricted to cleaner environments.

Examples of 'footprints' of seals with pumping features

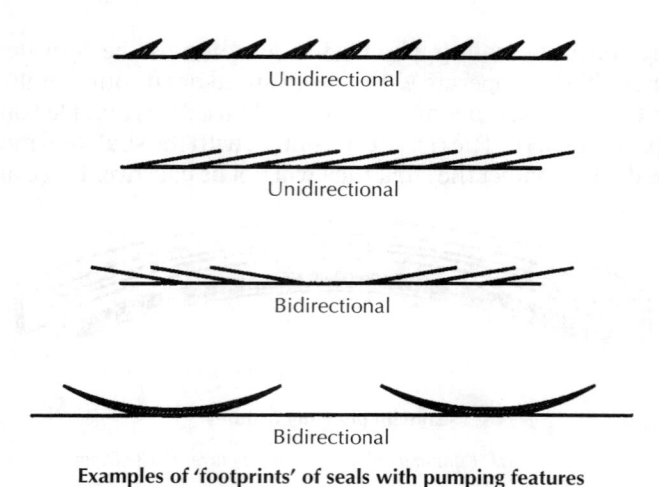

Figure 3.11 Bidirectional pumping aids. (Source: BHR Group)

3.2.2.2 Limits of performance

The performance parameters possible with lip seals depend on a combination of the seal design, flexibility, lip load, etc., but also the seal material and other factors such as the shaft. Any information provided is thus for general guidance.

A major factor is the heat generation under the lip. As the seal continuously runs on a localized section of the shaft, dispersion of the heat generated by friction is one of the key limiting considerations in the performance of the seal. The extent of this problem can be demonstrated by considering that a typical seal on a shaft of 50 or 60 mm diameter will have a power consumption of anything up to 100 W. This power is dissipated through the seal contact which will be of the order of 50 mm^2. The heat density is higher than that of an electric cooker hotplate. The effect of this is for the lubrication film under the seal to be at a considerably higher temperature than the bulk oil. Measurements for seals at a range of speeds are shown in Figure 3.12. This shows that selecting a seal material capable of operating satisfactorily at the bulk oil temperature will cause potentially serious problems as the seal lip may be up to 40°C hotter. This differential may be even higher in viscous oil, such as transmission applications, or when sealing grease.[7] With many typical engines and transmissions having bulk oil temperatures in the region of 100°C it can be seen that material selection can be critical. The selection of material can therefore have a direct effect on the performance limits of the seal.

Shaft surface speed will directly affect the amount of heat generated. The seal material characteristics will also contribute to the lubrication and hence thermal limit of a particular material. This is demonstrated by the difference in speed limits for the different elastomers, Figure 3.13. The materials with a higher temperature limit generally have a higher speed capability. This capability is demonstrated more broadly in Figure 3.14 where the performance limits of four different materials are illustrated, showing that the material properties affect both the shaft surface speed and temperature capabilities. These speed limits are for seals operating

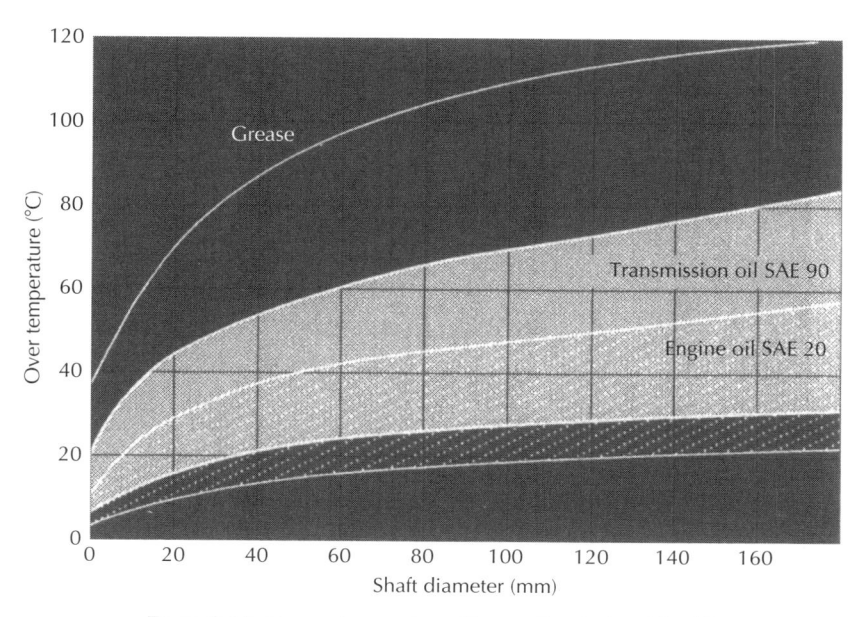

Figure 3.12 Lip over temperature. (Source: Freudenberg Simrit)

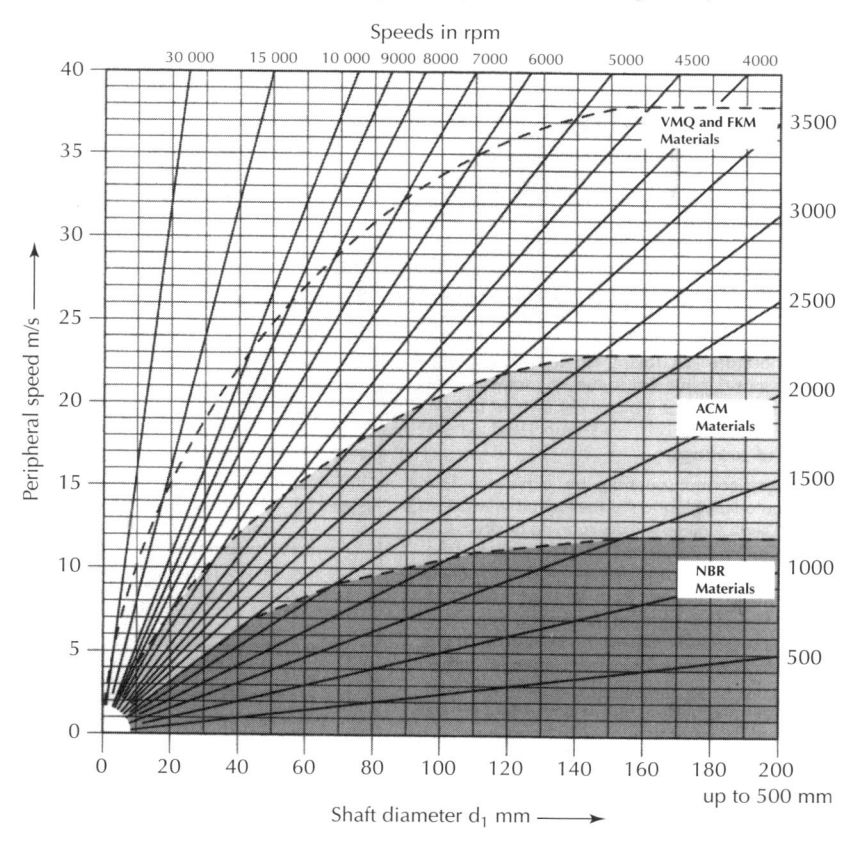

Figure 3.13 Shaft surface speed limits for elastomer lip seals. (Source: Trelleborg Sealing Solutions)

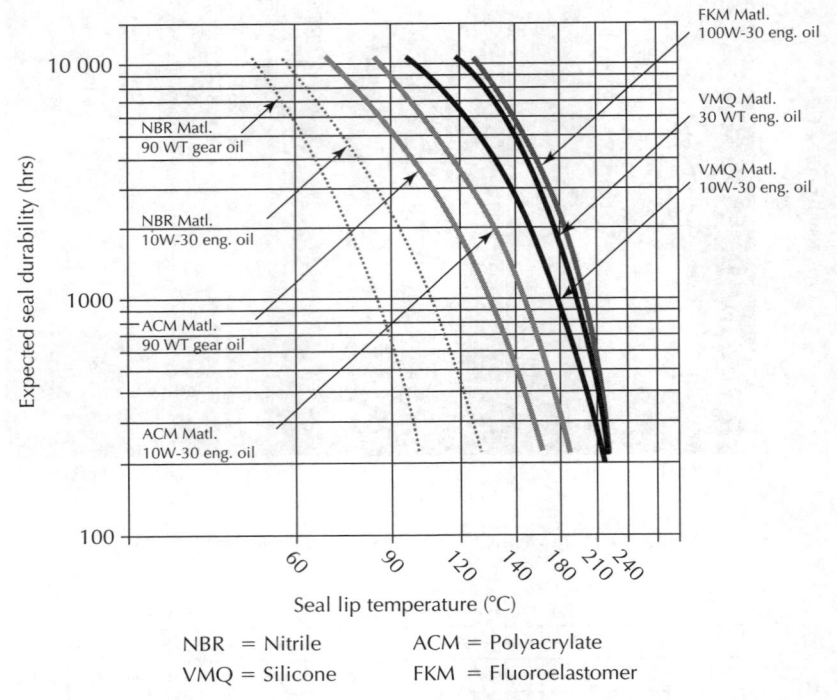

Figure 3.14 Performance limits of lip seal materials. (Source: Freudenberg Simrit)

without pressure. The performance limits when lip seals are pressurized are much more limited, as discussed in section 3.2.8. These performance limits are those suggested from experience and are based on the requirements for automotive or industrial applications. It is possible to use the seals outside these parameters but their life will much more limited. This will require testing to prove that the correct compromise of life and reliability is practicable. For example, in competition vehicles seals may be used at conditions that would not be considered possible in a standard automotive application, but in such cases a life of only a few hours will be required.

If a shaft is subjected to appreciable runout then the seal will have to flex dynamically to ensure that the lip follows the shaft. The ability of an individual seal design will depend on the lip design and material, as well as any temperature effects on the resilience of the elastomer. The generally accepted limits of shaft runout are shown in Figure 3.15. Again there is a choice to be made, as there can be a tradeoff of reliability against runout. Figure 3.16 demonstrates the benefits in extending seal life by decreasing the amount of eccentricity.

The viscosity of the sealed fluid will directly influence the friction generated in the lubrication film. Most of the data generally available is generated using fairly typical lubricants such as engine oil. The viscosity characteristics of these fluids are such that they will have a viscosity in the region of 5 to 10 cP at a typical

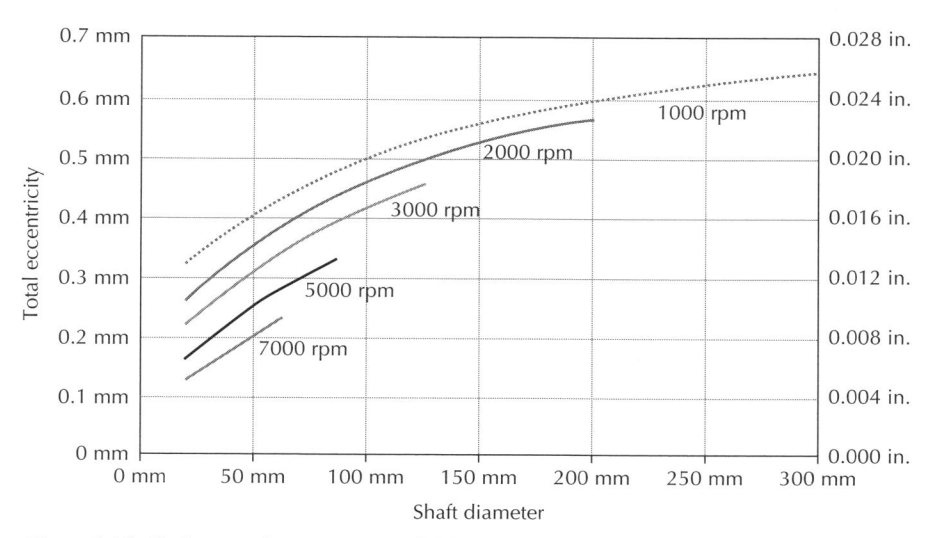

Figure 3.15 Shaft runout limits recommended for lip elastomer seals. (Source: Freudenberg Simrit)

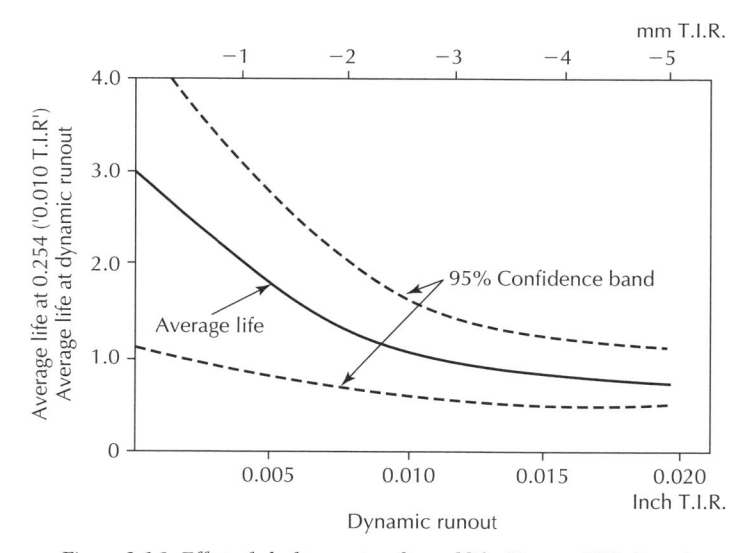

Figure 3.16 Effect of shaft runout on lip seal life. (Source: BHR Group)

operating temperature of 100°C. Higher viscosity fluids such as transmission lubricants will create higher viscous shear and hence increased friction. Examples of this effect are shown in Figure 3.17 where it can be seen that increasing the viscosity across the range of engine oils increases the power consumption by approximately 30%. Increasing the viscosity, such as with a transmission lubricant, will increase the power consumption still further.

A further problem occurs at low viscosity. Below 2 or 3 cP the lubricant film will become too thin to maintain separation of the seal and the shaft texture and

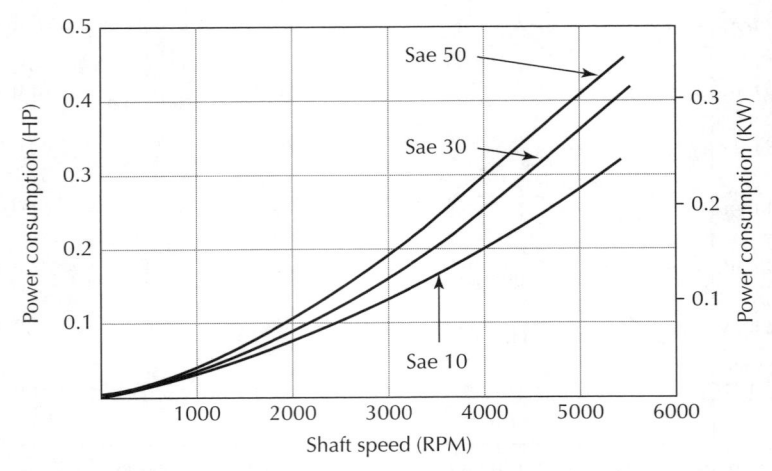

Figure 3.17 The effect of viscosity on seal power consumption. (Source: BHR Group)

boundary contact will occur. This will cause a large increase in friction and wear. For applications involving very low viscosity liquids, such as water-based lubricants, and water, then alternative seal designs and materials must be considered. If an elastomer lip seal is still to be used it will probably be of a specially developed material with improved boundary lubrication properties.[9] An alternative that is increasingly used is a lip seal manufactured with filled PTFE and these are discussed in section 3.2.3.

3.2.3 PTFE seals

Lip seals manufactured from filled PTFE have been available since the 1970s. The early designs replicated the construction of the original leather lip seals with a PTFE washer crimped into a spun or stamped metal housing, Figure 3.18(a). This style of seal is still widely used and is particularly suitable for manufacturing small quantities of seals. The sealing element starts out as a flat washer of filled PTFE which is then shaped to form the seal. Static sealing of the PTFE in the metal casing may be by means of either a flat washer, 3.18(a), or an O-ring, crimped behind the PTFE.

The use of PTFE lip seals has become more widespread, and this has led to the development of seals that can be produced more economically in higher volumes. In the bonded version, introduced in approximately 1990, one or more PTFE discs are retained in the main seal case through chemical bonding processes. This may either be bonding the PTFE directly to the metal case or, alternatively, by moulding elastomer around the case in such a way that it may be used to attach the PTFE disc securely to the case, Figure 3.18(b). The advantage of this compared to the mechanically retained device is manufacturing efficiency, cost reduction, and with some potential for improvement in performance. One disadvantage is that inherently high lip forces may still be required which may cause

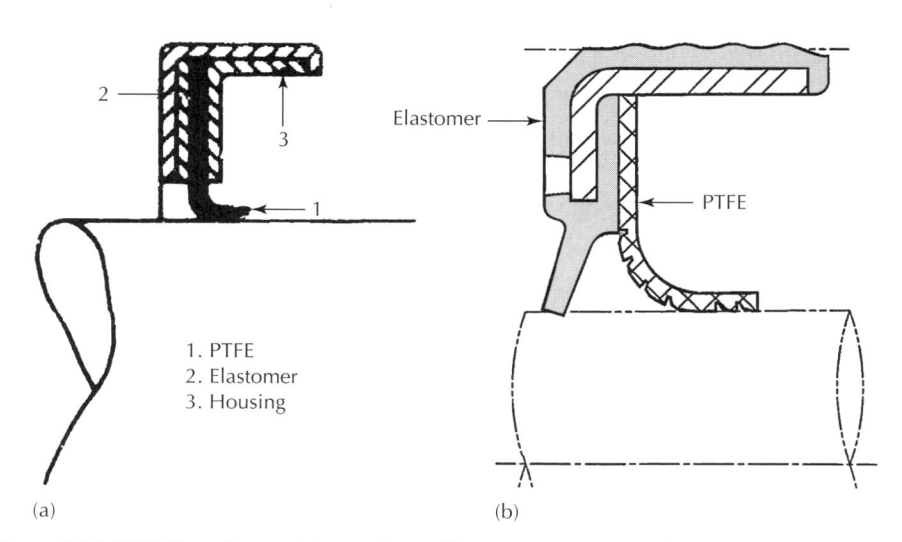

Figure 3.18 PTFE lip seal types: (a) manufactured by crimping a PTFE washer in a metal housing sealed by an elastomer washer; (b) PTFE bonded to an elastomer in a metal housing.

high levels of stress on the bonded area of the PTFE disc. A further disadvantage is that special moulding or bonding tools are required to manufacture these seals. The casing may have an elastomer backing and o.d. similar to an elastomer lip seal, which will facilitate fitting and sealing of the o.d. This construction also allows for the incorporation of dust excluders, etc.

The PTFE washer type seals do not have the same sealing mechanism as elastomer lip seals. The geometry does not present the same sharp lip edge with asymmetric angles as the elastomer seal. The belled plastic washer forms a PTFE bush running on the shaft. This does not provide the same inherent sealing mechanism as the elastomer seals but really forms a close fitting restrictor bush. A typical seal design will have a PTFE seal contact width of 2 or 3 mm. The sealing mechanism is enhanced in many applications by providing a spiral thread form in the rubbing face of the seal. This acts with the same mechanism as a viscoseal, section 3.6.6, with the seal returning fluid leaking under the bush to the oil side of the seal. To ensure that the seal can maintain some static sealing capability this spiral is often sliced rather than turned. However, for high volume automotive seals it is now common to stamp this feature into the surface of the PTFE. There have been attempts to improve the static sealing ability by including small 'dams' in the spiral feature but these have not proved to be effective showing similar leakage rates to a spiral without the features.[23] As the spiral feature is an integral part of the sealing mechanism these seals are essentially for one direction of rotation. This is different to the pumping aids on an elastomer seal described in section 3.2.2.1 which are an additional security external to the basic sealing function. If a PTFE washer design of seal is used without the spiral features then it can be expected to leak to a small extent, quoted leakage being between 0 and 0.4 g/hr.[10] The spiral design of seal is more likely to be used as a low pressure seal as a replacement for an elastomer lip seal, whereas the plain washer type is more prevalent as a pressurized lip seal, section 3.2.8.

The spiral can be an effective method of maintaining leak-free dynamic operation. It should also be noted that this effective fluid transfer will again potentially pump contaminant into the lip contact. Exclusion of contaminants from the seal lip area is therefore particularly important for this type of seal. A feature of a viscoseal is the tendency to pump air into the seal, particularly at the highest speeds. Introducing air to the high temperature oil film under the seal has been known to cause premature oil degradation and oil coke deposits in the seal area can inhibit the action of the spiral.

PTFE lip seals have become popular for two types of application:

- Where a lubricating liquid is not continuously available for the seal.
- Where the operating conditions are outside the range of elastomer seals, particularly temperature, fluid compatibility or the combination of speed and pressure.

The early applications of PTFE seals were areas where the use of an elastomer seal would not be considered viable due to the absence of a suitable lubricant, such as sealing dry gas, water or an aggressive chemical or solvent. In such applications the material of the casing and the elastomer casing seal also require careful consideration. The variety of potential applications require careful selection of the plastic and filler, as behaviour of the filler together with shaft texture is an important factor in achieving success in such situations, see section 5.6. The wear mechanism of plastics is very different in wet and dry situations.

It has now become quite common to use PTFE seals in lubricated conditions. This is particularly so in the automotive industry but also in other areas such as industrial transmissions. The combination of modern lubricants and increasing operating temperatures causes severe problems with the selection of suitable elastomers for some applications. We have seen in section 3.2.2.2 that seal lip temperatures may be in excess of 150°C. Combined with this modern lubricants have a significant proportion of additives to support the more highly rated machinery and longer oil change intervals that are now commonplace. These additives can be deleterious to the properties of the elastomer and significantly reduce the acceptable operating temperature of a material. In addition, lubricants from different suppliers can behave quite differently due to the use of different additive packages, as discussed in section 5.1. This can make selection of a suitable elastomer an almost impossible task and so the use of PTFE may be preferred. Another factor that can influence the decision to select a PTFE seal is if the equipment manufacturer includes a dry running requirement in the seal specification. This may sometimes be included as a seal on, for instance, a large gearbox casing, or an engine may run dry for some period on startup until oil splash has circulated, especially after a long period of storage such as being transported to site.

In addition to potentially higher leakage than elastomer seals PTFE seals have some further potential disadvantages that must be considered carefully. The PTFE does not have the inherent flexibility and resilience of elastomer and will therefore not follow shaft runout, etc. as effectively. It can be seen by comparing Figure 3.19 with Figure 3.15 that permissible runout for a standard PTFE washer type of seal is less than half that allowed for elastomer seals. The shaft surface finish required is finer, 0.1 to 0.4 μm Ra, and most experience indicates that a hardened shaft is required. Best results are obtained if the shaft has a surface hardness of 55–60 Rc, although 45 Rc may be used for low pressure applications and very clean conditions.

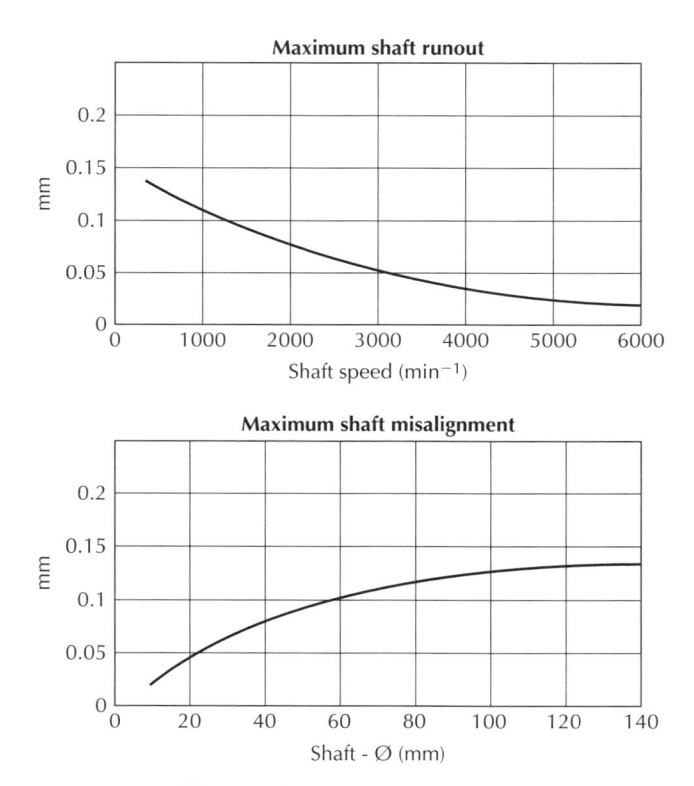

Figure 3.19 Permissible runout for washer type PTFE lip seals. (Source: Garlock)

PTFE seals with a spiral have better flexibility than the simple washer type and hence can operate with more runout.

In common with other plastic seals these lip seals require more careful handling on installation to ensure that they are not damaged or distorted. A shallow polished taper mandrel is required with a maximum of 15° taper; 10° is preferred. The taper and shaft must be scrupulously clean and free from scratches otherwise the seal is liable to leak immediately on startup.

Serious problems have occurred when equipment manufacturers have transferred a design from elastomer lip seals to PTFE without modifying the detail of the assembly or installation procedures to take account of the different requirements. These problems can include excessive rejects due to damage during assembly and excessive wear due to inadequate shaft specification.

3.2.4 PTFE lined elastomer seals

PTFE lined seals are manufactured by the bonding of a more flexible PTFE element to the contact surface of elastomer oil seals. This is carried out as one moulding/bonding process and the advantage compared to the PTFE washer seals is considerably improved dynamic seal performance on high speed shafts and significantly reduced specific lip force, leading to reduced frictional heating.

The disadvantage compared with other PTFE seal types is that in some cases the cost of specialized tooling may be higher. The care and handling precautions associated with PTFE seals, as discussed in section 3.2.3, also still apply.

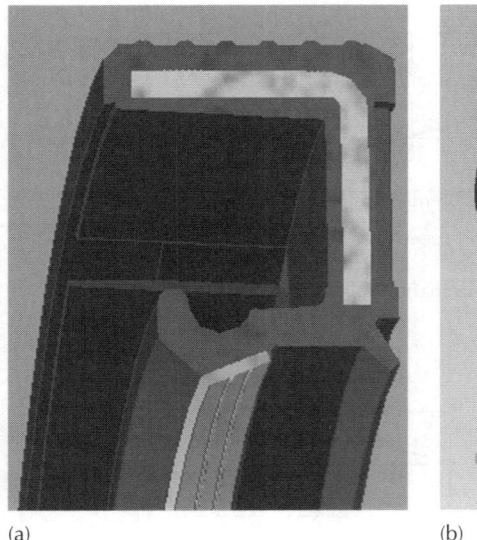

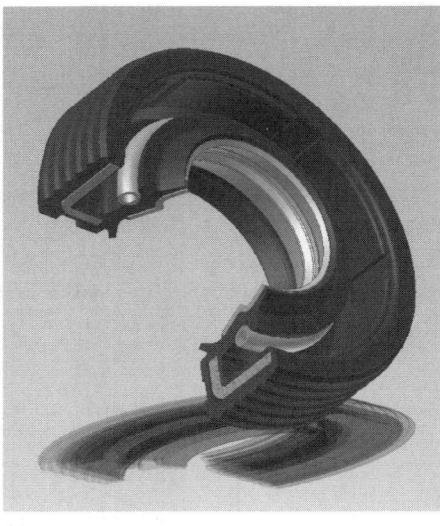

(a)   (b)

Figure 3.20 Elastomer lip seal with a PTFE insert on the atmospheric side of the seal lip. (Source: Race-Tec NAK)

3.2.4.1 Design and manufacture

PTFE lined seals (PL seals), Figure 3.20, are more specialized than most other elastomer and plastic rotary seals being essentially composite seals of PTFE and elastomer, or PTFE/elastomer and metal. The more common PL seals look like normal lip seals but have a bonded facing of PTFE material on the contact face of the seal lip(s).

To meet the highest performance demands the PL seal technologies require new parameters for the design of the seal lip. Although the differences are not very obvious to the casual observer, modified manufacturing processes and development of new materials are required.

Some suppliers who have recently introduced PL seals offer the PTFE bonding technology on standard low cost oil seal designs. Without the development of correct design parameters, such products only offer part of the performance gains achievable.

Advantages

Friction power loss   PL seals have been shown to run with typically 30–50% lower friction-related power losses compared to other PTFE lip seals and up to 80% in certain applications. Compared to standard elastomer seals the saving may be in the range 60–85% depending on the seal design and materials.

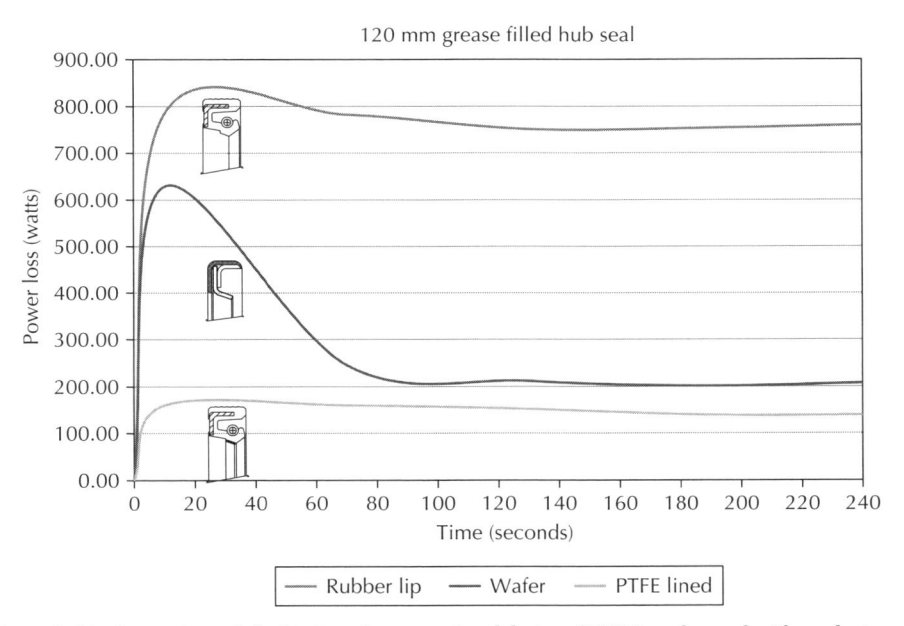

Figure 3.21 Comparison of the friction of a conventional design of PTFE washer seal with an elastomer seal with PTFE insert. (Source: Race-Tec NAK)

Lip forces   A comparison of overall lip loads shows that conventional washer type PTFE lip seals use significantly higher total lip forces than PL seals. This can have a number of practical disadvantages including the inability of the seal lip to follow the dynamic movements of the shaft.

Reduced shaft wear   As the grade of PTFE used may be selected according to the shaft material used, the rate of shaft wear can be minimized even on relatively soft shaft materials.

Flexibility in design   The PL seals can be applied to rotary, semi-rotary and reciprocating applications. Specialist geometries are used to suit the space envelope and packaging envelope.

Speed   These seals are used for Formula 1 applications which are regularly tested to 20 000 rev/min and to 25 000 rev/min for material developments.

Surface speeds up to 100 m/sec are possible. Some applications operate at 150 000–200 000 rev/min.

Vacuum   For some engine applications there is a requirement to secure the maximum possible vacuum within the crankcase of a race engine for performance reasons. These seals are able to maintain continuous dynamic contact with the shaft throughout the speed range even with high levels of vacuum.

Shaft runout and housing eccentricity   The inherent flexibility of the elastomer seal body combined with the low friction facing on the seal lip enable the PL seal lip to accept larger eccentricity between shaft and housing bore as well as larger amounts of dynamic shaft runout.

Temperature   PL seals have been shown to have lower underlip temperatures than elastomer lip seals and washer type PTFE seals, reducing the problems of shaft wear and thermal degradation of lubricants. As the elastomer material is not in contact with the shaft it does not have to withstand the underlip temperature. The temperature limit for the seal becomes the upper limit for the elastomer, typically 200°C for FKM and 250°C for VMQ/FVMQ.

The introduction of FFKM materials effectively gives the elastomer a higher application temperature range than the PTFE and so the PTFE becomes the limiting factor with an underlip temperature range of up to 280°C being possible. Premium seal suppliers are able to mould the FFKM and also bond it to metals and the PTFE.

Pressure   Pressures will increase the lip contact area in proportion to the pressure increasing friction. In applications where the pressure varies, the seal will relax at the lower pressure so that friction will remain proportional to the applied pressure.

Cost   Compared with elastomer oil seals the selling prices for PL seals reflect the difference in materials used and the additional processes required. In small batch quantities the PL seal will always require a mould so has a potentially higher initial cost than the washer type PTFE seal.

For medium batch quantities there is reasonable parity of price taking into account the tooling for the comparable seal types.

Storage   PL seals are supplied in the same form as elastomer oil seals, typically in paper-wrapped rolls, tubes, or individually sealed in plastic bags. This compares with typical washer type PTFE lip seals that often need to be stored on a mandrel, plastic or cardboard tubes, or are supplied mounted individually on a moulded 'keeper'. This creates some risk of seal damage when the seal is removed from the tube.

Continuing development   These seals continue to be developed to optimize sealing capabilities at higher speeds with negative pressure and to take account of shaft dynamics and vibration experienced at higher speeds. Seals are currently being developed which operate with less than 20% of the frictional torque of other seals. Another development is the use of duplex PTFE-lined sealing lips which are especially effective for engine applications.

These seals, when correctly designed, have made it possible to combine many of the advantages of elastomer and PTFE seals. They have proved to be highly effective in areas such as racing cars, where a very high performance is required for a relatively short life, but are also used in compressors and powder manufacturing machinery.

3.2.5 Excluders

The vast majority of lip seals are used to seal lubricant within a system. The sealing lip discussed in detail in section 3.2.2 is intended to work primarily with a clean lubricant. However, it is also necessary to keep the external environment away from the seal and ultimately the lubricant. A very significant part of the lip seal overall system is exclusion of contaminant. For sealed rolling bearings, as discussed in section 3.2.7, this is the overriding concern with retention of the lubricant a secondary consideration.

There are two factors to consider as a starting point with excluders for lip seals:

- As we have already discussed in section 3.2.2, it is well established that a lip seal will pump into the lubrication system. An inward facing seal will therefore potentially entrain dust. It is also more likely to entrain any liquid and pump it into the bearing or lubrication system. If the seal design is one with pumping features, section 3.2.2.1, this may increase any inward pumping of contaminant.
- The amount and type of contaminant to be excluded. The potential contaminants can be broadly classified as:

  Dry: Fine dust, sand, large particles
  Liquid: Typically water, aqueous solutions
  Slurry: Mud, slurry mixtures, etc.

Within any of these categories the contaminant may be of variable quantity and the types of solid will have varying degrees of abrasiveness. The exclusion may be required for both static and dynamic situations. A wide range of excluder types are used with lip seals with the selection depending on the severity of the application, operating conditions and life expectancy. They can be broadly classified as:

- Seal only.
- Radial dust lip.
- Multiple radial dust lips.
- Axial dust lips.

- Multiple radial and axial lips.
- Labyrinth and flinger features with any of the above.

Virtually any form of excluder involves providing a seal external to the lip seal. Because of the inward pumping action of the lip seal this immediately creates a potential problem as the volume between the seal and excluder will be gradually evacuated by the pumping action of the seal.

A method that was very common at one time was to use a pair of lip seals back to back. This can cause wear and high friction as the space between the seals is reduced to a vacuum.[1] It is therefore not advised to use this arrangement unless permanent lubricant supply to the interseal space is available, some form of venting is arranged, or precautions against the creation of a vacuum have been implemented.

The simplest form of excluder is to use a radial dust lip, moulded into the heel of the seal, Figure 3.22. These are intended to protect the main seal lip from a light dust environment. Two different philosophies are used in the approach to these dust lips:

- Contacting: As this will create a seal it can create a vacuum between seal and the wiper. In applications where an OEM uses dry assembly to provide rapid bedding-in then the wiper may itself create wear debris. The favoured approach with such a wiper is to ensure that it is lubricated and consider a small relief on the lip to allow venting to prevent a vacuum formation.
- Clearance: This will inherently provide reduced exclusion capability but will create less vacuum and hence reduce wear. A further potential problem is that during heat up/cool down cycles the clearance will allow some breathing of ambient air which may introduce dust and moisture. A clearance type of dust lip will also not be effective where there are very fine dust particles.

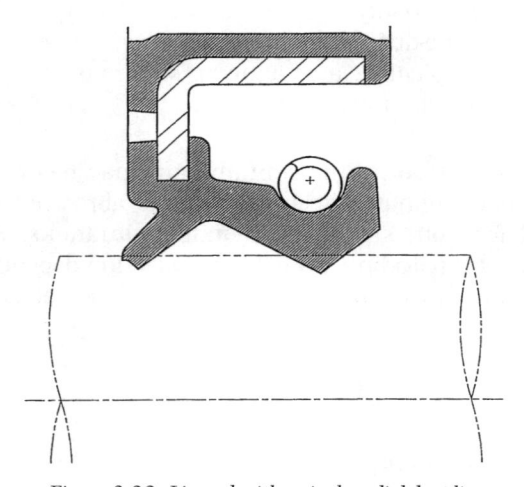

Figure 3.22 Lip seal with a single radial dust lip.

If used or installed incorrectly a seal with a dust lip can create almost as many problems as it solves and in some conditions it has been found that a seal without a dust lip is preferable.

An alternative is to use a felt dust lip. This will act as a filter and remove dust particles but will allow the space between the seal and dust lip to breathe, hence removing the problem of creating a vacuum between the seal and dust lip. Felt wipers can be effective in dry, dusty conditions but are not satisfactory for excluding water from the seal.

There is an increasing trend to use axial dust lips as part of the exclusion system. These will normally be in contact with a flange section of the shaft, probably as part of a cassette arrangement. Cassettes are discussed in more detail in section 3.2.6. An example is shown in Figure 3.23. The shaft flange will provide a centrifugal effect to any contaminant in the seal area and will disperse much of any contaminant present when the shaft is rotating. There is also the tendency to disperse any wear debris created from the dust lip which further helps to keep the seal area clear.

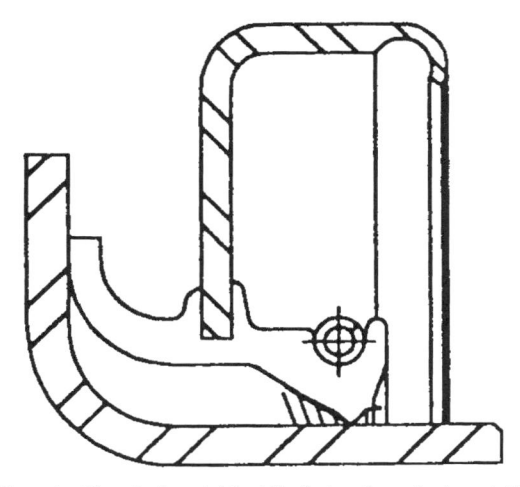

Figure 3.23 Lip seal with a single axial dust lip that seals against a rotating radial flange.

It is also possible to fit the wiper lip to the rotating flange, Figure 3.23. The rotation of the flexible sealing element will subject it to centrifugal force which will tend to lift the seal from contact during rotation. The excluder lip can therefore be designed to have a higher contact stress when static, to prevent ingress of contaminant, but will operate primarily as a centrifugal seal when the shaft is rotating at high speed. The provision of a contorted access route together with assistance from a centrifugal flinger has been shown to be more effective than the simple radial dust lip. Figure 3.24(a) shows three types of excluder arrangements and Figure 3.24(b) results of testing in a dusty environment. These results show that in the dust environment used for the tests the non-woven fabric wiper and the dust lip with a radial flange were effective whereas the radial dust lip rapidly became ineffective.

130   Seals and Sealing Handbook

(a)

| (A) Non-woven fabric dust lip type | (B) Dust lip, side lip and labyrinth type | (C) General dust lip type |

(b)

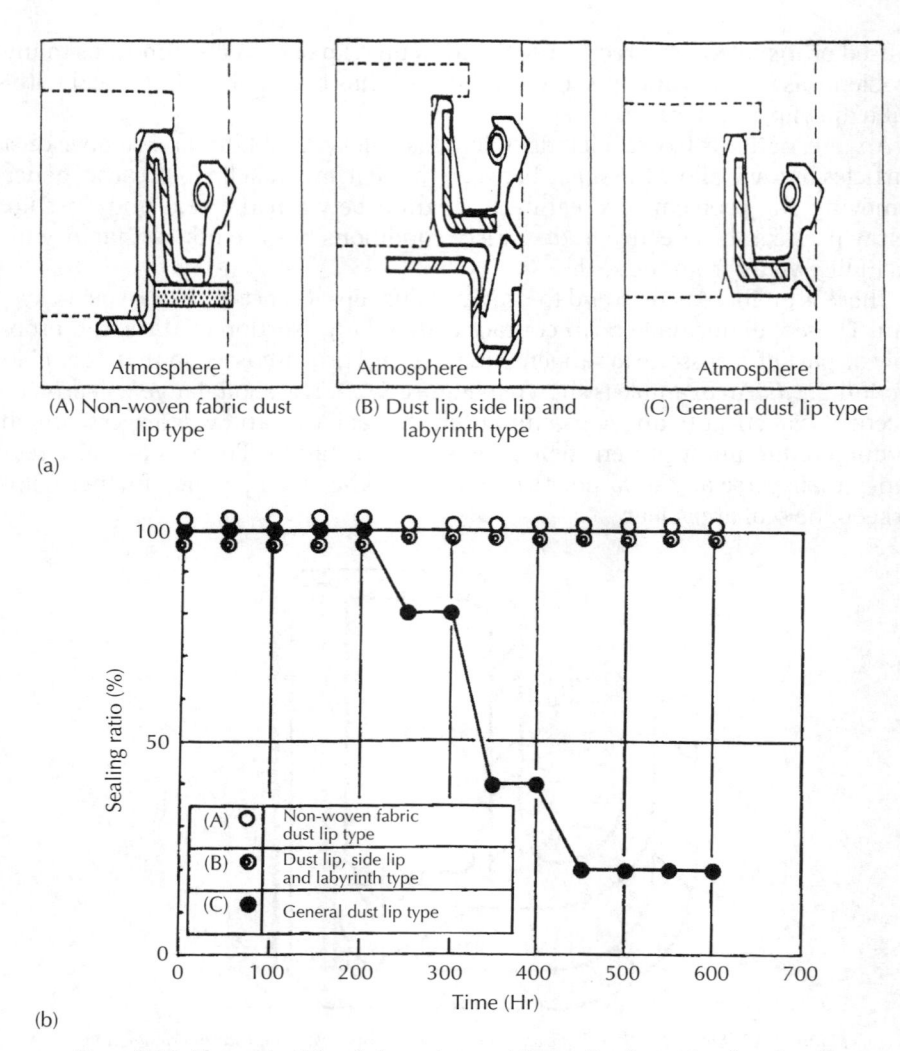

Figure 3.24 Three potential excluder systems and the results of a test in a dust chamber.

Excluder systems for lip seals have evolved into a wide variety of arrangements tailored to suit both the operating environment and constraints of the machine geometry. The techniques for higher speed seals will vary from those used at low speed. With high rotational speed it is possible to use centrifugal force as an expeller while the shaft is rotating. However, it may still be necessary to provide a static excluder. An example is four wheel drive vehicles which spend part of the time at high speed on the road and also operate at low speed in a dirty or wet environment. A seal and excluder system as shown in Figure 3.25 may be used. The rotating axial dust lip will provide a contact seal when stationary, or at low speed, and progressively reduce contact stress as speed increases.

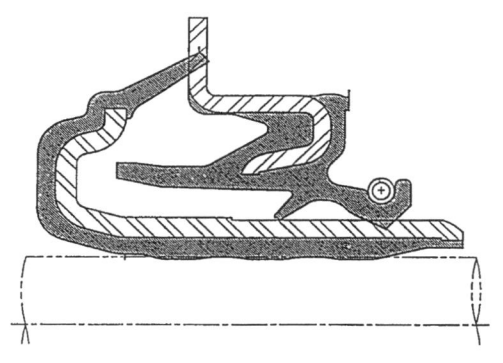

Figure 3.25 A seal and axial excluder that will permit high speed operation with positive exclusion at low speed. (Source: Freudenberg Simrit)

For slow speeds and very contaminated environments, such as agricultural and earthmoving equipment, more extensive exclusion systems are required. They will usually involve a multiple arrangement of dust lips combined with a labyrinth, to impede contaminant access, and radial flanges to provide centrifugal expulsion. An example is shown in Figure 3.26. The interspaces between the lips will be lubricated. The combination of a number of physical barriers, the labyrinth geometry and some radial faces to create an expeller provide a system that can give acceptable service before the contamination gains access to the seal lip. The heat generation with this type of sealing system will be a crucial factor and hence speed is limited. The actual speed limit will depend on the number and proximity of the seal lips and the material, but for reliability these designs are usually limited to surface speeds of less than 5 m/s. The actual speed in many of the applications for these seals will be below this speed.

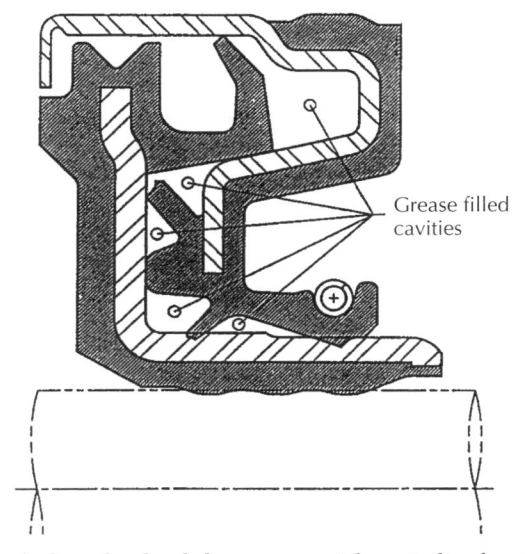

Figure 3.26 Example of a lip seal and excluder arrangement for agricultural or construction equipment with a combination of sacrificial excluder lips and radial flanges to provide some centrifugal exclusion. (Source: Freudenberg Simrit)

If the shaft has a large diameter the seal will probably have to operate with a significant amount of shaft movement, either radial or axial, and potentially at a high surface speed. One example where specialized seals have been developed is for the hydrodynamic bearings of rolling mills for metal production. The seal and excluder system is required to seal the oil in the bearing and exclude the contaminant that may be present in such an environment which may include liquid coolant used in the rolling process plus metal oxide slag, etc. removed during the rolling. A production example of a patented seal design that has been used for this purpose is shown in Figure 3.27. The main seal is fitted to the shaft on the neck of the roll and rotates with the shaft. An internal flinger returns the majority

Figure 3.27 Bearing seal and excluders designed for a steel rolling mill bearing to operate at high speed in wet contaminated conditions. (Source: Morgan Construction Co.)

of oil into the lubrication system. The internal lip seals the oil. Externally an axial wiper prevents liquid and debris passing down the radial face of the roll towards the seal. A labyrinth arrangement and external seal back up the axial seal in excluding contaminant. The pair of radial seal lips are designed to be flexible to cater for the radial and axial movements of the roll. A recent development of this seal has an alternative lip profile, Figure 3.28. The revised design maintains an effective sealing edge during radial displacement of the shaft and improved exclusion of contaminant.

Some applications may be too severe to provide reliable service with the excluder arrangements described above. In applications that involve continuous submersion in contaminant or continuous high speed in a contaminated environment then alternative sealing configurations may have to be considered. Bearings operating in aggressive slurry situations may be sealed by appropriate designs of mechanical face seal, section 3.4.4.8. Process plant bearings use both mechanical seals and labyrinths. High speed spindles also use a variety of labyrinth arrangements for applications such as machine tools. Potential seals are discussed in sections 3.6.2.2 and 3.6.2.3.

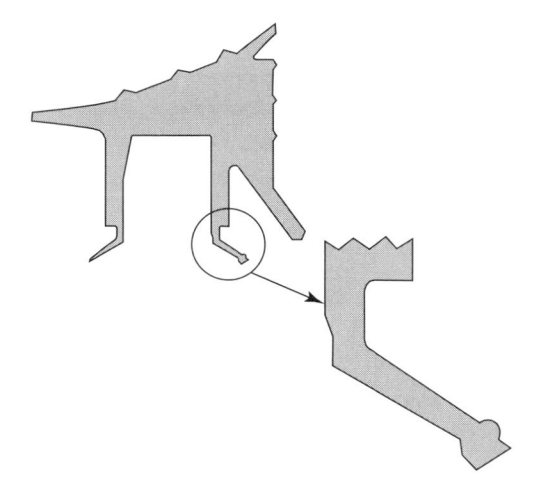

Figure 3.28 A modified version of the rolling mill seal with hinge points in the seal flange to control the deflection geometry with radial offset. (Source: Morgan Construction Co.)

3.2.6 Cassettes

Cassette is a term that has been adopted for an integrated seal and shaft sleeve assembly. It is analogous to what is known as a cartridge in mechanical seal terminology. Cassettes have evolved to include a seal and shaft sleeve assembly complete with many of the exclusion features described in section 3.2.5. Figure 3.29 is an example of a cassette arrangement for a single lip seal incorporating external dust lips. They have a number of potential benefits which should be considered at the design stage:

- The seal supplier is providing both the static and dynamic components of the seal in one controlled assembly.
- The unit is pre-assembled in controlled conditions so there is less opportunity for damage to the seal lip during transport, storage and assembly.
- The seal supplier provides the shaft surface texture and hardness required.
- A complex seal geometry with optimized excluder design incorporating radial flanges and axial sealing lips can be effectively assembled. Positioning stops can be designed into the cassette assembly.
- Machine shaft is only required to be machined to a surface finish suitable for static sealing.
- Reduced specification for assembly taper on the shaft.
- In the event of maintenance the shaft contact zone is replaced at the same time as the seal, eliminating the potential requirement to clean and remachine the shaft.

The major disadvantage of this approach is that a seal is required between the shaft and seal sleeve. This is usually achieved by an elastomer coating on the inside diameter of the sleeve, such as that shown in Figure 3.29. As the shaft is the primary method of dispersing the heat generated at the seal lip the reduced thermal

134 *Seals and Sealing Handbook*

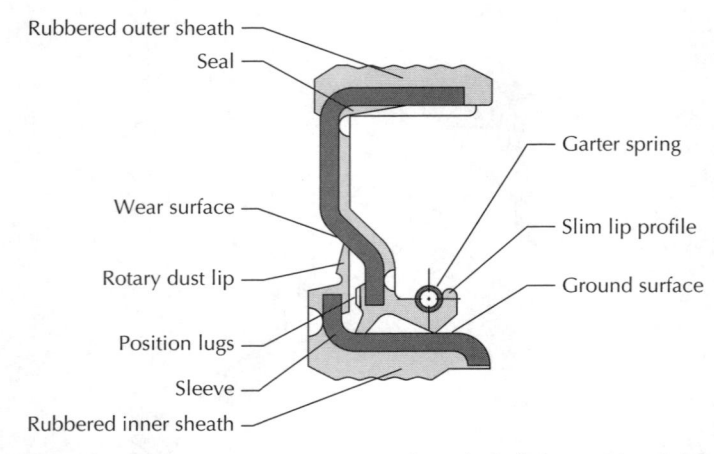

Figure 3.29 A simple cassette arrangement comprising lip seal, shaft sleeve with radial flange and dust excluders. (Source: Trelleborg Sealing Solutions)

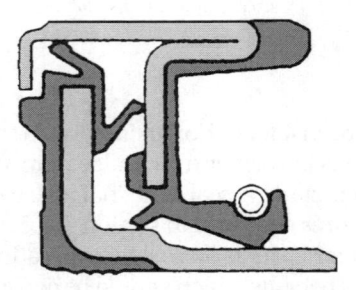

Figure 3.30 A cassette designed to provide improved heat conduction with metal-to-metal contact under the seal lip. (Source: Freudenberg Simrit)

conductivity of the elastomer layer will increase the inherent seal lip temperature thereby limiting the operating conditions compared to a seal running directly on a steel shaft. This effect can be reduced by using a sleeve that has an interference fit for the section of sleeve that is directly under the seal and an elastomer liner at the atmospheric heel of the sleeve, as shown in Figure 3.30. This arrangement will permit a 25% increase in shaft speed compared with a conventional cassette.

Seal manufacturers will also attempt to reduce the seal lip temperature by designing the seal to have a minimum practicable lip load and also incorporate pumping aids to achieve low friction.

As discussed above this approach has a number of benefits, particularly if incorporated at the initial design phase to take account of the simplification of manufacture and assembly. Once this approach is adopted it will then potentially be difficult to incorporate a standalone seal arrangement without major modifications.

3.2.7 V-ring seals

The V-ring is essentially an axial lip seal. The original seal of this type was developed by Forsheda AB during the 1960s but has now been widely replicated.

Rotary seals 135

The seal is a form of elastomeric lip seal which seals on a radial face rather than axially along the shaft. A typical seal is shown in Figure 3.31.

The seal is normally stretched to fit directly on the shaft with the elastomer tension then providing sufficient friction to drive the seal. It will rotate with the shaft and seals axially against a radial face. This radial face can be a bearing race, bearing housing, a stamped washer or even the metal casing of a lip seal. The seal fitting dimensions are such that a very light lip loading is created. The seal may therefore run dry in many applications with a low friction. Figure 3.32 provides

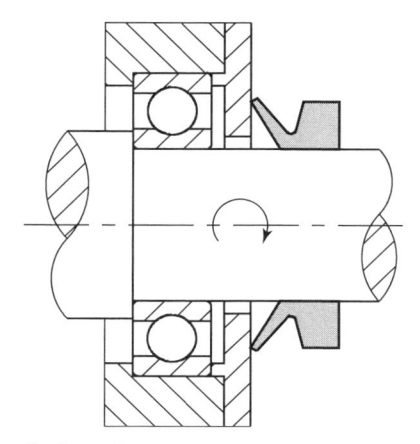

Figure 3.31 A standard V-seal arrangement. (Source: Trelleborg Sealing Solutions)

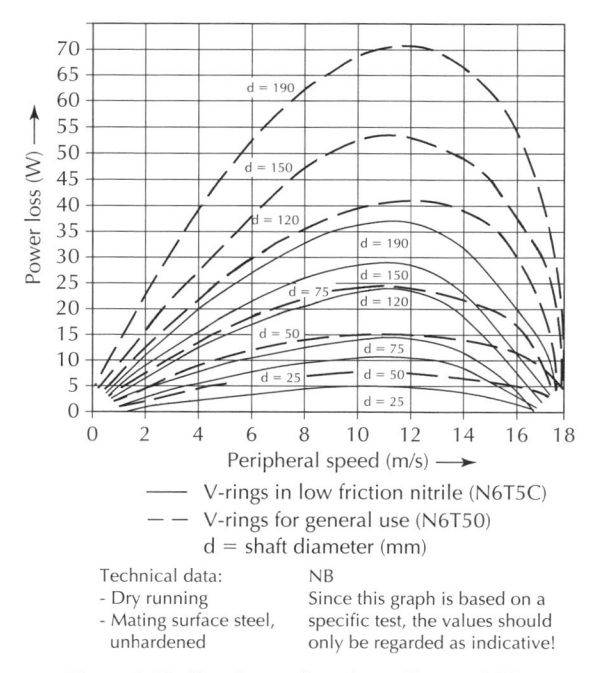

—— V-rings in low friction nitrile (N6T5C)
– – V-rings for general use (N6T50)
d = shaft diameter (mm)

Technical data:
- Dry running
- Mating surface steel, unhardened

NB
Since this graph is based on a specific test, the values should only be regarded as indicative!

Figure 3.32 V-seal power loss chart. (Source: SKF)

an indication of typical power loss from a V-seal. The configuration of the lip is such that as speed increases centrifugal force will tend to reduce the seal lip interference which enables the seal to operate at high speeds without excessive heat buildup. Above a speed of 15–20 m/s the seal lip has lifted off the counter-face and the seal then operates as a clearance seal.

The entire seal is subject to centrifugal force as the shaft rotates. Above certain speeds, which depend on diameter, some form of radial retention will be required. For general purpose sizes the limiting speed for operation without radial retention is 10–12 m/s. For very large sizes, such as 1–2 metre diameter, radial retention should always be used. This can be achieved by designing a suitable housing to retain the seal or using a clamping band, Figure 3.33.

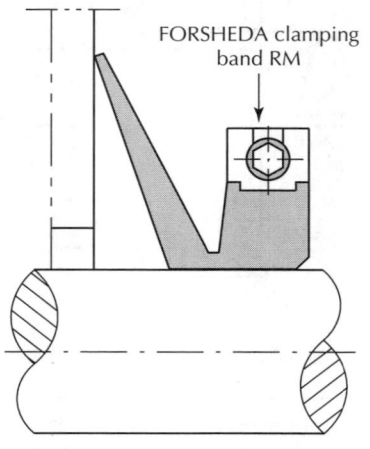

Figure 3.33 V-seal with a clamping band to prevent radial growth at high speed. (Source: Trelleborg Sealing Solutions)

Among the potential advantages of this design of seal compared with a radial lip seal are:

- Shaft surface finish less critical.
- Will operate with considerable runout and misalignment.
- Running surface is less critical.
- High speed capability.
- Axial shaft movement accommodated by seal flexing.

Typical surface requirements for these seals are:

Radial surface finish: 0.4–0.8 μm Ra
Flatness: 0.4 mm per 100 mm diameter

In certain high speed applications the seal can be mounted statically to operate against a radial face. In this arrangement the seal contact stress will not reduce as speed increases, so this would normally be used in liquid lubricated application.

V-ring seals are widely used as bearing seals in small high speed applications such as electric motors and other bearing seals in conditions of light contamination. For more arduous applications they are widely used as an excluder seal on the atmospheric side of a lip seal.

3.2.8 Bearing seals

The lip seals and V-ring seals discussed in sections 3.2.1 to 3.2.7 are generally fitted in a housing or on a shaft that may be adjacent to a bearing with the purpose of the seal being to either retain lubricant or assist with the exclusion of contaminant from the bearing. Seals are also fitted directly within rolling element bearings. They are normally more compact than a separate seal but can be adaptations of some of the seal designs already discussed.

Bearings with integral seals will be lubricated with grease on assembly by the bearing manufacture. Although it may be thought that the function of the bearing seals is to retain this grease their primary purpose is in fact the exclusion of contaminant. In the majority of cases the seal will be designed to seal the bearing from the external environment. It may be expected to leak a small amount of the grease, but in so doing lubricate the seal and assist with contaminant exclusion. The primary concern is that even minute amounts of solid debris or water will seriously reduce the life of the bearing.

A rolling element bearing is expected to support a shaft and allow rotation with a minimum of friction. The selection of a bearing seal presents an immediate compromise, minimizing friction while providing adequate exclusion. The bearings can be divided into three convenient major subgroups: shielded, labyrinth and sealed.

3.2.8.1 Shielded bearings

These are non-contacting seals that comprise some form of washer or plate, typically made from sheet steel, which provides a close clearance gap. An example is shown in Figure 3.34. The shield will be designed to form a long sealing gap with the land of the bearing inner ring shoulder, Figure 3.34(a), or a simple labyrinth seal using a recess in the inner ring shoulder, Figure 3.34(b).

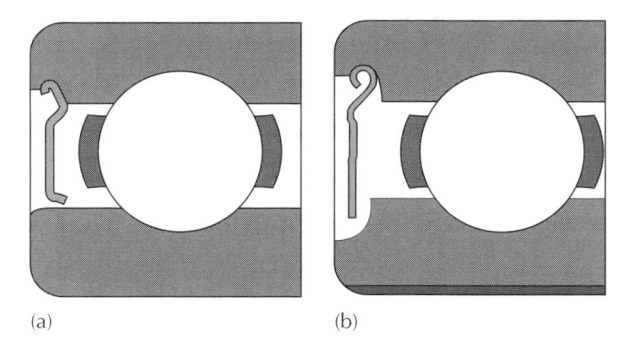

Figure 3.34 A rolling element bearing with integral shields to inhibit access of contaminant. (Source: SKF)

The bearing will be grease lubricated and the sealing will depend to some extent on the properties of the grease to fill the clearance gap and inhibit contaminant ingress. A shield offers the benefit of minimal friction and high speed capability but exclusion performance is limited. Applications are therefore limited to areas where there is only light contamination, such as small quantities of dust and moisture. They are only suitable for applications where there is no danger of water, steam, etc. coming into contact with the bearing. The water repellant properties of the grease will be critical if there is any opportunity for direct contact with water.

3.2.8.2 Simple labyrinth

A labyrinth bearing seal is an extension of the simple shield designs discussed in section 3.2.8.1. It will usually employ a pair of shields designed to provide a tortuous path into the bearing. If the external shield is fitted to the inner race it will then act as a flinger when rotating to expel moisture and debris. The complexity will depend on the application and also bearing design as more axial space is required than for a simple shield, Figure 3.35. Under normal bearing operating conditions the labyrinth will run with a small clearance. These seals will again be used with grease filled bearings and the grease may be used to fill the clearances of the seal on assembly to provide an improved barrier to ingress.

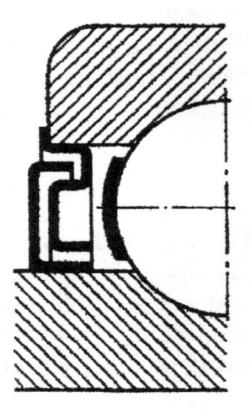

Figure 3.35 A labyrinth seal integral to a rolling bearing.

This type of seal will provide an improved resistance to ingress compared with a simple shield, especially during shaft rotation as the outer flange acts as a centrifugal expeller. This will repel a light spray of water and larger debris. The seal will not be impervious to continued exposure to water and large quantities of dust or resistant to immersion. It will also not be suitable for use in a steamy atmosphere as water vapour will be able to permeate into the bearing and contaminate the grease.

More complex labyrinths are used in specialized and some arduous environments. These are dealt with in more detail in section 3.6.

3.2.8.3 Sealed bearings

These employ a contact seal that is capable of providing a physical barrier, Figure 3.36. They may comprise a very simple looking design that is an elastomeric washer with a light lip contact on the inner race or a more complex design that has both a sealing lip and a wiper lip. In the vast majority of cases the seal

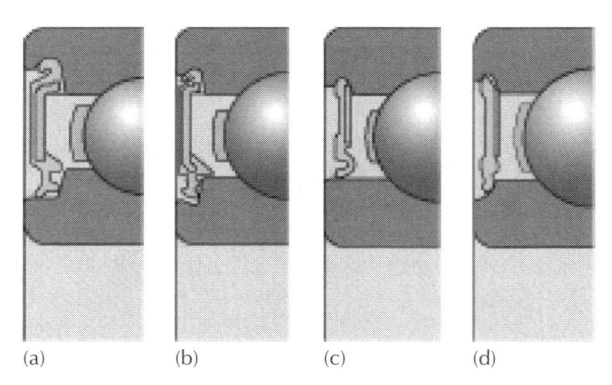

Figure 3.36 Examples of integral bearing seals. (Source: SKF)

is designed such that the lip is outward facing. This encourages the bearing grease under the lip to counteract ingress of contamination. A seal may be used with either oil or grease as the bearing lubricant. It is important on assembly that some of the bearing grease is used to lubricate the seal. The lubrication of such seals by grease has not been thoroughly researched but it is presumed that during operation some of the oil bleeds from the grease to lubricate the lip.[11]

The seal lip design may resemble either a conventional lip seal or a V-ring with either axial or radial sealing lips. The seal will often be reinforced by an integral sheet steel washer to provide rigidity. A major concern with these seals is the friction. It will increase the overall bearing friction but also generate local heat which will reduce the grease viscosity and potentially limit the bearing performance. A number of studies have been carried out to study the lip loading of various seal designs and the effect on torque and bearing performance.[12,13,14] However, the majority of work has involved empirical design based on test work and experience to provide an optimum design.

The seal may be designed to run against the inner ring shoulder (a), against a recess in the inner ring shoulder (b, c), or the lead-in at the sides of the inner ring raceway (d). If required, for example a bearing where the outer ring rotates, the seal may operate against the outer ring.[15]

Additional seal types have been developed for deep groove ball bearings, referred to as the low friction seal, Figure 3.37. These are designed to have very low contact pressure to fulfil the demand for adequate sealing and low friction operation.

These seals are used in bearings for arrangements where contamination is moderate and the presence of moisture or water spray, etc. cannot be ruled out, or where a long service life without maintenance is required.

3.2.8.4 Mechanical seals

There are two different types of compact mechanical seal that are specifically designed as bearing seals for aggressive applications. These are an elastomer energized metal seal for low speed highly abrasive environments and magnetic seals specifically designed for process pump bearings discussed in section 3.4.10.

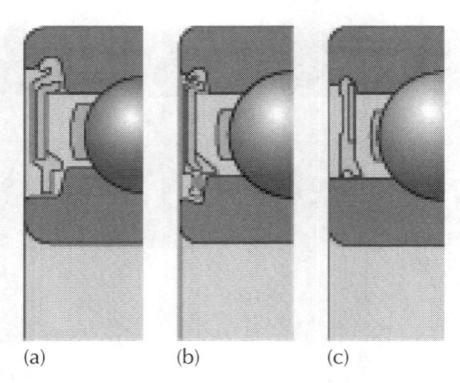

Figure 3.37 Deep groove ball bearings with low friction integral seals. (Source: SKF)

3.2.9 Lip seals for pressure

Standard lip seals as described in section 3.2.1 are only suitable for use in systems with a maximum pressure of up to 0.5 bar,[16] indeed some manufacturers recommend a maximum of 0.3 bar.[7] Pressure will distort the lip and both disturb the orientation of the seal lip angles and increase the contact area. This together with increased lip loading from the pressure will create additional friction and hence heating of the lip and limit seal performance, Figure 3.38(a). Figure 3.38(b)

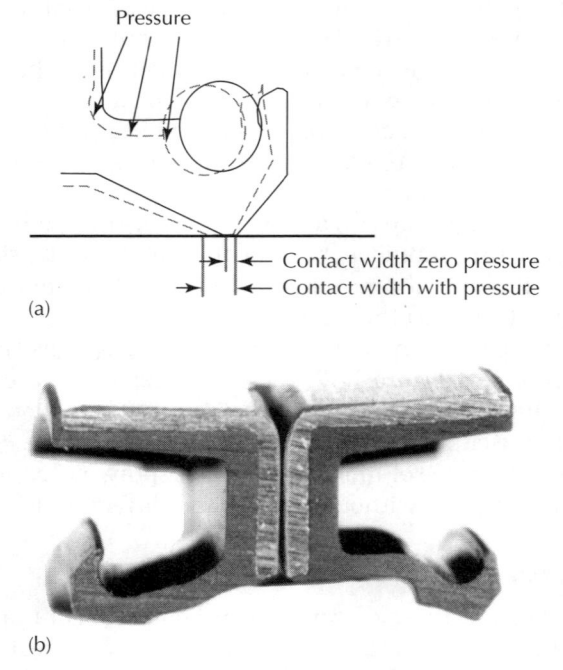

Figure 3.38 (a) Distortion of a lip seal by pressure. (b) Conventional lip seal that has been run at excessive pressure. (Source: (a) Freudenberg Simrit, (b) BHR Group)

shows a lip seal that has been run at excessive pressure, about 2.0 bar, and the subsequent distortion of the lip and excessive contact zone created.

Specific designs of lip seal are available for operation up to 10 bar. The reasons for selecting a lip seal for pressurized applications may include:

- Compact seal envelope.
- Economic seal assembly compared with alternatives.
- Conformability of elastomer can provide improved sealing at large diameters, e.g. in excess of 0.5 m, compared with alternatives.

Lip seals are popular for a number of specific applications where the benefits listed above are major considerations. These include:

- Shaft seals on fluid power hydraulic pumps and motors where they may seal a casing pressure of 1 to 7 bar.
- Vehicle steering column input shaft seals for hydraulic power steering racks. These seals are required to have low friction to avoid excessive steering effort and may operate at low pressure during most steering but up to 70 bar at full load. Shaft surface speeds are essentially very low.
- Ship stern shaft seals where a series of seals will seal externally between seal and bearing lubricant and internally between the lubricant and machinery space. The lubricant is typically pressurized to counteract the head of sea water on the external seal with the internal seal running at the pressure of the head of oil.
- Water turbine shafts under a variety of conditions.

To resist the distortion of the seal lip two alternatives can be considered:

- Standard lip seal design but with the use of a lip support, Figure 3.39.

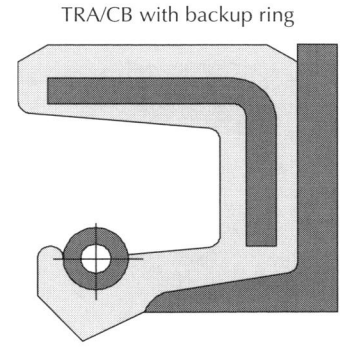

Figure 3.39 Example of lip seal with lip support for operation at pressure. (Source: Trelleborg Sealing Solutions)

- Revised seal design with a stiffer lip, Figure 3.40. A number of techniques are used to resist distortion including short, thicker elastomer section, extension of the metal casing into the flex section of the lip and redesign of the radial flange of the seal to support the lip.

The use of the support lip will reduce the potential of the seal to accommodate shaft runout, etc. and so it will require more attention to the installation tolerances. The stiffer lip designs will also reduce flexibility and require improved shaft alignment compared with standard seals.

The lip contact pressure will still be increased and this will limit the speed capability of the seals. As pressure is increased the permissible speed is reduced. The permissible speed will depend on both the seal design and also the material. Examples of recommendations of different manufacturers are given in Figures 3.41 and 3.42. There is some difference in capabilities of the individual designs from the suppliers. At higher pressures only very slow or intermittent speeds are permissible.

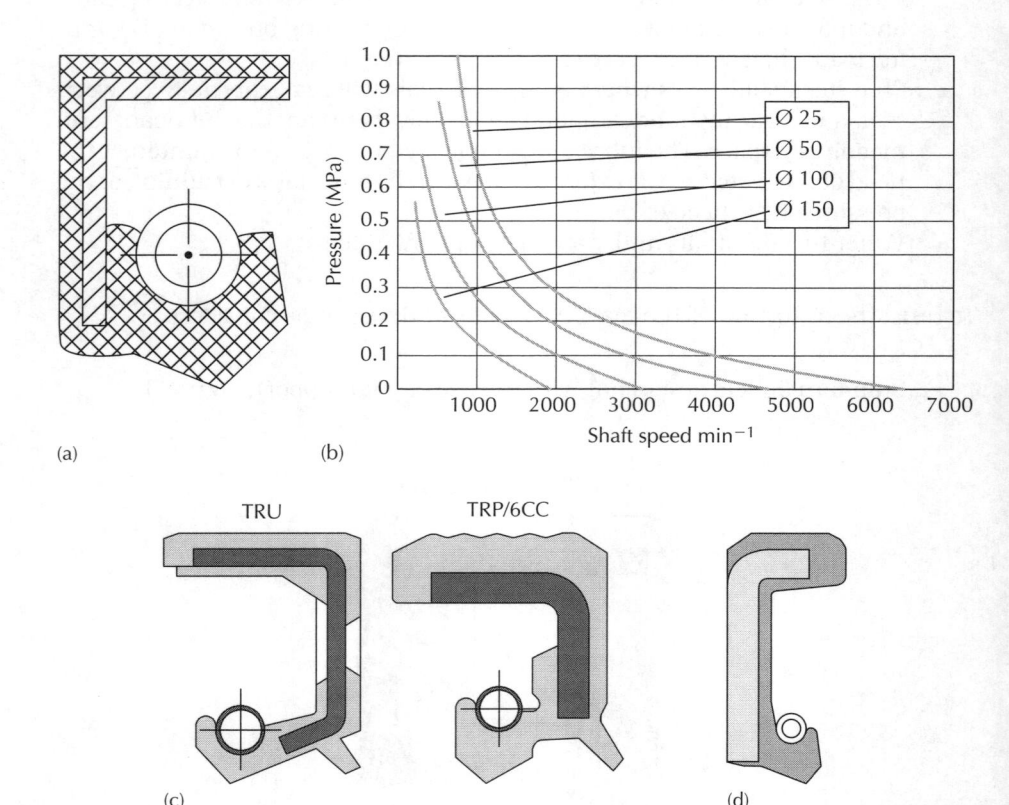

Figure 3.40 Lip seals designed to withstand distortion for operation in pressure systems. (Source: Trelleborg Sealing Solutions)

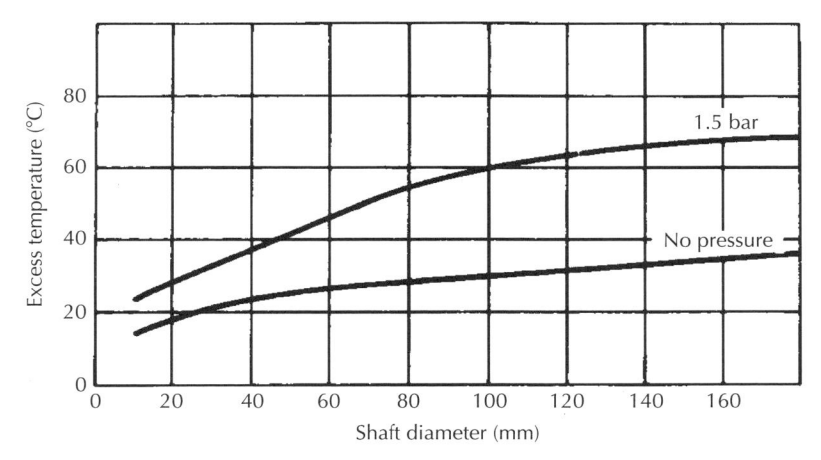

Figure 3.41 The effect of running of lip seal at 1.5 bar pressure on the temperature at the lip of a seal. (Source: Freudenberg Simrit)

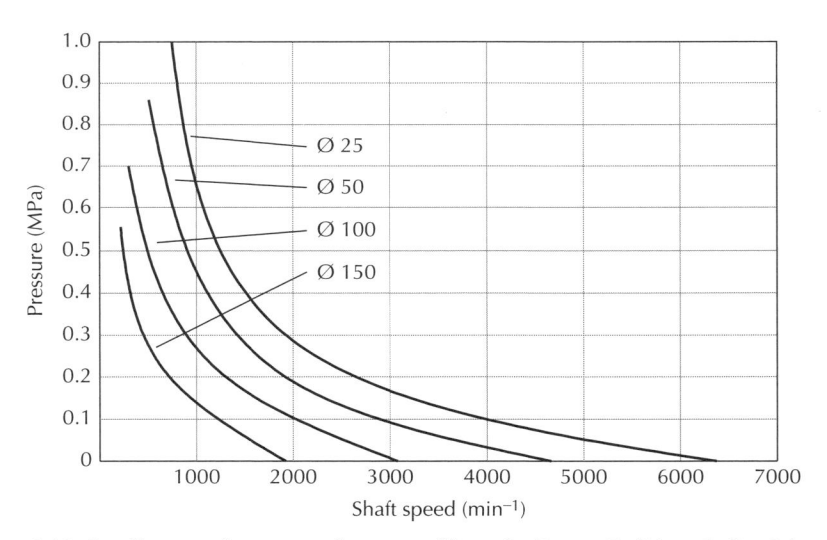

Figure 3.42 Speed/pressure limitations of pressurized lip seals. (Source: Trelleborg Sealing Solutions)

The ability of the elastomer material to operate at high temperature will also affect the performance of the seal. Figure 3.43 provides some information on anticipated life of nitrile and fluorocarbon seals at a range of speed/pressure combinations.

It is important to consider the resistance of the arrangement to dislodging of the seal by pressure and also the potential for static leakage around the seal. Normally it will be necessary to have a flange or circlip to retain the seal in the housing. A seal with an elastomer outside diameter is advisable to ensure an adequate static pressure seal.

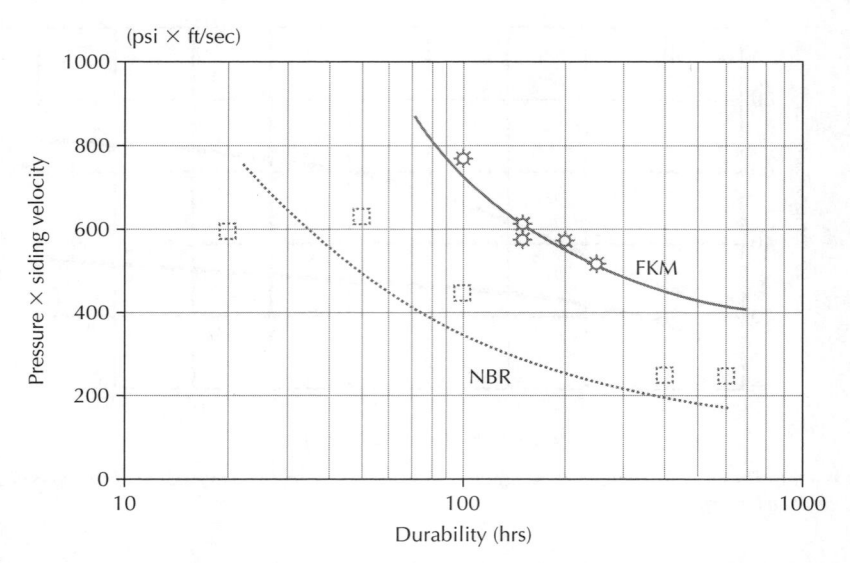

Figure 3.43 Life expectancy of fluorocarbon and nitrile lip seals when used in pressure applications. (Source: Freudenberg Simrit)

3.2.9.1 Redesign of lip angles for pressure

Although the stiffer profile of the lip seals designed for operation at elevated pressure will resist distortion, the lip design itself still promotes a relatively high lip load which will vary significantly with tolerances and the effect of pressure is still to increase the tendency of the seal to reverse pump the oil and reduce lubrication. A new approach is to design the lip to encourage lubrication, especially at lower pressure.[17] It includes an all-new, patented sealing edge profile with two approaches to reducing wear and increasing service life.

To improve lubrication in the seal zone, the profile of the sealing edge has been designed so that the contact surface angle to the shaft on the air side is larger than that on the oil side at zero pressure. At the same time, an artificial groove has been added to the sealing edge profile in the area where the most wear has been noted previously. This addition minimized the risk of wear and the buildup of damaging particles.

Figure 3.44 compares the old sealing edge profile and the new one. It shows both seals as they appear in a non-assembled state. The illustration clearly shows that the contact surface angle on the oil side of the old profile is much larger than that on the air side. However, the contact surface angle on the oil side of the new profile is much smaller than that on the air side. When assembled on the shaft, the sealing lip expands elastically and the sealing edge in this illustration turns in an anti-clockwise direction. This further increases the ratio of the contact surface angles on the old profile, but with the new profile it is balanced: the contact surface angles are nearly equal.

According to the hypotheses that explain the sealing mechanism, one would expect the new seal to constantly transport oil out past the seal. However, when

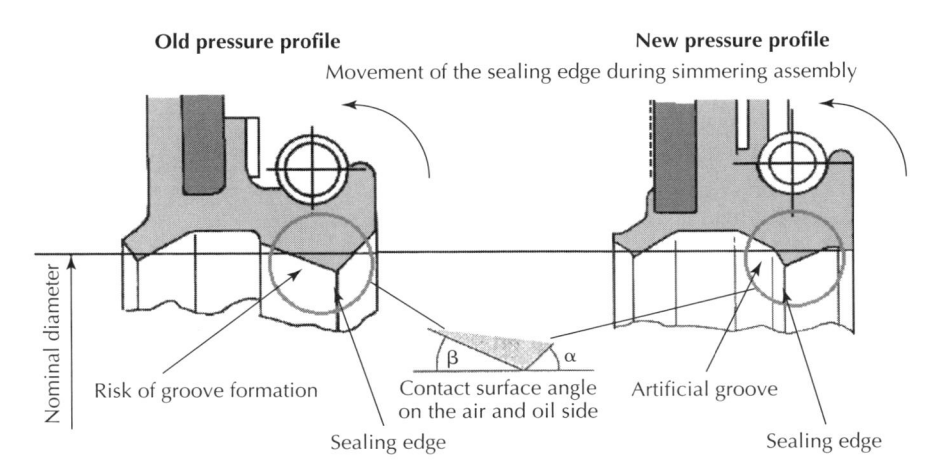

Figure 3.44 Comparison of the sealing edge profiles of a standard and new pressure radial shaft seal. (Source: Freudenberg Simrit)

in operation, the sealing lip geometry is affected by the pressure, which causes an additional rotation of the profile and therefore leads to a change in the contact surface angle.

At 3 bar, the calculated pressure distribution at the sealing edge is almost symmetrical. At 7 bar, the pressure gradient on the air side is lower than that on the oil side. The rotation of the sealing lip due to the contact pressure is such that α is larger than β. Hence, once the operating pressure exceeds approximately 3 bar, the seal will tend to return oil from the air side to the oil side.

In a hydraulic pump the pressure is usually in a permanent state of flux. This means that the sealing edge pumping direction of the new design of seal is always changing. At low pressure, the sealing edge transports from the inside to the outside while the opposite is the case at high pressure. This means that oil is constantly being exchanged in the seal contact zone, preventing the deposit of oil carbon and dry running, and reducing typical groove wear to a minimum.

This design of seal can therefore be considered for applications where there is a constantly varying pressure profile up to the pressure limits of the seal.

3.2.9.2 Stern tube seals

These are specialized seal systems specifically designed for sealing of oil lubricated stern tube bearings. Inboard the seals contain the lubricant within the bearing. Outboard they both contain the lubricant and exclude sea water and any contaminant from the bearing.

Due to the arduous operating conditions and high reliability required a multi-seal arrangement is employed, Figure 3.45. The inboard sealing will normally comprise a pair of seals with the main seal taking the pressure differential, the secondary seal being a backup that can be brought into operation in the event of problems with the primary seal.

Monitoring/lubricating oil tanks standard supply:

Forward seals
- 6 litres: up to and including size 380 mm
- 30 litres: size 400 mm and over

Aft seals
- 30 litres: size 400 mm and over

Lubrication for the forward seal includes the CCM system for enhanced oil circulation and cooling for sizes 500 mm and over.

System key

T1 Stern tube gravity tank with low level alarm switch (yard supply)

T2 Forward seal tank with low level alarm switch

T3 Aft seal tank with high level alarm switch

T4 Sump tank with low level alarm switch (yard supply)

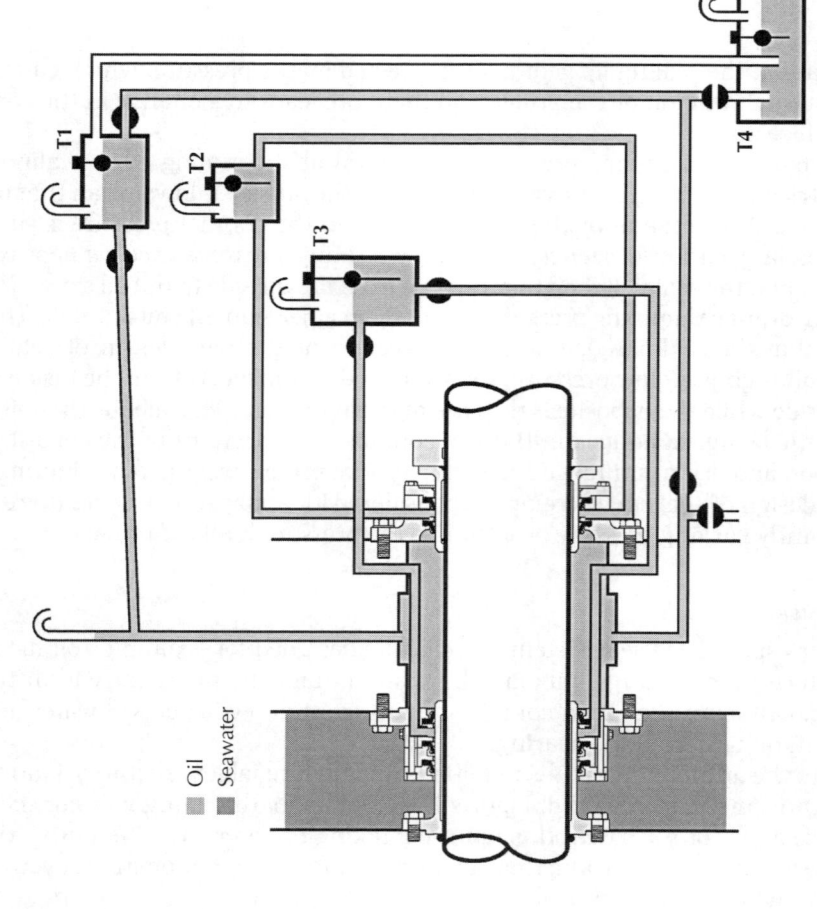

Figure 3.45 Stern tube sealing system for a ship with oil lubricated stern tube. (Source: Wartsila, DeepSea Seals)

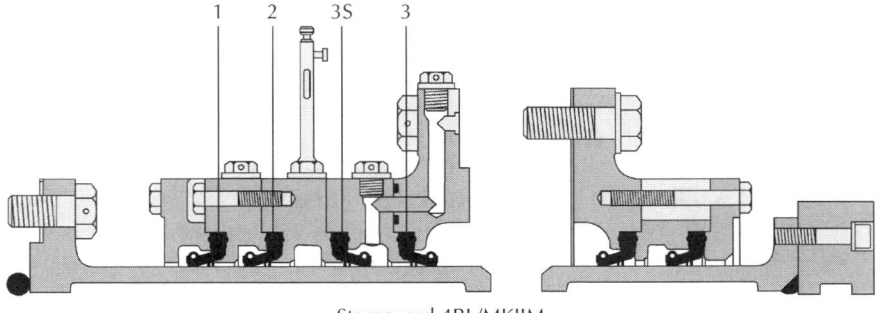

Figure 3.46 Aft sealing arrangement for an oil lubricated ship stern tube seal using a multi-barrier lip seal arrangement. (Source: Wartsila, DeepSea Seals)

The aft seal is a multi-barrier seal and will typically contain four seal rings, Figure 3.46. Two seal rings, 3 and 3S, face the stern tube oil side and the other two seal rings, 1 and 2, face the waterside, providing double security against seawater ingress. The aftermost seal ring 1 also functions as a dirt excluder. All four seal rings run on a shaft liner to avoid grooving of the shaft. The seal will usually be supplied as a cartridge, which includes the shaft liner and is ready for installation without any further assembly work to be carried out.

To provide optimum reliability and avoid potential pollution hazards from the stern tube oil the seal system will probably incorporate an active double security against oil spill. In normal operation only one oil seal ring 3 is active, and the other oil seal ring 3S is running with an equal, balanced pressure on both sides, thereby acting as a standby seal ring. This balance is achieved by connecting the chamber between seal rings via two pipes through the stern tube to a header tank. Because of the absence of pressure differential the second seal will be expected to have little wear and can be activated in an emergency. An example of an arrangement of this type of seal is the Wartsila Sternguard range.[18]

Materials are critical for stern tubes as the seals operate at pressure and at peripheral speeds up to 6 m/s. Fluorocarbon is widely used for the entire seal or for the seal lips to provide satisfactory long-term performance. Special materials have been developed to operate at the oil/water interface where lubrication is especially difficult.

Shaft liners are also critical because of the corrosion and wear problems. An example may be a seal liner that is made of high nickel chromium steel, offering the optimum combination of wear and corrosion resistance. Compared to the conventional materials it will reduce the risk of pitting corrosion and wear of the running surface. Ceramic coatings are also used to provide wear resistance but create the problem of higher underlip seal temperatures placing greater demands on material selection.

3.2.9.3 Plastic lip seals for pressure applications

Plastic lip seals, as described in section 3.2.3, can be used for a wide range of applications. The type of seal used for pressure applications will be the plain washer design, Figure 3.18. The spiral groove design will be prone to leakage under pressurized conditions. The additional strength of the PTFE material means that it will withstand pressure without excessive distortion. The washer type lip seal can be used up to a pressure of 30 bar. There is still the problem that exists with elastomer rotary seals of heat generation under the lip, so as pressure increases the speed capability decreases. The maximum combination of speed and pressure recommended for these seals is 100 bar × m/s.[10] So at 30 bar the speed would be limited to 3.3 m/s. A further potential limiting factor is the design of the hardware, particularly the shaft as this will influence dispersion of the frictional heat generated. Frequency of stop/starts and fluctuations in system pressure may also affect the performance as these will inhibit the bedding-in of the seal and accelerate wear.

There are two significant factors that can influence the selection of a PTFE lip seal for pressurized applications:

- The seal will tend to leak more than an elastomeric seal. Typical leakage rates may be between zero and 0.4 ml/h.[10] It may be necessary to arrange additional leakage collection measures such as a backup seal.
- As the seal is not dependent on the sealed fluid for lubrication it may be used to seal non-lubricating fluids such as water and mild chemicals or dry to seal gas. In this case the boundary lubrication properties of the seal and shaft combination are critical and selection and finish of the shaft is as important as the seal.

For non-lubricating liquids such as water it is important to have a hard shaft surface, 50–60 Rockwell C, a surface finish of 0.2–0.3 μm Ra and a polished texture. For dry operation a fine surface in the range 0.1–0.2 μm Ra is preferred. The reduced heat dissipation for dry applications must also be considered.

3.3 Alternative elastomer and plastic seals

For sealing rotary shafts in a pressurized system a lip seal has limitations, as discussed in section 3.2.9 above. Some of these limitations will apply to any seal design, most specifically the problem of achieving a compromise between frictional heat and sealing function. A number of seal designs have been developed for rotary applications. These seals generally offer the benefit of a compact seal envelope compared with alternatives such as mechanical seals or packing. The individual selection will depend on the seal life, leakage, permissible seal housing geometry and reliability requirements of each application.

3.3.1 O-rings

It is possible to use O-rings as a rotary seal and they are widely used as bearing seals for down hole drilling rock bits, Figure 3.47. However, in this application a compact seal with limited life expectancy is required. A specific property of

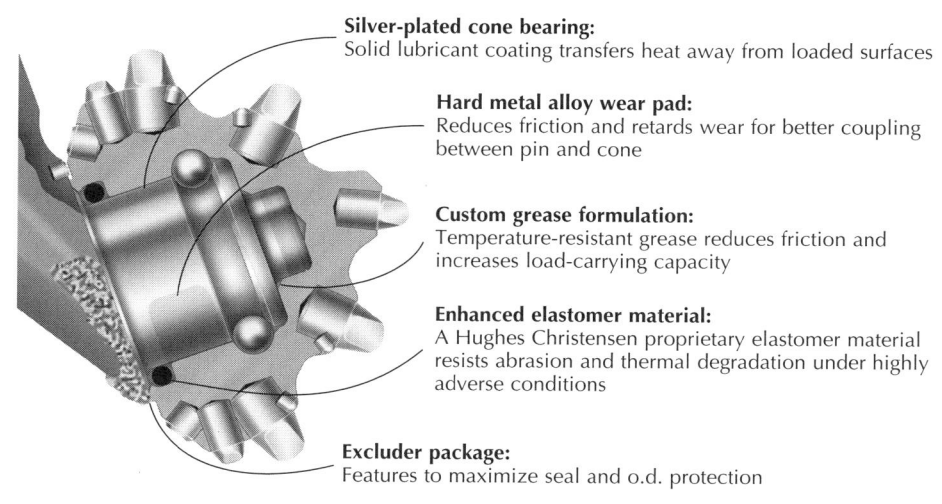

Silver-plated cone bearing:
Solid lubricant coating transfers heat away from loaded surfaces

Hard metal alloy wear pad:
Reduces friction and retards wear for better coupling between pin and cone

Custom grease formulation:
Temperature-resistant grease reduces friction and increases load-carrying capacity

Enhanced elastomer material:
A Hughes Christensen proprietary elastomer material resists abrasion and thermal degradation under highly adverse conditions

Excluder package:
Features to maximize seal and o.d. protection

Figure 3.47 O-rings being used as rotary seals to protect the bearings of an oilfield down hole rock bit. (Source: Hughes Christensen Co.)

elastomers can cause additional problems in highly loaded seals, this is what is known as the Gough–Joule effect. They showed that if elastomer under tension is heated it will tend to increase in modulus. In commercial elastomers this tends to be counteracted by the decrease in modulus with increasing temperature, but it can cause a transient increase in contact stress and hence friction. It is therefore usual to install O-rings for rotary applications with a small amount of circumferential compression to offset this effect.

The application of O-rings to rotary seals is limited to low speeds, of the order of 0.2 m/s, and it is necessary to provide a well-located shaft to housing assembly with good groove concentricity to avoid eccentric loading of the O-ring and consequent 'bunching' which will lead to failure. A compromise between large O-ring cross-section for stability and small cross-section for reduced heat generation add to the problems of selecting a suitable compromise.

However, O-rings are found in light duty rotary couplings where the rotation is intermittent and slow.

3.3.2 Elastomer fabric seals

For high pressure rotary unions rotating intermittently at low speed it is possible to use elastomer fabric seals. The sealing surface will often be of elastomer fabric to improve both wear resistance and lubrication. These seals are used to seal between adjacent high pressure passages in a union and so a small amount of leakage across the seal is permissible. An example of a union assembly is shown in Figure 3.48 and a seal in Figure 3.49. The typical operating limits of this type of seal are pressures of up to 400 bar, at room temperature, and a shaft speed of 0.2 m/s. The continuous pressure velocity (PV) rating will be of the order of 50 bar × m/s so at 400 bar the maximum speed will be only just over 0.1 m/s.

150　Seals and Sealing Handbook

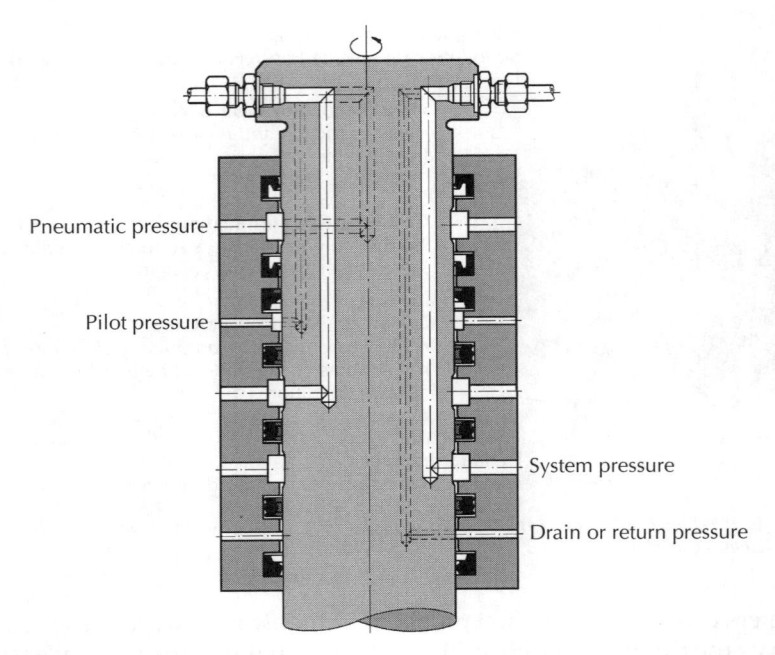

Figure 3.48 A hydraulic rotary union that uses elastomer or plastic rotary seals between the high pressure ports. (Source: Parker)

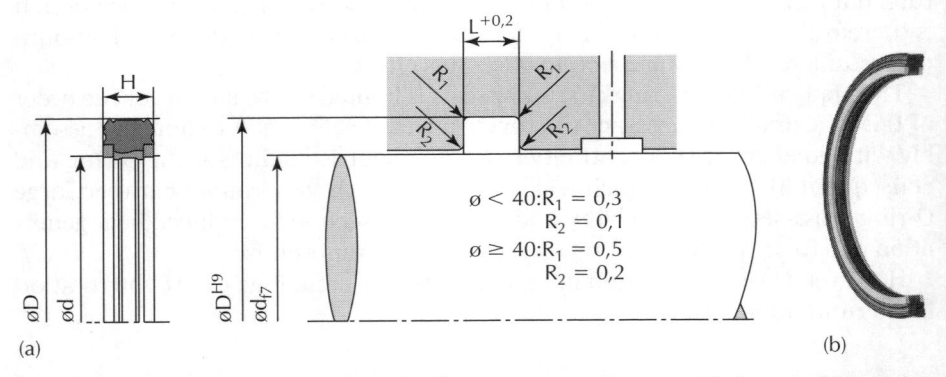

$\varnothing < 40: R_1 = 0{,}3$
$R_2 = 0{,}1$
$\varnothing \geq 40: R_1 = 0{,}5$
$R_2 = 0{,}2$

(a)　(b)

Figure 3.49 High pressure rotary seals for double acting pressure operation with elastomer fabric sealing surface. (Source: Parker)

3.3.3 Elastomer energized plastic seals

Many of the high pressure rotary applications such as rotary unions require compact seals between adjacent oil passages. A small amount of leakage is therefore acceptable and hence PTFE seals are now often selected in preference to elastomer fabric designs. Typical seals are shown in Figure 3.50. The circumferential grooves provide an increase in sealing contact stress and a reservoir for lubrication. Certain designs are available in which the grooves are not strictly circumferential

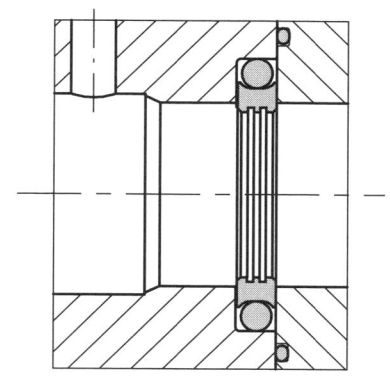

Figure 3.50 High pressure rotary seals for double acting pressure operation with PTFE dynamic component and elastomer energizer. (Source: Trelleborg Sealing Solutions and Freudenberg Simrit)

but have a wavy circumferential profile to improve lubrication, Figure 3.51, potentially at the expense of increased leakage.

Although some manufacturers design these seals to fit within grooves complying with standard dimensions specified for reciprocating seals, such as ISO 7425, these seals are not interchangeable with reciprocating seal designs that may have a similar overall appearance. The rotary seals are designed to have a lower overall seal contact loading on the shaft and an increased static contact area to prevent rotation of the seal in the housing. The application limits of this type of seal will be very dependent on the individual designs. Seal contact geometry, interference and material will all have a significant effect on the seal performance. Seals are available that will operate within the following areas, depending on individual designs:

- Pressure limit of 300–500 bar.
- Speed limit of 0.5 to 2.0 m/s.
- Combined PV limit of 25–40 bar × m/s (70 bar × m/s with wavy groove profile seals).

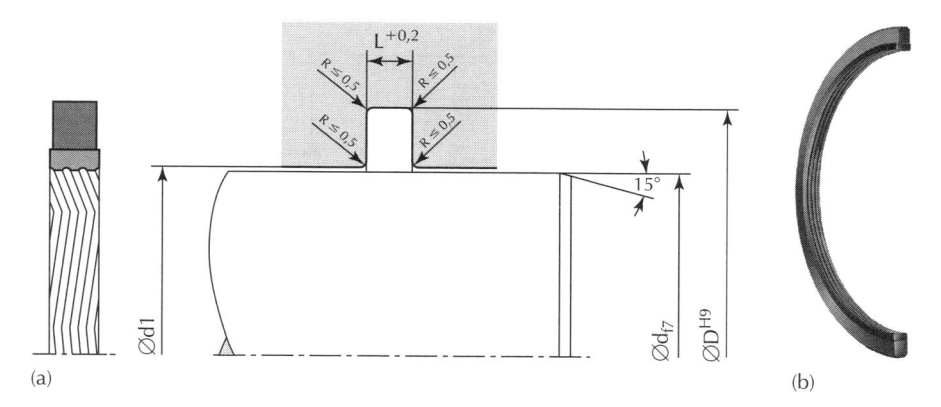

Figure 3.51 High pressure rotary seals for double acting pressure operation with PTFE dynamic component that has a wavy profile groove for enhanced lubrication. (Source: Parker)

Some supplier catalogues do not provide a PV limit but this is an important factor in the buildup of frictional heat with this type of seal and must be considered. If no limit is given, it must be assumed to be similar to those quoted above unless the supplier has some proven solution to avoid this problem.

These seals are specifically designed as high pressure double acting seals for sealing between the passages of a rotary joint and must therefore be expected to have a small amount of leakage. Quoted values are up to 0.4 ml/hr[10] but will probably be higher for seals with enhanced lubrication such as wavy grooves.

It must also be noted that the performance limits quoted by suppliers are usually at a moderate temperature and lower limits will apply as the temperature increases.

3.3.4 Spring energized PTFE seals

Spring energized U section plastic seals can be used for a wide range of rotary duties which can include virtually any liquid or gas at pressures of up to 200 bar. Speed is limited at these conditions but they provide a compact and fluid resistant seal that can operate with poor lubrication.

The construction is broadly similar to the designs discussed for both static seals, section 2.4, and reciprocating seals, Chapter 4. However, while they may appear to be superficially similar the detail of the seal design can be very different and they cannot be assumed to be interchangeable. The seals designed for rotary duties will be constructed to provide resistance to rotation in the seal housing, including during pressure, temperature and stop/start cycles. The seals will therefore usually be designed with a shorter dynamic lip and an additional location facility on the outside diameter. This has traditionally been achieved by using a U-seal but with a radial flange on the outside diameter, Figure 3.52. The radial flange is clamped by the seal housing on assembly to prevent rotation of the seal. An alternative method is to use a seal with an integral metal locking ring, which is an interference fit into the seal housing, Figure 3.53, so that location is achieved in a similar manner to that of a conventional rotary shaft lip seal.

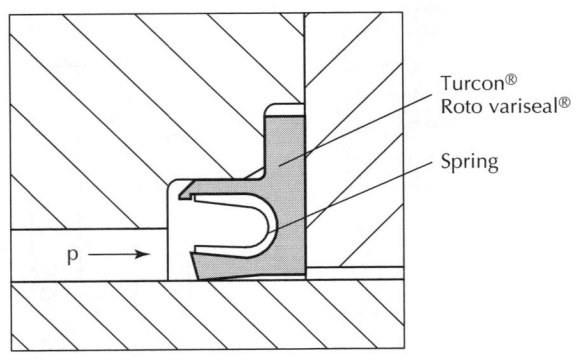

Figure 3.52 PTFE U-ring for rotary applications with a radial flange for positive location to prevent rotation. (Source: Trelleborg Sealing Solutions)

Figure 3.53 An alternative design of PTFE seal for rotary applications that uses a metal locking ring to provide the location. (Source: Balseal)

This positive location is important as there is a considerable risk of the seal rotating with the shaft and spinning in the housing, especially during cold starts. When first used the seal will warm up and bed into a sealing contact with the shaft. As it cools, when stopped the high thermal expansion coefficient of the plastic will tend to reduce the contact stress on the outside diameter and increase that around the shaft. When restarting there will then be a strong tendency for the seal to grip the shaft and spin in the housing if it is not restrained.

In addition seals designed for rotary applications will have a relatively flexible spring design so they will either be of the U section leaf spring or canted coil spring type. The selection is based upon the individual preference of individual manufacturers. Experience suggests that either type can give satisfactory performance, the selection of the correct overall design and plastic material type for the individual application are far more critical to performance.

As these seals do not have any elastomer components they offer the potential of wide fluid compatibility offered by PTFE, with the limitation usually being the metal of the spring, or occasionally the filler in the PTFE. They can therefore be used on a wide range of applications including oil, water, food, chemicals, pharmaceuticals, etc. As they also have dry running potential they can also be used to seal gases. If sterilization is required then it is also possible to obtain seals where the U section is additionally filled with silicone elastomer to provide a smooth seal geometry and avoid crevices.

The potential operating parameters for this style of seal are very wide. They can be used from low pressure, or vacuum, up to typically 200 bar, depending on the design. Speed capability is more modest, but up to 10 m/s is possible with low pressure and moderate temperature. The combination of speed and pressure, PV, is again important and a maximum of 50 bar $\times$ m/s is recommended at low temperatures for long-term service. An example of the speed $\times$ pressure limits for a range of temperatures recommended by one manufacturer is shown in Figure 3.54. This demonstrates the effect of increasing temperature to reduce the application limits of the seal.

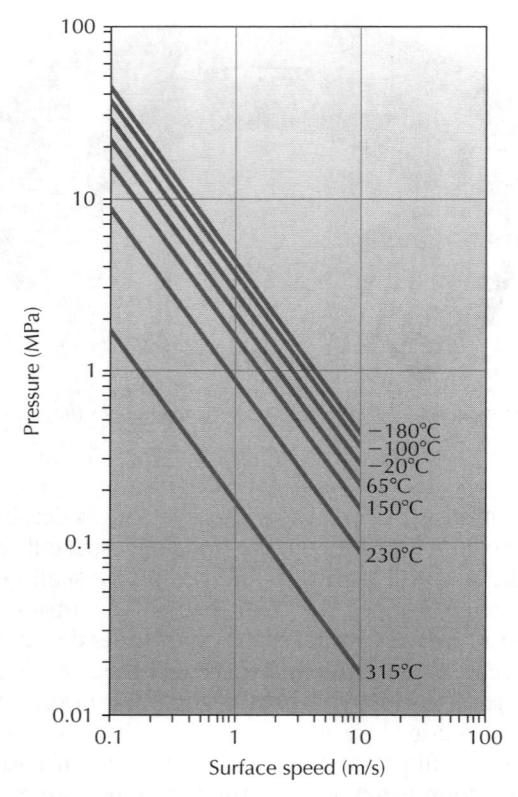

Figure 3.54 Pressure times velocity limits for plastic U-seals in rotary applications. (Source: Parker)

It is generally advisable to liaise closely with the seal supplier on selection of the seal material as this will be very dependent on the fluid to be sealed. While the majority of compounds are based on filled PTFE, virgin PTFE may be used for light duties, low friction or in particularly difficult fluid compatibility situations. In addition, ultrahigh molecular weight polyethylene is generally recommended for use in water. Spring materials will often be stainless steel on standard seals but corrosion resistant alloys such as Hastelloy or Inconel may be specified where additional corrosion resistance is required.

The shaft is again an important component in the overall seal installation. With rotary plastic seals in arduous applications the shaft can be absolutely crucial, making the difference between rapid wear and satisfactory performance.

The shaft should have a hardness of at least 55 Rockwell C which should be achieved by at least case hardening to ensure a satisfactory depth of hardening. If a hard coating is used considerable care is required. Chromium must have a high integrity and not be liable to cracking or peeling. Any form of flame or weld deposit coating must have a high integrity with excellent adhesion to the substrate and good conductivity. Any coating that comprises a combination of hard and soft materials is particularly unsuitable. To assist with removal of frictional heat a high conductivity is particularly beneficial. In critical situations a solid shaft will provide improved performance over a hollow shaft as the heat transfer is improved.

Typical shaft surface finish recommendations are:

Dry, low molecular weight gases: 0.1 μm Ra
Liquids: 0.2 μm Ra

As with other rotary seals a shaft lead should be avoided for optimum leakage control.

In addition it is necessary to consider the operating conditions:

- With a dry gas application the deposition of the transfer film is important to achieve satisfactory wear and friction. This can be achieved with a fine plunge ground finish to the specification above.
- With liquid applications it has been found that the transfer film will be washed away and very high wear rates can occur. Better performance can often be achieved by providing a polished surface, and so a fine plunge grind followed by a polishing operation is preferred. Particular care is required with any specialist hard coating to ensure that they are polished to provide a suitable surface that is free from sharp peaks in the texture.

3.3.5 Positive lubrication rotary seals

There are some specialist seals that are designed to be positively lubricated. They are designed to positively pump lubricant into the seal contact area, rather than pump it out as is the case with other designs such as the conventional lip seals. This can make it possible for a compact elastomer seal to operate reliably at a relatively high PV. The penalty for this positive lubrication is some leakage across the seal. However, there are situations where this may be acceptable. It may additionally be possible to have a tandem arrangement where a low pressure seal collects the leakage from the high pressure seal.

3.3.5.1 Positive lubrication seals for down hole and abrasive applications

A proprietary range of seals are available that were originally designed for some specific oilfield rotary seal duties operating in aggressive oil and mud environments. The seal is also expected to be very compact and many of the original equipment designs were based on using an O-ring as the rotary seal. The ideas for the seal were based on the observations that an O-ring mounted at an angle to the shaft could perform rather better as a rotary seal than when it is mounted at 90° to the shaft axis. This was shown to be due to the wiping action improving the lubrication under the seal.[19, 20] This mechanism was exploited to design a seal that would promote lubrication under the lip while excluding abrasives.[21] The primary application that was originally considered for these seals was in down hole equipment where a rotary seal is required to separate a lubricant from drilling mud. In most of these applications, which include high vibration and shock loads as well

156   Seals and Sealing Handbook

as the abrasive conditions, a small amount of lubricant leakage is normally acceptable provided the equipment can operate reliably for the specified time.

The seal is an elastomer interference seal which is fitted into a stationary housing. The inside diameter of the seal has an abrasive exclusion side which is an interference fit with the shaft, and the lubricant side which is relieved to provide a clearance with the shaft. The interface between the two sections of the seal comprises a wavy profile ridge, Figure 3.55. When the seal is stationary, the abrasive exclusion side of the seal is an interference fit on the shaft to provide a static seal. A typical installation is shown in Figure 3.56.

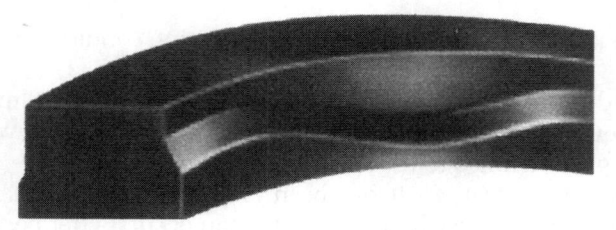

Figure 3.55 The geometry of positive lubrication seal showing the straight exclusion edge and the hydrodynamic wave for entrainment of lubricant. (Source: Kalsi)

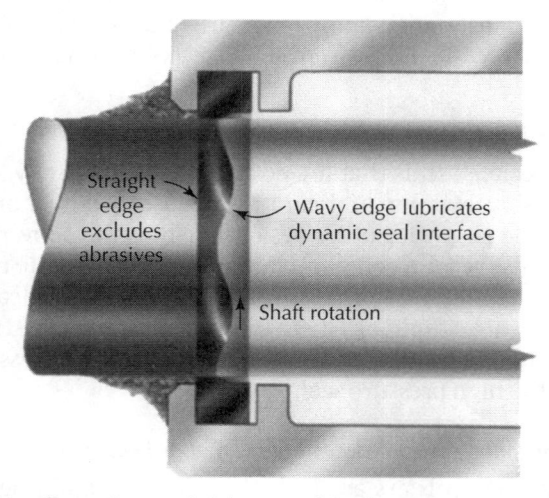

Figure 3.56 A seal installation showing the lubricant and the abrasive external to the seal. (Source: Kalsi)

When the shaft rotates the wavy profile on the oil side promotes hydrodynamic action which pumps lubricant under the seal. This positive lubrication feature permits the seal to work at relatively high rotational speeds and pressures in these aggressive conditions when compared with other compact designs of polymeric seals.

In down hole applications the seals will inevitably be exposed to a high ambient pressure which can typically range from 400 bar to over 1000 bar. This high ambient pressure in itself introduces problems with hydrodynamic lubrication. In addition the seals may also operate at a differential of up to 200 bar.

A number of variants on the basic seal design have been developed.

Wide footprint seal
This has been introduced to provide a longer life in abrasive conditions. The 50% wider footprint provides more material to accommodate wear and extrusion in severe applications. An optimized profile to provide hydrodynamic lubrication with higher interference is also used to assist with debris exclusion and provide compensation for wear.

Axially constrained seal
This seal is designed to reduce wear caused by abrasive entrainment in applications with zero pressure differential or where pressure reversal may occur. The axial constraint prevents movement of the seal which could entrain debris.

Enhanced lubrication seal
This seal has a different profile for the dynamic sealing lip, Figure 3.57, and was developed to improve lubrication with low viscosity lubricants, especially on the thin stainless steel sleeves that are used in some down hole tools. With an ISO 32 viscosity lubricant the torque is approximately half that of the standard seal. This can permit a longer life or higher speeds for the same conditions. Additional testing is under way with the potential to gradually extend this design to a wider range of the seal types available.

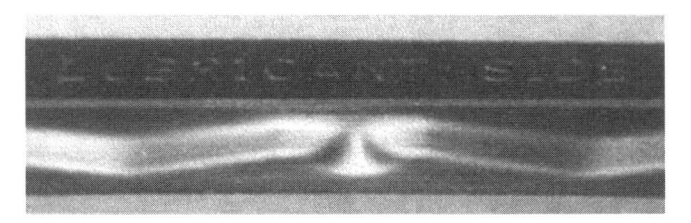

Figure 3.57 The enhanced lubrication seal design provides additional pumping into the sealing zone. (Source: Kalsi)

Optimized static lip
These seals have a higher interference and revised geometry on the lip of the seal adjacent to the abrasives. This is intended to reduce entrainment of abrasives and ensure that the seal is not contaminated while stationary.

Composite seals
Material options include seals that can be a composite arrangement to improve either temperature capabilities or wear resistance. A composite FKM/FEPM seal uses the fluorocarbon for the body of the seal and a dynamic seal layer of FEPM for improved wear resistance.

Dual hardness seals are also used with a harder elastomer on the dynamic seal surface to handle differential pressures in excess of 200 bar. The seal can then offer the wear and extrusion resistance of the hard material with lower interference force and improved resilience from a lower hardness body material.

A unidirectional version of the seal is also available with a hydrodynamic geometry designed to provide a higher pumping rate permissible in many applications such as slurry handling pumps. It is intended for pressurized lubricant retention and abrasive exclusion and can be used with very low viscosity fluids, including water.

The typical housing arrangement is similar to that which may be expected for a large cross-section O-ring or other solid elastomer seal. As the seals generate the hydrodynamic regime from new, and are usually expected to provide a good static seal, the shaft surface finish is relatively fine, compared to those for conventional rotary shaft lip seals. A surface finish of 0.05–0.2 μm Ra is required. As with other rotary seals discussed, a machining process that does not create a lead is preferable to avoid pumping generated by the shaft surface texture.

The static housing surface is rather different to most seals. To prevent rotation of the seal in the housing a ribbed surface finish achieved by knurling can be used or a surface texture of up to 2.8 μm Ra achieved by grit blasting.

These seals were originally developed for down hole drilling motors. The sealed bearing assembly has to be protected from the drilling mud and debris. It provides a positive flow of lubricant into the seal interface which creates a finite loss of lubricant. This permits the motor to operate for up to 500 hours until it is withdrawn for a drilling bit change. The seals are now used on a range of arduous oilfield duties. It should be noted that very often a relatively short life of a few hundred hours is required, but high reliability is required for the period of operation as this may be down hole where failure costs will be very high. They can be adapted to cater for a number of otherwise adverse conditions such as poor radial shaft location or axial float.

3.3.5.2 Positive lubrication seals for water turbine applications

A different design of proprietary seal is available as a cartridge seal assembly for the shafts of water turbines, particularly Francis and Kaplan designs.

A pair of sealing elements are fitted back to back within a cartridge, with flush water introduced between them, typically at 2 bar above the sealed pressure. The seals are manufactured from elastomer with a fabric reinforced running sleeve for the dynamic surface on the inside diameter, Figure 3.58(a).

The operation of the seal is shown in Figure 3.58(b). In the relaxed, unpressurized state, the ported pressure face (A) allows fluid to the outside diameter of the housing. The atmosphere face (B) does not have any ports and forms a static seal that prevents the fluid from continuing around the back of the seal to atmosphere. The seal is designed to have a small clearance on the inside diameter along the shaft, so some fluid can pass between the shaft and the running sleeve.

The pressure distribution on the outside and inside of the seal produce a turning moment (M) that tilts the seal to bring the atmospheric edge of the seal sleeve into close proximity to the shaft, controlling the leakage. A hydrostatic pressure field is created with a distribution approximating as shown in the energized diagram.[25] Dynamic contact between the atmospheric end of the seal and shaft is prevented by the hydrodynamic fluid film created by shaft rotation.

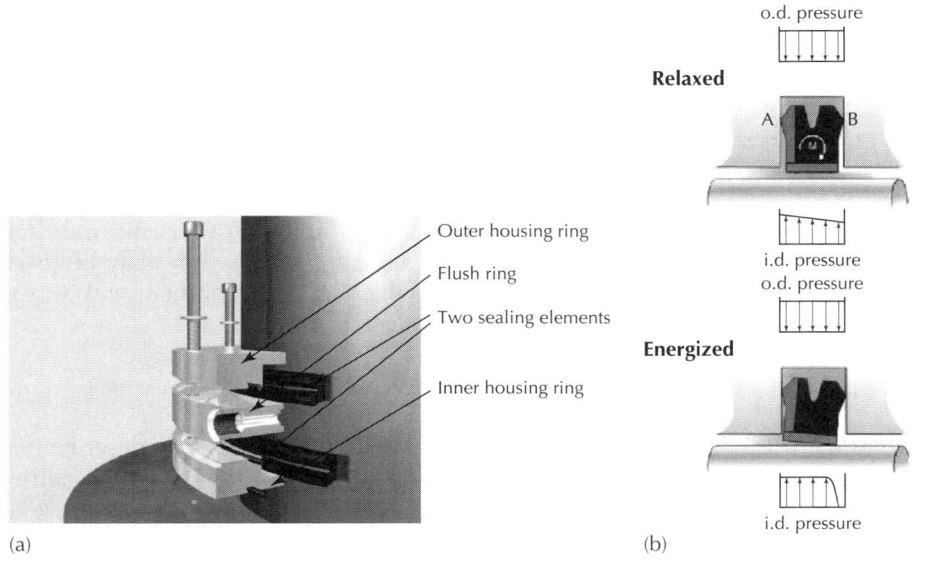

Figure 3.58 Seal for water turbine applications that uses the pressure distribution to provide hydrostatic lubrication: (a) a seal cartridge arrangement; (b) the deflection of the seal due to pressure which creates the sealing action. (Source: James Walker)

An equilibrium force occurs between the closing and opening forces providing a seal that is effectively pressure balanced with a low level of controlled leakage.

This seal provides an arrangement that is relatively economic, convenient to fit, does not need any subsequent adjustment and operates with low power consumption. It can be used up to a pressure of 10 bar and surface speed of 10 m/s.

3.4 Mechanical seals

3.4.1 Introduction

A mechanical seal is a device that uses a pair of ostensibly flat radial faces, one stationary and the other rotating, to form a dynamic seal. The faces are held in sealing contact, usually by a combination of the force of a spring and the system pressure. The face materials are selected to be a compatible tribological pair when operating in the sealed fluid. One of the faces will be well located either in the machine housing or on the shaft and the other face, loaded by the spring, will be permitted to float sufficiently to maintain the required sealing contact and face orientation. The selection of either a static or rotating floating component depends on the application and operating parameters.

This type of seal is now the primary method for providing the rotary shaft seal for containing the product in most types of rotary machines such as centrifugal pumps, rotary compressors, mixers and other process equipment. There are a wide variety of designs to suit individual applications and also facilitate the appropriate production volumes. To seal high pressures, difficult or hazardous fluids two or more seals may be used in a variety of configurations which are discussed in more detail in the relevant sections. The production requirements can vary from the very high volumes for domestic equipment, washing machines and dishwashers, or automotive water pumps to special one-off designs for large high pressure pumps or compressors.

This section describes the basic operating mechanism of the seals and then discusses the generic types and typical applications where they may be used. Further useful information covering the application of mechanical seals in the process industries is also included in Appendices 1 and 2.

3.4.2 Basic design

Mechanical seals are divided into three very broad categories that describe two important ancillary functions: the method of statically loading the faces together and the secondary, quasi static sealing of the floating component. These are generally termed:

- Elastomer bellows seals.
- Pusher seals.
- Metal bellows seals.

In elastomer bellows and pusher seals the static loading is provided by some form of springs. The most common types are either a large single coil spring or a series of smaller springs. The secondary sealing is provided by either an elastomer bellows or an O-ring respectively.

Metal bellows seals provide both the sealing and the spring through the bellows.

The basic operating method and key features of typical seals is discussed with reference to a pusher seal and a metal bellows design, Figure 3.59.

A typical pusher type mechanical seal consists of a rotating face, a stationary face and what are generally known as secondary sealing elements plus additional components to adapt the seal to the pump such as a flange and a shaft sleeve. The stationary face is seated in a flange which is bolted to the pump cover. For the majority of conventional pusher designs the rotating face can move in the axial direction and is kept in place by a spring holder and one or more springs. The rotating parts are installed either onto a shaft sleeve or directly on the shaft. Seals that are hydraulically balanced and/or of a cartridge design will normally be on a sleeve; low pressure seals may fit directly on the shaft. The O-ring, or other elastomeric seal, that can move axially with the rotating face is generally known by a name such as 'dynamic secondary seal'. The secondary sealing elements are often elastomer but sometimes PTFE or other materials may also be used.

A bellows type mechanical seal is similar in overall concept, but uses a metal bellows to achieve the flexibility in the design. A bellows seal avoids the use of a 'dynamic secondary seal' and only requires static secondary sealing. This provides

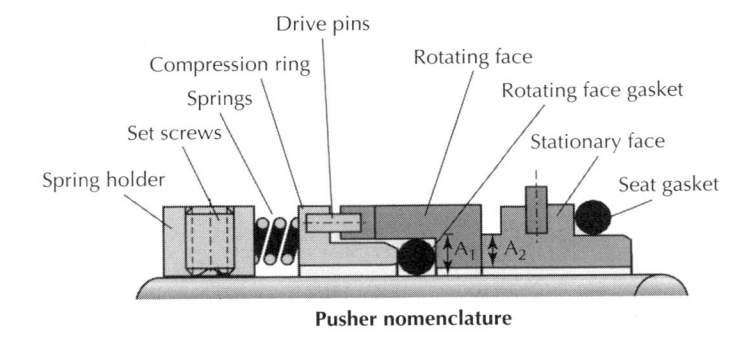

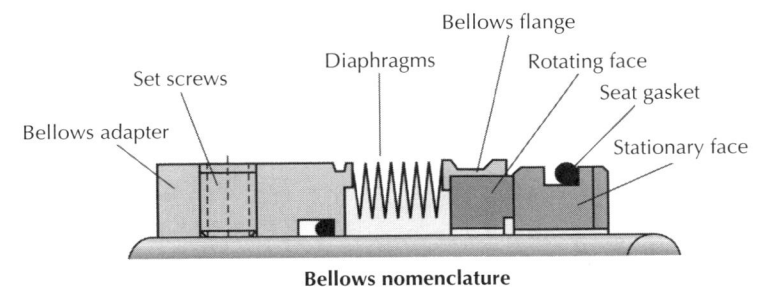

Figure 3.59 Typical pusher and metal bellows seals. (Source: Flowserve)

more choice in the selection of the secondary sealing material and geometry which is a particular advantage at high temperature. The majority of metal bellows seals use a welded bellows but some designs use a hydroform bellows. A further potential benefit in some applications is that a metal bellows seal does not require the use of any drive pins, lugs or other device to transmit the torque to the floating seal face.

3.4.3 Method of operation

Seals are expected to operate with low leakage and with acceptable reliability for the application. This reliable operation is achieved by maintaining a minimal but sufficiently thick film between the faces. This fluid film is typically below 1 μm, of a similar order to the maximum peaks of the surface roughness. The design of the seal is a compromise between obtaining sufficient thickness of fluid film to ensure adequate lubrication and long seal life and on the other hand keeping it as thin as possible to minimize leakage. As a starting point the seal faces are lapped flat usually to within one or two light bands for regular size seals. This represents a flatness somewhere between 0.3 and 1.0 μm depending on the light source used.

The floating seal face represented on the left of Figure 3.60 can be seen as a piston which is subject to a number of forces to keep it in balance.

162 *Seals and Sealing Handbook*

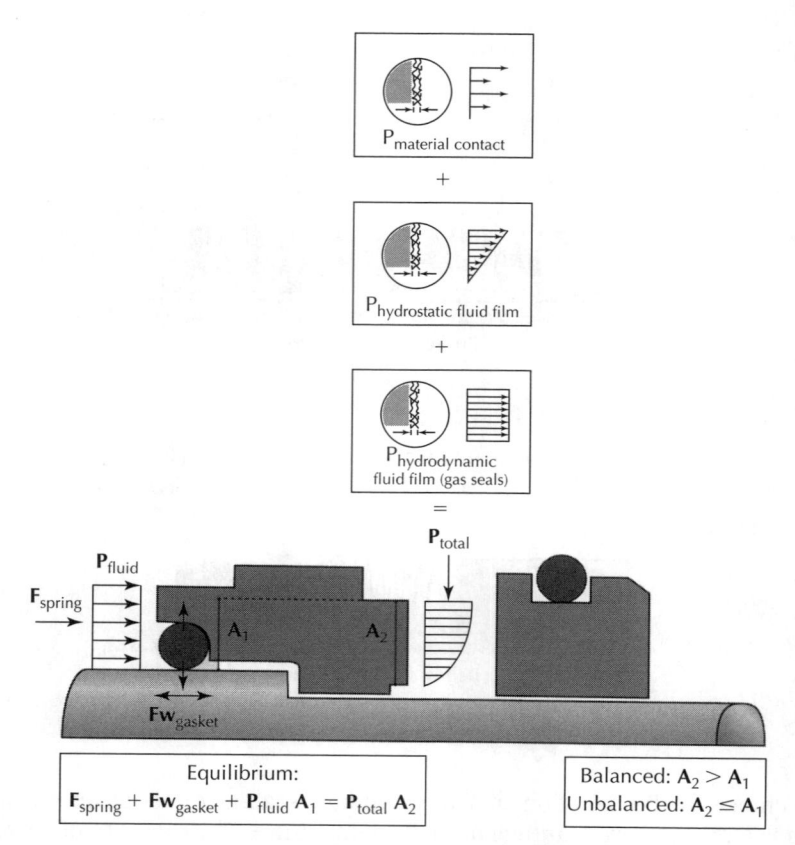

Figure 3.60 Hydraulic forces. (Source: Flowserve)

The closing forces are the spring force and system pressure, F_{spring} and P_{fluid}. The opening forces are more complex and include:

- Mechanical contact between the faces. This will be minimal or zero during normal operation.
- Hydrostatic pressure from fluid that is present between the faces.
- Hydrodynamic pressures generated during rotation.

An additional force which may be in either direction and thus considered as a potential dead-band effect is the friction or hysteresis of the secondary seal. This can be very difficult to quantify.

An important factor in the pressure forces is what is known as the hydraulic balance of the seal. In Figure 3.59 the pusher seal has pressure acting over area A_1 which is reacted by face pressure over area A_2. It can be seen that A_1 is greater than A_2 and hence the face pressure is nominally greater than the system pressure. This is known as an unbalanced seal and will only be suitable for low pressures. In Figure 3.60 it can be seen that the steps in the shaft and seal face give a seal face area that is larger than the pressure area. This is known as a balanced

Rotary seals 163

seal. It allows the designer to both offset the system pressure but also achieve better control of the deflections discussed below.

The hydrostatic pressure in the seal faces is dependent on the actual orientation of the seal faces, Figure 3.61. Assuming that at rest the faces are aligned in parallel radial orientation then for the majority of seal designs any pressure forces, which usually act on the outside diameter, will distort the faces to create a divergent fluid film with most of the load support at the outside diameter close to the sealed fluid. This will tend to create a seal which has minimal film support and may run in a boundary lubricated condition. It must be remembered that the diagrams are greatly magnified and that in reality deflections of a few tenths of a micron are crucial to the correct operation of the seal. Conversely generation of frictional heat at the seal faces will cause a thermal deflection which tends to open the faces to the sealed fluid. The sealed pressure can then penetrate between the faces and provide load support to maintain some face separation.

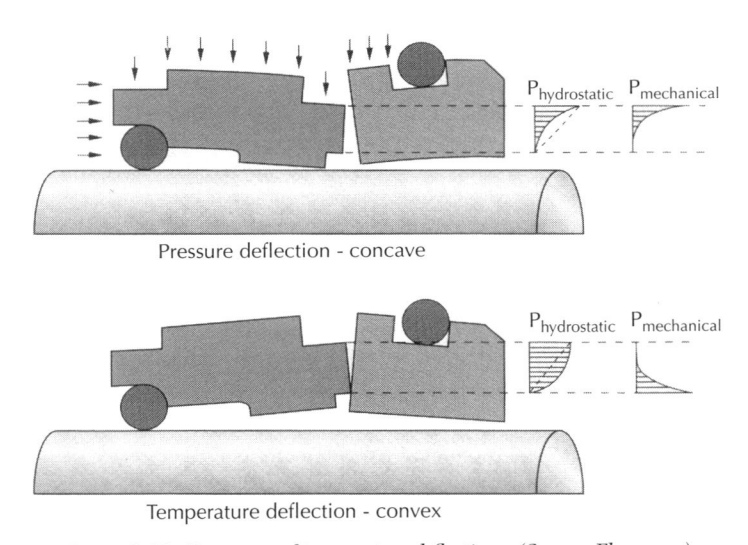

Figure 3.61 Pressure and temperature deflections. (Source: Flowserve)

Both the total deflections and the relative magnitude of the thermal and pressure components are crucial to the successful operation of the seal. The actual values are dependent on a large number of variables including:

- Face geometry.
- Method of mounting and support to react the forces.
- Face thermal conductivity.
- Face elastic modulus.
- Face thermal expansion coefficient.
- Pressure differential across the seal.
- Area of face exposed to process liquid.
- Flow regime of liquid in the seal chamber.

164 *Seals and Sealing Handbook*

- Flow rate of any cooling in the seal chamber.
- Thermal and heat transfer properties of the liquid.

It can be seen that this is a very complex issue with several interdependent variables. It is only since the advent of powerful FEA software that all the relevant parameters could be thoroughly analysed, Figure 3.62. For a thorough analysis manufacturers may now use an integrated package that analyses the mechanical distortions, lubrication conditions and cooling flow around the seal. The behaviour of mechanical seals has probably been the subject of more research effort than any other area of sealing technology, and Lebeck[26] has provided a comprehensive review of this subject.

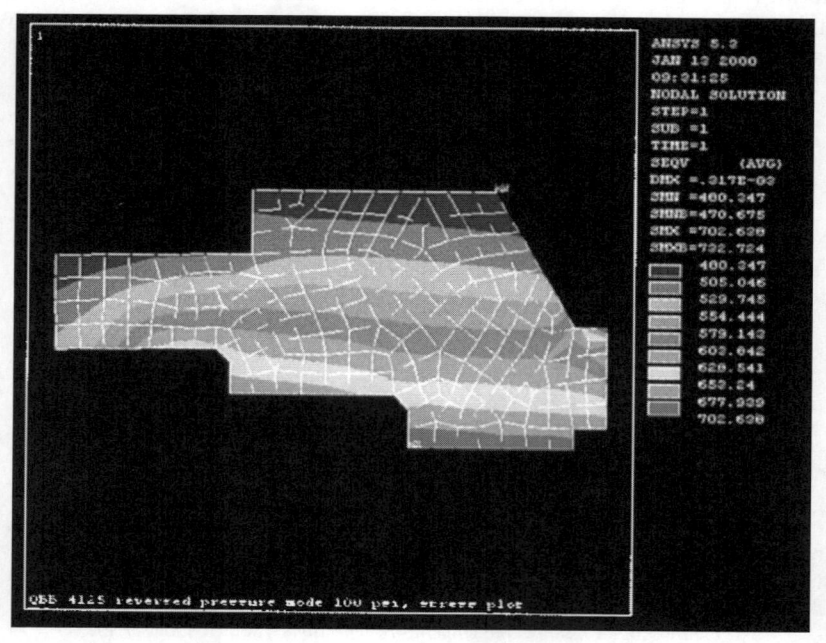

Figure 3.62 Example of the FEA analysis of a mechanical seal face. (Source: Flowserve)

The degree to which a seal relies on hydrostatic load support will depend both on the basic seal type and the intended application. High pressure seals will be carefully designed to deflect sufficiently to provide the desired support but without allowing excessive leakage. A significant contribution to the design of modern seals has been the use of materials such as silicon carbide which have both a high elastic modulus and high thermal conductivity. The deflections due to both pressure and temperature can therefore be reduced by an order of magnitude compared with metals.

In a plain face seal any hydrodynamic lubrication is predominantly generated by circumferential waviness of the seal faces.[27] This may not be present when the seal is assembled but is generated by various factors including discontinuities in the geometry such as drive slots and location holes, uneven spring loading and the method of face retention. It is not a major factor in high pressure seals or those operating on low viscosity fluids but can be significant for low pressure general purpose seals.

There are two classes of seal that depend on hydrodynamic, or aerodynamic, lubrication, those with enhanced or positive lubrication features, section 3.4.4.10, and dry gas seals, section 3.4.4.9.

Most of the seal designs that have been extensively analysed are those for the more arduous duties where high reliability is required. It is probable that most low pressure general purpose seals operate with a combination of boundary and hydrodynamic lubrication, with considerable reliance on the material properties to provide reliability during transient conditions.

Sections 3.4.1 to 3.4.3 have been prepared with assistance from Flowserve, Flow Solutions Division, whose assistance is gratefully acknowledged.

3.4.4 Basic seal types

3.4.4.1 Elastomer bellows, high production volume

The highest production volumes of mechanical seals are those manufactured for domestic washing machines and automotive water pumps. These are surprisingly difficult duties. In washing machines and dishwashers the seal can run dry for extended periods and be subject to sudden temperature changes at the start of a cycle. Automotive water pump seals are subject to an often variable operating cycle and run in a boiling liquid and at higher engine speeds may also be in liquid starved conditions.[28]

The design of these seals is a compromise between price, production engineering of the seal and pump assembly and the expected seal life required. A car water pump seal may be expected to provide 2000 hours' life and that for a truck engine over 10 000 hours.

These seals are usually of the stationary floating face design with a single coil spring surrounding the elastomer bellows. They will also normally be supplied by the seal manufacturer as a cartridge assembly, or in automotive industry terminology a cassette, complete with a sleeve ready to be assembled directly onto the water pump shaft, Figure 3.63.

The face materials have been progressively improved to provide longer life. A moulded carbon is typically used for the softer face on the stationary component. The original concept of using the cast iron impeller boss as the hard face was replaced by an alumina ceramic insert but it is now most common to use silicon carbide for the rotating face. Some heavy duty truck water pump seals will use a pair of silicon carbide faces.[29]

A major factor with the design of these seals is achieving satisfactory cooling to the seal faces, compatible with the high volume seal design. Many designs crimp the elastomer in a shroud with an elastomer sleeve as a seal, but this can inhibit heat transfer. A variety of arrangements are used to achieve an acceptable compromise, Figure 3.64.

This type of seal is manufactured in very high volumes for automotive and similar applications. Often these will be specifically designed for the individual

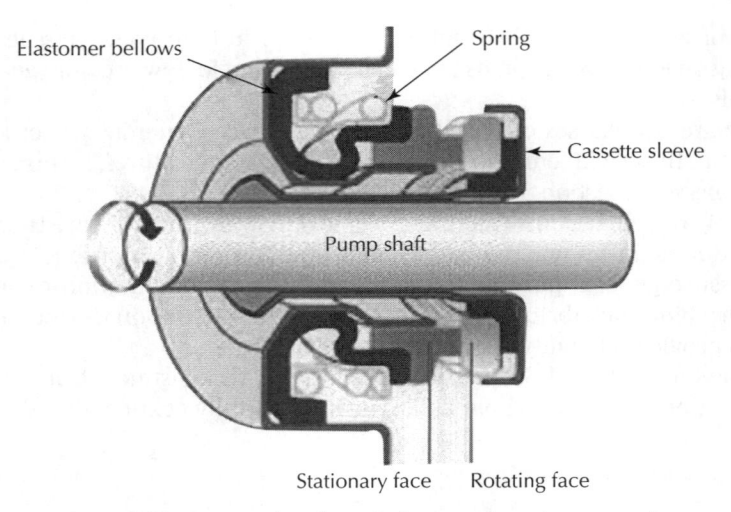

Figure 3.63 Cross-section of a typical automotive water pump seal.

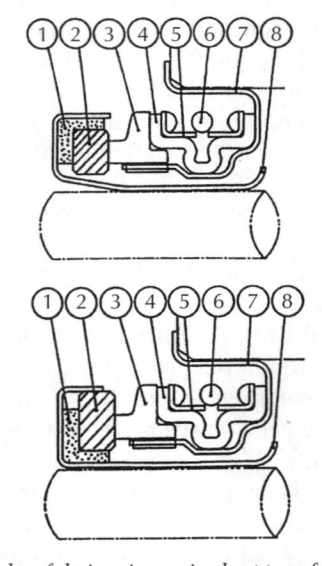

Figure 3.64 Examples of designs improving heat transfer from the seal faces.

customer. However, the basic concept provides a compact and very economic seal that should be considered if a relatively high volume of seals for a low pressure duty on a small shaft in the range 12 to 20 mm is contemplated. This same basic type of seal is found on high volume light to medium pump duties such as swimming pool and spa pool pumps and other domestic or light industrial water pump duties. Depending on the application the seal faces may vary from high duty combination of carbon or silicon carbide against silicon carbide to alumina ceramic against a graphite loaded phenolic for lower temperatures.

3.4.4.2 Elastomer bellows, industrial

Industrial elastomer bellows seals are normally of a rotating spring design, Figure 3.65. They may be of a narrow cross-section design, Figure 3.65(a), which will allow fitting within a narrow bore seal housing, or a more compact length arrangement with the spring on the outside of the bellows, Figure 3.65(b), which provides a shorter axial length. The second type, 3.65(b) permits a shorter overall design of pump providing the seal housing is designed to accept the seal. The torque transmission is normally through the interference of the elastomer bellows on the shaft. This provides both an economic installation with no set screws, etc. and allows rotation in either direction independent of the spring. The metal components of the seal will normally be manufactured by an appropriate volume technique such as stamping or spinning and will only be machined for low volume sizes. Hence the cost of such seals can be very size dependent as high volume sizes will be ready tooled for economic manufacture. This type of seal has been shown to provide acceptable performance with higher amounts of runout than many other basic design types[30] so are well worth considering if the operating parameters permit. The temperature, pressure and fluid resistance are usually limited by the elastomer bellows. Application up to 10 or even 14 bar may be quoted but distortion of the bellows and the fact that these are typically unbalanced seals tends to limit the maximum operating envelope. The introduction of

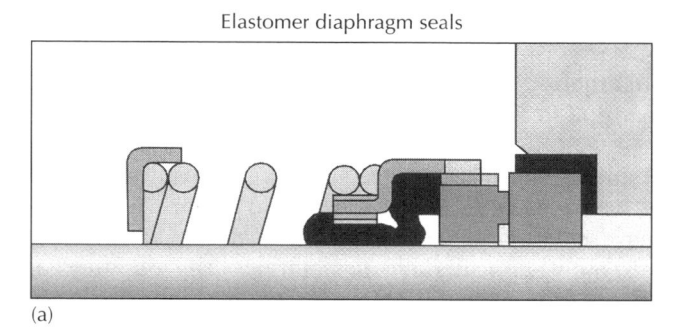

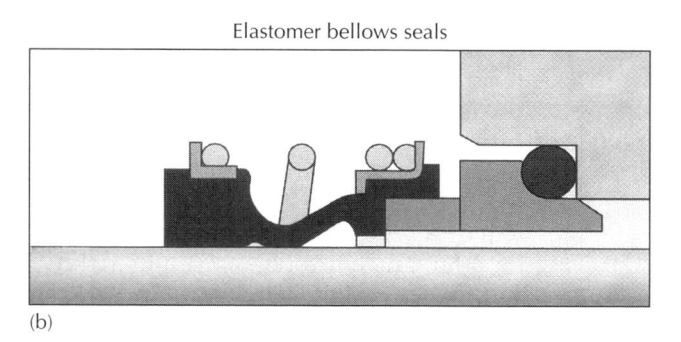

Figure 3.65 Industrial elastomer bellows seals: (a) narrow seal housing type; (b) compact axial length design.

seals with high temperature resistance elastomers has extended the operating range of these seals so that temperatures in the range 150 to 180°C can be contemplated depending on the liquid involved.

These seals will be the primary option for many general purpose industrial duties where the pressure is below 10 bar. A range of elastomer bellows are available to suit the individual fluid resistance required, but the fluid resistance of both the elastomer and the metal will prevent their application in the most aggressive chemicals and special seals are required as discussed in section 3.4.4.6.

3.4.4.3 Pusher seals

This is a general term that is applied to mechanical seals that employ an O-ring or other elastomer or plastic seal as the floating secondary seal. They can vary from simple unbalanced designs that would be used in applications comparable to those discussed for elastomer bellows seals, section 3.4.4.2, up to very specialist high pressure, high duty seals. A general purpose unbalanced pusher seal is shown in Figure 3.90, with hydraulically balanced designs discussed in more detail below. The term pusher derived from the fact that the spring force pushes the seal along the shaft or sleeve to provide the sealing contact. The most common type of secondary seal for the floating component is an elastomer O-ring.

Other seal designs are used, such as U-rings in some seals. Alternative seals, often made from PTFE, were popular as these provided improved fluid and temperature resistance. As seal emission requirements have become more stringent it has been found that the PTFE will not provide sufficient sealing integrity. It has also been shown experimentally that they do not seal adequately in the presence of shaft misalignment.[30] Unless the fluid is very clean they also cause problems with shaft wear due to fretting damage. PTFE seals are therefore relatively rare and have been replaced by O-rings manufactured from the more fluid resistant perfluoroelastomers that are now available. Alternative materials such as graphite rings have also been tried but again do not provide sufficient resilience for the dynamic oscillations experienced by the seal.

Pusher seals allow considerable flexibility to the designer and are often selected for light hydrocarbons, high pressures and high speeds. The seal design can be made relatively strong and the damping provided by the O-ring can be an advantage.

The disadvantages of a pusher seal are the temperature resistance of the O-rings and the potential for the floating action to be inhibited by accumulation of leakage products on the atmospheric side of the seal, often known as hang-up. For higher temperatures a metal bellows will be used, section 3.4.4.4.

There are three general spring configurations, multi spring, single spring and wave spring.

Multi spring designs employ a series of small coil springs around the circumference of the seal, Figure 3.66. The advantages of multi springs include compact axial length, more even face loading, better resistance to centrifugal force and better balance for high speeds. Disadvantages include more prone to clogging by solids in the product, less corrosion resistance from small cross-section springs and short springs reduce both installation setting tolerance and axial

A – Seat/mating ring   B – Face/primary ring   C – Spring
D – Sleeve   E – Retainer
(a)

(b)

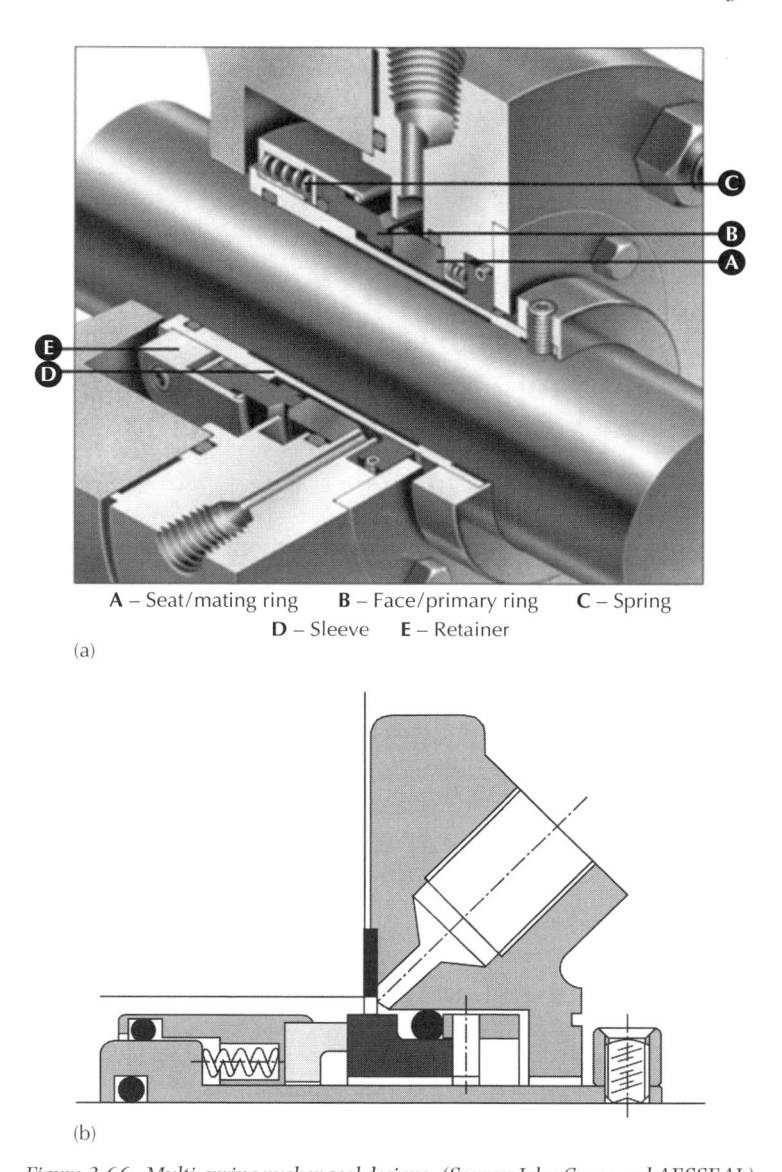

Figure 3.66 Multi-spring pusher seal designs. (Source: John Crane and AESSEAL)

shaft float capabilities. To overcome the problems of corrosion and clogging, some designs arrange the springs outside the product space, but this can demand even smaller springs which in turn further limits the installation requirements.

A large single spring, Figure 3.67, offers improved resistance to clogging and corrosion in the fluid and having a longer compression length is less sensitive to installation variables and can withstand more shaft axial float if fitted to a large or multistage pump. The limitations are the extra axial space required, reduced speed capability due to growth of the spring at high speeds and the uneven spring force

170 *Seals and Sealing Handbook*

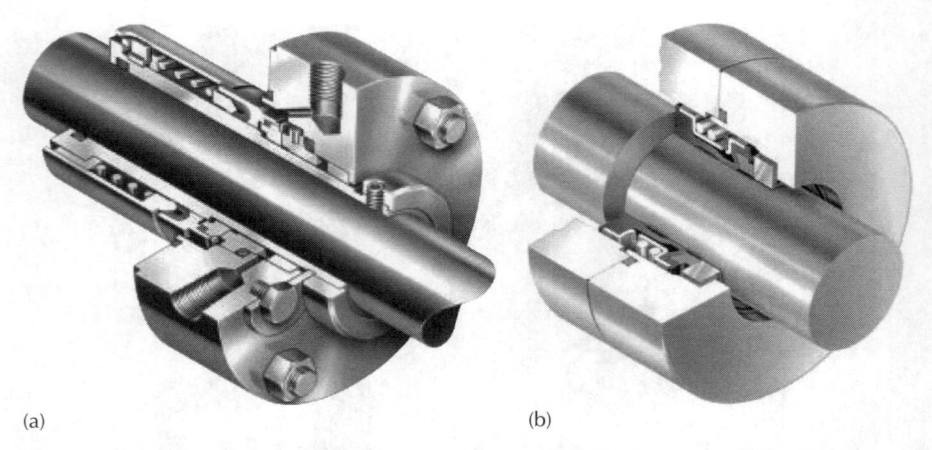

(a)　　　　　　　　　　　　　　　　(b)

Figure 3.67 Pusher seals with a large single spring. (Source: Flowserve and John Crane)

at the end of the coil can create sufficient distortion to make the face loading uneven. This can be a particular problem with applications such as boiling liquids.

Wavy springs provide the most compact arrangement, Figure 3.68, where axial space is limited. The short axial travel gain places a premium on the installation of the seal and they will not be suitable if there is appreciable shaft axial movement. They are widely used to give a compact seal arrangement in aerospace applications and are also used in some industrial pumps where axial space is limited.

Pusher seals cover the widest range of seal designs and applications and different types will be covered in the section on application examples, section 3.4.6.

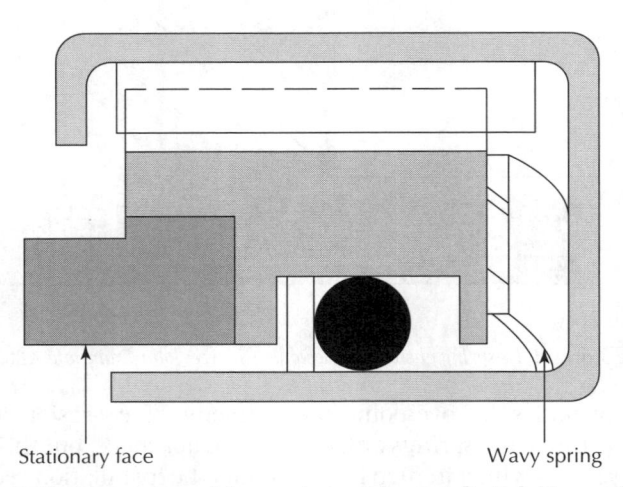

Stationary face　　　　　　　　　　　　Wavy spring

Figure 3.68 Wavy spring seal for use where axial space is limited. (Source: Eaton)

3.4.4.4 Welded metal bellows seals

Metal bellows seals are used to overcome several of the problems discussed concerning pusher designs. The absence of a dynamic elastomer secondary seal eliminates the possibility of 'hang-up', caused by buildup of deposits or seal swell. A rotating bellows has the benefit of 'self-cleaning action', disposing of particles or solids through centrifugal force. As there are only static secondary seals it is possible to use a material such as graphite foil which has higher temperature capabilities and a good range of chemical resistance. As there are fewer components they can also be easier to install.

The most common design of metal bellows seal is the welded bellows type, Figure 3.69. To manufacture the bellows a series of annular metal discs with the wave profile are edge welded alternately on the outside and inside diameters to

A - Set screws
B - Socket head cap screws
C - Compression ring
D - Flexible graphite packing
E - Bellows assembly
F - Insert/primary ring

Sealol® welded metal bellows

(a)

(b)

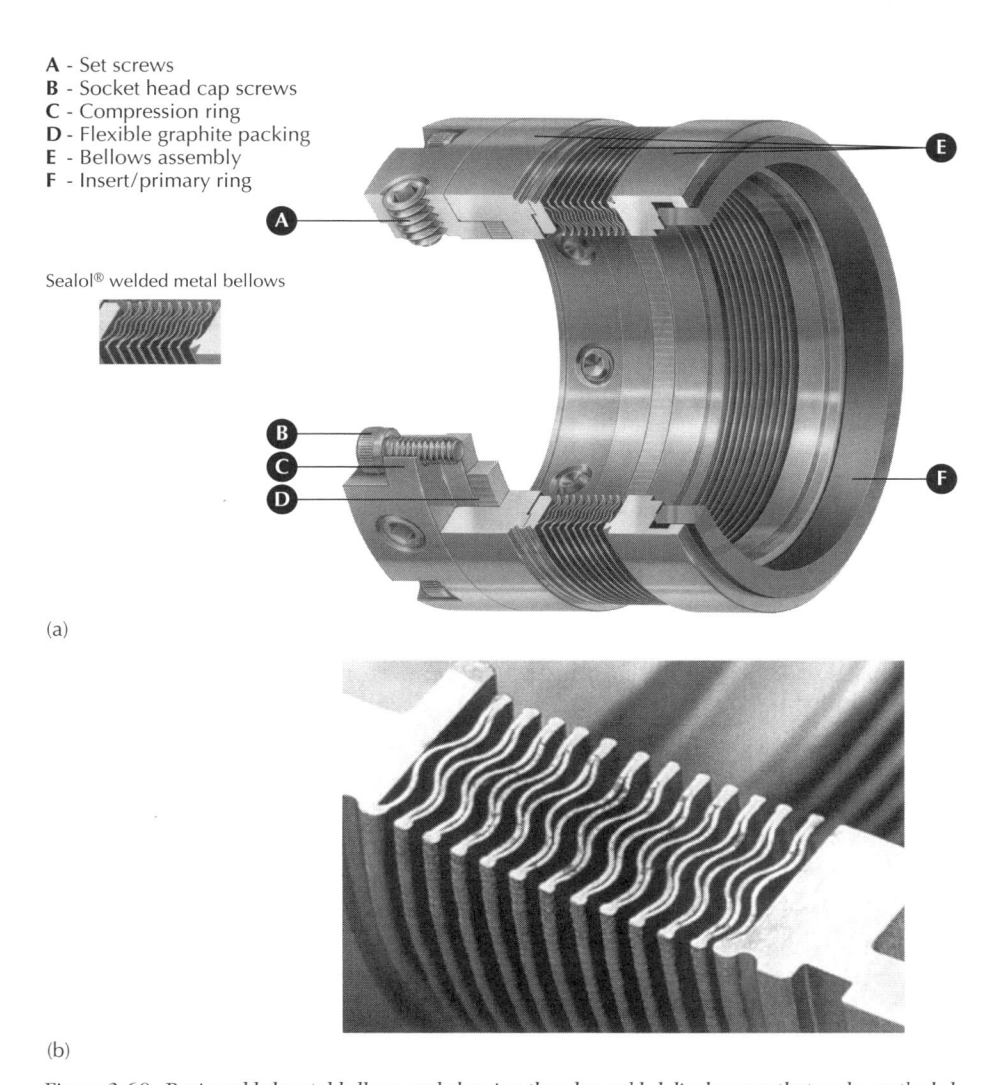

Figure 3.69 Basic welded metal bellows seal showing the edge welded diaphragms that make up the bellows. (Source: (a) John Crane, (b) Flowserve)

172 *Seals and Sealing Handbook*

make up the bellows stack. To provide the necessary flexibility the metal plates are typically about 0.15 mm thick. The welding process is also very critical as it must provide a high integrity emission-proof weld without impairing the metallurgical properties of the diaphragm plates. The detail design of the bellows is critical to reliable assembly and operation. For instance the bellows designed by one of the originators of metal bellows seals, Sealol, now part of John Crane, uses a plate shape called a nesting ripple. It has a three-sweep radius which is designed to allow the bellows to be flexed repeatedly without the metal being stressed beyond its endurance limit. They also have a 45° tilt edge at the bellows inside diameter to disperse stresses.

The bellows are often manufactured from 316 stainless steel, as a standard item, but high nickel alloys are widely used for corrosion resistance and high temperature service. The bellows manufacture is an intricate process, usually with high cost alloys, and a high degree of quality control is required. Metal bellows seals are therefore generally more costly than standard pusher designs, although as volumes have increased bellows manufacture has become more automated and the price differential decreased.

The advantages of bellows seals as mentioned above are high temperature capability and freedom from hang-up. They also have a number of technical disadvantages in addition to the extra cost:

- Limited pressure range. The thin metallic bellows limits the pressure usually to a maximum of somewhere between 20 and 30 bar depending on design.
- Solids deposits in the bellows will create high stress on the welds and can cause failure. A rotating bellows is normally used if solids are present to minimize this problem.
- The floating component does not have the damping, provided by the O-ring material and friction, of a pusher seal. In conditions of low lubricity such as light hydrocarbons or hot water, vibrations can be set up by an unstable fluid film that can cause short-term leakage and in the longer term bellows failure.
- Due to the manufacturing process non-standard sizes are very expensive to manufacture and so a non-catalogue size of seal will probably involve significant extra cost and time compared with a one-off pusher seal.

3.4.4.5 Formed metal bellows

The formed metal bellows was originally introduced to provide metal bellows seal advantages across a wider range of applications. The bellows is manufactured from a tube that is then formed into the bellows. To provide the strength and flexibility required a double ply bellows is used, Figure 3.70. The formed bellows offers similar benefits to a welded bellows by eliminating the sliding secondary seal. A further benefit is that the shape of the bellows gives an open geometry that is not liable to clog with solids. It is therefore applicable to solids laden applications such as pulp and paper, chemical processing, food processing and wastewater treatment. The formed bellows design is also relatively easy to clean and so it is suitable for food applications where in-place cleaning is required. Special food compatible designs are available, Figure 3.71.

Rotary seals 173

A - Seat/mating ring
B - Face/primary ring
C - Drive ring
D - Bellows

Metal bellows technology

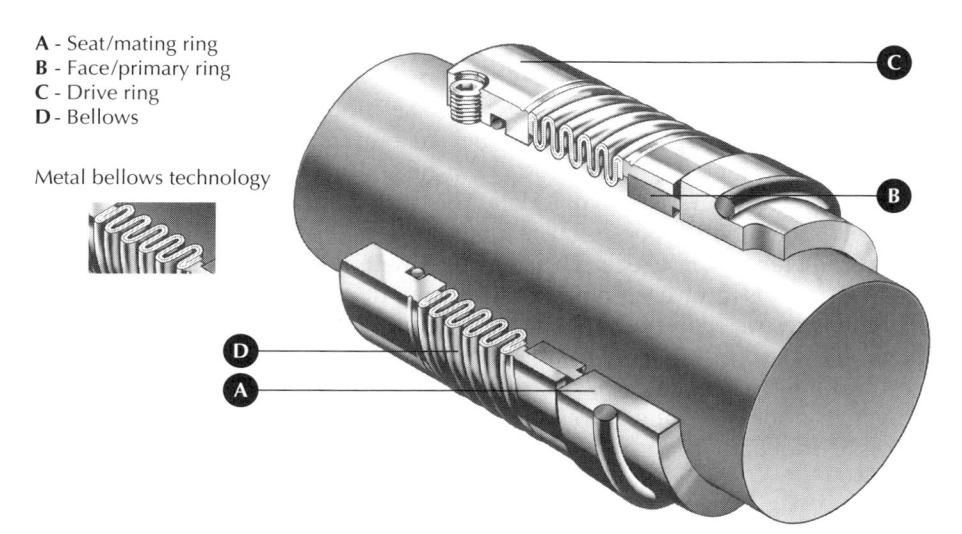

Figure 3.70 Formed metal bellows seal. (Source: John Crane)

Figure 3.71 Hygienic seal design incorporating a formed metal bellows. (Source: John Crane)

Application limits for these seals are lower than welded metal bellows, typically 260°C and 16 bar, although special designs can be obtained up to 30 bar. A limitation, which also impacts on the cost of the seals, is the selection of bellows material. In order to provide a suitable material compatible with the seal duties and also the forming process it has been necessary to standardize on high nickel alloy. One of these seals will also be longer than a welded metal bellows seal, taking up a similar axial length to a single spring seal.

This is a seal design that should be considered as a potential option for applications involving some solids where a smooth profile is an advantage and especially for hygienic duties.

3.4.4.6 PTFE bellows

For highly corrosive duties where contact with any metal has to be avoided then seals manufactured with PTFE bellows can be used, Figure 3.72. The seal is mounted externally so that springs and metal clamp components are remote from the chemical product. It is also necessary for the seal face materials to be manufactured from corrosion resistant materials. The combination of external mounting, PTFE bellows and restrictions on face materials severely limits the potential operating parameters of these seals. The usual face materials are alumina ceramic and carbon. The seals receive very little cooling from the pumped liquid and the pressure limit is restricted by both the outside mounting and the strength of the PTFE bellows. Typical operating limits may be restricted to 8 m/s and 13 bar. These seals also have limited resistance to overpressure which may be caused by pressure surges with rapidly closing valves. Precautions are necessary to ensure that the bellows are not damaged.

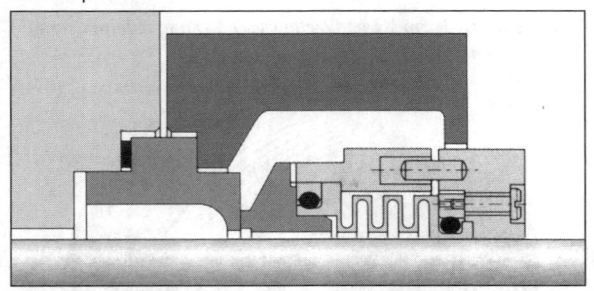

Figure 3.72 Mechanical seal for highly corrosive applications with PTFE bellows. These seals are mounted externally so that any metallic components are removed from the product liquid. (Source: Flowserve)

3.4.4.7 Alternative chemical seals

Pusher seals may also be mounted externally, Figure 3.73, in highly corrosive duties and will provide higher speed and pressure capability than the PTFE bellows design, at the expense of reduced chemical resistance. The face materials and O-ring can be manufactured in highly corrosion resistant materials and these are the prime areas of the seal in contact with the liquid. Any outside seal is still liable to opening if pressure surges occur so this may be a consideration.

To provide an internally mounted seal that can be used in highly corrosive services a range of seals have been developed that have all non-metallic wetted components that are specifically designed for non-metallic pumps. These are available as a cartridge arrangement which is mounted on a carbon filled PTFE sleeve, Figure 3.74. A liner of the filled PTFE also protects the inside face of the gland plate. Using resin impregnated carbon and reaction bonded silicon carbide seal faces together with appropriate fluoro- or perfluoroelastomers this seal is resistant to most chemicals. The O-ring grooves are specifically designed to compensate for the high compression set than can occur with the most highly chemical resistant fluoropolymers, section 5.1.4.17.

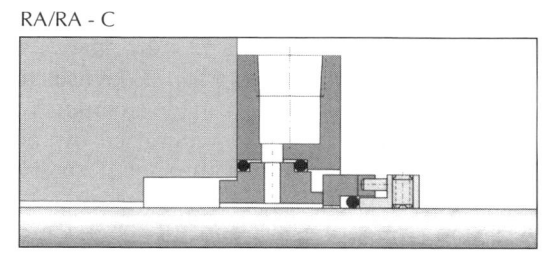

Figure 3.73 External pusher seal for corrosive duties allows higher operating speeds and pressures. (Source: Flowserve)

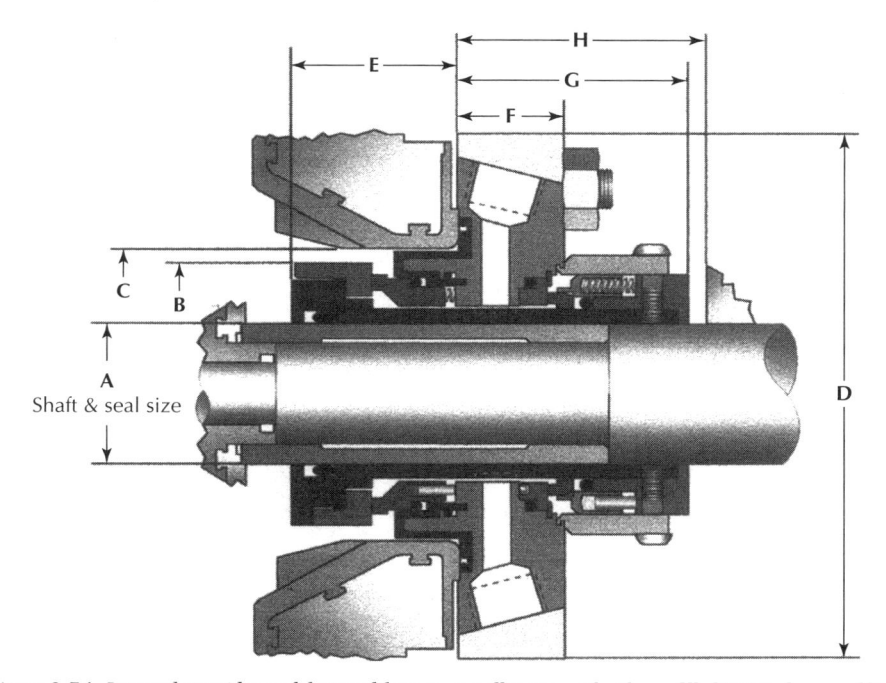

Figure 3.74 Internal cartridge seal designed for non-metallic pumps that has a filled PTFE sleeve and liner for the gland plate. (Source: Flowserve)

3.4.4.8 Bearing seals

Compact versions of mechanical seals are used to seal bearing assemblies in arduous environments where more conventional bearing seals, such as elastomeric lip seals, do not provide either sufficient life or wear resistance. There are two quite different configurations that are designed for very different applications, one for process pumps and the other for earthmoving machinery and similar abrasive environments.

Pump bearing seals
Elastomer lip seals are no longer considered adequate to meet the reliability requirements to seal the bearings of modern process pumps. In order to provide improved

isolation compared with a labyrinth seal, which can 'breathe' to inhale water vapour and exhale oil mist, magnetic mechanical seals can be used. A major factor in many plants is resistance to water jets from wash hoses used to clean the plant.

Magnetically energized mechanical seals provide a compact arrangement that can be considered, Figure 3.75. The seal is energized by magnets, instead of springs, to minimize space. It will fit into the same space formerly occupied by a lip seal or labyrinth. It has a fixed rotary seal face on the shaft, so no shaft fretting damage occurs. The sealing function is provided by the mechanical seal faces, so equipment breathing will not occur and washwater can be excluded. The magnetic seal design also operates equally well in horizontal or vertical applications. A phosphor bronze shroud is incorporated that equally spaces the magnets to ensure uniform seal face loading. As the seal faces are not limited to materials which need to be magnetic, the product can be supplied in conventional mechanical seal face materials. These are antimony-impregnated carbon versus solid tungsten carbide, and tungsten carbide versus bronze-impregnated PTFE.

It is also possible to hermetically seal the bearing chamber if required. A sealed expansion unit, with integral diaphragm, is offered for some applications. This is used to seal the breather port orifice, in the bearing chamber, which is sometimes the other source of moisture ingress. The seal can also be used in dusty environments using an air purge.

Seals of this type can operate in the marginal lubrication conditions found in bearing chambers including flooded, oil mist, greased or dry running.

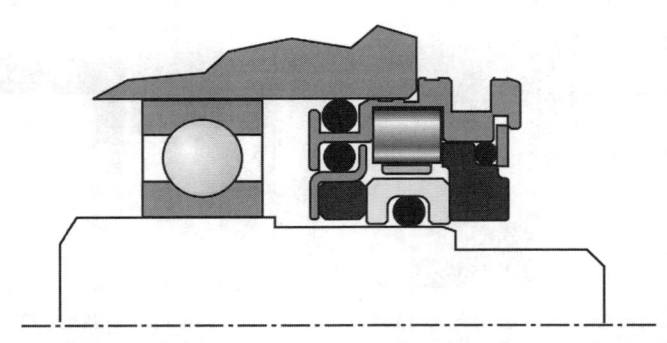

Figure 3.75 Magnetically energized mechanical seal designed to be used as a seal for bearing housings on process pumps in refineries and petrochemical plant. (Source: AESSEAL)

Earthmoving equipment bearings

Where high bearing reliability is required in wet and abrasive laden environments a range of metallic mechanical face seals have been developed. They were originally developed for use on tracked vehicles where effective protection against abrasive media such as dust, sand, mud, stones or earth for the bearings of the tread, support and guide rollers and for wheel or axle hubs is required. The use of these seals has extended to a wide range of off-road vehicles in components where effective sealing of lubricant is required.

A pair of alloy steel seal faces which both have a similar geometry are each supported on an O-ring, Figure 3.76. The lubricant is on the inside of the seal with the

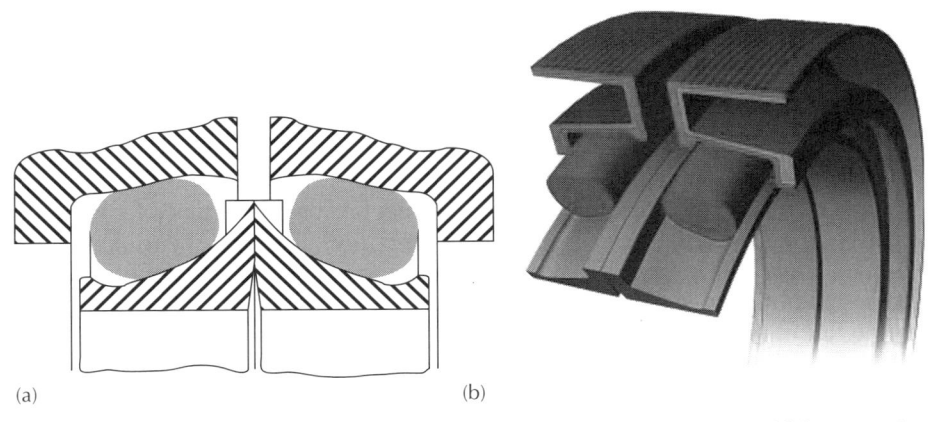

Figure 3.76 (a) A mechanical seal with a pair of steel seal faces used as a bearing and lubricant seal on earthmoving equipment. (b) With pre-manufactured adaptors for press fit. (Source: FTL Seals Technology)

outside diameter exposed to the environment. The seal rings are designed in such a way that they have a narrow parallel contact band at the outside diameter and open out from the seal contact with a tapered clearance towards the centre axis, forming a wedge-shaped gap. This gap has the following functions:[33]

- Encourages lubricant towards the seal faces by capillary action and centrifugal force.
- Provides adequate lubrication and cooling and prevents the potential for cold welding of the seal faces.
- As wear increases, the sealing face graduates towards the centre axis; the seal therefore has large wear reserves which virtually only end when the inside diameter is reached.
- The traces of lubricant that become visible at the outer gap indicate that the seal is working satisfactorily.

The O-ring has a number of functions, it provides the axial load on the metallic sealing faces, transmits the frictional torque to the housing components and provides a static seal between each seal ring and the housing bores.

The seal faces of the metallic rings are lapped to a radial width of about 2.5 mm. Operating limits of these seals are:

- Peripheral speeds of up to 10 m/sec with adequate oil lubrication and cooling flow.
- Speeds up to 3 m/s with grease lubrication.
- Pressure up to 3 bar.

To ensure reliable operation it is necessary to ensure that the frictional heat generated at the seal faces is effectively dissipated by ensuring adequate lubrication. Continuous lubrication is necessary to prevent pick up or galling. These seals are most successful in relatively viscous lubricant such as transmission lubricants.

Corrosion during stationary periods must be avoided. Overpressurization, such as relubrication using a grease gun, must also be avoided.

The correct machining of the housing profile to the specified surface finish is important to ensure correct operation of the O-rings and avoid either slippage or leakage. Where it is not possible to machine the housing profile, adaptor rings can be used as shown in Figure 3.76(b).

A specially adapted version of this type of seal has been developed for the sealing of bearings on down hole rotating equipment, specifically the bearings of rock bits used to drill oil wells. It has been found that the conventional double floating face design of this type of seal does not function reliably in the oscillating pressure regime found down hole. A design with a single floating face, Figure 3.77, has

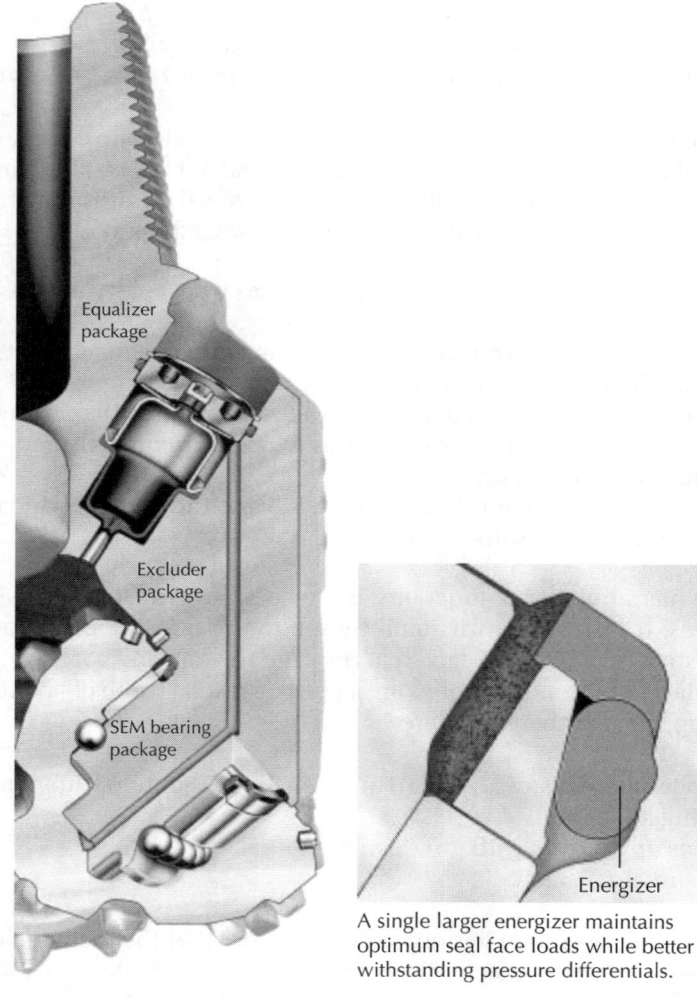

A single larger energizer maintains optimum seal face loads while better withstanding pressure differentials.

Figure 3.77 A modified version of the metal face seal for earthmoving equipment developed to have a single floating face and used for sealing the bearings of down hole rock bits. (Source: Hughes Christensen Co.)

been proved to provide reliable sealing against the ingress of drilling mud and loss of lubricant enabling considerable extensions in the operating life of drill bits.

3.4.4.9 Dry gas seals

Dry gas seals have developed over a period of 40 years. Major developments occurred when machining techniques were introduced that could provide the seal face features that are now employed to develop aerodynamic lift. The majority of new rotary compressors are now fitted with dry gas seals. Applications are also broadening into turbines and as the speed capabilities increase they provide the opportunity to reduce the leakage of gas or vapour in a number of high speed rotary applications.

Gas seals are basically an adaptation of pusher seals that have a series of exceptionally fine laser machined grooves on the outer radial portion of the rotating seal face. The most common form of these grooves is a spiral profile, Figure 3.78. As rotation commences the grooves entrain gas and create aerodynamic lift to separate the faces. The spiral profile will only provide lift in one direction of rotation. Alternatives such as a T groove profile, Figure 3.79, will provide lift in both directions of rotation. The plain face section of the seal face, termed the sealing dam, will provide the static sealing when the seal is stationary and act as the controlling clearance when the seal is rotating. The pressure generated by the grooves is throttled across the sealing dam and the combination of lift and restoring force establishes a seal fluid film which may be in the range of two to ten microns depending on the seal design, gas and speed of rotation. The design of the seal with respect to lift, face distortions and restoring force are crucial to maintaining the few microns of gas film that provides non-contacting operation. The design of the seal to provide a stiff gas film is important so that it can accommodate transient conditions such as pressure changes, axial shaft movements, etc. As the seal face gap closes the force that it will absorb increases, to create an equilibrium, Figure 3.80.[34]

Figure 3.78 Spiral groove pattern on a dry gas seal face. (Source: John Crane)

180   *Seals and Sealing Handbook*

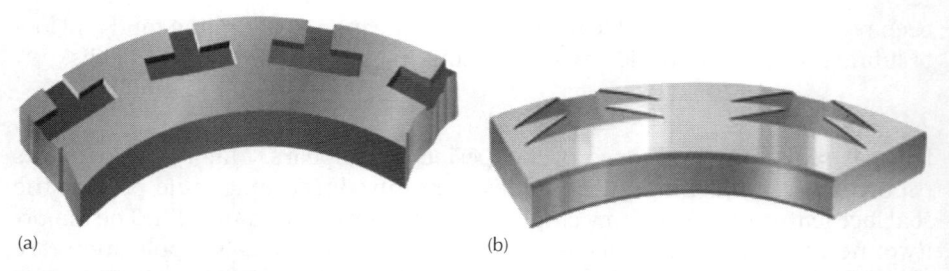

(a)   (b)

Figure 3.79 Dry gas seal groove pattern profiles to provide bidirectional rotation capability. (Source: Flowserve T and John Crane Chevron)

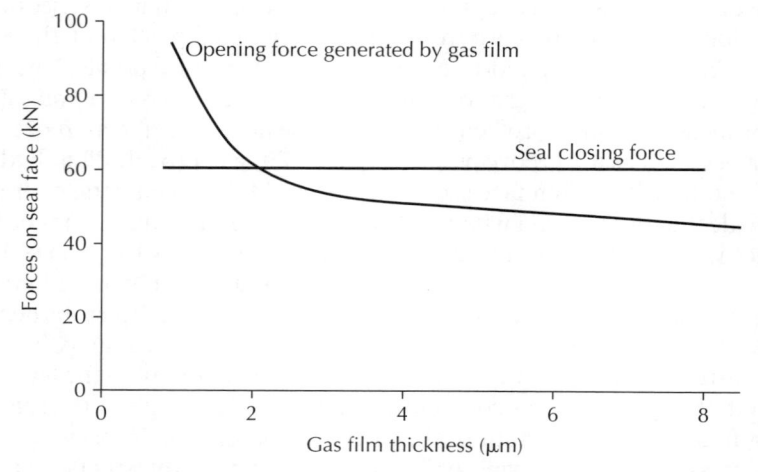

Figure 3.80 Gas film support vs film thickness for a typical dry gas seal.

The selection of the spiral groove design or a bidirectional design such as the T or chevron pattern groove is dependent on a number of factors. Spiral grooves are said to provide a stiffer and smaller gas film giving lower leakage and the capability to deal with transients. T and chevron grooves provide for bidirectional rotation. This is a potential benefit if the machine is likely to be subject to any tendency to reverse roll when stopped and also allow identical seals to be fitted to both ends of a machine. These benefits are very dependent on the actual machine involved and method of operation. The bidirectional design also potentially provides more flexibility in the use of spares.

To enable a dry gas seal to operate reliably under changing conditions the seal's design very carefully takes account of the gas pressure forces, thermal gradients and the restraint of the supporting structure. The cross-sectional profile is normally designed to provide a relatively short axial length and deep radial profile which allows the seal faces to adapt the radial profile to maintain the sealing gap. Seals are available for turbo compressors and turbines that can operate at speeds in excess of 200 m/s and pressures can be over 400 bar. The seals may be used singly in non-hazardous situations but the majority of seals are used in dual arrangements designed for individual circumstances. These are discussed further in section 3.4.6.15.

The use of dry gas seals has also expanded into a number of seal arrangements for liquid sealing. They are used directly as dry gas seals, as a backup seal to a standard liquid seal and also as a gas barrier seal for hazardous or toxic products. These are discussed further in the section 3.4.6.8 on application examples.

3.4.4.10 Enhanced face lubrication

A number of seal designs use face profiles that have been developed in parallel with the dry gas seal technology. There have been designs intended to improve seal reliability in applications where the pumped fluid provides very little lubrication to the faces, such as boiler feed pumps, for many years.[35, 36] The techniques applied included either introducing grooves or cutouts in the outer section of one seal face to introduce liquid to the faces,[35] or to promote the formation of a wavy seal face profile to encourage hydrodynamic lubrication.[36] The advent of detailed design and analysis plus the availability of manufacturing techniques such as laser machining have allowed the effective exploitation of the two basic philosophies. Laser machining is used both to provide micro grooves to seal faces and also to manufacture faces with a machined waviness.[37] Typical examples of both types are shown in Figure 3.81.

Flowserve   Crane

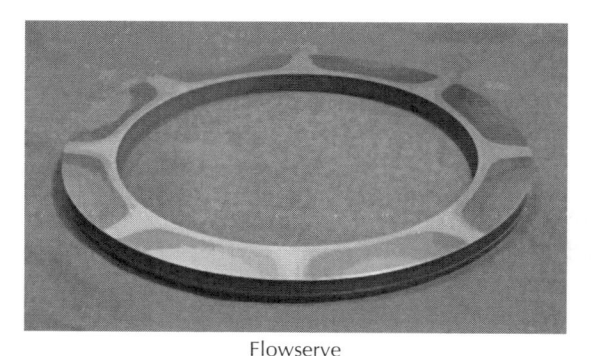

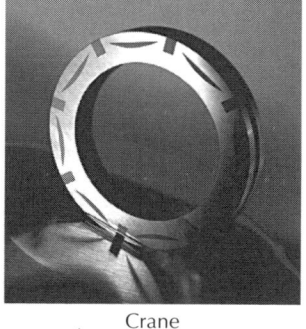

Figure 3.81 Mechanical seals with laser machined faces to enhance the face lubrication: (a) wavy face profile; (b) laser machined grooves. (Source: (a) Flowserve, (b) Crane)

These seals may be used in two quite distinct ways to promote seal lubrication and improve reliability:

- With very low viscosity liquids, especially those that are being pumped close to their boiling point, there is a severe problem of maintaining a stable lubricant film in the seal without either vaporization which can lead to instability and severe damage[38] or boundary contact and subsequent high wear and overheating. These problems are particularly acute on applications such as boiler feed and circulating pumps and with light hydrocarbons. In these cases designs such as those shown in Figure 3.81 will promote lubrication of

the seal face and provide significantly improved performance.[37] The leakage will be expected to be slightly higher than when a plain face seal is used but on low viscosity liquids will still be extremely low. This type of seal will normally be used on a clean duty.
- For arduous duties with the potential for solids in the pumped product, or other potential problems such as solidification of the product around the seal faces an alternative technique may be used. The laser machined lubrication features are provided at the inside diameter. They are then used in a dual seal arrangement to pump a clean fluid into the seal interface. This provides a clean, cool fluid to lubricate the seal faces and exclude the pumped product. Figure 3.82 shows how this arrangement operates. The buffer liquid pressure is provided only by a static head and is significantly below that of the pumped product. Seals of this type have now been developed to generate significant pressures across the seal face to enable them to operate in high pressure pump duties.[39] A particular feature of this type of arrangement is the ability to provide a consistent lubricant film to the seal faces across the full range of pump operating duties and is better able to survive transient events such as running against closed valve conditions.

The advantages can include:

- The technology is non-contacting and therefore the usual operating limits of contacting seals do not apply.
- The power consumed is lower than a standard dual arrangement.
- The positive flow of clean fluid into the seal chamber provides a cleaner sealing environment within the seal chamber.

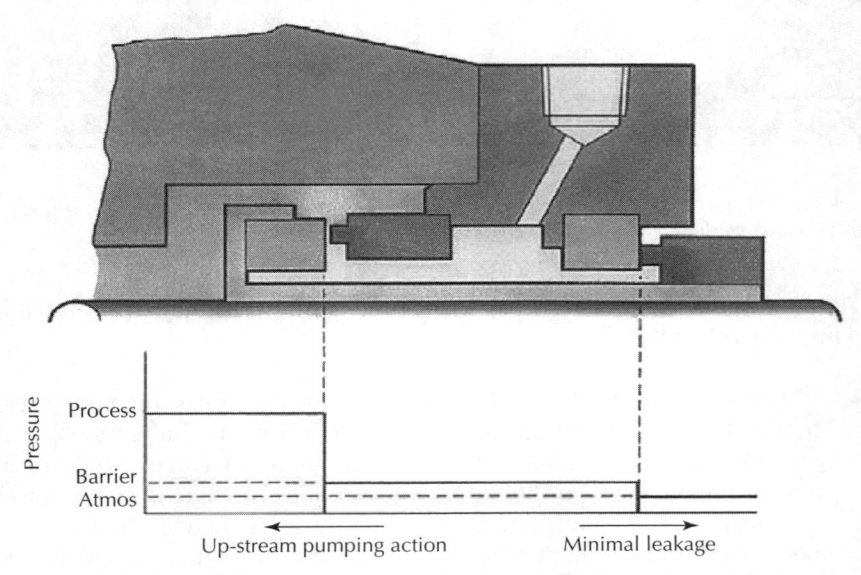

Figure 3.82 Active seal lift to promote lubrication by the buffer liquid. (Source: John Crane)

- In services where the process pressure is variable, or where pressure spikes are likely, active lift constantly regulates against this varying pressure.
- Seal leakage to atmosphere is significantly reduced when compared to a pressurized dual seal.
- An upgrade of a single or multiple seal is feasible when process changes have shortened seal life.

3.4.5 Seal housing design

The initial application of mechanical seals was in the narrow parallel bore housings designed for packing. On a typical pump this would allow a radial space of 9 to 12 mm. However, even in the early days of the use of mechanical seals, it was realized that a stuffing box was not necessarily the optimum geometry to obtain the best performance from a mechanical seal and alternatives were proposed, and in some cases incorporated into pump designs.[40]

The constraint of the narrow parallel bore creates two problems. First, it restricts the design of the seal to a slim profile and severely inhibits the opportunities for the seal designer to optimize the seal geometry with regard to achieving optimum face presentation together with satisfactory spring and torque transfer arrangements. Second, the liquid flow around the seal is very limited which inhibits cooling.

A comprehensive research project to investigate a wide range of housing designs has been carried out.[16–19] The variables included parallel bores of varying radial clearance around the seal, positive and negative tapers, and combination taper and parallel bores. In addition, a variety of flush and dead-ended arrangements with and without neck bushes were included. To ensure that the results were broadly applicable, representative examples of the commonly used seal geometries were included.

This work showed that the narrow confines of a stuffing box could cause a stagnant region that not only impaired heat transfer and dispersion of gas, but could actually entrain gas and vapour into a vortex close to the seal faces with most seal designs. An example is shown in Figure 3.83. A further problem with many typical balanced seal designs is the step in the outside diameter to provide the nose of the balanced seal face. This creates a pocket that can entrain a vortex of predominantly vapour to inhibit heat transfer, Figure 3.84.

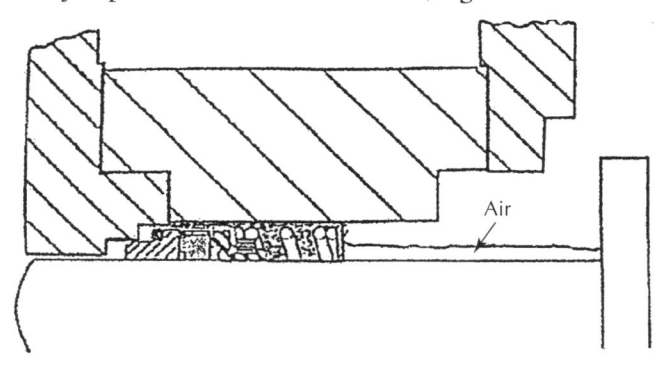

Figure 3.83 Narrow bore seal housing entrains vapour without circulation. (Source: BHR Group)

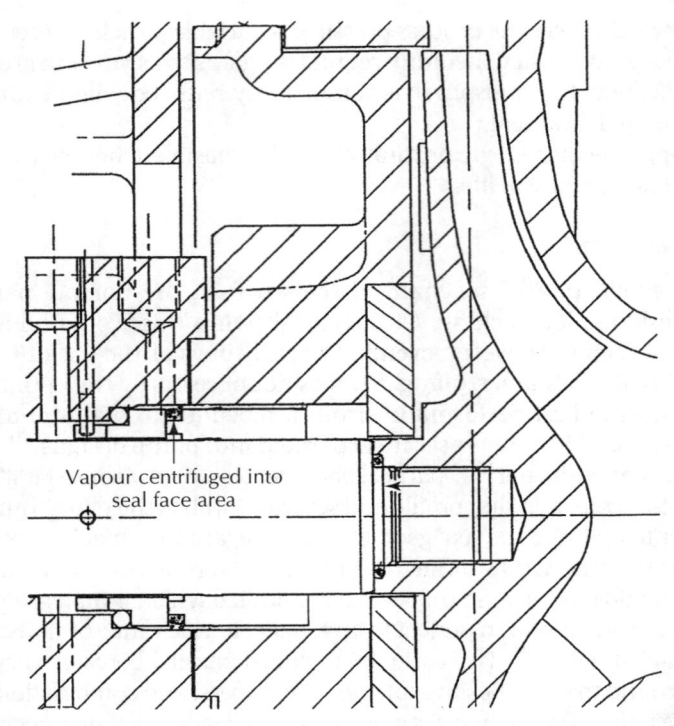

Figure 3.84 Seal outside profile can cause entrainment of vapour close to the faces. (Source: BHR Group)

The only effective method of removing this entrained vapour in a parallel bore is to provide an adequate circulation that will disturb the vortex and create an axial flow in the seal housing. Both the circulation quantity and position of the ports in the seal plate can be important. The seal design also becomes a factor, as a result of the small radial gap between the seal and the housing wall. A single exposed spring design can act as a viscous pump, and thereby help to entrain additional vapour around the seal face.

A larger radial clearance around the seal did not appear to be as beneficial as expected. Either a single or a number of very stable, cylindrical, Taylor vortices could form in the seal chamber, Figure 3.85.[42] These could often require a

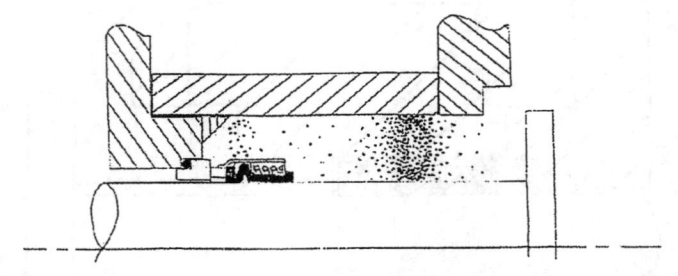

Figure 3.85 A wide parallel bore can create stable toroidal vortices that can lead to erosion. (Source: BHR Group)

considerable circulation flow to disturb them. Circumstantial evidence has supported the view that these were not desirable, as heavy erosion of seal chambers had been experienced. A wide variety of tapered housings were also tested. In common with other work, these were seen to be beneficial in helping with the removal of both vapour and solids. However, it was still found that under some circumstances a stable vortex could form in the housing, and gradually accumulate vapour. It was found that introduction of one or more axial strakes, often known in the industry as a 'vortex breaker', would cause sufficient disturbance to oscillate the vortex and assist with the removal of vapour from the seal chamber, Figure 3.86. Testing on an actual pump installation[44] helped to confirm the findings of the test rig work, but also illustrated that the impeller and pump geometry is a further significant variable. Balanced impellers worked more successfully than a number of observers had expected, as they can often allow the efficient flushing of vapour from the seal chamber. Other designs, such as back vanes, did not always create the level of circulation behind the impeller that had been expected. More recently computational fluid dynamics (CFD) work[45] has provided similar results to the earlier experimental work.

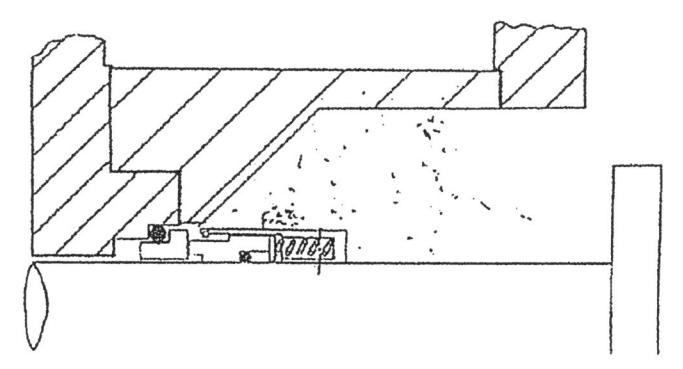

Figure 3.86 A tapered seal housing which has a cone starting close to the seal faces can provide the optimum circulation and dispersion of vortex flow. (Source: BHR Group)

Progressively seal chamber design has evolved to provide an improved environment for the seal. A number of the pump standards now specify a much larger radial space or in some cases a tapered housing. A number of devices have also been introduced that promote flow in the seal chamber, as demonstrated in Figure 3.87. The use of vortex breaker devices is extremely varied and can be application dependent. Most research suggests that a number of small devices in the seal chamber provide the optimum dispersion without causing gross turbulence. Some seal manufacturers provide a seal housing insert to help avoid the formation of erosive vortices, Figure 3.88.

This is an area that still requires some further investigation to provide optimized guidelines. The potential pump purchaser should consider that the selection of an optimum housing can save the additional capital and running costs of providing a circulation flow to the seal housing.

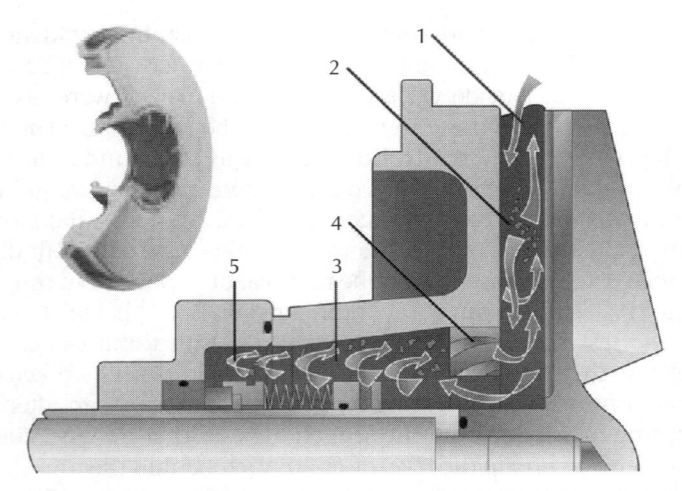

Figure 3.87 Parallel bore with a flow inducing device. (Source: Gould Pumps Inc, ITT Corp.)

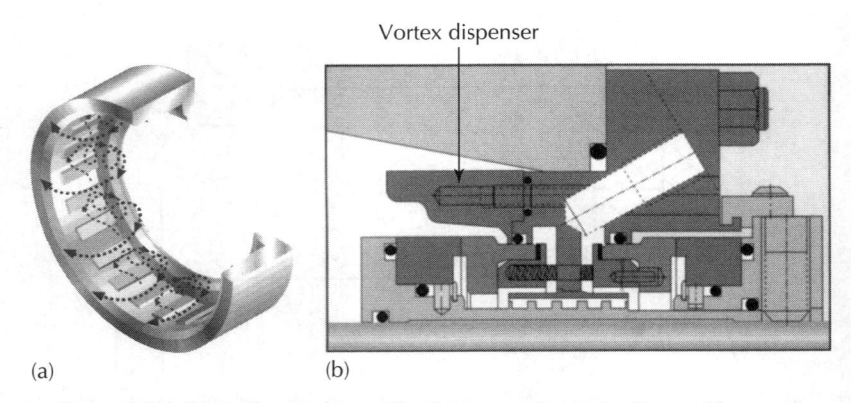

Figure 3.88 Vortex dispersing insert for slurry pump housings. (Source: Flowserve)

3.4.6 Typical applications and seal arrangements

3.4.6.1 Dead ended single seal

Dead ended is the general term used for seal installations where no external flush or cooling supply is provided to the seal. All the cooling and dispersion of any solids is achieved entirely by the internal flow in the pump seal housing, Figure 3.89. This type of arrangement is used where the pressures and temperatures are moderate, up to potentially 15 bar and 150°C maximum. It is also necessary for the pumped liquid to have adequate heat transfer properties and be neither a fire hazard nor toxic.

Such single dead ended arrangements are widely used on water pumps. These can vary from automotive water pumps, as discussed in section 3.4.4.1, to many industrial and building services water pumps. Many of these seals will be elastomer bellows designs. For small sizes, up to 19 or 25 mm they are most

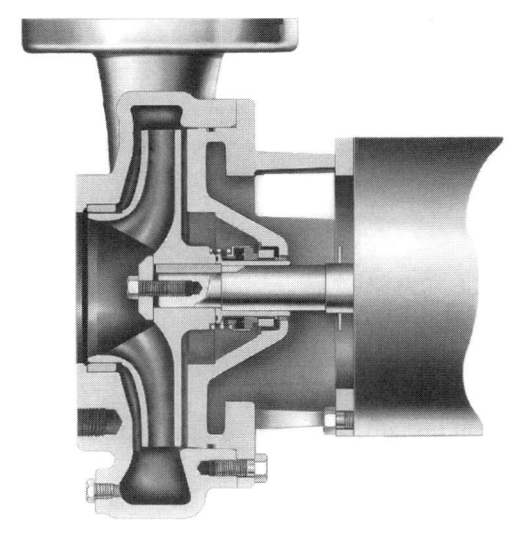

Figure 3.89 Typical dead ended seal arrangement on a general purpose water pump. (Source: Gould Pumps Inc, ITT Corp.)

likely to be produced in high volumes and be a stationary bellows design comparable with the automotive seal types.

For larger pumps the seals may be of the industrial elastomer bellows design, section 3.4.4.2. These are used very widely for applications such as building services where they may be used on hot water up to 10 bar and 150°C. To achieve long-term reliability with a dead ended arrangement the design of the pump and specifically the seal housing is critical. A pump design which ensures good circulation around the seal and especially thorough flushing of any vapour buildup from around the seal will help to promote a reliable seal operation, section 3.4.5. It is important to note that a narrow parallel bore seal housing as used for packing, if used dead ended, will entrain vapour that will not be readily dispersed.[41] For high temperature operation it is also necessary for the liquid around the seal to be sufficiently above boiling point to avoid vaporization around the seal or instability in the seal faces. The pump design and particularly the impeller type can change the seal housing pressure from close to suction to approaching that of the discharge pressure and this must also be considered.

A pusher seal, such as a simple unbalanced design, Figure 3.90, may also be used for these duties. The disadvantages of a pusher design can be a positive drive such as a grub screw is usually required whereas the bellows seal may use the grip of the bellows on the shaft, and the long-term ability of the O-ring to slide on the shaft compared with the flexing of a bellows. If the application demands a high duty elastomer this can be a very significant cost factor with an elastomer bellows in which case a pusher seal may be selected. An elastomer bellows seal requires moulds for the bellows and is usually based around high volume metal forming which means that considerable investment is required before manufacture. As the pusher seal is manufactured from turned components it is more viable for lower volume applications.

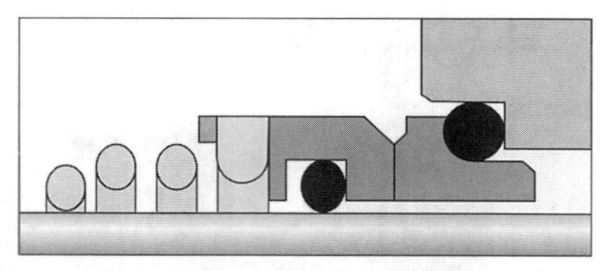

Figure 3.90 Simple unbalanced pusher seal for general purpose duties as an alternative to an elastomer bellows.

At higher pressures and temperatures a balanced pusher may be used to provide higher performance without circulation. This will be very dependent on an adequate design of seal housing. Such arrangements are permitted for refinery duties[46] but are not currently widely used.

3.4.6.2 Cartridge assembly

For many years mechanical seals were provided as a set of component parts, Figure 3.91, and the user would assemble the seal into the pump or other machine to be sealed. Having lapped faces flat to less than a micron and with performance dependent on clean assembly and correct installation of springs, O-rings and other components this was left to the pump supplier or maintenance fitter. There is the further problem – it is not possible to pressure test the seal until after it is installed in the machine, which can then involve an expensive rebuild if some damage occurred on assembly.

It is now common practice for seals to be supplied as a cartridge unit, Figure 3.92. The complete seal is assembled on a sleeve together with the flange for mounting on the pump, often known as the seal plate. The complete seal assembly is carried out by the seal supplier, or by a specialist service workshop, and can be pressure tested before delivery. The supply and installation of appropriate face and O-ring materials, etc. for the duty can be certified by the seal supplier.

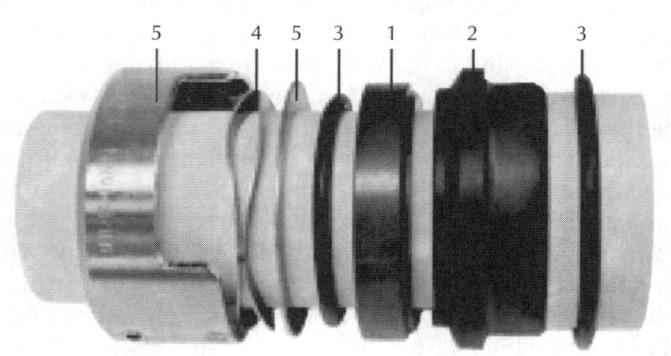

1 – Rotating seal face   2 – Stationary face   3 – Secondary seals
4 – Spring   5 – Other metal components

Figure 3.91 Individual seal component parts ready to be assembled by the user. (Source: Burgmann)

Rotary seals 189

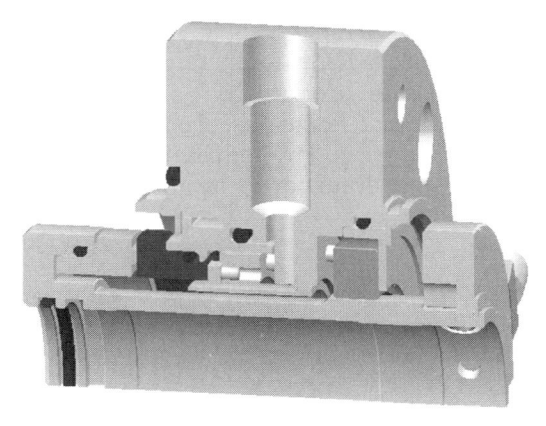

Figure 3.92 Typical single seal cartridge assembly as supplied directly by the seal manufacturer ready for direct fiiting. (Source: AESSEAL)

Some form of retaining clips are fitted to locate the sleeve and seal plate such that the correct spring or bellows compression is achieved.

End users have usually reported a significant improvement in seal reliability when converting to the use of cartridge seals which more than offsets the higher initial purchase cost. Cartridge seals are mandated for refinery pumps by ISO 21049/API 682[46] as part of the emphasis on achieving high reliability service.

They are also used by the automotive industry for water pump seals, Figure 3.93, where they are more familiarly known as a cassette. In this case the reduced assembly time and operations will also be a significant factor. The reduced assembly time is also a factor for industrial seals as less production or maintenance staff time will be involved compared to assembly with individual seal components.

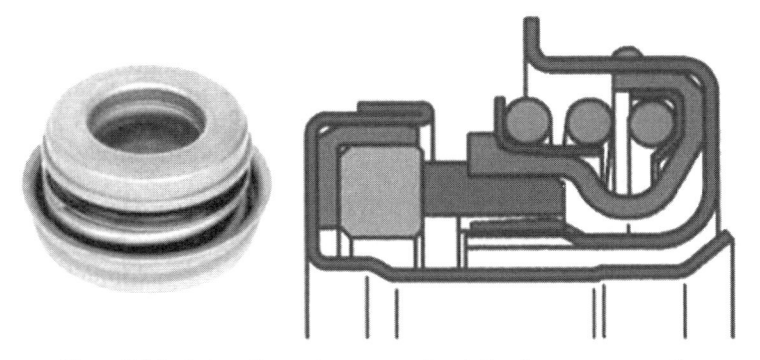

Figure 3.93 Automotive water pump seal cartridge, known as a cassette.

Cartridge arrangements do present some challenges to the seal manufacturer, which must also be considered by the pump manufacturer. A fairly obvious potential problem is that there must be sufficient radial space to allow for a sleeve as well as the seal. This is where the narrow parallel bore provided for packing is a particular problem. Providing a seal on a sleeve and still leaving some radial gap for liquid circulation can create a radially shallow seal that is difficult to optimize. Providing a seal plate that has facilities for bolting to a wide range of pumps and also providing the ports for a range of auxiliary services can create considerable packaging problems. A standard design of seal plate for a single seal will usually be provided with four drilled ports. Two for circulation to and/or from the seal housing plus quench flow and drain ports external to the seal, Figure 3.94.

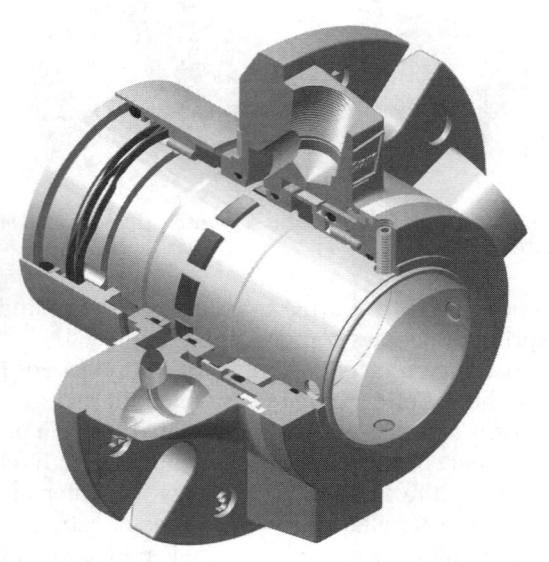

Figure 3.94 Seal or gland plate for a cartridge seal showing the cutouts for a variety of bolt arrangements and the ports for auxiliary pipework services. (Source: John Crane)

3.4.6.3 Single seal plus circulation

As operational parameters create higher temperatures or an increase in vapour around the seal it is necessary to increase the fluid flow and cooling. The contributory factors include increases in temperature, pressure or speed and alternatively if the pumped liquid has less heat transfer capability. Hence mineral oil is likely to create higher temperatures than water even though it may have superior lubrication characteristics.

The first step that is taken to alleviate the situation is to provide some additional circulation of the pumped liquid around the seal faces. The possibility of designing the seal housing to accomplish this has been discussed in section 3.4.5. The more traditional method is to provide a supply from the pump discharge, either from the back of the impeller or in the discharge pipe, and circulate this to the seal housing, Figure 3.95. It is possible to provide such a facility internally within the pump and this has been used both on industrial pumps and also on diesel engine coolant pumps,[47] Figure 3.96.

Circulation can also be taken from the seal chamber and circulated back to the pump suction. This can have a different effect as this will reduce the pressure around the seal to something close to suction pressure whereas circulation from discharge will raise the pressure. The actual implications will also depend on

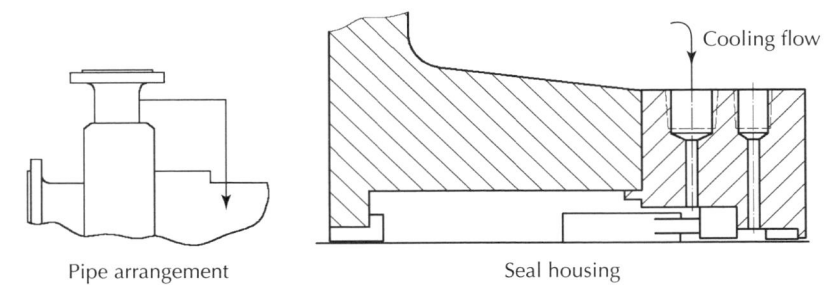

Figure 3.95 Circulation from pump discharge to the seal housing to assist with cooling the seal.

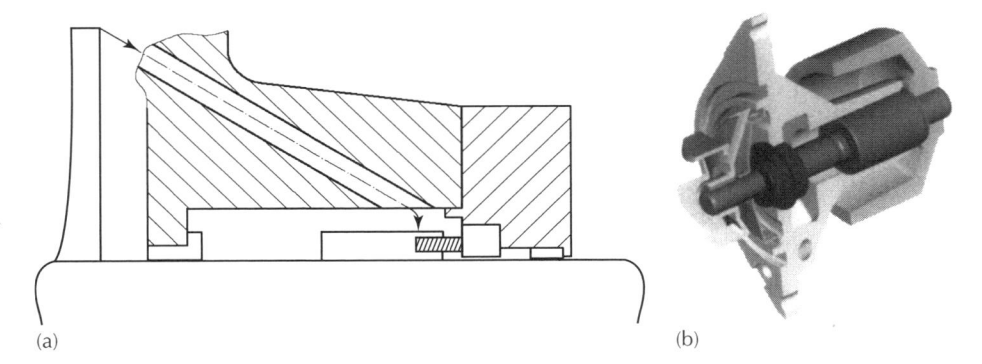

Figure 3.96 Internal circulation of high pressure from impeller o.d. (Source: Sealing Technology)

the impeller type as this will affect the seal housing pressure. This arrangement is particularly recommended for vertical pumps as it can help to ensure that any gas or vapour buildup in the seal housing is rapidly removed on startup, Figure 3.97.[46]

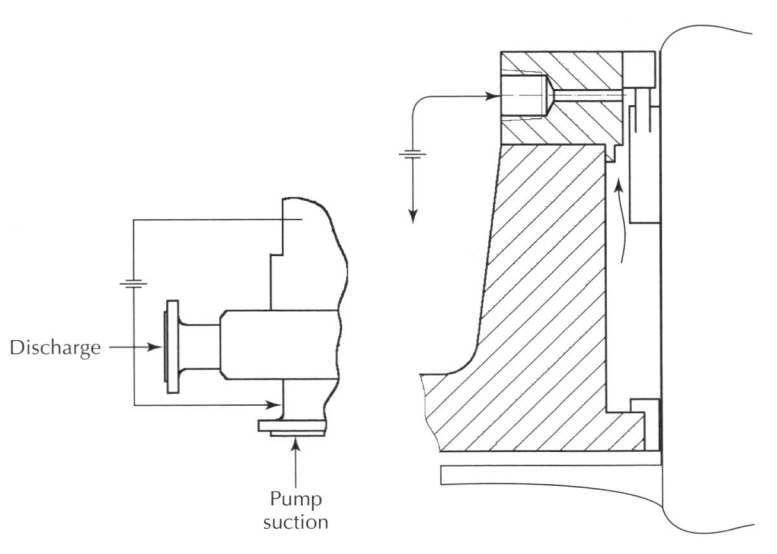

Figure 3.97 Vertical pump with circulation from seal housing back to suction, API Plan 13.

It is important to take into consideration that any circulation will be a diversion of a proportion of the pump flow and as such a loss of efficiency. Some throttling is therefore necessary to avoid excessive flow. Design of the porting is also important to ensure that adequate flow close to the seal is obtained to disperse vapour and provide cooling without causing erosion of the faces or side loading to the floating component.

Direct circulation of the pumped liquid will provide cooler running and assist with vapour dispersal. This will not be sufficient in cases of high temperature, abrasive solids or for viscous solids. A variety of additional auxiliary facilities are then required as described below.

3.4.6.4 Single seal plus circulation plus auxiliaries to API plans

A wide variety of auxiliary facilities are used to improve the environment in the seal housing to facilitate reliable operation of mechanical seals. The most comprehensive description of these is provided in the industrial standards for refinery seals[46] where the various configurations are referred to as 'plans', each having a unique Plan Number. For full details it is necessary to refer to the complete standard. Some examples are given below.

Plan 12, Figure 3.98, involves recirculation from the pump discharge through a strainer and flow control orifice to the seal. The purpose of the strainer is to remove occasional particles. Strainers are not normally recommended because blockage of the strainer will restrict or shut off the circulation and cause seal failure. This arrangement is not recommended to achieve high reliability because of potential problems with blocked strainers.

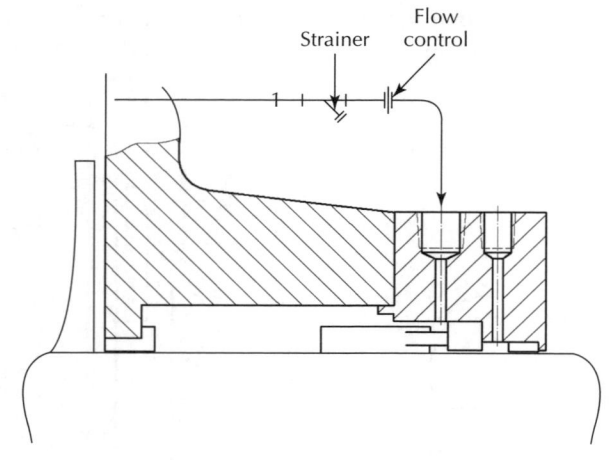

Figure 3.98 Plan 12.

Plan 13 is designed specifically for vertical pumps.

Plan 14, Figure 3.99, is the provision of recirculation from pump discharge through a flow control orifice to the seal and simultaneously from the seal chamber, with additional flow control orifice if required, to the pump suction.

This allows fluid to enter the seal chamber and provide cooling while continually venting back to suction and so it will reduce the pressure in the seal chamber compared with Plan 11 or 12.

Plan 21, Figure 3.100, is used where the temperature in the seal chamber is unacceptable and provides recirculation from pump discharge through a flow control orifice and cooler, then into the seal chamber.

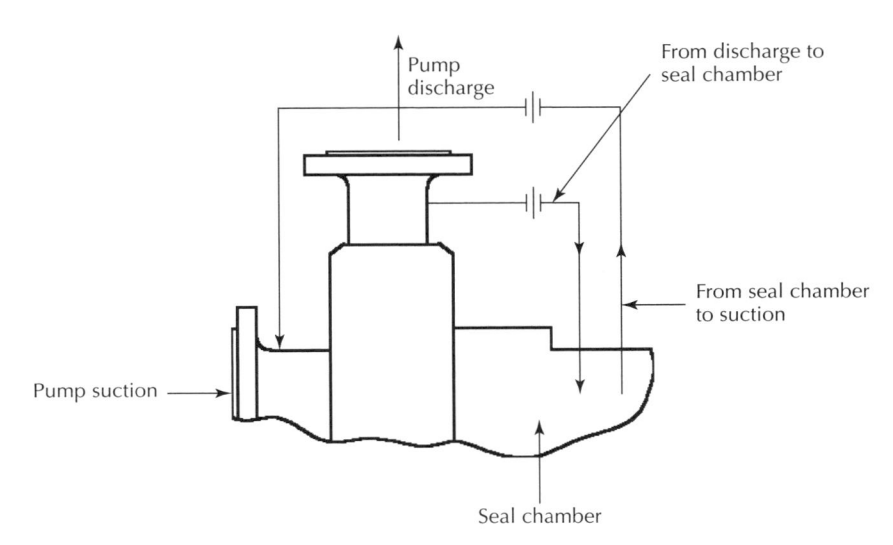

Figure 3.99 Plan 14.

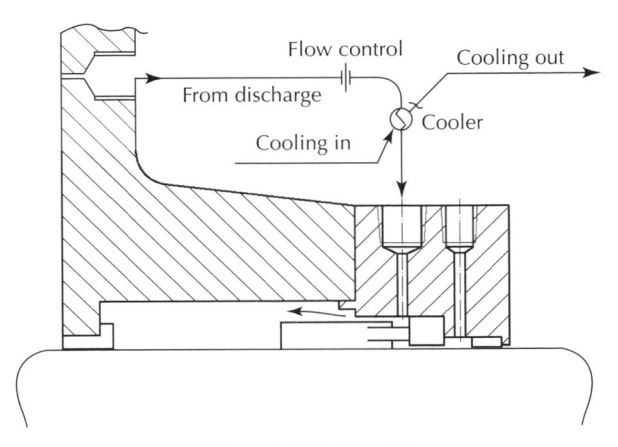

Figure 3.100 Plan 21.

Plan 23, Figure 3.101, is used for high temperature and usually high pressure services where the pumped liquid is well above atmospheric boiling point. Recirculation is provided from a pumping ring in the seal chamber through a

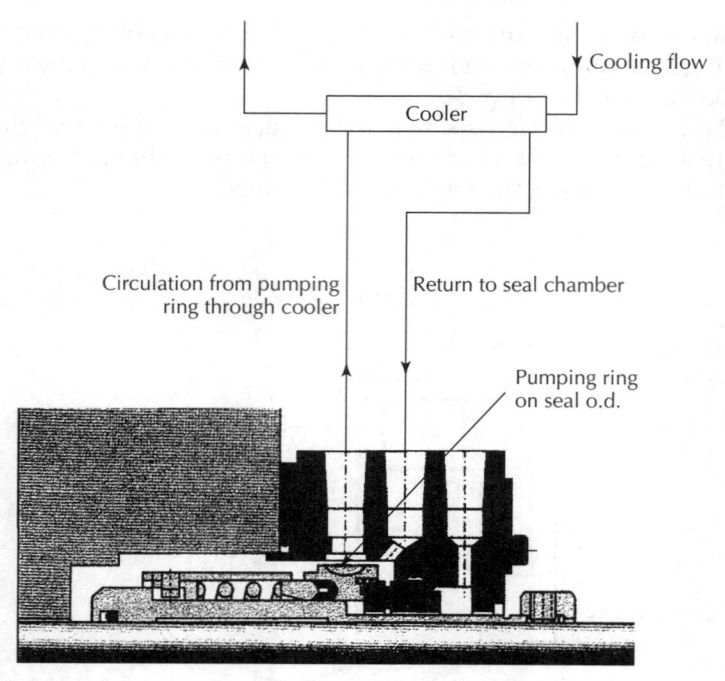

Figure 3.101 Plan 23. (Source: Flowserve)

cooler and back into the seal chamber. A close clearance bush or other restrictor is usually provided in the neck of the seal chamber to minimize liquid flow between the pump volute and seal chamber. This minimizes the heat load on the cooler by cooling only the small amount of liquid that is recirculated and enables the seal temperature to be well below the pumped liquid temperature. This arrangement is widely used on boiler feed water duties to reduce the seal operating temperature to well below the atmospheric boiling point of the water.

Where the product contains abrasive solids, Plan 31 may be used, Figure 3.102. Recirculation is taken from the pump discharge through a cyclone separator. A cleaner fluid is then delivered to the seal chamber and the underflow containing the solids is returned to the pump suction line. This arrangement should be used with some care as cyclone separators are very dependent on the pressure differential and flow rate for optimum performance. The design of the pipework installation and variations in pump flow conditions may considerably affect the efficiency of the separator.

If recirculating the pumped liquid will not provide suitable conditions for the seal then in Plan 32 flush is injected into the seal chamber from an external source, Figure 3.103. Considerable care is required in selecting the source of seal flush. If a less viscous liquid is used it may be liable to vaporize in the seal

Rotary seals 195

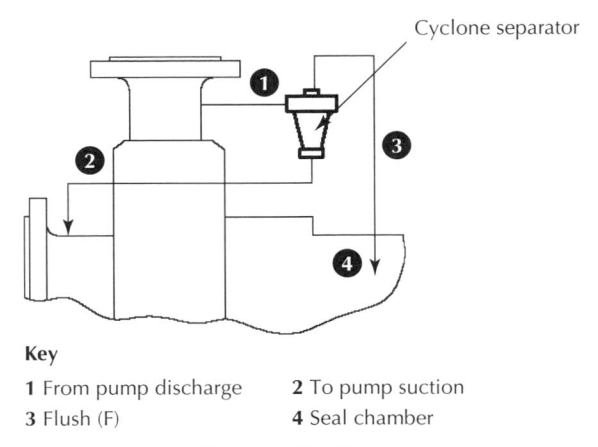

Key

1 From pump discharge   **2** To pump suction
3 Flush (F)   **4** Seal chamber

Figure 3.102 Plan 31.

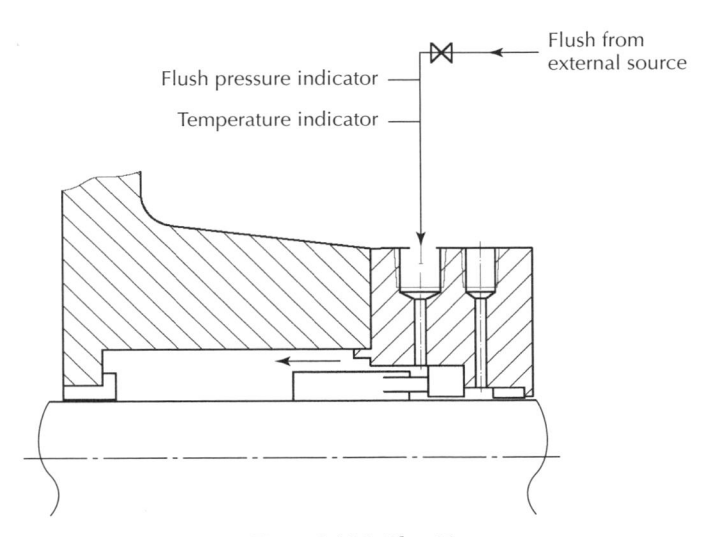

Figure 3.103 Plan 32.

chamber. Compatibility with the product is also necessary to avoid contamination of the liquid being pumped with the injected flush.

A number of these facilities may be used in combination depending on the circumstances. For instance in Plan 41, Figure 3.104, recirculation from pump discharge is first passed through a cyclone separator. The clean fluid then flows to a cooler and then to the seal chamber. The solids laden underflow from the cyclone is returned to the pump suction line.

The selection of a suitable arrangement depends very much on a combination of seal type, pump type, characteristics of the pumped liquid, user experience and preferences together with any regulatory requirements.

196 *Seals and Sealing Handbook*

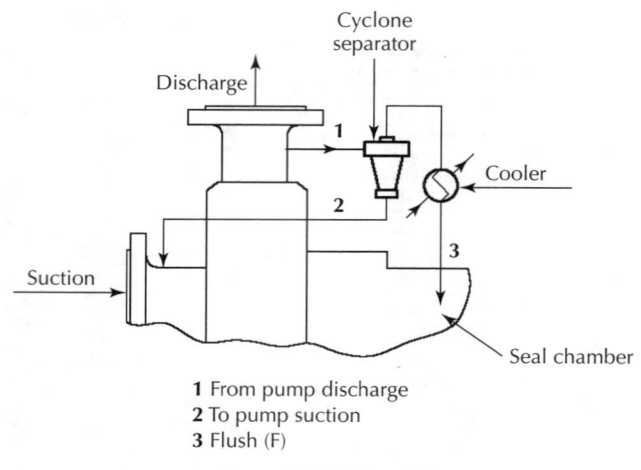

1 From pump discharge
2 To pump suction
3 Flush (F)

Figure 3.104 Plan 41.

3.4.6.5 Backup arrangements for single seals

A number of backup arrangements are used on the atmospheric side of single seals. There are two major reasons for the use of a backup arrangement: to alter the external environment of the seal, for instance to flush away leakage deposits, generally known as a quench supply, and to limit or prevent leakage in the event of primary seal failure.

Quench

A quench provision of steam may be used to flush away leakage deposits on hydrocarbon duties. Small quantities of hydrocarbon leakage tend to carbonize as they evaporate across the seal face leaving a heavy carbonaceous residue on the atmospheric side of the seal. The steam will soften and remove this residue.

On pumps handling aqueous suspensions or solutions a solid residue can form as any leakage occurs so a quench of water may be used to dissolve or flush away the deposits. These systems are recognized as Plan 62, Figure 3.105. In refinery duties the normal method is the use of steam. A further example for slurries is discussed in section 3.4.6.12.

The design of the backup must take into account the containment of any quench fluid and also any safety requirements to restrict leakage to atmosphere within acceptable limits.

Where the consideration is limited to the quench fluid it is normal to use a close clearance throttle bush for steam and a light duty seal such as a lip seal for quench water or other non-hazardous liquid. The throttle bush will be manufactured from a compatible material that will accept some dry running contact with the shaft. They also need to be non-sparking and avoid static buildup so may be made of carbon or bronze. When used in this way a disadvantage is the leakage of steam to the atmosphere. This is also adjacent to the pump bearings and places a premium on the efficiency of the bearing seals.

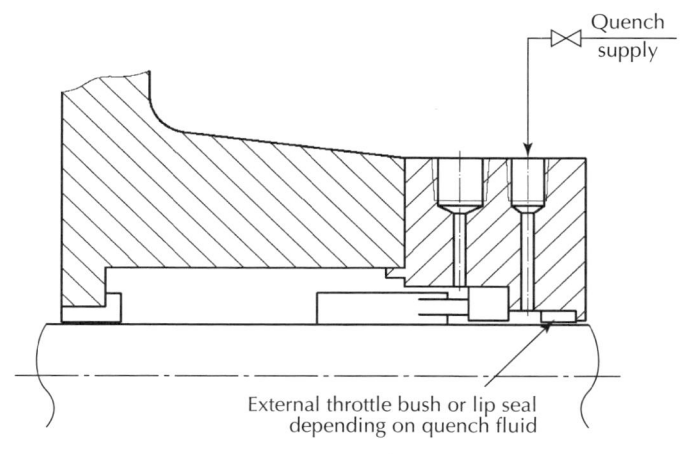

Figure 3.105 Plan 62.

Leakage containment
Most mechanical seals suffer from gradual deterioration over a period of time with gradually increasing leakage. There is, however, the possibility of sudden leakage if a component failure occurs, such as the fracture of a carbon or ceramic face material due to thermal or mechanical shock caused by dry running or other malfunction. Any leakage containment device must take into account the sealed conditions to be restrained and the potential hazards. The close clearance throttle bush can therefore only be considered where the potential for total seal failure is low and it is acceptable to have a considerable quantity of leakage for a short period of time if it does occur. The actual quantities of leakage will be very dependent on clearance of the bush, viscosity of the liquid, boiling point and pressure together with the capacity of any drain facilities.

To provide effective containment both of any small amounts of long-term leakage and a high degree of leakage control in the event of seal failure backup mechanical seals are used. These will be used if seal failure will cause an unacceptable danger or level of emissions. The seals may be dry contact seals or dry gas lift type.

With either seal type the interseal space will be vented either to flare stack or a recovery system depending on the application and location. A sensor and alarm system will be installed to warn of primary seal leakage above a predetermined amount. The selection between a dry running contact seal or a dry gas lift-off design will depend on the application and the potential conditions to which the backup seal will be exposed during primary seal failure.

Dry contact seals resemble a conventional wet mechanical seal but have lightly loaded faces and carefully selected mating faces to provide reliable running in dry conditions. The capabilities for containing vaporized leakage will be limited, typically to the flare system pressure. The overall speed and pressure are also limited to flooded conditions of the order of 20 m/s and 20 bar. They are therefore not suitable for light hydrocarbons that could flash and cause a high pressure vapour in the interseal chamber or for high speed and pressure applications.

The dry lift-off design can be used to higher operating conditions, typically 30 m/s and 40 bar. As the dry running conditions are not limited it can be used for light hydrocarbon duties where sudden seal leakage may create a high pressure vapour condition. Such seal designs also offer the benefit of lower face friction and hence reduced temperature and power consumption compared with a dry contacting seal. The dry lift-off design can also be used with a buffer gas as part of a dual seal system as discussed in section 3.4.6.8.

3.4.6.6 Multiple seal systems

It is quite common for a pair, or sometimes even three, seals to be used as a system for a variety of reasons. Common terminology is for double, tandem and dual seals and these will usually refer to specific arrangements.

Dual seals is a general term within API 682/ISO 21049[46] that applies when two mechanical seals are used with provision of a fluid service between the seals. The barrier, or buffer, space between the seals may be either pressurized or unpressurized using either liquid or gas. Each is covered by specific arrangements and within the standard these are known as Arrangement 2 with the buffer space at less than system pressure and Arrangement 3 where the barrier fluid is above the seal chamber pressure. Although the terms barrier fluid and buffer fluid are often used interchangeably they do have specific meanings within the dual seal terminology; see the Glossary on Mechanical Seal terms, section 7.1.3. Seals are normally used in what is termed a face-to-back configuration for both Arrangement 2 and 3. This means that both seals face in the same direction, Figure 3.106.

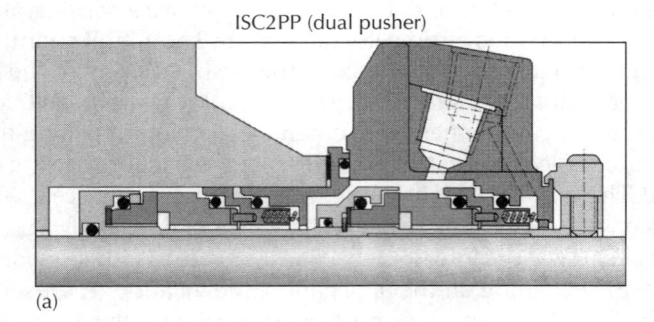

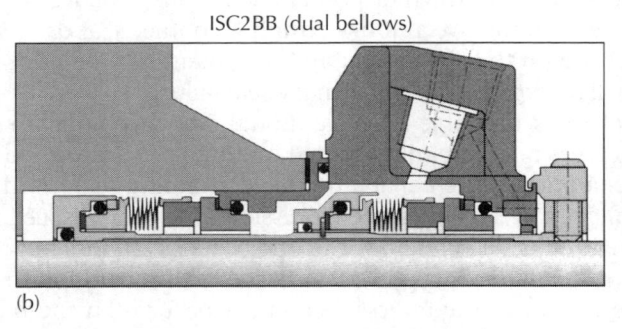

Figure 3.106 Dual seal arrangements: (a) pair of pusher seals; (b) metal bellows seals. (Source: Flowserve)

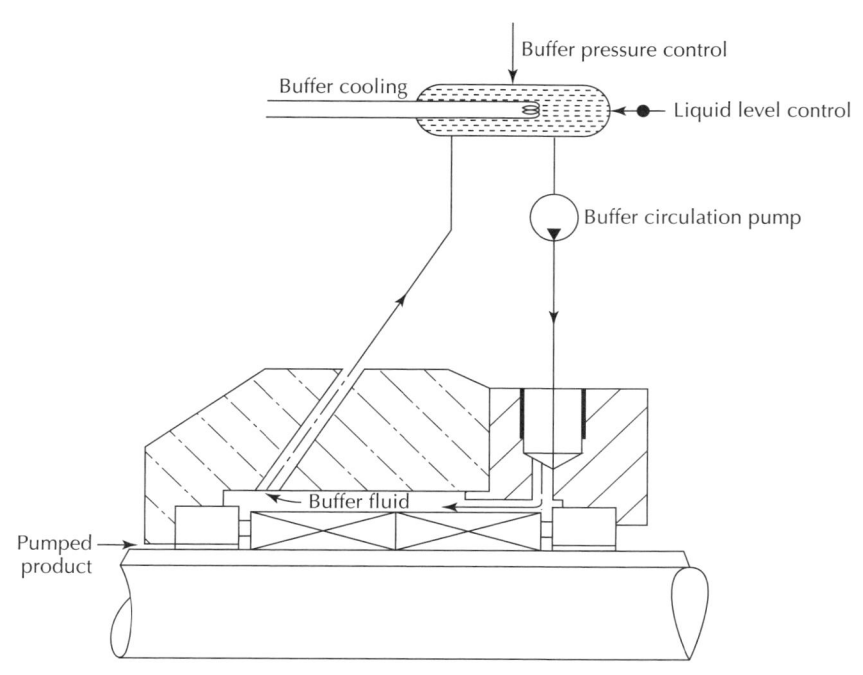

Figure 3.107 Traditional back-to-back double seal arrangement that has been superseded by dual seals as shown in Figure 3.106 for refinery duties.

This configuration has two distinct advantages over the traditional 'back-to-back' seal arrangement, Figure 3.107, previously used for pressurized double seals and these are discussed in the section on double seals. A similar seal arrangement can therefore be used for either Arrangement 2 or 3 and will also survive without serious contamination of the barrier liquid if the pressure supply fails in an Arrangement 3 system.

The use of dual seals, such as API Arrangement 2, with a low or unpressurized buffer is familiarly known as a tandem arrangement. The predominant use of this system is as a backup or containment system to the primary seal. Within a process plant system this may be applied as Plan 52, Figure 3.109, where an external reservoir will provide buffer liquid for lubrication of the outer seal. During normal operation the circulation is maintained by an internal pumping ring that is a designed-in feature of the seal. The reservoir is usually continuously vented to a vapour recovery system and is maintained at a pressure less than the pressure in the seal chamber. The system will be fitted with suitable alarms and trips to alert staff in the event of a rise in the buffer pressure and close the vent to contain the leakage. The outboard seal will be designed to operate at the full system pressure until it is possible to shut down and replace the seals. Individual company policy or local environmental regulations may stipulate a maximum period for use of the backup seal.

This system is not universally favoured as problems can occur if the outboard seal has operated for a long period at low pressure. It may then not operate reliably

200 *Seals and Sealing Handbook*

- Conventional seal with a single balance line

Double balanced seal

(a)

- Top: balance line with pressure on the o.d. of the seal.

- Bottom: balance line with pressure on the seal i.d.

(b)

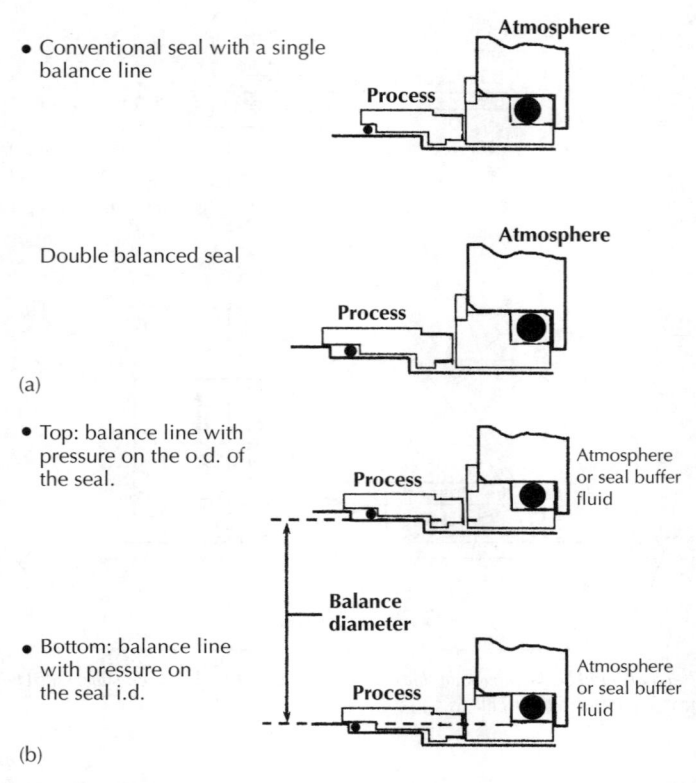

Figure 3.108 Double balance line feature allows seal to work with pressure from either direction. (Source: Flowserve)

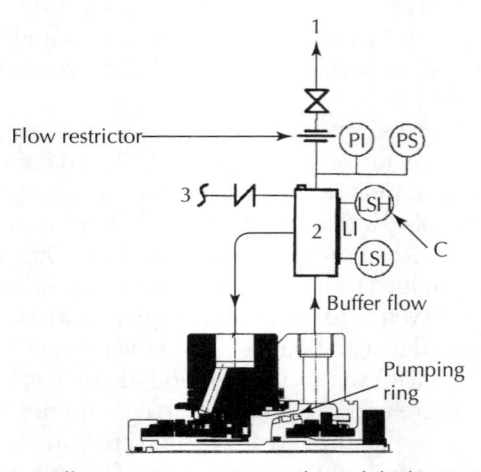

1 To leakage collection system
2 Reservoir
3 Makeup buffer fluid

LSH Level switch high
LSL Level switch low
LI Level indicator

PI Pressure indicator
PS Pressure switch

Figure 3.109 Plan 52. (Source: AESSEAL)

Rotary seals 201

if suddenly exposed to pump pressure and high temperature. Heat generation of the backup seal also has to be considered.[48]

An increasingly popular alternative to the tandem liquid seals is to use a dry gas seal as the outboard seal, as discussed for backup seals in section 3.4.6.5, but in a tandem arrangement a controlled buffer gas is provided. A typical system is Plan 72, Figure 3.110. A coalescing filter (FIL) is used to ensure solids and/or liquids which might be present in the buffer gas do not contaminate the seals. A pressure indicator (PI) is provided together with a low pressure and high flow switches (PSL and FSH respectively). An externally supplied gas buffer feeds a pressure control valve (PCV) which is used to limit the buffer gas pressure applied to the containment seal to lower the process-side pressure of the inner seal and prevent reverse pressurization of the inboard seal. The buffer gas may be used alone to dilute the seal leakage or it may be used in conjunction with additional

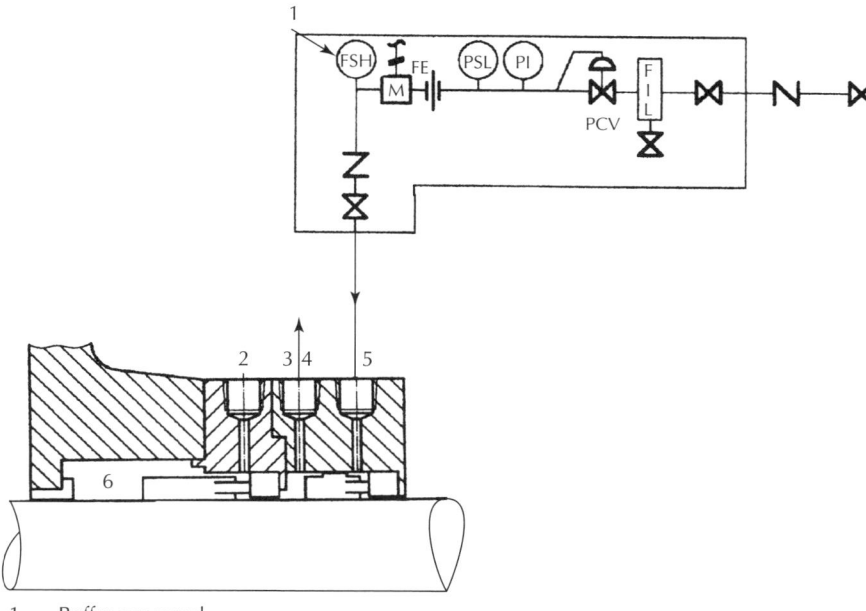

1	Buffer gas panel
2	Flush (F)
3	Containment Seal Vent (CSV)
4	Containment Seal Drain (CSD)
5	Gas Buffer Inlet (GBI)
6	Seal chamber
FE	Flow meter (magnetic type shown)
FIL	Coalescing filter, used to remove solids and/or liquids from buffer gas
PCV	Pressure Control Valve, used to limit buffer gas pressure to prevent reverse pressurization of inner seal
PI	Pressure Indicator
PSL	Pressure Switch Low
FSH	Flow Switch High

Figure 3.110 Plan 72.

facilities such as those specified in Plan 75 or 76 to help sweep leakage into a closed collection system. Plan 75 provides a drain and collection chamber for liquid leakage, while Plan 76 provides a leakage collection chamber for vapour products if the pumped fluid vaporizes at atmospheric conditions.

The dry gas seal used for this arrangement may be either a wavy face, or with laser machined grooves depending on supplier and customer requirements. The seals will be designed to operate reliably on liquid for a satisfactory period of time with minimal leakage within the statutory requirements. This type of seal will have much lower heat generation than the liquid types and the system auxiliaries and controls are often much simpler. For these reasons it is now very often the preferred option for a tandem arrangement.

A tandem seal may be specified either to provide a high degree of local containment, often to comply with emission requirements, but also for applications such as unattended plant on pumping stations, etc. where it can provide a capability for extended running after a primary seal failure until it is possible for maintenance staff to attend.

An alternative use of this type of seal arrangement is where the pressure is beyond the capability of a single seal. By controlling the buffer space to an intermediate pressure a higher overall pressure can be reliably sealed. An example of this technique is that used for the seals on the boiler circulating pumps for nuclear power plant. Those employed on pressurized water reactors operate at up to 170 bar.[49] A series arrangement of three seals is used, Figure 3.111. A carefully

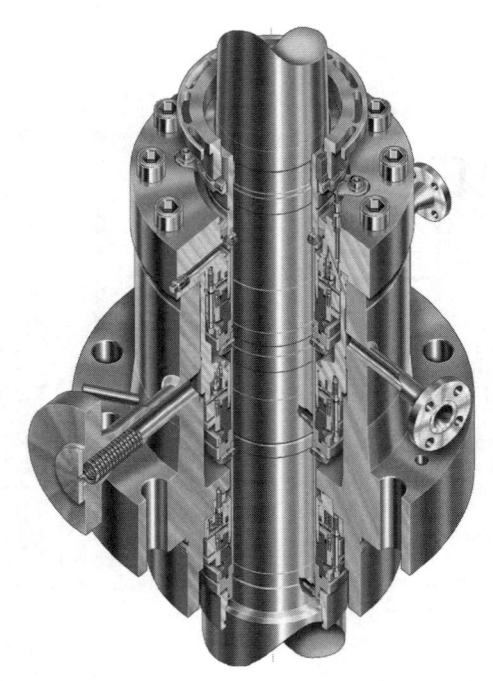

Figure 3.111 Three seals used in series to break down the pressure of a pressurized water reactor circulating pump from 170 bar. (Source: Flowserve)

regulated bleed flow is designed so that each seal operates at one third of the pressure. These seals have been very highly developed for the duty and seal life of up to 15 years has been obtained. Any one seal is designed to be capable of operating at the full system pressure in the event of a system malfunction to provide sufficient time for a controlled shutdown of the pump. A similar philosophy is employed for boiling water reactors[50] which operate at 70 bar and 280°C. In this case two seals are used in tandem with a throttle ring controlling the relative pressure of the inboard and outboard seals.

3.4.6.7 Double seals

It is fairly common terminology to refer to a dual arrangement with barrier space above the sealed pressure, such as API Arrangement 3, Figure 3.112, as a double seal. However, this may not be a universal understanding so must be used with care. With the barrier space at a higher pressure than the seal chamber the lubricant for the inboard seal will be the fluid supplied to the barrier space. This

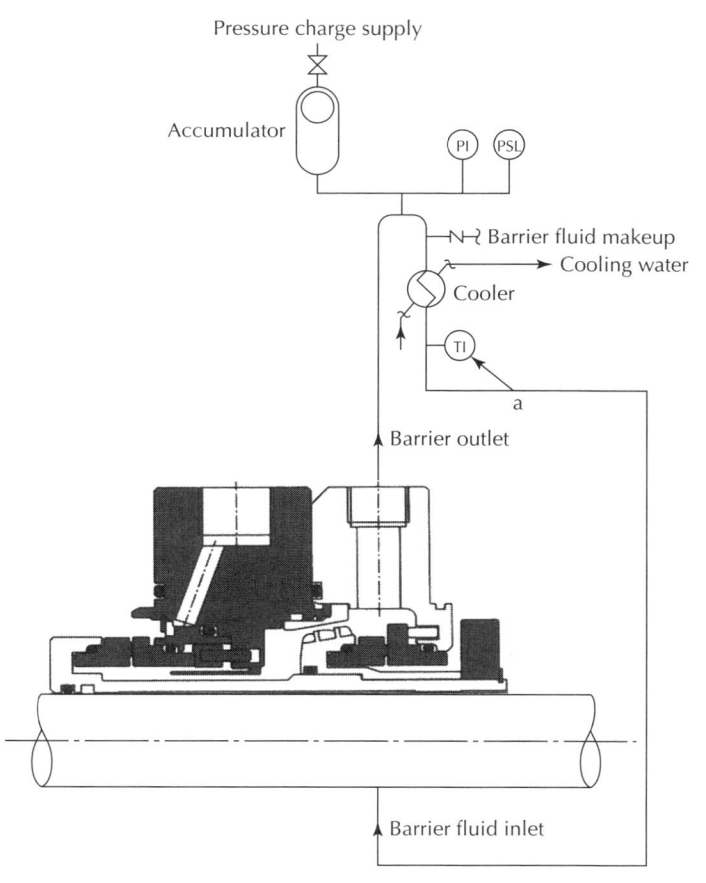

Figure 3.112 An API Arrangement 3 Dual Seal configuration designed to have a pressure between the seals in excess of the seal chamber pressure. (Source: AESSEAL)

arrangement will be used when it is considered necessary to exclude the pumped product from the seal faces. This may be because of one or more of the following:

- Solids content.
- Vaporization hazard.
- Toxic contents.
- Potential for polymerization.
- High viscosity.
- Very low viscosity or mixed phase.
- Provision of an inert barrier.
- To provide a hygiene barrier between the product and atmosphere.

The traditional method of providing a double seal of this type was to have a pair of similar seals back to back, Figure 3.107. This design has two problems that can lead to failure in practice.

- If the barrier fluid pressure falls below the seal chamber pressure the process pressure will tend to lift off the seal and process fluid will leak into the barrier system. In most cases this will lead to failure of the barrier system. This can cause a time-consuming shutdown if the barrier system has to be cleaned.
- As the inboard seal, between the barrier and process fluids, faces the process any wear of the seal will cause it to move towards the process fluid. If the fluid is viscous or contains solids the secondary seal on the sleeve will be tending to move onto a contaminated shaft sleeve area and this can cause the seal to stick or leak.

With the dual arrangement, Figure 3.106 or 3.110, both these factors are designed out of the seal. The seals have a double balance feature, Figure 3.108. If the barrier fluid pressure should fall below that of the seal chamber the pressure on the inboard seal reverses. With this type of seal when the O-ring moves to the other end of the groove under reverse pressure the seal pressure balance is changed such that the seal will not be opened. This therefore provides a much more reliable configuration as short-term loss of the barrier pressure should not cause failure of the complete system.

As the inboard seal wears, any movement of the floating component will tend to slide the O-ring into the zone of the relatively clean and benign barrier fluid.

However, both seal configurations still require a liquid barrier system that involves a method of pressurizing and circulating the barrier liquid with appropriate temperature and level controls and associated safety trips.

3.4.6.8 Gas barrier double seals

An alternative to a liquid double seal arrangement is to use a pressurized gas barrier. A pair of gas seals is used in either a back-to-back or face-to-face configuration, Figure 3.113. An inert gas such as nitrogen is provided as the buffer

GF-200

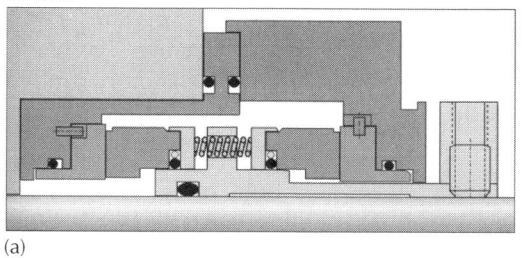

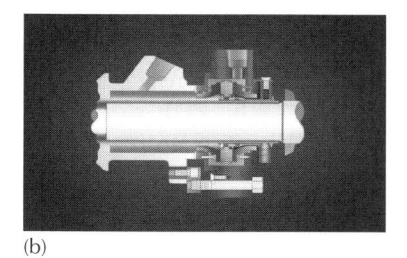

(a) (b)

Figure 3.113 Pressurized gas barrier double seal configurations for difficult liquids: (a) back-to-back arrangement used in a wide bore pump housing which provides adequate axial length; (b) a face-to-face arrangement that reduces the axial length required can be used with a narrow bore pump design on lower pressure duties. (Source: (a) Flowserve, (b) Sealing Technology)

fluid at a pressure higher than the liquid in the seal chamber. When the pump starts the gas seals will lift off and operate on a film of buffer gas. Some leakage of the inert gas into the process liquid will occur. A relatively simple gas supply system is required, Figure 3.114, compared with a liquid buffer system. The gas supply is fed directly to the interseal space with no need for circulation. A vent port is provided to depressurize during shutdowns. Care is required to ensure that the pump is correctly vented on startup as buffer gas may accumulate in the pump while stationary. It is necessary to ensure that the inert gas is compatible with the process system. Otherwise this arrangement provides a potentially reliable system which provides practically total freedom from harmful emissions and lower frictional losses than a liquid seal system. It has been used quite extensively as a retrofit option on pumps handling difficult liquids.[51]

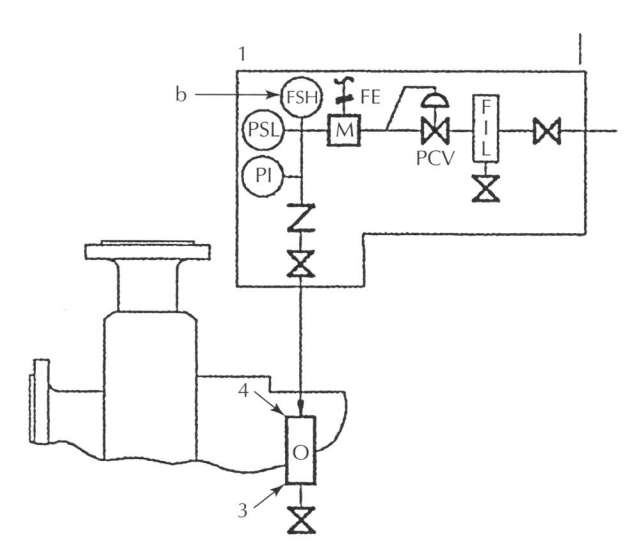

Figure 3.114 The gas supply system recommended for a pressurized gas barrier double seal arrangement, API/ISO Plan 74. (Source: Sealing Technology)

The back-to-back option permits the use of higher duty seals and is typically rated to 35 bar. It is also available at fairly large sizes up to 150 mm. The face-to-face configuration provides a reduced axial length and is therefore suitable for retrofitting to narrow bore seal chambers, but pressure rating is more limited, typically 14 bar. The single large rotating face also limits the size that is generally available to 75 mm.

3.4.6.9 High temperature

The preferred design for very high temperatures is to use a metal bellows seal. With the use of suitable secondary seals these can be used up to 400°C, but the bellows limits pressure to typically 20 bar. To achieve high temperature operation the elastomer secondary seals cannot be used. The most common substitute is to use an exfoliated graphite packing ring. This does not provide the same flexibility and resilience as an elastomer and so the seal must be designed to accommodate the required flexibility in other ways. For this reason a stationary bellows design is often preferred, Figure 3.115(a). The flexibility to cater for any misalignment vibration, etc. is provided by the bellows. A disadvantage of this arrangement is that fine solids in the process fluid can accumulate in the convolutions and restrict movement or cause excessive stress. If there are problems with solids then a rotating bellows design may be used, Figure 3.69. The rotation of the bellows helps to expel any solids, but as the stationary face is retained on a graphite secondary seal it has less flexibility than a typical elastomer sealed stationary and hence the seal is more sensitive to misalignment than a stationary bellows design.

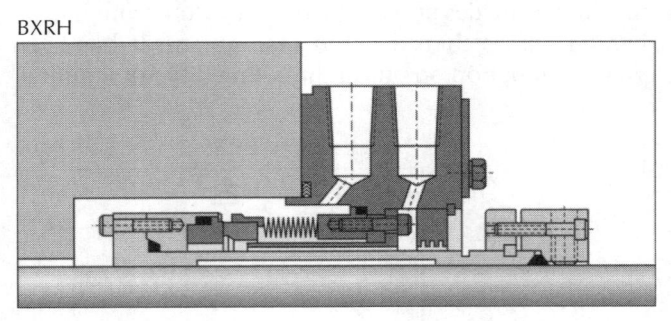

Figure 3.115 A stationary metal bellows seal designed for high temperature operation with exfoliated graphite secondary seals. (Source: Flowserve)

3.4.6.10 High pressure, high temperature

As metal bellows seals are limited to typically 20 bar if it is required to seal a high temperature above this pressure it is necessary to use a pusher seal design. With the use of high temperature elastomers it may be possible to use the seals up to 250°C, depending on the characteristics of the process fluid, including the atmospheric boiling point, potential for solidification, etc. Depending on the specific fluid characteristics and safety requirements the usual options are to provide either a single seal with special coolant circuit, such as that described for API Plan 23, or a dual seal with a coolant buffer system as described in section 3.4.6.7.

3.4.6.11 High speed

At high rotational speeds, such as in excess of a surface speed of 30 m/s, or approximately 6000 rev/min on a 100 mm shaft, the design of the seal to cater for the speed becomes important. It is necessary for the rotating mass of the seal to be dynamically balanced and be as compact as feasible to minimize centrifugal forces. It is also necessary to provide a stable lubricating film in the seal face with minimal friction to provide stable seal operation with minimum heat generation.

A conventional rotating floating component will create high out-of-balance and potential instability. High speed seals are therefore designed with a stationary floating component with the rotor clamped directly to the shaft. This may be a one piece ring, Figure 3.116(a), or a specialist designed rotating component with a contained seal face, Figure 3.116(b). This arrangement can involve a higher rotating mass but provides the opportunity to control the seal face distortions more accurately and hence control the lubrication and leakage. A pusher seal type is often favoured as the O-ring provides some damping and prevents high speed excitation of the sprung components. Metal bellows seals are used, where it is necessary to seal high temperature or aggressive fluid, but will normally incorporate a friction damper on the bellows to maintain stable operation, Figure 3.117.

It is also necessary to ensure a stable lubricating film to minimize friction and avoid instability caused by problems such as vaporization in the film. Many high speed seal designs incorporate some type of enhanced lubrication features such as wavy faces or pumping grooves to ensure a stable film.

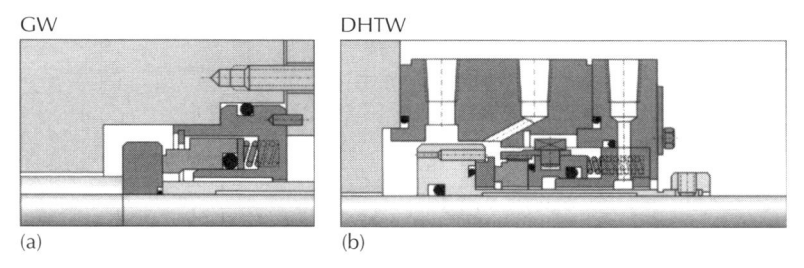

Figure 3.116 (a) A high speed seal design with the rotating face clamped directly by the shaft sleeve. (b) An alternative high speed seal with specialist designed rotor and contained seal face that offers the potential benefit of improved face distortion control.

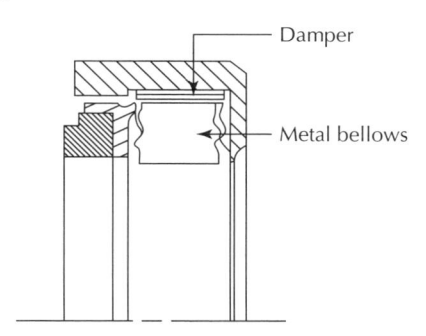

Figure 3.117 A high speed metal bellows seal with a friction damper on the bellows to prevent bellows instability. (Source: Eaton)

3.4.6.12 Slurry seals

There are two major problems when mechanical seals are exposed to slurries. Abrasive slurry will cause excessive wear and any slurry will cause clogging of the springs, bellows and other floating components of a conventional seal. Seals have been developed specifically for slurry duties that address these problems. They are designed such that the components exposed to the slurry have as smooth a profile as practicable and either use a spring design that has a smooth profile or remove the springs to be remote from the slurry, Figure 3.118.

The seal faces will be a hard/hard combination with either a pair of silicon carbide or tungsten carbide faces. Many of the designs also have the seal faces relatively exposed close to the largest diameter of the seal, as, for example, in Figure 3.118(b), which helps to promote cooling of the seal faces. A suitable tapered seal housing has also been shown to be successful at promoting fluid circulation around slurry seals[31] and this facility is included within the cartridge for some seals, Figure 3.118(a).

Either of these designs is capable of operating on concentrated slurry without any flushing to the product side of the seal. This removes the requirement to provide pressurized fluid to the pump and also avoids dilution of the pumped product. They have proved to be highly reliable when installed in the appropriate pump and housing arrangement.

However, slurry seal installations may include a low pressure water quench externally to the seal to flush away leakage deposits that could clog the seal.

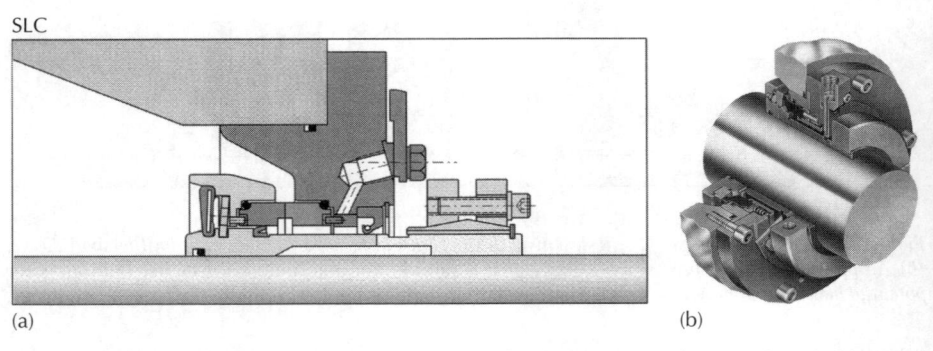

Figure 3.118 Examples of mechanical seals for slurry duties: (a) a spring design with smooth external profile; (b) external springs remote from the product. (Source: (a) Flowserve, (b) John Crane)

3.4.6.13 Mixer seals

Seals for mixers have to deal with some specific conditions that are outside those normally encountered with other seal applications. The range of applications is very wide, so there is no single sealing solution. Mixers are used in a wide range of industries such as chemicals, food, pharmaceuticals, hygiene products, oil refining and biotechnology. Vessels may be relatively small in specialist plant but in major production facilities may be 10 000 or 20 000 litres. Depending on the mixing requirements the mixer shaft may enter the bottom or side of the vessel, in which case it will be flooded with the mixed products, or more often will enter the top of the vessel so that the seal is operating in the gas space above the mixing process.

Shaft rotational speed is often much slower than a pump. However, the long shaft with long overhang to the mixing impellers can cause very large shaft runout and vibration. In a large vessel removal of the shaft will be extremely difficult so designs that avoid this necessity are also required.

A range of possibilities exist with respect to the seal requirements:

- If no special precautions are required then a single seal with dry running capability may be used. This may be a contact seal designed for dry running, or a lift-off design that will run on a gas film when the mixer is operating. The selection will depend on parameters such as the speed and pressure involved.
- Lubrication of the seal may be required either because of the speed and pressure or because of the internal mixer environment. In this case a double seal with lubricating buffer may be used. Precautions will be necessary to ensure that the buffer is compatible with the product.
- If hazardous products are involved, either because of flammability or toxicity then a suitable low or zero emission dual seal arrangement will be required.
- In pharmaceutical, food and biotech applications hygiene will be a major consideration and a steam or biocide flush may be required, or a high temperature barrier adjacent to the seal. The detail design of the components exposed to the product must be assessed for external profile and drainability, etc.[32]
- If a temperature sensitive product is involved cooling may be required to ensure that seal frictional heat does not degrade the product. Figure 3.119 shows a seal arrangement that can be used to provide either heating or cooling in the seal area depending on the application.
- To minimize the effects of shaft runout and vibration the seals may be supplied as an integral cartridge assembly with shaft bearings that mount on the top flange of the vessel, Figure 3.120.
- Where high product cleanliness is required it may be necessary to ensure that no wear debris, etc. can fall into the product and include a debris collection well below the seal[32] as shown in Figure 3.121.

3.4.6.14 Large diameter seals

A few applications employ especially large diameter seals. Examples are ship stern shafts and large pumps and turbines in, for example, hydroelectric power generation. Shaft sizes may extend from 0.5 metre to in excess of 1.0 metre.

210 *Seals and Sealing Handbook*

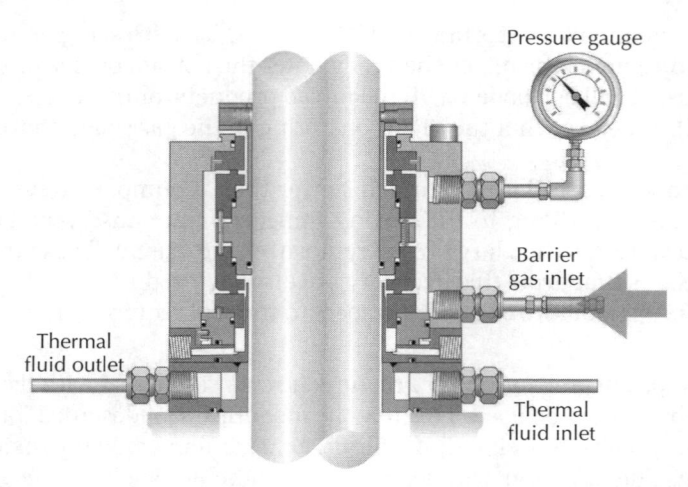

Figure 3.119 A mixer seal assembly that uses a dual gas barrier seal arrangement with a thermal fluid chamber between the seal and the mixer to control the temperature for either hygiene or product quality requirements. (Source: Sealing Technology)

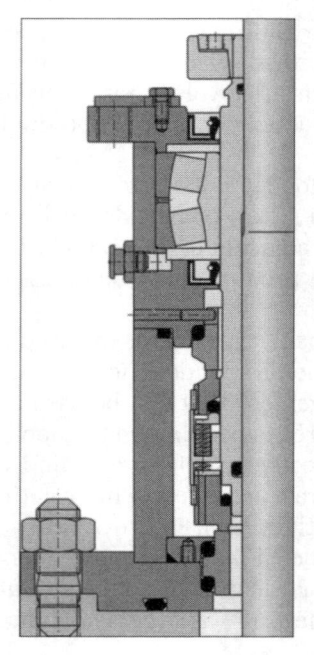

Figure 3.120 A double seal arrangement incorporating shaft bearings for a top entry mixer. (Source: Flowserve)

Mechanical seals are used as the aft seal on marine stern tubes. They may be used directly to seal the stern tube bearing oil from seawater as an alternative to the lip seal arrangements discussed in section 3.2.9.2, or to provide a more secure anti-pollution option they may seal the seawater with a separate lip seal used to contain the bearing oil.

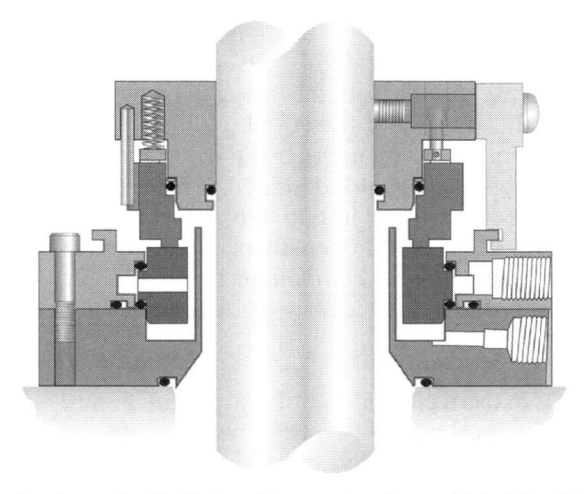

Figure 3.121 A single mixer seal with debris well to prevent seal wear debris falling into the mixer product. (Source: Sealing Technology)

This latter system helps to eliminate oil loss from the outboard seal, even when it is fouled or badly damaged. It is also less susceptible to wear and ageing. An example is shown in Figure 3.122. In the outboard seal assembly, water is excluded by a radial face seal unit which surrounds and encloses an oil sealing unit comprising a single elastomeric lip seal. This lip seal runs on a chromium steel liner which rotates with the shaft. A second lip seal, positioned forward of the oil seal, acts as a backup seal if water excess should penetrate the drain space.

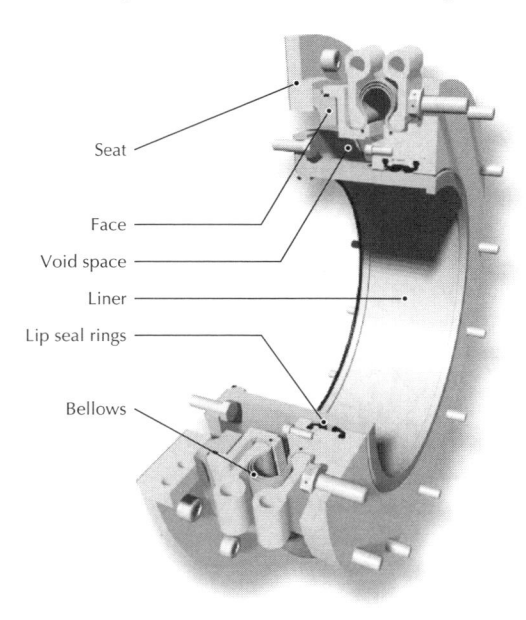

Figure 3.122 Ship stern shaft seal with large elastomer bellows to provide flexibility. (Source: Wartsila)

This drain space not only allows any oil or water passing to freely drain inboard, but also forms a 'coffer-dam' between the seawater and the oil. It is then possible to operate the stern shaft bearing at a low oil pressure of about 0.2 bar, irrespective of the draught of the vessel.

Mechanical seals can be used for either the water barrier seal discussed above or as the water to oil seal. They are designed to be serviced in situ with all the wearing components constructed in two halves. Some designs use a bellows, as shown in Figure 3.122. This may consist of either three, four or six sections, depending on seal size. The seat component is machinable and the secondary rubber components are split. If necessary, the face component can be refurbished under controlled conditions.

A more recent development, intended to provide extended operating periods without removal, is to use a pusher type of seal design, Figure 3.123. The purpose is to allow the use of hard materials for the face and seat components to reduce face and seat wear caused by abrasive particles in the seawater. A chromium steel liner is used on the inside diameter of the face carrier to minimize wear caused by the secondary seal. The purpose of this design is to achieve up to 10 years' use without maintenance. A further advantage claimed for this type of seal is that the sealing faces are positioned further forward than is possible with the bellows type of seal. This means more effective protection against fishing lines and nets, thus reducing the chance of the seal being damaged during operation.

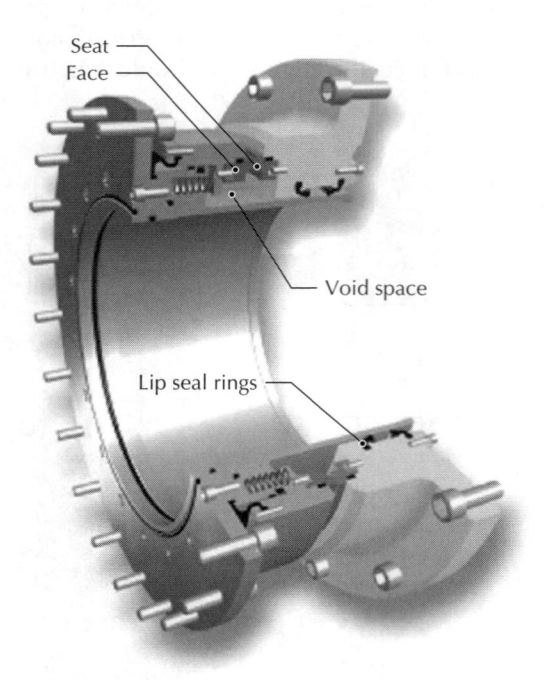

Figure 3.123 A more recent development using a pusher seal design for a stern shaft. This may be fitted with hard faces. (Source: Wartsila)

Another application for large diameter mechanical seals is on water turbines, including pump storage schemes where the turbine will also operate as a pump. Similar seals may also be used on very large cooling water pumps.

The seals for these applications have been specifically developed for the application and incorporate a number of specialist features. An example is shown in Figure 3.124. They will be designed individually for each application and are fully split for ease of installation and maintenance. In particular the seals on pump turbines can see particularly difficult operating conditions.[52] It is normal practice on such installations to run the machine up to synchronous speed before startup, and to reduce the auxiliary motor power the pump chamber will be filled with compressed air. The machines may also be run for extended periods on air for power factor improvement, or when in standby mode for potential sudden power requirements on the grid.

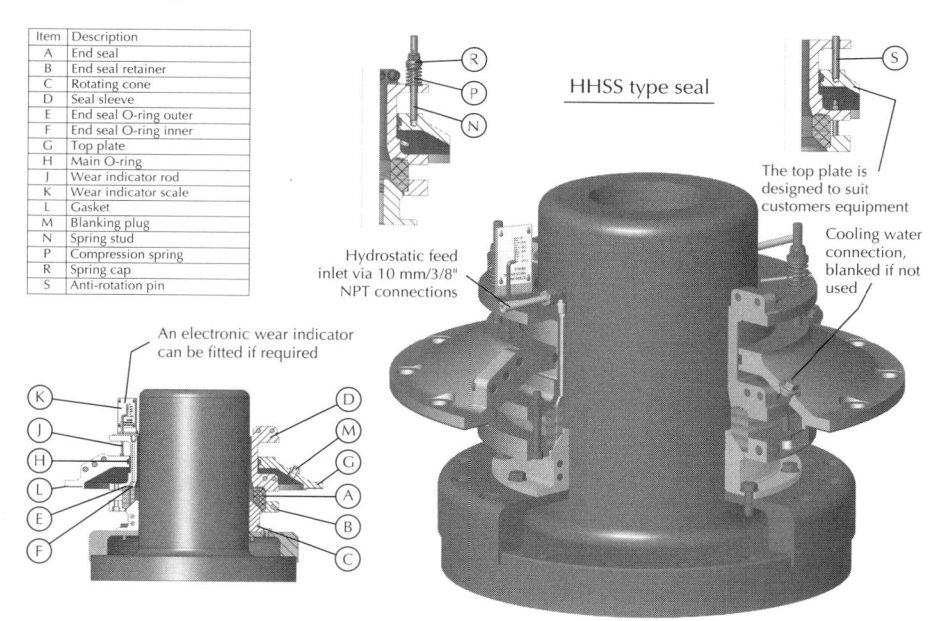

Item	Description
A	End seal
B	End seal retainer
C	Rotating cone
D	Seal sleeve
E	End seal O-ring outer
F	End seal O-ring inner
G	Top plate
H	Main O-ring
J	Wear indicator rod
K	Wear indicator scale
L	Gasket
M	Blanking plug
N	Spring stud
P	Compression spring
R	Spring cap
S	Anti-rotation pin

Figure 3.124 A large diameter mechanical seal for water turbine applications that uses a cone face geometry with a composite material for the conformable seal face. (Source: SM Seals)

The main wearing face is made from a composite material that is designed to accommodate damage from debris and regenerate mutually conforming seal faces if damaged by abrasive particles. The design can be adapted to suit either horizontal or vertical shafts and is capable of operating at circumferential velocities up to 30 m/sec and pressures up to 10 bar. With hydrostatic feed these parameters can be increased but the majority of water turbine applications are in the range 5–10 m/s.

In the example in Figure 3.124 the seal hard face cone (C) is secured to the shaft, as is the rotating face. The composite fibre end seal (A) is located in the sleeve (D) and is secured in position by the retaining ring (B). Face loading is achieved by the springs (P) and turbine water pressure. The wear indicator system (J/K) can be linked mechanically to a gauge panel or alternatively to an electronic digital display unit.

The composite seal face does not have the self-lubricating properties of carbon and so it is necessary to provide cooling water, which can be unfiltered penstock water, to the seal whenever the shaft is rotating. Some variants such as high speed or higher pressure seals may have a hydrostatic feed which will require the use of filtered water.

The seals are designed to permit adjustment of the seal face loading after installation to optimize the seal running conditions with respect to leakage, seal face temperatures and wear. This is achieved by providing adjustment to the seal springs. If there is not the facility to access the springs for adjustment pneumatic cylinders may be used for face loading. An alternative available for vertical shafts is to use a series of weights to load the seal.

The incorporation of the wear indicator permits maintenance to be planned, and seal life in excess of 10 years is possible. Smooth contours and no springs in the water prevent clogging in silty conditions. The composite wearing face has the advantage for large diameter shafts that it can be dropped without damage unlike carbon. It can also withstand vibration and abrasives without permanent damage.

An alternative design of the water turbine seal uses a more conventional looking radial seal face geometry, Figure 3.125. This can be used in either a vertical or horizontal shaft machine. In this design the face loading is remotely controlled by

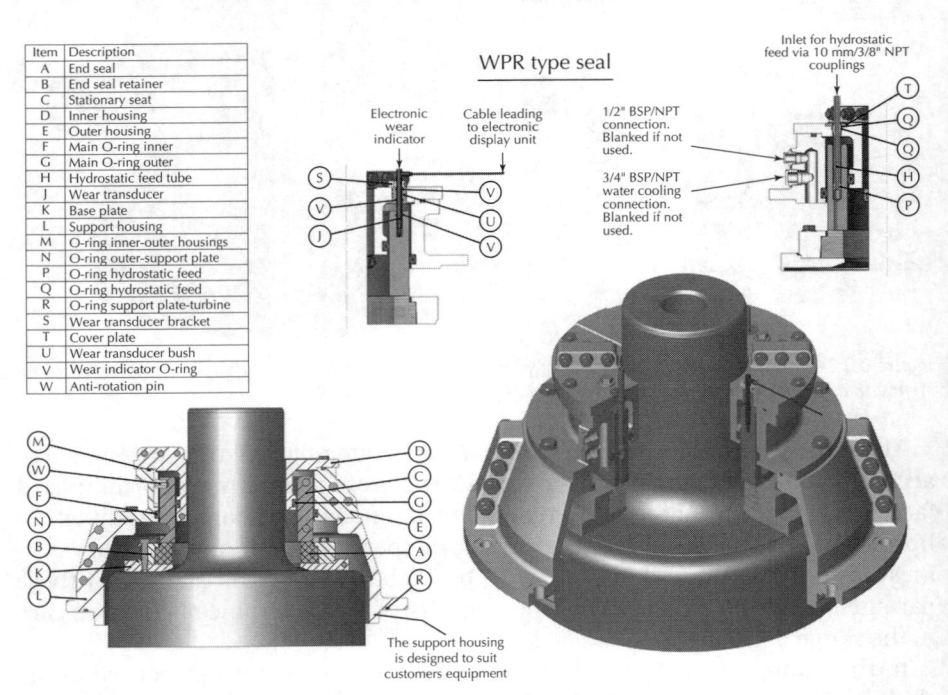

Item	Description
A	End seal
B	End seal retainer
C	Stationary seat
D	Inner housing
E	Outer housing
F	Main O-ring inner
G	Main O-ring outer
H	Hydrostatic feed tube
J	Wear transducer
K	Base plate
L	Support housing
M	O-ring inner-outer housings
N	O-ring outer-support plate
P	O-ring hydrostatic feed
Q	O-ring hydrostatic feed
R	O-ring support plate-turbine
S	Wear transducer bracket
T	Cover plate
U	Wear transducer bush
V	Wear indicator O-ring
W	Anti-rotation pin

Figure 3.125 An alternative design of water turbine seal that has a radial face configuration and uses hydraulic water pressure for the face loading. (Source: SM Seals)

water pressure and so it is particularly suited to turbines where there is no access to the seal for adjustment after installation. They can be used within similar operating limits to the cone face seal.

In this design the composite fibre end seal (A) is secured to the turbine shaft as is the rotating face. The stationary seat (piston) (C) is loaded by water pressure in the operating chamber, and slides axially between the O-rings (F and G) in the inner (D) and outer (E) housings, to compensate for wear. The operating chamber is pressurized by connecting it to either a header tank, the chamber around the seals via a control device or a separate water source through a custom built control circuit through a regulator depending on the application. The additional facilities such as the wear indicator (J) are also fitted to this seal.

3.4.6.15 Dry gas seal arrangements

A variety of arrangements are used with dry gas seals depending on the duty, type of gas and the design of the machine.

Single seal
A single gas seal can be used for non-toxic, non-flammable gas duties where a small amount of leakage of the process gas is acceptable. The arrangement will comprise a single dry gas seal with a labyrinth inboard between the seal and the process gas, Figure 3.126. The labyrinth provides a throttle which helps to protect the mechanical seal from contaminants within the process. It is normal also to inject some clean dry sealing gas, which can be process gas, between the labyrinth and mechanical seal. Most of this will flow through the labyrinth into the process, but some will be lost as leakage through the mechanical seal. This will normally be piped away to a flare or other collection system.

Outboard of the dry gas seal there will normally be some form of barrier seal to separate the gas seal from the shaft bearings. It is important to prevent migration of the bearing lubricating oil mist into the gas seal where it can cause contamination. This is achieved by providing a further separation gas supply to the space between a pair of elements of an oil barrier seal. The oil seal is often a segmented ring seal, section 3.6.7, or potentially a labyrinth, section 3.6.2, and in some cases a lubricated mechanical seal.

If improved protection from process liquids and contamination is required a segmented ring seal may also be used on the product side of the dry gas seal instead of a labyrinth. This provides reduced flow across this barrier seal but with the potential disadvantage of being a wearing component.

The single gas seal offers the major benefit of a simple arrangement with a minimum of auxiliary support systems. The major disadvantage is the absence of any backup system should the primary seal fail. This is a major safety issue that precludes the use of a single seal in many process applications. The segmented seal on the process side may be considered to offer a better backup sealing than a labyrinth in the event of primary seal failure, but will still only provide limited effectiveness. Guidance on the selection of safety systems is provided in API 617.[53]

(a)

(b)

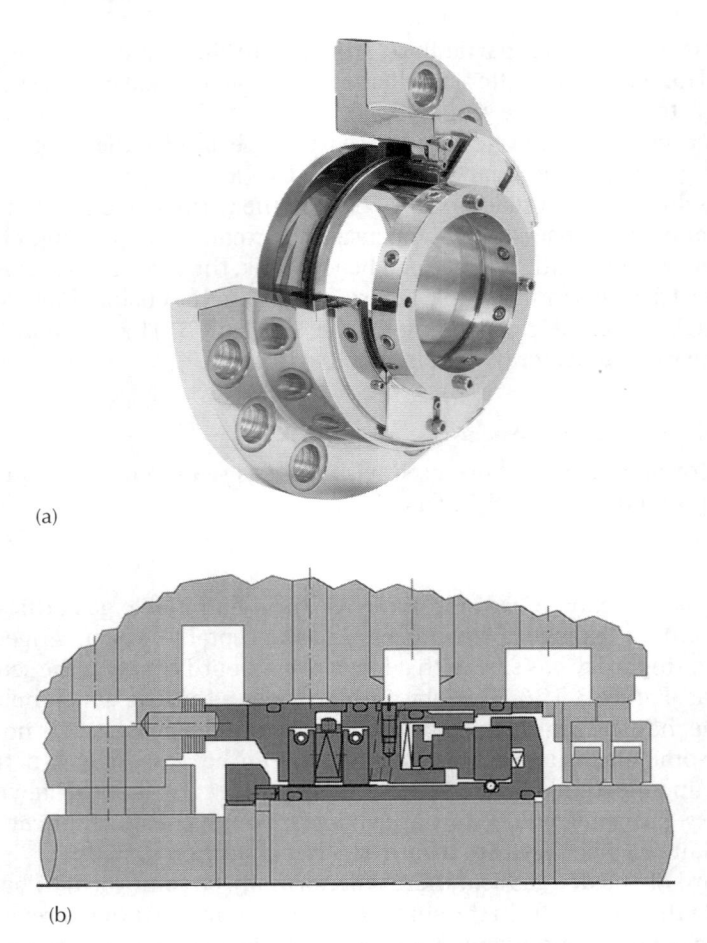

Figure 3.126 Single gas seal arrangements for non-hazardous duties: (a) a conventional design of turbo compressor seal, (b) a compact, narrow face seal design plus segmented oil barrier seal used for the restricted housings of a screw compressor. (Source: (a) John Crane, (b) Flowserve)

Tandem seals
This is the seal arrangement that is most commonly used in process and gas transmission duties as it provides the necessary level of backup in the event of problems with the primary seal. A fairly standard arrangement is for a pair of seals in tandem on a single cartridge, Figure 3.127. The primary seal will contain the process gas pressure with a small amount of leakage into the buffer space which would then be piped away to flare or other leakage receiver. The secondary seal contains the leakage in normal operation and acts as a backup in the event of failure of the primary seal.

There will again be an injection of clean dry gas inboard of the primary seal to maintain clean conditions at the seals, and a separation gas to isolate the gas seals from potential contamination by bearing oil mist. The separation gas, and any small amounts of process gas that leak across the secondary seal are also piped away to flare. This system is widely used as it provides the backup necessary

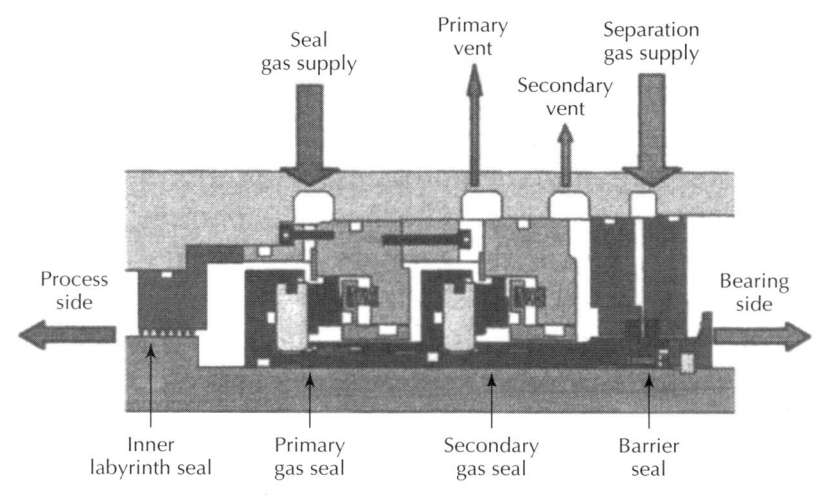

Figure 3.127 Tandem gas seal arrangement. (Source: John Crane)

to deal with the potential failure of the primary seal. It has the disadvantages of more complex auxiliary support systems, with monitoring, controls, alarms and shutdowns, plus increased cost together with requiring more axial space and associated increases in rotating mass for the compressor.

A variation of the tandem arrangement is to use an intermediate labyrinth, Figure 3.128. The purpose of this arrangement is to reduce the possibility of leakage of process gas into the secondary vent system, between the outboard seal and the oil barrier seal. An inert or otherwise acceptable gas is fed to the labyrinth at a pressure slightly higher than the intermediate buffer pressure between the seals. This assures preferential flow of this buffer gas through the labyrinth to both the

Figure 3.128 Tandem gas seal with intermediate labyrinth to provide increased protection from process gas leakage. (Source: John Crane)

primary and secondary vents. This system may be used if leakage to the secondary vent is unacceptable, either because the vent is to atmosphere, or because of a toxic or other hazard associated with the gas. This arrangement has a number of disadvantages compared with the more standard tandem design. The auxiliary support system becomes even more complex with additional gas supply and controls and associated alarm and shutdown systems. More axial space is required which can impact on compressor design and the additional gas supply makes it more difficult to monitor the condition of the primary pressure seal by measuring leakage flow.

Seals for these arrangements are available to operate at speeds up to 200 m/s and pressure from 200 to 400 bar depending on the design.

Back-to-back seals
The double opposed configuration, Figure 3.129, is similar in principle to that used for liquids using API 682 Plan 74, described in section 3.4.6.8, Figure 3.113(b).

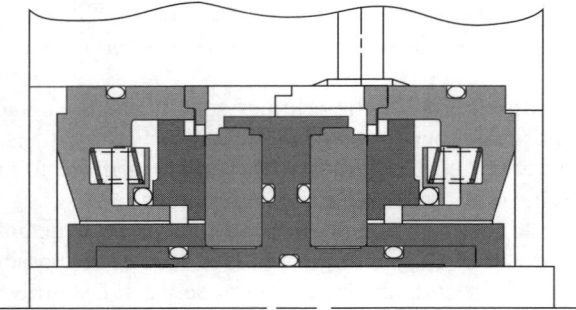

*Double seal arrangement where hazardous gas is not permissible to leak into atmosphere.

(a)

(b)

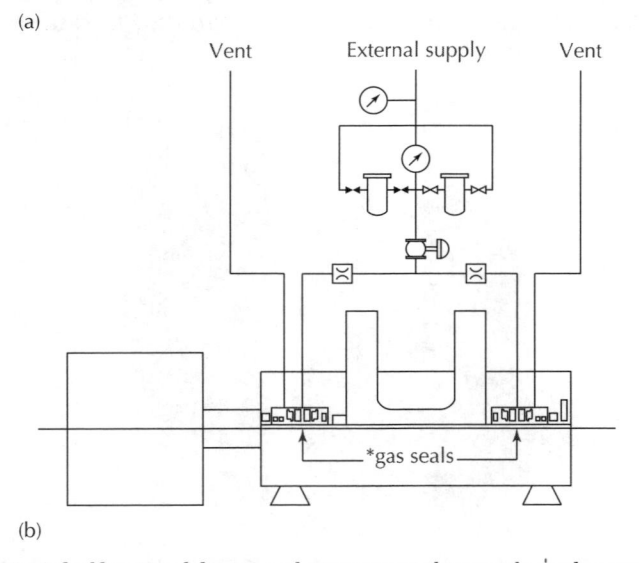

Figure 3.129 A double opposed dry gas seal arrangement that may be used to avoid a vent to flare and escape of emissions to atmosphere on lower pressure applications. (Source: John Crane)

A seal buffer gas is supplied between the pair of seals at a pressure higher than the process gas pressure at the seal. The buffer gas, which will normally be an inert gas such as nitrogen then leaks both to atmosphere and into the process gas across the inboard labyrinth. For optimum reliability it is still preferable to supply a clean process flush gas between the labyrinth and seals to prevent contaminated process gas accessing the seal area.

This arrangement is used where it is required to prevent leakage of process gas to eliminate emissions or if the gas cannot be vented to flare. This is done so as to comply with environmental regulations or because of a toxicity hazard. It is also useful if the process gas is contaminated with particles as the process gas is positively flushed away from the process side seal. It can also be used for very low pressure applications where a tandem arrangement is not suitable.

This system has a number of advantages compared with a typical tandem arrangement:

- The auxiliary support system is relatively simple, see API Plan 74, with significantly reduced piping and controls.
- The nitrogen supply is usually very clean and dry which reduces the probability of contamination of the seal by any solids or liquids that may be present in process flush gas.
- Overall gas leakage is low.

The potential disadvantages to consider include:

- Because of the opposed seal arrangement and limited cooling available from the small flow of buffer gas, the pressure limit is lower than a tandem seal. This may vary depending on seal supplier, detail of the seal design and the application but a typical maximum is 40 bar, subject to confirmation of operating conditions.
- An inert gas supply at a suitable pressure is required.
- Some inert sealing gas will leak into the process gas. This has to be acceptable to the process.

3.5 Compression packing for rotary shafts and valves

3.5.1 Introduction

Compression packing is often known as 'packing', 'gland packing' and also 'soft packing'. The concept dates back to the beginning of the industrial revolution. Although it has often been written off as being outdated the progressive development of materials has transformed the performance of the concept and particularly in the area of industrial valves, it is still the most popular method of sealing.

This section describes the general principles of packing to achieve a seal and the application to rotary shafts.

The term compression packing is probably the best description of this form of seal as they are created by using a series of rings of a soft compliant material that is compressed axially from the seal. Most packing is manufactured by braiding a fibrous material into a square cross-section cord that is then cut to length to form the packing. An example is shown in Figure 3.130.

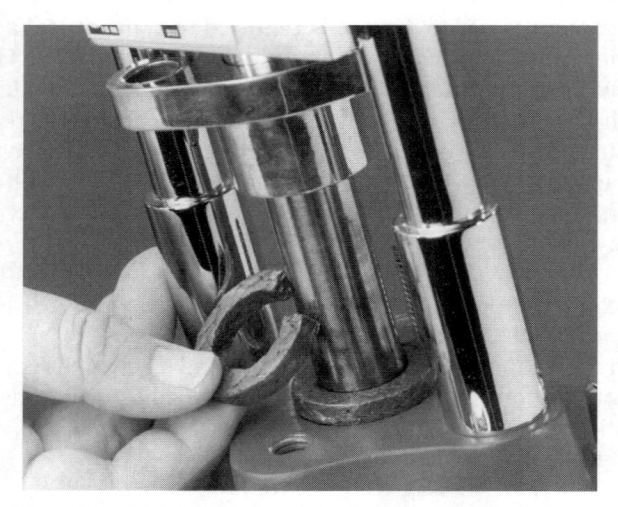

Figure 3.130 Photo of packing being fitted. (Source: James Walker)

The traditional materials for manufacturing packings were natural fibres, such as cotton, flax and hemp for use at low temperatures with non-aggressive fluids and asbestos for high temperature and to provide chemical resistance. Over the last 30 years asbestos has been replaced and a number of synthetic fibres and materials have been progressively developed. The majority of packings are now manufactured from PTFE, Aramid fibres or graphite. A number of packings that are manufactured from a mix of synthetic fibres to take advantage of the individual properties they provide have also been introduced. However, some of the natural fibres are also still in regular use, particularly for water applications and also provide some of the most durable materials, as discussed in section 3.5.6. Materials such as glass fibres are also used for chemical and temperature resistance.

Until the middle of the 20th century compression packing was the almost universal method of sealing rotary and semi-rotary shafts. In many areas it has now been replaced by usually lower leakage and often more compact types of seal. The majority of rotary pumps are sealed by mechanical seals. At lower pressures various forms of elastomer and plastic seals are used. For many valves, including such common examples as domestic taps, it is common to use an O-ring or a plastic U-ring depending on the application. However, packing continues to be popular for a number of applications and for process plant valves it can be the preferred method of sealing to achieve low emission levels, Appendix 1. On pumps packing is generally restricted to relatively low temperatures and low value liquids where there is no danger from VOC pollution. They are to be found on water service pumps, solids handling pumps and also other rotary machines, particularly if slow speeds and fluids containing solids or fibrous particles are involved. In some industries where there is regular pump maintenance due to wear, such as mineral handling, then the use of packing that can last for a similar period of operation as replaceable pump components is a viable option. Packing is also preferred by some authorities for emergency service duties such as building fire water as it is considered to provide the benefit of being able to function without sudden failure, and can be adjusted to provide satisfactory short-term emergency service if required. With the availability of modern mechanical seals with standby arrangements this perceived benefit of packing is probably less convincing than in the past. A major benefit of packing compared with a standard mechanical seal, or typical elastomer and plastic rotary seals, is that it does not require a major machine stripdown for replacement. Releasing the gland follower will allow removal of the old packing and fitting of new rings. There are few other seals that can offer this benefit, split mechanical seals are available but will be a much more expensive option, and elastomer lip seals are available as split seals but usually only in large sizes.

3.5.2 Method of operation

A typical compression packing seal arrangement will be made up from four or five turns of packing fitted in what is often referred to as the gland housing or stuffing box, Figure 3.131. The free packing will be a light interference fit in the gland, but will be relatively easy to push into place. It is then axially compressed by the gland follower which creates a radial expansion of the packing to create a sealing force. The amount of axial compression required will vary considerably depending on the packing and application. For a pump shaft rotating at 1500 or 3000 rev/min a very light loading will be required, as described in section 3.5.4 on fitting. For very slow speeds and valve glands a relatively high loading may be applied to give low leakage with specially designed packing.

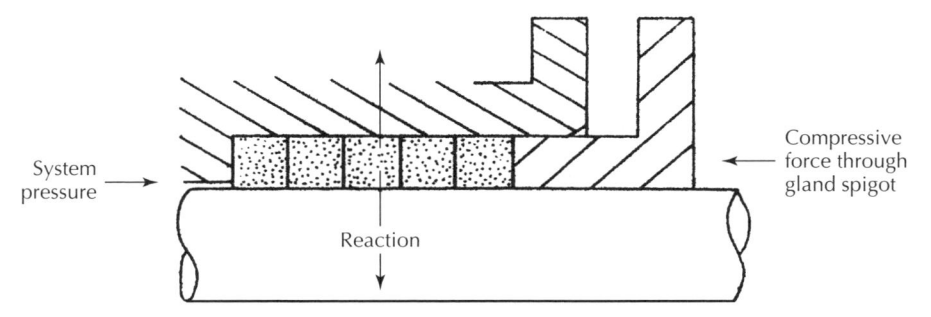

Figure 3.131 Cross-section of simple gland arrangement.

The method of operation of the packing will also be different depending on the application. Where a light gland loading is used, as described above for a high speed shaft, then it has been shown that most of the actual sealing occurs with pressure drop across the last turn of packing.[57] The lubricating film will be present under the inner packing rings. At low speeds, with a higher gland loading, or at low pressure then the pressure drop will be more progressive along the length of the gland. In this second case the action of the packing is more analogous to a very close clearance bush or a fibrous labyrinth seal.

The former case suggests that the once common practice of fitting anything up to eight turns of packing on a pump is counterproductive. It requires a longer shaft, which is both additional expense and causes rotational stability problems. Any attempt to reduce leakage by tightening the packing will cause additional friction and dry running of the atmospheric end packing rings that are withstanding the pressure forces, further accelerating wear. Optimum performance can be achieved with three or four turns of packing, although five turns is quite common to allow the use of a lantern ring. Considerable success has been achieved in some applications with two turns of packing as much less heat is generated.

Packing lubrication may be derived from three sources. The material may have some self-lubricating properties, as in the case of PTFE and graphite-based materials. The fibres and core of the material will be impregnated with lubricant during manufacture. Also, the sealed fluid will be the source of potential lubrication and cooling.

For higher speeds such as centrifugal pumps the flow of sealed fluid becomes important. Although it is possible in some cases to achieve virtually zero visible leakage from pump packing this requires extremely careful packing and running-in of the packing followed by uninterrupted steady state running. In practice this is not a practical situation, as every time the pump stops and restarts, or there is a change of the pumped conditions, bedding-in will have to restart. It is therefore normal for packings to operate with 60 to 100 drops per minute of leakage. Hence their application can only be considered where this level of leakage is acceptable. To obtain satisfactory performance with a pump packing several steps are necessary. The first is to ensure that a good quality packing has been purchased. The selection of the material fibres, lubricant impregnation and braiding process are all critical to the final performance. A low cost packing from an unknown source may be deficient in all three of these important areas. It is then necessary to both fit the packing carefully and pay close attention to the running-in process, section 3.5.4.1.

Low speed applications and valve packing, where the shaft movements are intermittent and relatively slow, will depend almost entirely on the properties of the packing material and impregnated lubricants to permit a very low leakage seal. The most up-to-date valve packings designed for process valve applications can provide extremely low leakage.[58] For many purposes the low levels of leakage achievable with correctly applied premium specialist packings are considered to be comparable to that achieved with metal bellows sealed valve stems, Appendix 1.

3.5.3 Packed gland arrangements

A simple gland arrangement for a low pressure application will consist of three or four turns of packing in a stuffing box bore, Figure 3.132. This can be used for

Rotary seals 223

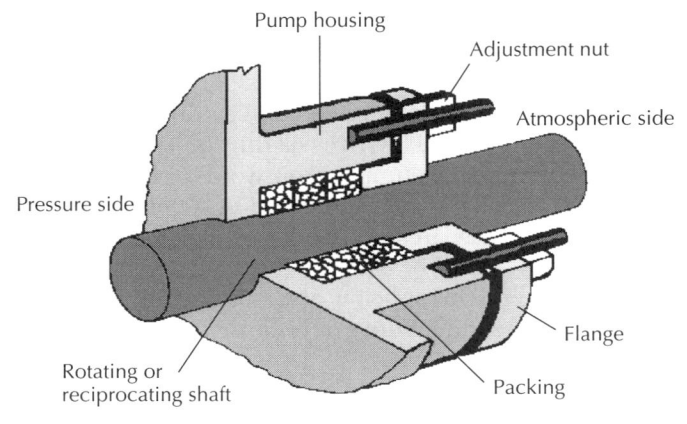

Figure 3.132 Isometric arrangement. (Source: BHR Group)

clean water applications where no addition of clean fluid or cooling is required. Such arrangements have been used successfully up to 40 to 60 bar, but very careful fitting and adjustment of the packing is required. It would more normally be used at low pressure, less than 10 bar.

A variety of stuffing box configurations are used on process pumps and other machinery to either improve packing life or provide a safer working environment.

A typical packed stuffing box arrangement is shown in Figure 3.133. In this case there are five rings of packing with a lantern ring between the second and

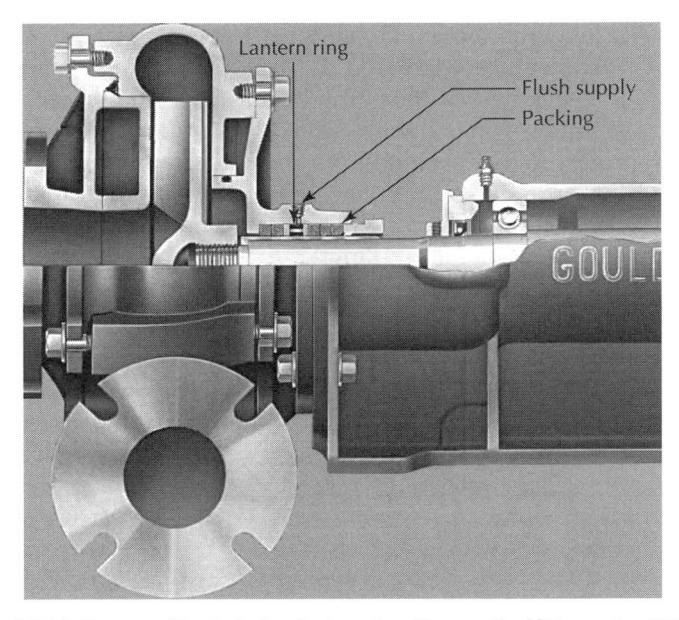

Figure 3.133 Pump packing including lantern ring. (Source: Gould Pumps Inc, ITT Corp)

third rings of packing. The lantern ring is used for the injection of a lubricating, cooling or flushing liquid. This will be supplied at slightly above the pump pressure to ensure a flow of liquid from the lantern ring into the pump. The outer three rings of packing will seal the pressure of the lantern ring pressure to atmosphere but will operate on clean cool liquid. There will thus be a small flow of the lantern ring injection liquid both into the pump and outwards as leakage.

With a clean pressurized liquid the lantern ring is not required, as discussed above. It may, however, be used in a number of applications:

- If the stuffing box pressure is below atmospheric, either because the pump is operating on a suction lift duty or because back pumping vanes are fitted to the impeller.
- An abrasive or crystallizing liquid or solution is being pumped.
- An intermittent duty such as a sump pump where the pump may run dry for some periods.
- A high temperature application where some cooling is required.

With this arrangement the flow through the lantern ring into the pump will be limited to a similar value to the leakage from the pump so it will not provide a high volume of flushing or cooling flow. It will, however, be sufficient to ensure lubrication in the event of dry running and remove soluble abrasives that may otherwise crystallize in the seal area. In some situations, such as suction lift it may be satisfactory to provide the lantern ring flow from the pump discharge, similar to a mechanical seal Plan 11 configuration. If the lantern ring is used to provide a clean liquid, or the pump may run dry on occasions then it will be necessary to provide an alternative supply to the pump.

In highly abrasive situations, such as slurry handling pumps, an alternative arrangement is used, Figure 3.134. In this situation a clean water supply is provided between the packing and the pump impeller. The pump will usually have impeller back pumping vanes fitted and will pump the clean water into the pumped flow, protecting the packing from the abrasive slurry. The back pumping vanes can be very effective and packing in this situation may often work under vacuum even with considerable pump suction head.[31] This configuration is very dependent on the flushing liquid supply for reliable operation and a

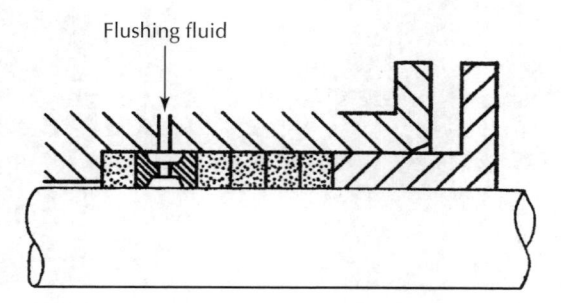

Figure 3.134 Slurry pump with flush at impeller end.

relatively high volume of water flow is required, of the order of one to five litres per minute depending on pump size and slurry density. A pressure of 0.5 to 1.0 bar above stuffing box pressure is recommended to ensure continuous flow. A flow meter and pressure gauge are also therefore required if the flow is to be regulated effectively. If the water flow is interrupted the packing will probably run dry and wear very quickly. Due to the effectiveness of the back pumping vanes the failure may not be evident until the pump stops. Such complications, and the logistics of supplying a reliable water supply to often remotely located pumps, has increased the interest in the application of mechanical seals to slurry applications, section 3.4.6.12.

3.5.4 Fitting and using packing

The fitting of packing is covered in some detail in Appendix 1. Traditionally one of the most difficult aspects of packing replacement was removal of the used packing. This is much less of a problem with modern synthetic packing materials. There is still the potential when overhauling old machinery that it may contain packing made from asbestos. If the presence of asbestos-based packing is suspected then removal should be undertaken with appropriate precautions, and the packing should be thoroughly wetted to avoid creation of harmful dust.[59] If the packing is badly worn and disintegrated careful inspection is required to ensure that all the old packing is removed.

It is important to ensure that the correct size of packing is used. Techniques such as flattening oversize packing or excessively compressing undersize rings do not provide a reliable seal. It is quite possible to find that budget packing is undersize and this should be avoided.

It is then important to cut it to size accurately from the reel of packing. This should be carried out using either a mandrel of the correct size to represent the shaft diameter or a cutting jig with a scale that may be provided by a packing manufacturer. It is sometimes suggested that techniques such as a 45° scarf cut are essential for good performance. This is much less important than accurate and clean cutting of the join. Depending on circumstances and also the type of packing fibre it can be difficult to cut an accurate and repeatable scarf join, in which case a straight butt join cut accurately will be eminently preferable. A scarf join should not be used unless some form of jig is available to ensure the packing is cut accurately at 45° to the correct length and with the join perpendicular to the shaft.

An alternative that can avoid the problems of providing accurately cut packing is to purchase precut rings from the packing manufacturer. Where a limited number of sizes are used this will both reduce fitting time and avoid the waste associated with reels of packing. Some surveys suggest that more packing is wasted from reels of packing than is ever used, as many partly used reels are discarded rather than returned to stores.

Packing rings must be fitted carefully one at a time. Care is required with scarf joins to ensure the correct overlap. The joins in the packing should also be at least 90° apart on successive rings.

3.5.4.1 Tightening the packing

This depends very much on the application. For a pump or other shafts where the speed is sufficient to generate any appreciable frictional heating, in excess of 1.0 m/s, the packing should be left with the gland nuts little more than finger tight. It has proved possible in the laboratory to achieve very low consistent leakage by tightening the packing moderately and then releasing it for some period of an hour or so to relax, before starting.[60] However, this is not often a viable method for industrial applications, unless the machine is to be left on standby, in which case it should be considered.

For these applications the packing must be sufficiently slack for a slight dribble of leakage to occur when started. Most packing suppliers will provide a running-in guide. This will normally advise gentle tightening of the packing by one sixth of a turn on the nuts to reduce leakage to approximately one drop per second at 15 or 20 minute intervals. This must be performed with considerable care as it is very easy to tighten sufficiently to overheat the packing. It is often preferable to wait some considerable period of time, say two or three hours, before any adjustments are made, as it can be found that the packing will bed in and run with reduced leakage. Thereafter very gentle adjustments should be made, often even less than one flat on the nuts, and reasonable time allowed for bedding-in to reduce the leakage before further adjustments are made. It should be noted that some packing materials and constructions are relatively dense and lacking in resilience which can make them extremely sensitive to gland adjustments. This is particularly true of some Aramid, yellow fibre, and also graphite filament packing.

In many situations access to the gland close to the rotating shaft to adjust the gland is not allowed for health and safety reasons which can make the use of packing unsatisfactory. Where access is possible ring type spanners are advisable as they are less likely to inadvertently catch the shaft, an event which can throw the spanner a considerable distance.

For valves and other shafts with very low speed or intermittent movement then the approach is to provide a higher gland loading to inhibit leakage and use the lubricating properties of the packing.

For process valves where low emission packing is used the packing supplier will probably provide information on the packing compression. The value will depend on the type of packing, density and configuration. It is necessary to check that the valve operates freely as the packing is tightened.

For slow speed applications the packing can be adjusted to minimize leakage consistent with not creating excessive friction and overheating.

For both valves and other low speed applications live loading of the packing can be an effective method of maintaining low leakage performance with minimum attention.

3.5.4.2 Live loading

Live loading is the term applied when some form of spring energization is included in the gland bolting system. It is normally achieved by using Belleville disc spring washers between the gland nuts and the gland follower. This provides additional

Rotary seals 227

compliance in the bolting applied to the gland follower. This will ensure satisfactory loading over a long period, but it is also important for maintaining satisfactory loading during temperature and pressure cycles and can avoid the use of excessive loading when new. It is typically necessary to use live loading to achieve a valve packing performance that will comply with the various emission requirements within EPA, ISO and TA-Luft, etc., Appendix 1. The live loading will accommodate a wide range of variables that could affect the packing performance such as vibration, differential thermal expansion, packing relaxation, and bolt creep. A diagrammatic live loaded arrangement is shown in Figure 3.135.

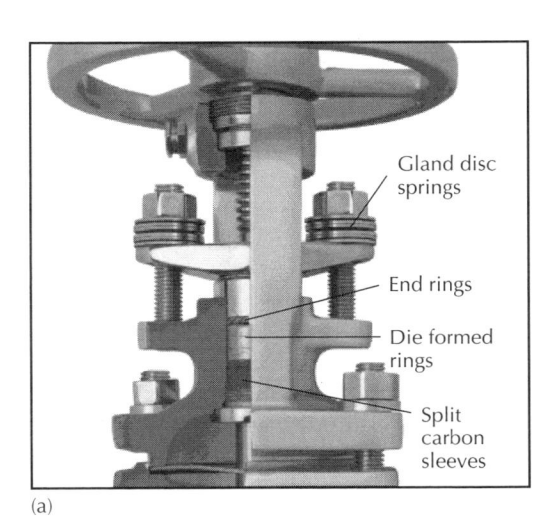

Figure 3.135 Live loading, standard + enclosed, e.g. Latty.

Belleville washers offer a number of benefits for this type of application including compactness, high spring force at low deflection, absence of set or fatigue under normal loads, a straight line load/deflection curve and simple adjustment by the addition or removal of individual springs. They come in a wide variety of materials from the relatively standard 17-7PH stainless steel, to alternative stainless and alloy steels, phosphor bronze and high nickel alloys depending on application.[61]

The selection of springs and the spring force required will require consideration of the operating environment and information from the packing supplier on the recommended packing stress for the type of packing to be used. Some suppliers offer a system with enclosure, or contained unit. With a matched packing set and spring combination the enclosure can be tightened to give a prescribed loading to the packing.[75] A further claimed advantage of this arrangement is that it is then tamperproof.

The use of live loading on pump packing where the speed is higher is comparatively rare. The packing load is much lower and can be critical to individual applications. The selection of a suitable load and loading method is therefore more difficult and it can often cause a situation where continuous wear occurs and packing life is reduced.

228 *Seals and Sealing Handbook*

For large rotary machines with slow rotational speed, such as driers and some mixers, again live loading can be a benefit, especially if access for adjustments is difficult or not possible. The loading method and compression load required will require careful consideration in liaison with the packing supplier.

3.5.5 Packing types

Compression packing is constructed in various forms and from a wide variety of materials. The most familiar industrial construction is the braided fibre material. Other constructions can include extruded cord, wrapped tape and expanded graphite tape.

Extruded cord is the simplest form of packing, usually manufactured from an unfilled plastic. It is not particularly compliant and will only be used for very simple installations such as cold water taps.

With braided or plaited packing a series of sets of fibres are interwoven by machine to form what is usually a square section packing, Figure 3.136. The packing may consist of a single fibre material or a selection of fibres to provide the optimum benefits from each fibre. This is covered in more detail in the description of individual packings in section 3.5.6, but a combination may include Aramid to provide strength and wear resistance and graphite PTFE fibres for reduced friction and additional compliance. The fibres can be lubricated before braiding and also as a final operation depending on the type of lubricant and the application. The leading manufacturers use a variety of specialist techniques to incorporate lubricants in the fibres. There is considerable benefit from well-impregnated lubricant that will be released over a long period of time. A wide variety of lubricants are used depending on the application and to cater for special requirements such as food or water regulations.

Figure 3.136 Braiding machine. (Source: James Walker)

A number of braided fibre packings are also available with a core of elastomer, often silicone. This provides additional resilience to improve sealing with shaft vibration and more compliance during adjustment. The properties of the elastomer must be considered with respect to fluid compatibility and temperature resistance. Temperature at the elastomer may not be as high as that at the dynamic surface. The silicone elastomer will have a similar maximum temperature to most of the synthetic fibre materials used for general purpose dynamic packing.

Expanded or exfoliated graphite material is used to manufacture packing rings. Tape material is wound into a mould and is then compressed into a packing ring. Woven tape materials may also be used, Figure 3.137. The standard rings are a square section. Packing sets for low emission valve applications may be of a variety of geometries which are not square to obtain an improved radial loading, Figure 3.140. Considerable work has been undertaken over a long period to evaluate the optimum density of graphite packing rings.[62] There is a compromise to be achieved between having a soft conformable material and a packing ring that will resist permeation of high temperature gas or steam. The density recommended may vary depending on the application. It is also the practice in some applications to use packing rings of different densities in one set, the high density providing the extrusion resistance and the lower density being conformable to give a good seal on the stem. Appendix 1 describes the possibilities for low emission valve stem packing.

Figure 3.137 Expanded graphite packing rings. (Source: James Walker)

Some packings are manufactured by wrapping a tape, such as PTFE, over a core of braided material or an elastomer core. The primary application is semi-static duties such as furnace doors, hatch covers, etc.

3.5.5.1 Packing types and application areas for rotary shafts

Many types of packing are equally suitable for rotary, static/valve or reciprocating applications. This section highlights the materials and constructions that are especially suitable for rotary applications. The packing types and materials specific to valve applications are discussed in section 3.5.5.2 and types preferred for reciprocating applications in section 4.6.1.

Cotton

Cotton packing is used particularly for cold water applications, but is also useful for other liquids such as oil. Main applications are the water and marine industries. It is also easy to cut, fit and handle compared with some synthetic fibres. Water quality grades are available depending on lubricant. It is particularly useful for pumps that have soft shaft material.

Application conditions:

Media in the range:	pH 6–8
Operating temperature:	−40 to +90°C
Maximum shaft speed:	7 m/s
Maximum pressure:	10 bar

Flax

Flax can provide a dense but flexible packing, and is used in the marine and water industries particularly for seawater, wastewater and other applications involving water with suspended solids. Lubricants can be mineral-based grease and oil, tallow and impregnants such as mica or PTFE dispersion depending on the application. The flexibility of the natural fibres can allow a more controlled response to gland adjustments than many synthetic fibres.

Application conditions vary depending on the lubricant additives but are within the range:

Media in the range:	pH 5–10
Operating temperature:	−40 to +95°C
Maximum shaft speed:	4 to 9 m/s depending on lubricant and construction
Maximum pressure:	20 bar

Ramie

Ramie is an extremely durable, rot resistant fibre and is particularly used in reciprocating pumps. It can also be used for rotary shafts and is suitable for fluids such as cellulose slurry, brine circulation, cooling water systems, and with fluids that crystallize or contain suspended solids. The limitations are primarily the chemical and temperature range when compared to synthetic fibres.

Application conditions:

Media in the range:	pH 4–11
Operating temperature:	−30 to +120°C
Maximum shaft speed:	15 m/s
Maximum pressure:	20 bar

PTFE

A variety of lubricated PTFE packing types may be used for rotating shafts. Many packings are combined with graphite to provide both lubrication and heat dispersion benefits. They may be manufactured from either expanded PTFE, such as Gore GFO, or a solid PTFE fibre depending on the application. PTFE dispersion may also be used to provide a denser packing to resist leakage, but with the potential disadvantage of being more sensitive to gland adjustment. An elastomeric core may also be used which enables the packing to absorb misalignment.

A range of lubricants and additives may be incorporated with the packing during manufacture to provide compatibility with chemicals, food and potable water. These packings can be used with aggressive chemicals and at temperatures beyond the range of the natural fibres. For highly corrosive products PTFE will be the primary selection. The potential leakage must of course be considered when selecting packing for a highly corrosive rotary shaft application.

Application conditions:

Media in the range:	pH 0–14 but not strong oxidizing agents and molten alkali metals
Operating temperature:	−100 to +260°C
Maximum shaft speed:	10 to 20 m/s depending on construction and the PTFE fibre. The highest speeds are generally achievable with expanded PTFE with entrapped lubricant
Maximum pressure:	10 to 20 bar depending on material

Aromatic polymer fibre

The most familiar fibre used for these packings is the yellow Aramid fibre, but alternatives are also manufactured. These are extremely strong and abrasion resistant fibres. They are therefore used in applications on reciprocating pumps that handle highly abrasive slurries or aggressive chemical solutions in the mineral, pulp and paper, wastewater and chemical processing industries. They may also be used on liquids that are liable to solidify such as tar and bitumen. The strength and abrasion resistance of these fibres means that they can cause excessive shaft wear and be excessively sensitive to gland adjustment. The manufacture of the packing is therefore very important. Variants may include special lubricant treatments, incorporation of a flexible elastomer core and a combination of Aramid fibres with expanded PTFE/graphite fibre, Figure 3.138. This latter improves both the adjustability and the lubrication while reducing shaft wear. It should be noted that hardened shafts, coating or sleeve are essential with the majority of Aramid packings. Some more recent fibre developments provide improved performance over the more familiar yellow Aramid fibre.

Application conditions will be very dependent on the packing construction and lubricants used. An elastomer core for instance may limit the fluid and temperature resistance:

Media in the range:	pH 2–13
Operating temperature:	−50 to +250°C (280° for special fibres)
Maximum shaft speed:	20 m/s
Maximum pressure:	25 bar

Figure 3.138 Hornet packing. (Source: James Walker)

Glass

General purpose glass fibre packings are manufactured with a combination with other fibres which aids the manufacture and lubrication of the packing. The application limit of these packings is limited by the additional fibres used. They offer a wide chemical compatibility but potentially limited temperature range. The major benefit is as a general purpose replacement packing. Dependent on the fibre mix they can offer low shaft wear and also be potable water compatible.

Application conditions:

Media in the range:	pH 0–14 (dependent on fibre mix)
Operating temperature:	−50 to +130 or 280°C (dependent on fibre mix)
Maximum shaft speed:	12 m/s
Maximum pressure:	20 bar (dependent on fibre mix)

Specialist glass packings will use a glass fibre yarn with high temperature lubricants and graphite impregnation. This will primarily be used if high temperature resistance is required, but the dynamic properties are more limited.

Application conditions:

Media in the range:	pH 4–10
Operating temperature:	−40 to +350°C
Maximum shaft speed:	10 m/s
Maximum pressure:	10 bar

Graphite filament

These are typically manufactured from graphite yarn, impregnated with PTFE dispersion and graphite powder. The benefits are a high temperature and speed capability together with wide chemical resistance. The packing is expensive and also very sensitive to gland adjustment so can be difficult to use. This has proved to make it difficult to apply reliably in some situations. However, it can be used in a wide variety of applications so has the possibility of being a standard packing across a site if this is important. Application conditions:

Media in the range:	pH 0–14 (excluding strong oxidizing agents, molten alkali metals, fluorine gas and fluorine compounds)
Operating temperature:	−50 to +400°C
Maximum shaft speed:	20 m/s
Maximum pressure:	25 bar

Metallic foil
Metallic foil packings are manufactured using a foil, usually aluminium or lead, which is coated with lubricant such as oil and graphite. The foil is usually crinkled, then twisted and folded over a core of lubricated yarn to provide flexibility. This type of packing was widely used for oil duties before mechanical seals became popular. A glass yarn core can provide temperature resistance. Packings made entirely from foil without a fibre core will have poor flexibility. This will make adjustment difficult and limit their ability to cope with shaft misalignment, etc.

Application conditions:

Media in the range:	pH 6–8 (aluminium) excluding steam and corrosive agents
	pH 4–10 (lead)
Operating temperature:	−70 to +540°C (aluminium), +260°C (lead)
Maximum shaft speed:	7 m/s (aluminium), 20 m/s (lead)
Maximum pressure:	20 bar

3.5.5.2 Packing types for valves

Valve stem motion may be rotary, linear or a combination of the two depending on the type of valve. The discussion of valve packing types is included adjacent to the rotary packing for the convenience of the reader as much of the information is applicable to each category.

The general description of compression packing gland arrangements is discussed in sections 3.5.2, 3.5.3 and 3.5.4. A wide selection of seal types are used for valves depending on the type of valve and application. Compression packing is widely used on rising stem valves whether they are purely linear or rotary/linear in action. Purely rotary action valve stems such as ball valves may use packing or an O-ring or other elastomer or plastic configuration depending on the intended application.

Valve applications vary extremely widely from room temperature water through the many oils and lubricants to high temperature water and steam to aggressive chemicals and hot VOCs. While there is the possibility of using one almost universal material and style of packing this would involve the use of a premium product which is not necessary in a high proportion of applications.

Many of the packings discussed in section 3.5.5.1 can be used for valve packing especially in low duty applications where there is no leakage or emission hazard. They can generally be used within the application limits for temperature and fluid compatibility in section 3.5.5.1 but as they are essentially static or very slow moving the pressure limits can be those applied to the reciprocating applications in section 4.5.1.

The packings that are essentially designed for dynamic duties will not provide the very low leakage levels that are achieved with specialist valve packing. The braided fibre construction designed to have some resilience and inclusion of lubricants for dynamic use mean that there are more inherent leakage paths in the packing. The gland load applied for valve applications will cause the embedded lubricants to 'bleed' with time and hence some gland follow-up will be required.

However, for general purpose maintenance of plant valves and for many service and other duties on non-hazardous fluids the use of dynamic packings that are in use on machinery in the plant is both technically satisfactory and makes for considerable simplification of maintenance, purchase, stocking and fitting procedures. The recommended areas for the use of general purpose packing and when specialist valve packing will be required is presented in Appendix 1.

There has been extensive research over more than a decade to both produce valve packing with extremely low levels of leakage and also devise procedures to measure these low emission levels reliably.[64–73] This has led to the development of packing sets that have been shown to produce extremely low emission levels, and for most practical purposes are considered to offer a similar level of integrity to that provided by a metal bellows sealed stem, Appendix 1.[71]

These packings are all based around the use of expanded graphite material. This has been found to provide both the low emission and wear resistance to provide the required performance over a long period. It has a wide temperature range and can be used in all but the most aggressive chemicals, where a PTFE packing may then be used. Guidance on the application areas of the two materials is provided in Appendix 1.

The initial use of graphite as a valve packing involved the use of compressed graphite rings. It was found that these could suffer excessive wear and be subject to localized scoring. It was also found that by using a turn of braided carbon packing at the top and bottom of a stuffing box the packing would operate far more satisfactorily. This was thought originally to be due to removal of graphite wear particles by the braided packing. Subsequent work showed that the real benefit is from the deposition of lubricants from the braided packing.[69] An example of a packing set of this type is shown in Figure 3.139. To reduce emission levels a number of suppliers have introduced packing sets that use radially chamfered rings of packing to promote radial energization of the appropriate rings and hence conformability to the shaft. Two examples are shown in Figure 3.140. The most recent developments are to use braided graphite ribbon, which when combined with suitable lubricants and reinforcing have proved to be suitable for both on–off and control valves.

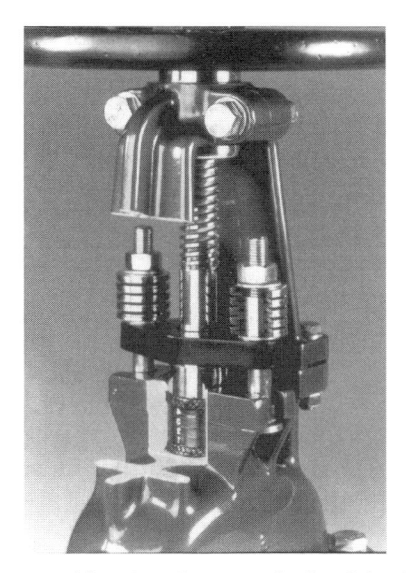

Figure 3.139 Valve packing set, exfoliated graphite rings plus braided carbon. (Source: James Walker)

(a) (b)

Figure 3.140 Low emission valve packing sets manufactured from graphite rings. (Source: Garlock)

Some examples of valve packing materials include the following:

Compressed exfoliated graphite fibre rings
The rings are moulded from expanded graphite foil. Suppliers may also incorporate lubricants or low friction coatings. The rings may be endless for optimum performance, or split to facilitate fitting.

These packing rings are often used in conjunction with braided packing rings, as described above in this section. This will depend on the individual material and the application. The lubricants or coatings supplied by some suppliers can reduce the requirement for braided rings, but may also limit the temperature.

They are also supplied as tailor-made low emission sets with radially chamfered rings, Figure 3.140(a).

Media in the range:	Liquids or gases pH 0–14 except strong oxidizing media
Operating temperature:	−250 to +450°C in oxidizing media +600°C in steam, +1000°C non-oxidizing Temperature limit also dependent on lubricants used so may be lower than the limits of the graphite
Maximum pressure:	150 to 500 bar dependent on design and arrangement

To achieve highest pressures the valve design and anti-extrusion design will be critical.

Braided graphite ribbon

A range of valve packings are manufactured using woven constructions of exfoliated graphite ribbon, Figure 3.137. These can now be used for many applications that may previously have required the use of the moulded graphite foil rings. As they can be readily used in the form of rings cut from a roll there is less necessity to have stocks of packing rings to suit a range of valve sizes. The construction may vary depending on the intended application. The density of the weave, lubrication and reinforcing can be varied to provide high pressure resistance or improved lubrication. Packings are available that may be reinforced with corrosion resistant wire, this may be stainless steel or a high nickel alloy. These packings are intended solely for valve duties and are not recommended for use in pumps.

Media in the range:	Liquids or gases pH 0–14 except strong oxidizing media
Operating temperature:	−250 to +450°C in oxidizing media +600°C in steam, +1000°C non-oxidizing atmosphere Temperature limit also dependent on lubricants used so may be lower than the limits of the graphite
Maximum pressure:	150 to 500 bar dependent on design and arrangement

To achieve highest pressures the valve design and anti-extrusion design will be critical.

Control valve packings

Process control valves have traditionally been sealed by V-ring sets manufactured from filled PTFE. These have been used to provide lower friction and improved sealing compared with traditional woven compression packings.[73] However, the PTFE does not provide sufficient resilience and recovery to provide effective emission control especially where temperature cycling occurs. To improve the emission control perfluoroelastomer rings have been included into the stack of PTFE. This requires separate lubrication to achieve low friction and also a persistent lubricant for sealing gases. The temperature range is also relatively restricted by the elastomer.

Packing manufacturers have developed graphite packing that is suitable for use in control valve applications. This can be a graphite foil type with the incorporation of PTFE,[68] or it may be a woven graphite ribbon material with specialized lubricants and reinforcing.[72] The temperature range of the packing will be dependent on the method of lubrication. Those incorporating PTFE are likely to have a lower maximum temperature than those with alternative inert lubricants.

Media in the range:	Liquids or gases pH 0–14 except strong oxidizing media
Operating temperature:	-250 to $+400°C$ in oxidizing media. Temperature limit also dependent on lubricants used so may be lower than the limits of the graphite
Maximum pressure:	200 bar dependent on valve design

Emissions measurement

Correctly applied the graphite valve packings described above can provide extremely low emission levels. However, the measurement of emissions requires specialist equipment and procedures to ensure reliable measurement.[65,66] A number of standards are used to qualify valves fitted with low emission packing. These include: Draft ISO 15848, Parts 1 to 3, VDI 2440 (also referred to as TA-Luft), Shell MESC SPE 77/312 and Environmental Protection Agency (EPA) Method 21. There is still considerable debate on the comparative relevance of the differing test methods. It has been recognized that it is ultimately necessary to coordinate these standards into something that is acceptable to the majority of manufacturers and users, but this may take some time.[74] Further details of the standards are included in section 7.2

3.6 Clearance seals

3.6.1 Introduction

Clearance seals are very widely used on rotary shafts. There are a wide variety of seal types which are discussed in the sections of this chapter. A very obvious problem with a clearance seal is that it does not in isolation provide a complete barrier to leakage. However, they are used for a number of reasons including:

- Very low friction.
- High speed.
- Absence of wear, especially if operating in gas at high speed.
- Capability to deal with contamination with minimal wear.
- Provide an effective throttle or as part of a pressure breakdown for protection of a main contacting seal.

The seals may vary from quite simple, very low cost, high volume products to very highly engineered designs for turbo machinery and very high speed spindles.

3.6.2 Labyrinth seals

As implied by the name, a labyrinth seal functions by providing a contorted path to inhibit the leakage. With a rotating shaft the seal can be designed to work by two methods:

- Rotating radial faces can cause centrifugal separation of liquid or solids from air.
- A series of restrictions followed by a clear volume creates expansion of a gas and hence reduces the pressure.

There are thus two main areas where labyrinth seals may be used. The first is for the protection of bearings in environments where liquid contamination is a problem and the second is as a pressure seal on turbo machinery.

3.6.2.1 General purpose bearing seals

A variety of shields and simple moulded elastomer or plastic devices are used to protect rolling element bearings. Some of these are briefly described in section 3.2.5. Metal shields may be used to act as centrifugal expellers to remove moisture and occasional solids from the area of a bearing, Figure 3.141. Simple labyrinths are also used as shown in Figure 3.35. These provide only very limited protection and will not prevent either moisture or fine dust from reaching the bearing. Neither design should be used in either a moist, steamy or dusty environment.[76, 77]

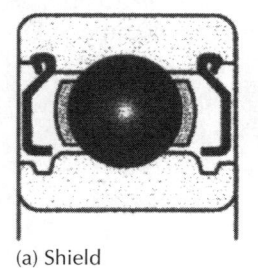

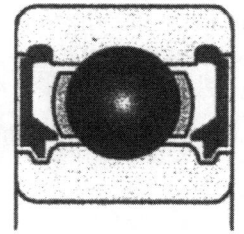

(a) Shield   (b) Non-contact seal

Figure 3.141 (a) A shielded bearing is effectively a very simple labyrinth. (b) A simple moulded labyrinth seal for very light duty protection of bearings from large particles of contaminant.

3.6.2.2 Pump bearing excluders

Labyrinth seals are widely used to protect the bearings on process pumps and other similar continuously running machinery such as fans, mixers, etc. These are required to provide effective protection to the bearings in a process plant environment. A typical life requirement can be 40 000 hours, so a sealing system that is effectively non-wearing is attractive. The seals are required to retain the lubricant in the bearing housing and exclude external contaminant that may be dust, steam or water spray. These seals provide a contorted path between the bearing and atmosphere, together with a series of rotating radial faces to create centrifugal separation of liquids and any solid contaminant. Drain passages can be used to ensure that the separated liquid is either drained back to the sump, in the case of bearing oil, or that potential contaminants are removed externally, Figure 3.142.

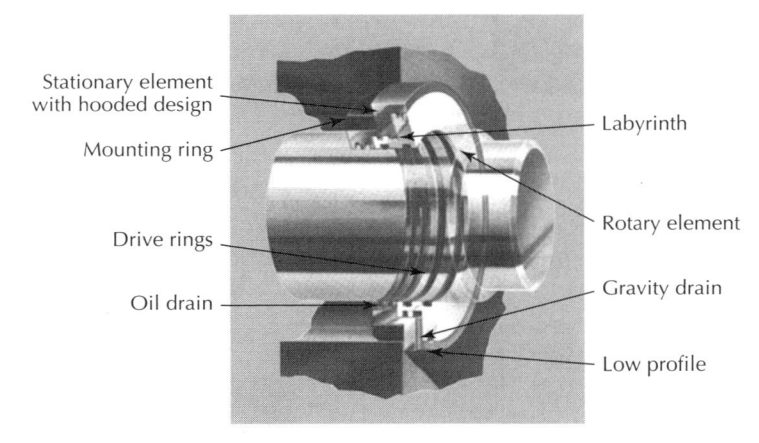

Figure 3.142 Process plant bearing protectors provide a labyrinth with radial expeller faces and drains to return liquid to the appropriate side of the seal for lubricant retention and contaminant removal.

The external access to the labyrinth passage may be shielded, such that there is an axial overlap on the outside diameter of the rotating part of the seal, Figure 3.143. This provides additional protection to prevent access of any external liquid that is being expelled by the centrifugal action of the rotor.

A problem with this type of seal is that it is primarily designed to provide dynamic liquid sealing, either lubricant retention or expulsion of contaminants. If the shaft is stationary this function is not available. This can be a particular problem as when stationary the bearing housing may cool down and ingest moist air, leading to contamination of the lubricant. Some labyrinth type bearing protectors use an O-ring arrangement. This is designed to lift off when the shaft is rotating, and seal when stationary, thus providing static sealing protection.

One design is widely used in the process industry for this purpose. This is shown in Figure 3.143(a). The O-ring (3) is stretched by the centrifugal force when the shaft rotates and lifts off to float in the seal groove. It will sit down to seal the labyrinth gap when the shaft stops. The O-ring has generally been found to exhibit minimal wear even after tens of thousands of hours in operation.

240 *Seals and Sealing Handbook*

Inpro/seal® VBXX-D

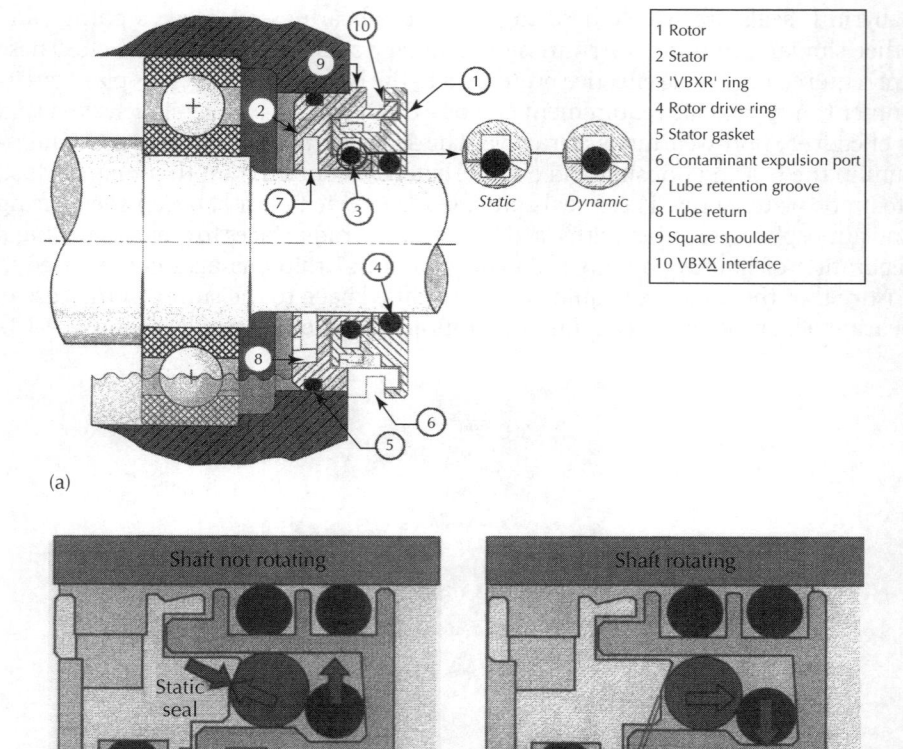

1 Rotor
2 Stator
3 'VBXR' ring
4 Rotor drive ring
5 Stator gasket
6 Contaminant expulsion port
7 Lube retention groove
8 Lube return
9 Square shoulder
10 VBXX interface

Figure 3.143 (a) Inpro seal O-ring labyrinth seal with automatic O-ring designed to seal when the shaft is stationery. (b) An alternative design. The O-ring expands under centrifugal force to release the static seal. (Source: Inpro)

An alternative design is shown in Figure 3.143(b) moisture ingress under bearing chamber breathing conditions has been addressed through the 'Arknian' shut-off valve. This valve comprises an axial energizing member, which automatically adjusts the contact force on the primary shut-off valve. In the equipment idle condition, the elastomer applies a radial load on the upper quadrant of the primary shut-off valve. When the equipment is operating, the energizing member is subjected to centrifugal forces which urge it radially outwards and disengage it from the primary shut-off valve. This allows the primary shut-off valve to move away from the stator, thus creating a micro-gap.

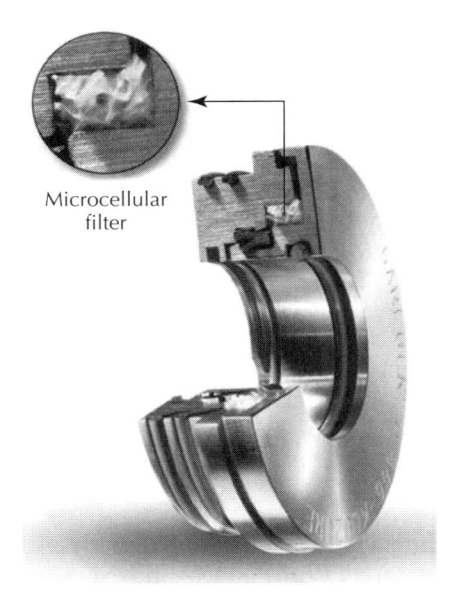

Figure 3.144 Labyrinth bearing seal with a porous insert to act as a filter. (Source: Garlock Klozure)

This gap permits the bearing chamber to breathe. As the equipment stops, the energizing member again takes up the idle position and axially moves the primary shut-off valve to make a static seal with the stator. The shut-off valve member is manufactured from a wear-resistant material with a low coefficient of friction.

The ingress of solid contaminant can also be addressed by the use of a small porous filter element in the seal, Figure 3.144. The effectiveness of the filter will presumably be somewhat affected by progressive wear as it is fitted between the rotor and the stator.

This design of seal depends on good alignment of the shaft and housing to provide a small clearance between the rotor and stator of the seal. This is not always possible, as in, for example, self-aligning pillow blocks. Specific designs are available that align the seal with the shaft, Figure 3.145.

These designs are widely used for bearing protection and lubricant retention in a wide range of process industry and other continuous duty applications. They may not be suitable in all cases and will not be a suitable selection if for instance there is excessive water or solid contaminant in the area of the bearing housing. In such cases a positive contact seal, such as a mechanical seal may be required. Typical design options that are used for bearing protection are discussed in section 3.4.4.8.

3.6.2.3 Machine tool spindles

The bearings of machine tools are expected to rotate at high speed and with high accuracy in an environment that includes a spray of cutting fluid and machine

242 *Seals and Sealing Handbook*

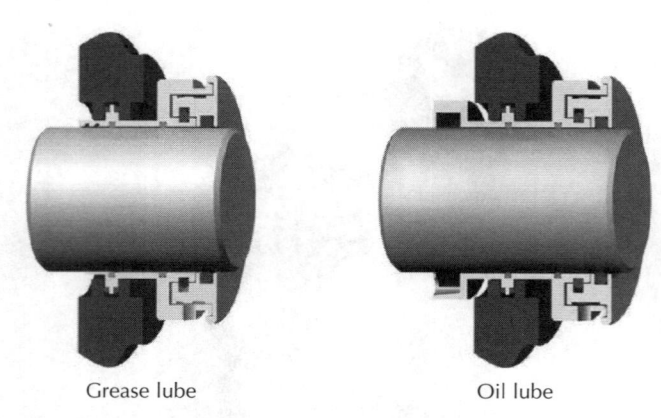

Grease lube   Oil lube

Figure 3.145 A labyrinth arrangement for self-aligning bearings which allows the seal to remain aligned with the shaft. (Source: Inpro)

tool chippings or grinding dust. An effective seal with extremely low power consumption is therefore required. While lip seal and excluder arrangements may be considered for low speed spindles, many modern machine tools are operating at high spindle speeds, from 5000 to 100 000 rev/min and also often in a virtually continuous production environment. It is therefore necessary to have a seal reliability similar to that discussed for process pump bearings in section 3.6.2.2, to protect high precision bearings operating in an environment of continuous coolant spray and machining chippings or swarf.

Simple labyrinths, Figure 3.35, may be used in some spindles but for improved protection a shrouded radial labyrinth, similar in concept to those used for pump bearings, Figure 3.143, may be used. To provide improved protection against the ingress of contaminant it is also common to provide an air purge to the seal. This is intended to positively exclude coolant spray and machine tool chippings, Figure 3.146.[78] This type of arrangement is preferred by a number of machine tool manufacturers to provide effective contaminant exclusion. The use of the air purge is also effective in preventing the breathing of contaminant when the machine cools down. Particular precautions to this effect are necessary in certain cutting operations.

An additional level of safety can be provided by including a positive contact seal that will operate if the air supply fails. Figure 3.147 shows a labyrinth with air purge and an elastomer V-ring.[79] The air purge will lift the seal and vent to atmosphere removing contaminant. If the air supply fails the V-ring will act as a contact seal to exclude contaminant. This arrangement can also help to make the spindle seal less sensitive to the orientation of the machining head as it is better able to be flooded by coolant.

3.6.2.4 Turbo machinery labyrinths

The labyrinths used in turbo machinery are a very different concept to those used for bearing protection. These are used to seal the pressure typically between

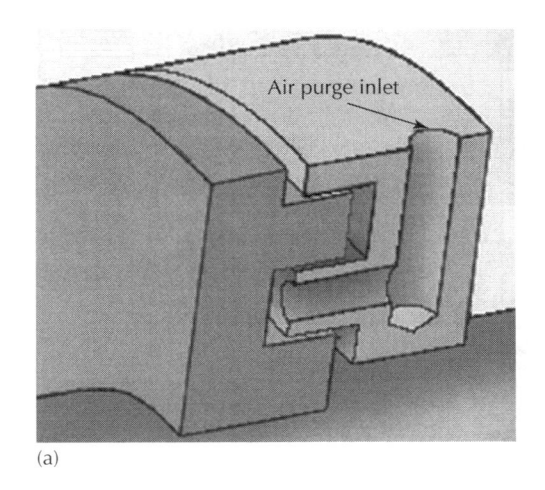

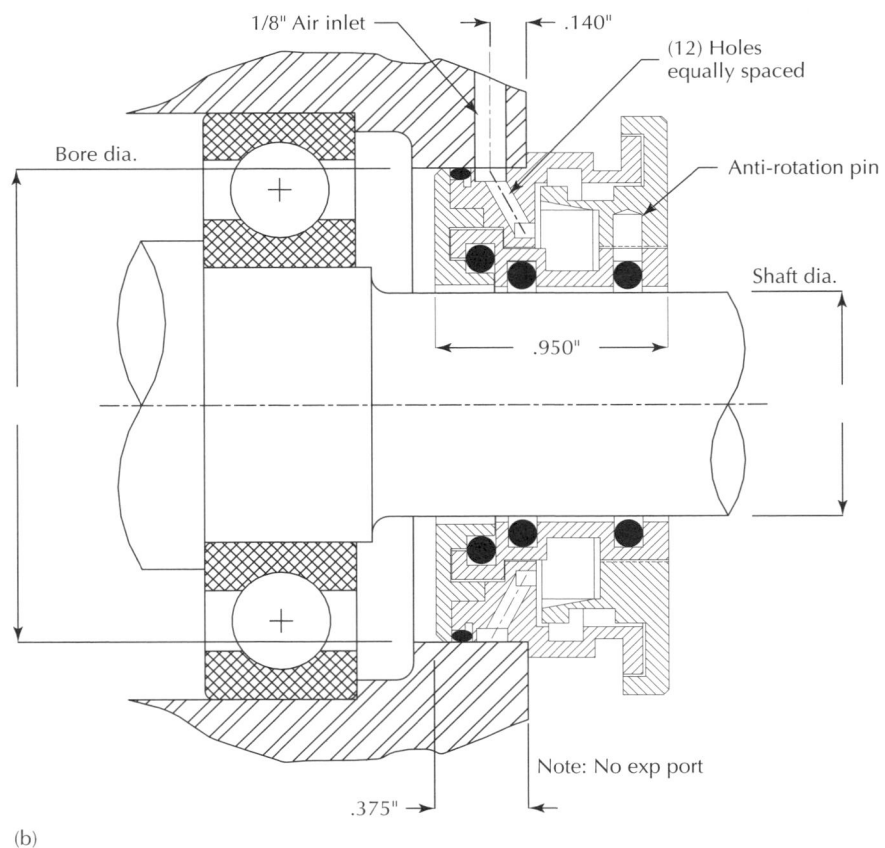

Figure 3.146 A labyrinth for machine tool spindles that includes an air purge for additional protection. (Source: Inpro)

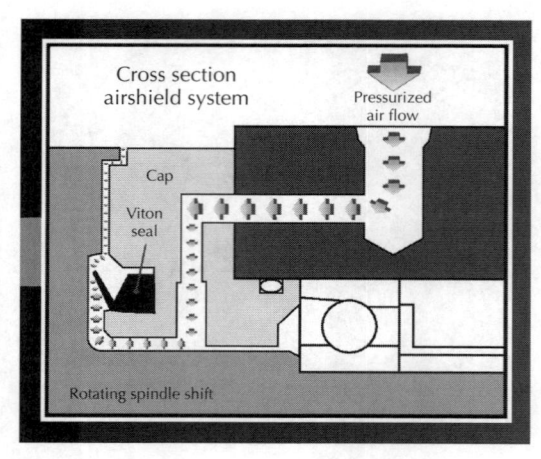

Figure 3.147 An air purge labyrinth with a V-ring to act as a positive shut of seal when there is no air pressure. (Source: Setco)

stages or sections of a machine. This may be a steam or gas turbine or a rotary gas compressor. Predominantly the use of these seals is now concentrated on the internal seals within turbo machinery. Although they were traditionally used as the seal to atmosphere on equipment such as steam turbines, this function can now be accomplished with much lower leakage and a more compact seal arrangement using either a dry gas seal or one of the advanced face seal designs such as wavy faces, depending on the actual seal duty requirement and steam condition.

A series of labyrinth teeth are arranged to provide a close clearance gap between the rotor and stator of the machine, Figure 3.148. A pressure drop will then occur across each gap. The configuration may be either a straight-through, sometimes known as a see-through design, or a stepped arrangement. The stepped design offers a more complex path, and hence the offer of better pressure reduction as it prevents any gas by-passing the expansion volume between the teeth. The additional leakage for the straight-through design is considered to be some 20–30% higher for equivalent dimensions. The disadvantages of a stepped design include more complex manufacture and also the requirement for split housing components which could be both a manufacturing and assembly disadvantage. There is also a limit to the design if it is necessary to accommodate considerable differential expansion between the shaft and casing during startup and shutdown. The staggered design presents further manufacturing difficulties and is therefore even less common.

The teeth of the seal are designed to be a close clearance with the counterface, and with large rotating machinery the possibility of some shaft instability or other runout must be accommodated especially during startup, shutdown or if an off-design running condition occurs. For this reason one part of the labyrinth must be a wearing material. This will depend on the application as the material must be capable of providing sufficient wear resistance and be compatible

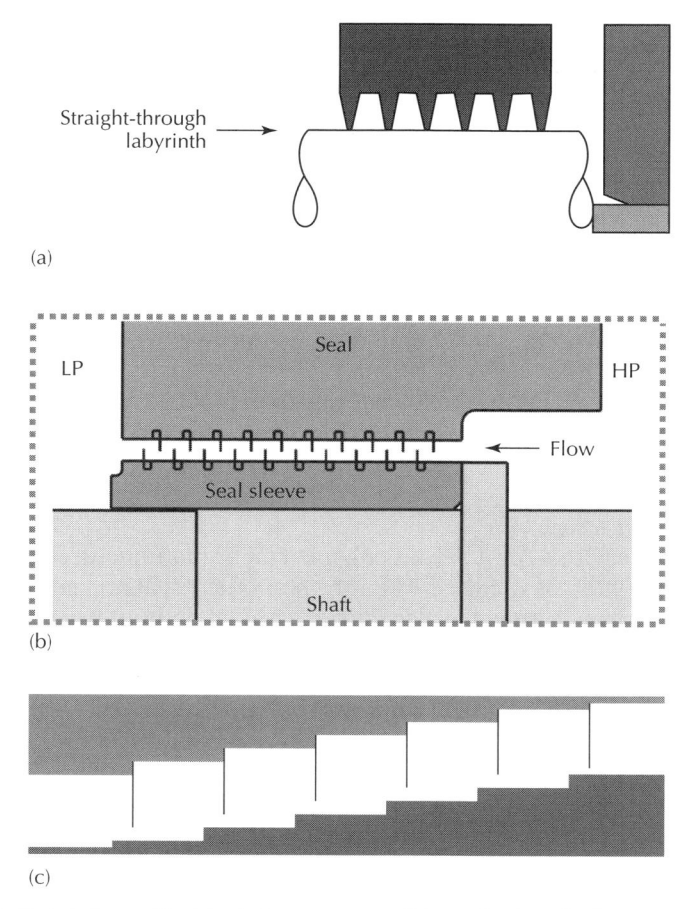

Figure 3.148 Basic turbo machine labyrinth arrangements ((a) straight-through, (b) staggered, (c) stepped).

with the temperature and operating fluid. The potential materials are discussed later in this section. In a conventional straight-through labyrinth it is common to provide the teeth in the housing, where they are readily replaceable when they have worn.

Calculation of leakage
There are a considerable number of publications that have offered calculations of labyrinth seal leakage.[80, 81] As it is a leakage of gas through an annular gap, it may at first appear that a knowledge of the gas, or vapour characteristics and seal dimensions, should enable a progressive calculation of the pressure breakdown. However, it has been found these calculations using bulk flow models[81] are not entirely satisfactory, and it is considered that a computational fluid dynamics (CFD) model is necessary to make a satisfactory calculation of leakage.[82] A major factor in the leakage that is not considered by the models is the rotation of the shaft which creates a rotordynamic component to the gas flow.

The preparation of a complex CFD model will therefore be necessary before an accurate leakage calculation is feasible.

There is considerable interest in improving the performance of the seals in turbo machinery as it has a direct effect on the performance. Any leakage between stages or across balance drums, etc. represents a loss of performance. Also as pressures are increased to uprate efficiency a conventional labyrinth will become longer with consequent undesirable effects on machine size, cost and also rotordynamic stability. To illustrate the importance of this area there may be up to 50 gas path sealing locations in a gas turbine. It has been estimated that in a large turbofan engine seal deterioration can lead to a 1% increase in fuel consumption per year. In smaller engines the internal flow can account for 17% of the power.[92]

Reducing the clearance in these seals has therefore become a priority. Modifications to labyrinth seals have included two variations, rub-tolerant polymers and abradable seals.[83]

Rub-tolerant polymer seals

Labyrinths made with a rub-tolerant polymer can be manufactured with reduced clearances. If contact is made between the stationary seal and smooth rotating shaft, the stationary teeth will deflect during contact without wear or damage to the rotor or seals. These are generally manufactured from thermoplastics, with good fluid and temperature resistance. The thermoplastic matrix materials are tougher and offer the potential of improved hot/wet resistance.

Abradable seals

The use of abradable seals is one way of accomplishing reduced clearances, and limits the risk of damage to the rotating or stationary member if a rub occurs. The concept of an abradable seal was first used in aircraft engines. However, over the years, the use of abradable seals has become more common in centrifugal compressors. In conventional seal applications, the stationary static aluminium labyrinth contracts the gas as it flows through the close clearance gaps under the teeth, and then expands it between the teeth. This repeated contraction and expansion reduces the flow of the gas and lowers its flow rate. Leakage through the seal is proportional to the clearance, Figure 3.149(a). For an abradable seal, the labyrinth teeth are now on the rotating component, and the smooth abradable seal is the stationary component, Figure 3.149(b). In the case of the abradable seal much tighter clearances can be achieved, since if contact should occur between rotor and seal, the rotating teeth will cut into the seal and remove the abradable material, leaving grooved markings where contact had been made. The rotating teeth can be made an integral part of the impeller, Figure 3.150, for both shaft and impeller eye seals. A crucial factor from the leakage aspect is that the abradable wear does not increase the clearance so leakage is not affected to any great extent.

Another application for abradable seals in centrifugal compressors is in the balance piston. The balance piston, Figure 3.151, rotates with the shaft and, if

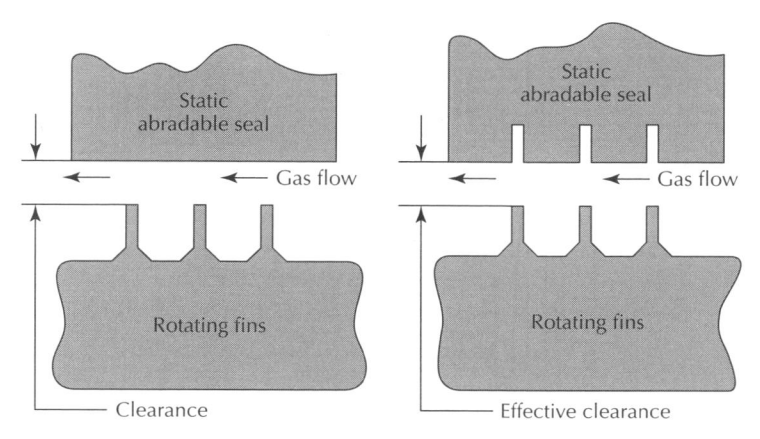

Figure 3.149 Example of wear of a conventional labyrinth with an abradable design. (Source: Sealing Technology)

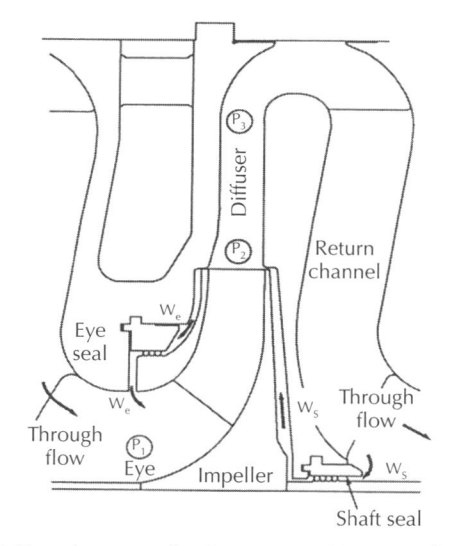

Figure 3.150 Abradable seals on centrifugal compressor. (Source: Sealing Technology, Elliott)

contact does occur with the abradable seal, there is negligible effect. The efficiency gains that have been achieved using abradable seals in these applications are shown in Table 3.1.[83]

Materials for labyrinth seals
The traditional labyrinth, Figure 3.149(a), requires a material for the teeth that will allow them to wear when rubbed by the rotor without damage to the main machine rotor. They would therefore be manufactured from a suitable material such as bronze or aluminium.

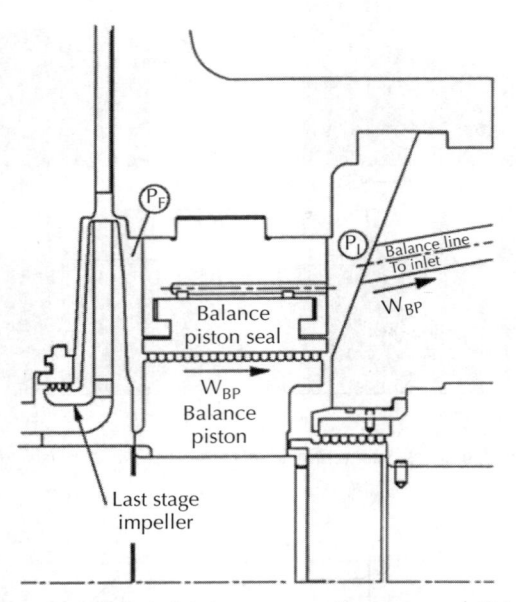

Figure 3.151 The balance piston labyrinth is important to the compressor efficiency and thrust bearing loading. (Source: Sealing Technology, Elliott)

Table 3.1 Efficiency gains using abradable labyrinth seals

Inlet flow (ICFM)	Impeller eye and shaft seal clearance reduction (%)	Stage efficiency gain (%)
1000	24	1.1–2.4
10 000	21–22	0.6–1.5
100 000	20–21	0.3–1.1

Inlet flow (ICFM)	Balance piston seal clearance reduction (%)	Total efficiency gain stage & balance piston (%)
1000	33	5.9
10 000	33	1.5
100 000	36	0.9

(Source: Sealing Technology Elliott)

For abradable seals it is necessary to select the material carefully to ensure both abradability but also fluid compatibility, erosion resistance and ensure that galling of the rotor does not occur. Nickel graphite abradable seals were originally developed for gas turbines and have also been used extensively in various applications for centrifugal compressors. However, to achieve optimum abradability and erosion resistance, special control of the hardness needs to be achieved. Silicon aluminium polyester abradable material has also been used in various centrifugal compressor applications, with considerable success. For lower temperature duties filled TFE and also silicone elastomer have been tested. Examples of candidate materials from Dowson,[83] are shown in Table 3.2.

Table 3.2 Temperature limits abradable materials for compressors

Abradable material	Minimum temp.	Maximum temp.
Mica-filled tetrafluoroethylene	−150°F (−100°C)	350°F (180°C)
Nickel graphite blended powder	−320°F (−196°C)	900°F (480°C)
Aluminium powder alloy containing silicon and polyester resin	−320°F (−196°C)	650°F (340°C)
Silicon elastomer containing hollow glass microspheres	−100°F (−73°C)	500°F (260°C)
Nickel chromium powder with lucite polymer	−320°F (−196°C)	1200°F (650°C)

(Source: Sealing Technology, Elliott)

For rub-tolerant polymer seals a number of filled high duty plastics are being tested, including polyamide-imide (PAI) and PEEK. Materials such as carbon fibre filled PEEK are showing potential use for this application at temperatures up to 340°C.

A material combination being investigated for gas turbines is a plasma sprayed boron nitride–silicon/aluminium powder. This is used in conjunction with labyrinth teeth manufactured from a titanium/aluminium/vanadium alloy.

Rotordynamics
The presence of swirl within the labyrinth was mentioned previously in this section in connection with the problems of calculating the leakage. A serious effect of this swirl is also the contribution to instability of the rotor. This problem was first identified in 1980.[84] As attempts are made to reduce rotor mass, and increase operating pressures, this can become an increasing problem. The fitting of 'swirl breaks' before a labyrinth is considered essential.[82] However, the design of labyrinths to include devices to interrupt the swirl have also been developed by major compressor and turbine manufacturers. Two examples are shown in Figure 3.152.[85, 87]

To reduce leakage further and also overcome the stability problems experienced using labyrinths a number of alternative designs have been developed such as hole slot and honeycomb seals, section 3.6.3.

3.6.3 Honeycomb and hole slot seals

Honeycomb and hole slot or hole pattern seals have been developed as an alternative to labyrinth seals. A particular benefit is that they inhibit the rotordynamic swirl and can therefore be used as a damper to stabilize the machine rotor.

Honeycomb seals are used in gas turbines as well as steam turbines and rotary compressors. The design and the materials will vary because of the very different operating conditions. For gas turbines the seals may also be designed to

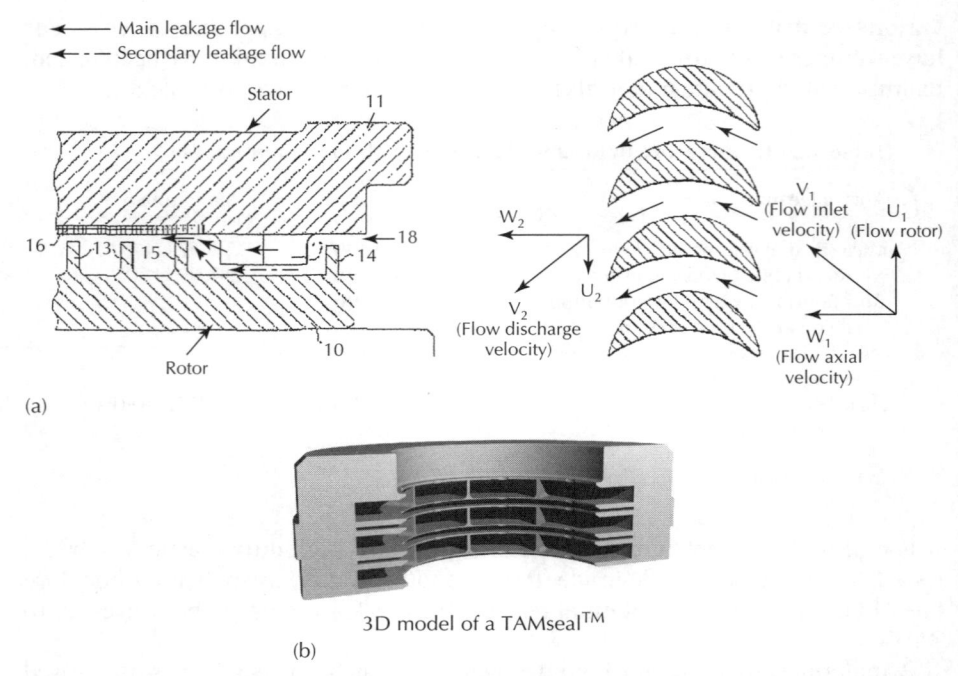

Figure 3.152 Examples of swirl reducers in labyrinth seals: (a) Elliott Patent; (b) Sulzer. (Source: Elliott Patent No. WO 2004/113770 and Sealing Technology, Sulzer)

be abradable as the operating temperatures and rate of temperature change create much more significant variations in clearance between rotor and stator.

An example of a gas turbine honeycomb pattern is shown in Figure 3.153. These are manufactured from high temperature alloys, such as Hastelloy X and Haynes 214, N07214.[86] The honeycomb is manufactured from thin metal foil with thicknesses in the range 70–130 μm. The degradation of the honeycomb material with time, and the effect of the high temperatures, in the range 700–1100°C, on the abradable properties of the alloy are prime concerns in the development of these seals.

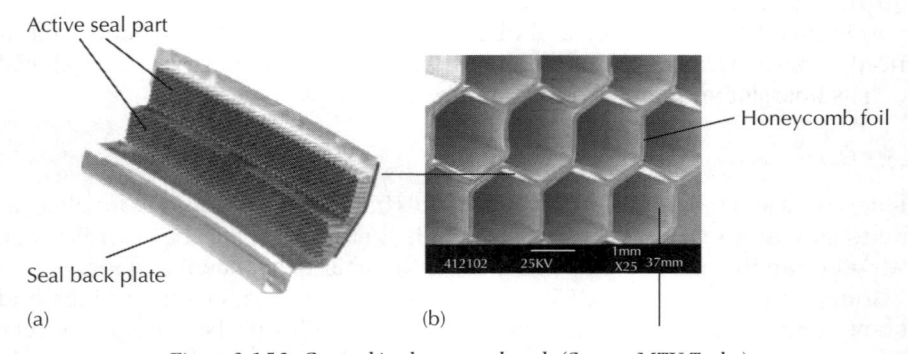

Figure 3.153 Gas turbine honeycomb seal. (Source: MTU Turbo)

In compressors and steam turbines the honeycombs are manufactured from aluminium or steel and have not been primarily designed to be abradable, although an alternative approach is being developed as discussed below. A typical size of hole, for metal honeycombs, is in the region of 3 mm with a similar depth. However, the depth can be an important factor affecting the damping performance of the seal with a difference as small as 0.2 mm affecting the rotor stability.[82]

A hole slot or hole pattern seal is similar in principle to a honeycomb, but is manufactured by drilling the required hole pattern in a metal bush. This provides a more rigid component which will have handling and maintenance advantages for industrial compressors and steam turbines. A typical industrial compressor hole slot seal is shown in Figure 3.154.

Figure 3.154 Hole slot seal for an industrial gas compressor. (Source: Sealing Technology, Elliott)

The relative leakage rates achieved with the different designs of seal measured by Childs[82] are shown in Figure 3.155. These measurements are all made at the same value of clearance and show that labyrinth and hole slot seals have a similar leakage rate. In practice labyrinth seals are used at closer clearance than the other seals so would provide a further reduction in leakage. A concern with honeycomb seals in compressor duties is that they may become filled with debris entrained in the gas flow. The worst case scenario of this is that they would convert to a smooth bush which, as Figure 3.155 shows, does have a significantly higher leakage.

One approach to combine the relative benefits of both seal types is to use labyrinth seals in tandem with a honeycomb, Figure 3.156. The rotor is provided with additional damping which permits stable operation and closer clearances for the labyrinth.[87]

Development is under way to investigate alternative approaches to the manufacture of honeycomb like seals. These offer the potential benefit of being rub-tolerant honeycomb seals. These include PEEK polymer and silicone elastomer material. For both cases, since the cavities in the materials are not honeycomb in shape, these seals will be called rub-tolerant 'cellular' seals. The cellular structure is manufactured by a special process that is being patented. The cellular structure consists of 1.25 mm diameter cells with a wall thickness of 0.125 mm. In this process the cells can be varied to any depth/diameter or randomly selected

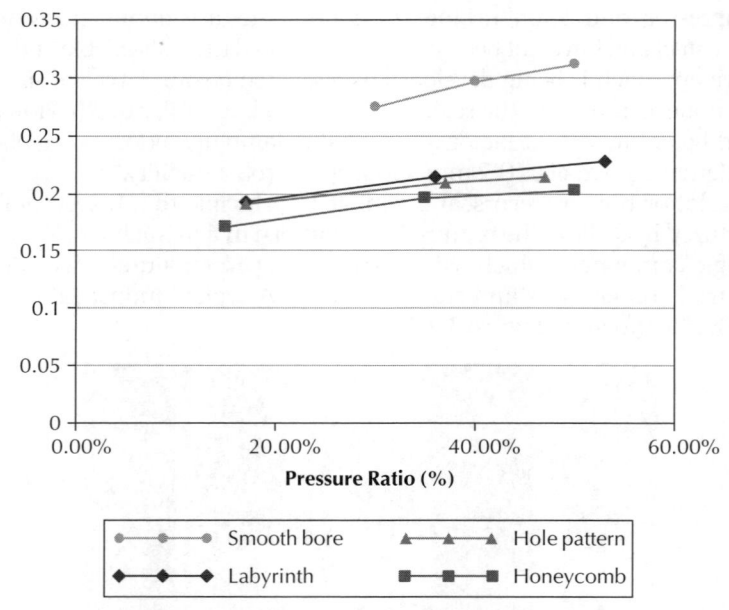

Figure 3.155 A comparison of leakage rates for honeycomb, hole pattern and labyrinth seals at 0.10 mm clearance compared with a smooth bore clearance. (Source: Dasa Childs)

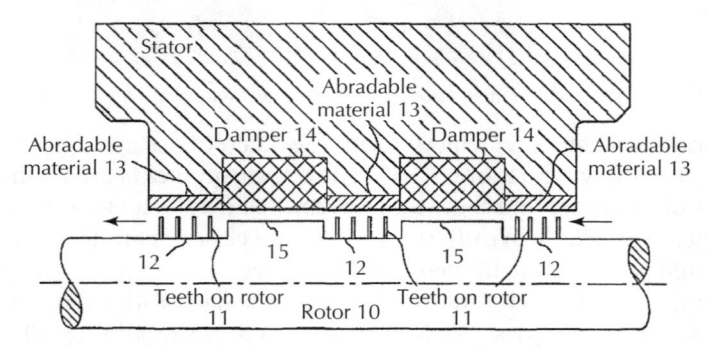

Figure 3.156 An arrangement to use abradable labyrinth seals and honeycombs in series to provide shaft damping. (Source: Patent WO 2004/113771)

alternative depths. An example of this structure is shown in Figure 3.157. This approach offers potential manufacturing benefits compared with stainless steel honeycomb structures and the opportunity to vary the hole depth as required in the structure of the seal.

3.6.4 Brush seals

Brush seals comprise a series of bristles in a housing which rub on the shaft. The housing or support plate is a relatively close clearance from the shaft to support the bristles, but sufficiently clear to avoid shaft contact. Brush seals have been used in gas turbines for up to 25 years but the development is still continuing

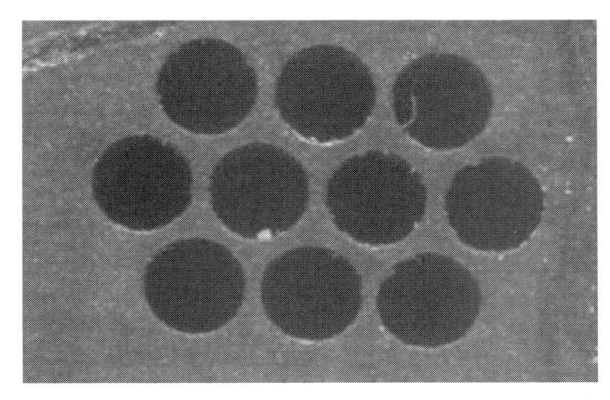

Figure 3.157 A prototype cellular seal manufactured from polymer to produce an abradable seal for compressors. (Source: Sealing Technology)

rapidly. This development is both improving performance and widening the scope of potential applications.

There are different methods of manufacture, which are shown in Figure 3.158. The bristles are arranged so that they are at an angle to the shaft, typically 30 to 45° and either a very slight clearance or light interference when cold, depending on design and application. The bristles will be very fine, 0.5–0.8 mm diameter, and for gas turbines manufactured from nickel alloy, cobalt alloy or stainless steel. During operation the bristles rub lightly on the shaft and form a curtain effect to inhibit leakage. They are likely to cause shaft wear, so a suitable hard coating is required.

Brush seals offer a number of benefits when compared to labyrinths or the other clearance seals discussed above. The leakage is considerably reduced and they are extremely compact. A single brush seal may be approximately 5 mm wide, saving considerable axial space. Leakage rates are typically only 20% of that experienced for a labyrinth, so combined with the reduced axial space requirement it is possible to obtain considerable gains in machine efficiency, size and weight. In addition a brush seal is less affected by shaft runout, the bristles will deflect to accommodate the relative radial movements between the rotor and turbine casing without inhibiting performance, although the permitted runout is constrained by the support plate for the brush and the force required to flex the bristles.

Typical performance parameters for gas turbine brush seals are speed to 350 m/s, temperature to 600°C and a pressure drop per seal of 20 bar. It has been found that brush seals should not be used in tandem as the pressure per stage becomes indeterminate and erratic performance can occur.

If a brush is subjected to overpressure it can distort and lock the bristles. Stiction in the bristles may also cause erratic operation on some occasions.

The pressure and performance limits of a brush seal are dictated by the compromise between maintaining sufficient clearance between the back plate and the rotor while ensuring that there is sufficient support for the bristles. The friction created by the bristles rubbing on the rotor is also a negative feature compared with the pure clearance seals such as labyrinths and honeycombs.

254 *Seals and Sealing Handbook*

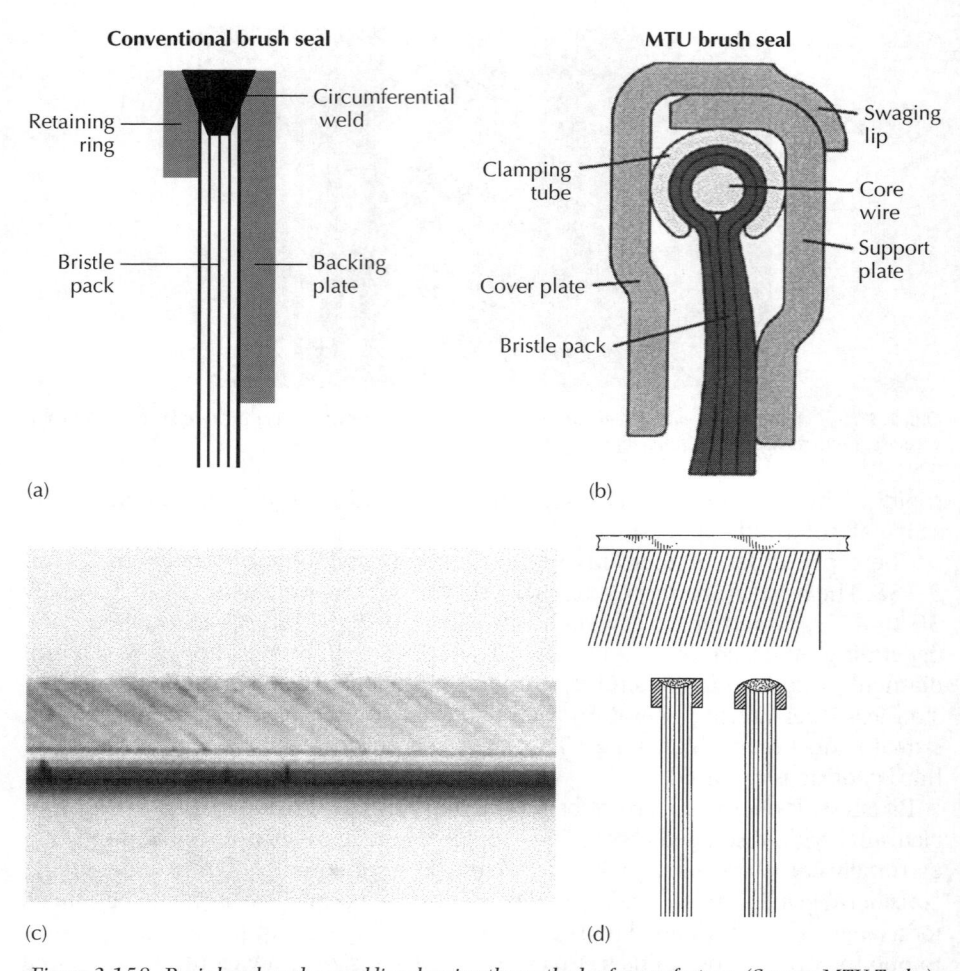

Conventional brush seal **MTU brush seal**

(a) (b) (c) (d)

Figure 3.158 Basic brush seal assemblies showing the methods of manufacture. (Source: MTU Turbo)

A particular problem with brush seals is that they can be seriously damaged if any reverse rotation occurs. This is quite likely with for instance an idle gas turbine that may windmill in the reverse direction.

An approach to the solution of this problem, and also to reduce or effectively eliminate wear of the bristles, is to fit hydrodynamic shoes between the bristles and the shaft. Pairs of axially spaced bristle bundles are mounted in the machine housing and the lower ends of the bristles are connected to a shoe, Figure 3.159. As the shaft rotates a hydrodynamic film is created between the shoe and the shaft. Each of the shoes is connected at discrete points to the end of the seal bristles such that the leading edge of the shoe is oriented to have less contact with the rotor than the trailing edge of the shoe. In operation, the shoes function very similarly to a tilting pad bearing shoe. Prior to rotation of the rotor, the shoe is in contact with the rotor surface. Because the leading edge of the shoe has less contact with the rotor

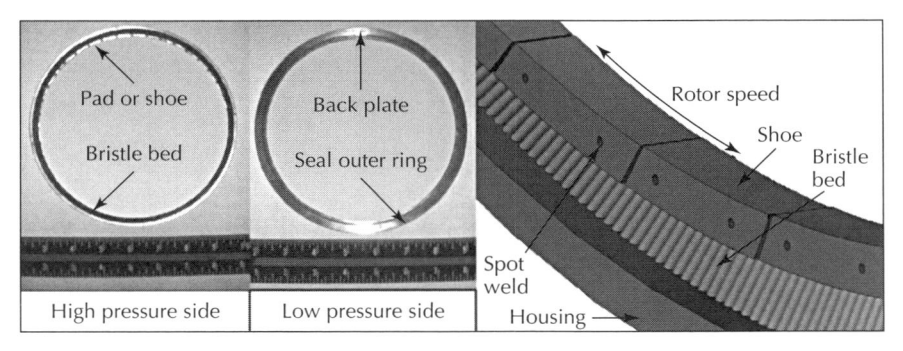

Figure 3.159 A shoed brush seal offers potential benefits of limited reverse rotation capability, reduced leakage and less wear. (Source: Sealing Technology)

or stator than its trailing edge, when the rotor begins to rotate a hydrodynamic wedge is created which lifts the shoe slightly off the surface of the rotor. Consequently, the shoe floats over the rotor at a design gap. Tests have shown that these seals can operate at a very small effective gap, equivalent to a labyrinth with 0.04 mm clearance, which provides extremely low leakage.[88]

A number of other developments continue either to extend the range of applications or to facilitate retrofitting of brushes to older machinery to improve performance.

One example is a brush seal that is intended to fit within a labyrinth, Figure 3.160. This seal is intended to fit different labyrinth designs, and be applicable for both new and as a retrofit to existing seals. The retainer of the brush seal is fitted in a slot formed in the labyrinth seal. The brush seal does not have separate front and back plates. The first tooth of the labyrinth seal forms the front plate of the brush seal, and a second, adjacent tooth of the labyrinth forms the back plate.

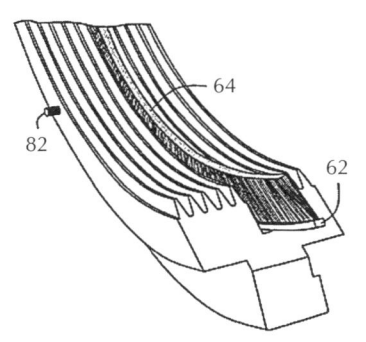

Figure 3.160 A brush seal retrofitted to a labyrinth to reduce the leakage. (Source: Patent WO 2004/023008)

Another development is a brush seal with the sealing element consisting of Kevlar fibres, very densely packed with up to 6000 fibres per millimetre, Figure 3.161. The seal housing is made from a PEEK polymer compound with a similar

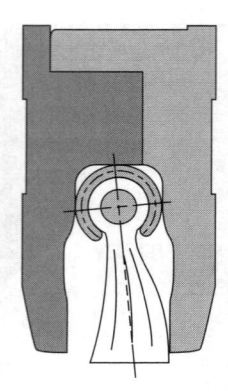

Figure 3.161 The brush seal from MTU Aero Engines, manufactured from non-metallic materials. Cross-section of the non-metallic brush seal, showing the PEEK housing and aramid fibre brushes. (Source: MTU Turbo)

thermal expansion coefficient to steel. The main benefits that are offered by this type of seal, compared with the conventional metallic brush seals, are:

- Leakage cut by a factor of two.
- Seal price up to 50% less than other brush seals.
- Considerably reduced heat generation on the surface of the rotor.

The seal can be used up to 200°C and pressure differentials up to 20 bar per seal element. The seal itself can be designed to accommodate the requirements of the respective application. Seal size ranges from 20 mm to 600 mm in diameter, but above 250 mm the seal housing is manufactured from metal. Potential areas of application include sealing of bearing chambers, gearboxes, industrial turbo-compressors, motor spindles for machine tools and similar equipment.

3.6.5 Leaf seals

In an effort to improve on the performance of brush seals a new development is leaf, or lamella, seals. The factors that these seals are designed to address are:

- Improved axial stiffness to increase pressure capability.
- Avoidance of bristle locking.
- Reduction of friction and wear by designing in aerodynamic lubrication.
- Improved deflection capability to permit operation at increased shaft runout.

A leaf seal is made up from a series of leaf elements disposed around the shaft, Figure 3.162. These leaves are compliant in a radial direction but stiff in an axial direction. It is not necessary to provide a supportive backing ring as for bristles of a brush so removing the tendency for bristle seal elements to lock in position as pressure is applied.

(a) (b)

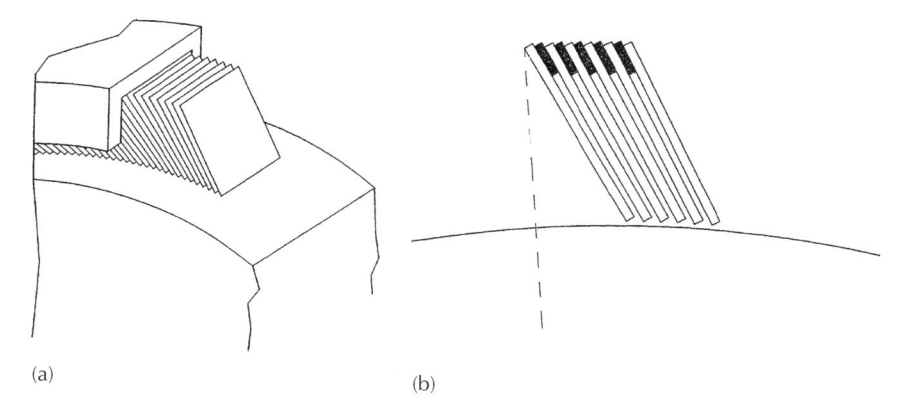

Figure 3.162 *A schematic of a leaf seal construction. (Source: Patent WO 2005/103535)*

The seals are manufactured by brazing a series of leaf elements in a carrier. Original designs used plain leaves. An interleaf clearance is provided by a spacing member in order to allow provision of an air film under the surface of the leaf for lubrication. This air film is caused by pressure in the spacing between the leaf seal element. The air film is formed between leaf seal element tip edges and the rotor. Alternatively, each seal element can be tapered from its mounting end in order to create spacing at the seal end and so the desired lubrication and leakage behaviour. The manufacturing methods for these seals to provide the necessary gaps between the leaves is quite complex, and does not provide an optimum seal.

The focus of development for these seals is to devise satisfactory manufacturing methods and also optimize the provision of gas to either the upper or lower side of the leaf to provide either pressure loading to seal or increased gas for aerodynamic lift.

The manufacture is facilitated by correct selection of the alloys and brazing materials and methods. The gas flow over or under the leaf is controlled by selection of the thickness of the distance pieces between the leaf foils.[89] The presence of aerodynamic lift can be encouraged by forming blisters on the leaf to create a void on the undersurface for an air supply.

An alternative approach is to manufacture the leaf seal elements with a recess that matches a similar recess in the mating leaf.[90] This method permits a controlled gas supply to the dynamic seal interface without providing the gaps between the elements. This system can be tailored to provide either lift or pressure loading to the leaf as required. This will depend on the position of the seal in the turbine, whether it is in a low or high pressure region. The number of channels in a leaf can also be varied depending on requirements.

A number of variations that provide similar configurations are also being developed as the requirements for more efficient gas turbines grow. The finger seal is a comparable approach to the leaf seal, but with a shoe in contact with the shaft to create a somewhat similar concept to the shoe described for the brush seal, except that they are individual for each finger, Figure 3.163.[91]

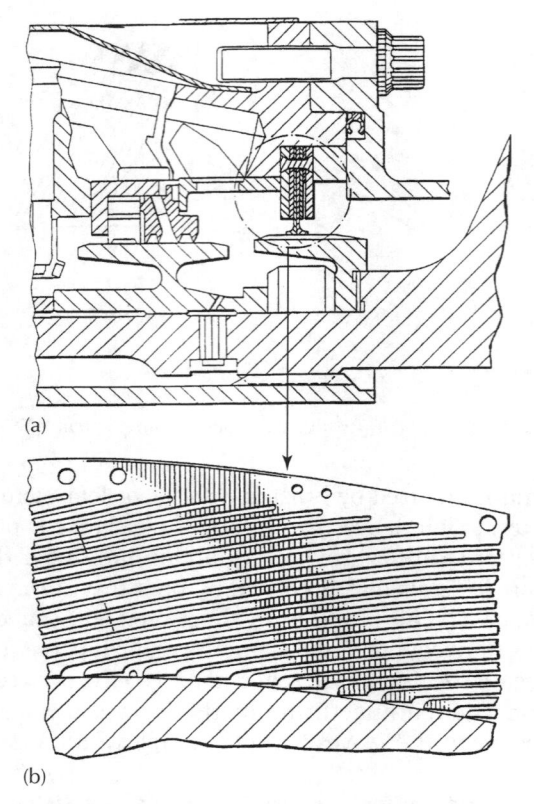

Figure 3.163 A finger seal configuration for gas turbines.

3.6.6 Viscoseals

A viscoseal, also known as a windback seal, is one of the very few clearance seals that have the capability to effectively seal a liquid. It is therefore able to provide a low friction and reliable seal. However, there are also two distinct disadvantages. The seal is only effective when the shaft is rotating and close clearances are required to provide a seal. It is also important to consider that the seal will only be effective at a liquid/gas interface. If there is liquid on both sides it will pump the low pressure liquid into the seal.

The seal can be created either as a helical screw form in a cylindrical surface or as a spiral on a radial face. The most commonly discussed arrangement is the cylindrical form. The seal depends on the viscous flow generated between the rotating shaft and stationary housing. The relative movement of the plain surface past the thread form will create a viscous flow and hence generate a pressure dependent on the clearance, thread dimensions, speed and fluid viscosity. As efficient operation requires an accurate thread form it is most usual to provide the thread on the outside diameter of the shaft, Figure 3.164.

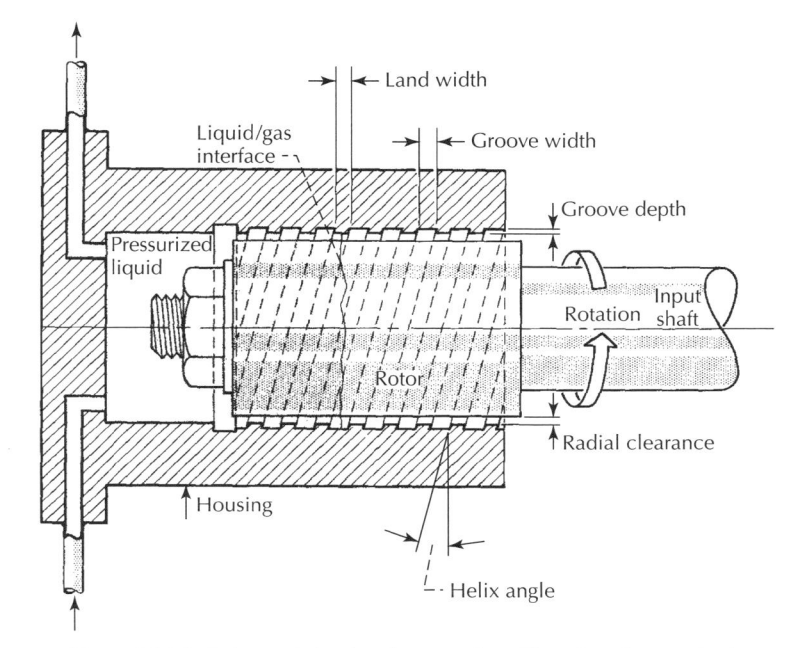

Figure 3.164 A viscoseal showing the screw thread in a close clearance bush.

A considerable amount of research was carried out on these seals during the 1960s and 1970s to provide information for the NASA space programme.[93, 94] The actual flow patterns in the screw thread and clearance are very complex. For a liquid such as water or medium viscosity oil the flow in the seal will be laminar which provides a simpler solution. With a viscous fluid and high speed the flow becomes turbulent. This improves the sealing capability but is more difficult to calculate.

Experiments have shown that the leakage calculations for the laminar case are reasonably accurate.[95]

This depends on the calculation of a 'sealing coefficient' to define the pressure limit of the seal. The coefficient is calculated from the geometry of the thread, Figure 3.165.

$$\Lambda = \frac{6\eta VL}{c^2 \cdot \Delta P} \quad \text{or} \quad \frac{\pi NDL\eta}{10\, c^2 \cdot \Delta P}, \quad (N \text{ r.p.m.})$$

There are two important geometric ratios that affect the seal performance and are used to calculate the sealing coefficient. These are:

The clearance ratio, which is the ratio of the clearance gap to the thread depth, $\beta = (h + c)/c$.

The land ratio, which is the ratio of the land width to the thread pitch, $\gamma = b/(a + b)$.

$$\Lambda = \beta(\beta - 1)\tan\alpha \cdot \left[1 + \frac{E/(\beta - 1)^3}{\gamma(1 - \gamma)\sin^2\alpha}\right]$$

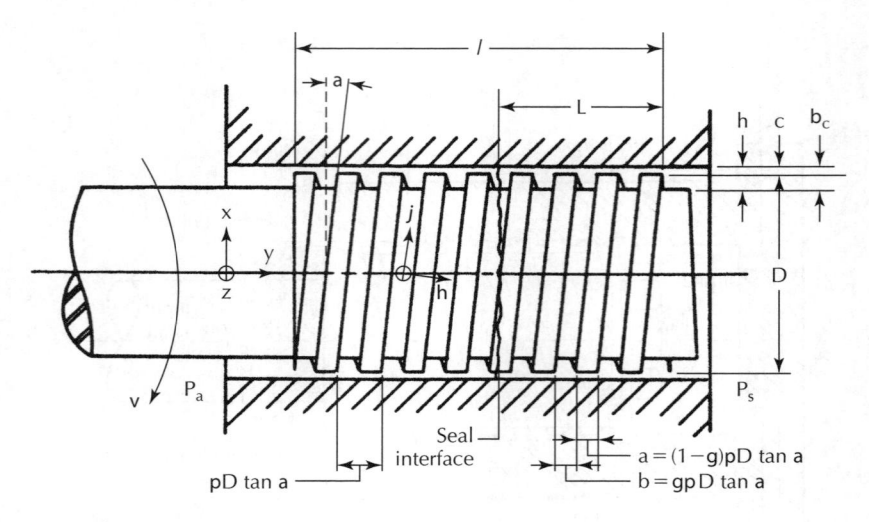

Figure 3.165 Viscoseal geometry nomenclature.

P_a = Ambient pressure P_s = System pressure

Groove depth, h
Land width, a
Groove width, b
Clearance between land and housing bore, c
Helix angle, a
Seal diameter, D
Seal length, L

Two important geometric ratios also have to be defined, these are:
the 'clearance ratio'   b = (h + c)/c,
and the 'land ratio'   g = b/(a+b).

Theoretically the pressure limit should be very dependent on the eccentricity, but experiments at up to 60% of maximum eccentricity have shown that this is not a major factor up to this value. However, it can be seen that the pressure capability is dependent on the square of the clearance, so it is obviously important to keep this to a minimum. In practice it is found that a clearance in the region of 0.02 mm is required to provide a practical seal that will operate at pressure with a low viscosity liquid, and preferably less than 0.01 mm.

It can be seen that the inter-relationship of the various screw thread parameters is quite complex, Figure 3.166.[93] In practice it can be seen from Figure 3.166 that a low clearance ratio is required, which means that the thread depth must be machined to a ratio of between two and three times the clearance, requiring a thread depth of less than 0.1 mm. The optimum land ratio is achieved if *a* and *b* are equal. To provide an acceptable land ratio and optimize the helix angle, it will usually be found that a multi-start thread is required.

The problem of providing a static seal remains. One method attempted to provide a static seal has been to incorporate a mechanical seal in series. This can then rely on the pressure generated in the viscoseal to lift it off,[95] or use alternative means such as speed operated pilot to pressurize the seal open. Any such

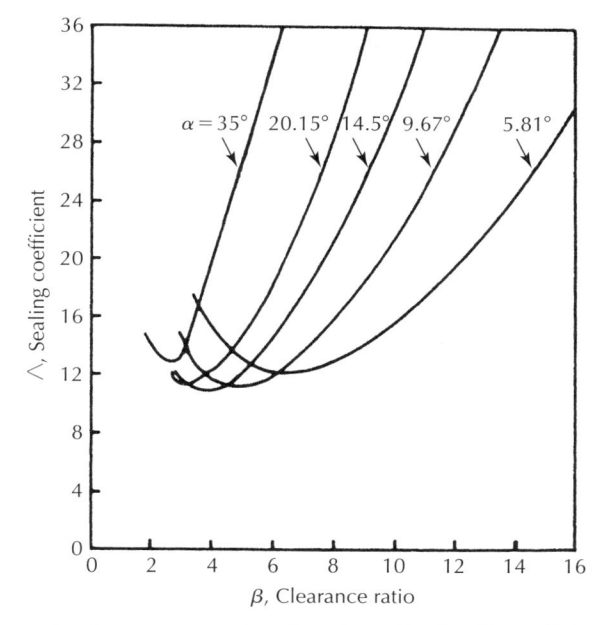

Figure 3.166 Viscoseal performance parameters, the relationship of sealing coefficient to clearance ratio. (Source: BHR Group)

means is an additional complexity and prone to malfunction, especially after a long period of continuous running. In effect, for virtually any general purpose sealing requirement, much of this technology has been replaced by the mechanical seal lift-off geometries discussed in section 3.4.

In practice a further problem has also been experienced. This is gas ingestion. As the seal will be designed with reserve capacity the liquid to gas interface will be part way along the length of the seal. At this interface the low pressure liquid is mixed with the atmospheric gas. This can lead to the formation of a foam like mixture that decreases the effective viscosity. This can cause a breakdown of the sealing function whereby both leakage of liquid to atmosphere and gas ingestion to the liquid will occur. When this occurs increasing the speed of the seal will make the leakage worse. Some methods for reducing this problem have been suggested, but often at the expense of further complication to the seal design.[96] Some work suggests that gas ingestion is considerably reduced if the rotor is located concentrically with the housing and success has been achieved with floating housing arrangements.[97] A further contributor to avoiding gas ingestion suggested by the same authors is to provide a section of plain shaft without screw thread in the length of the seal.

Viscoseals have also been investigated for the sealing of crankshafts on diesel engines. In this low pressure application a wider clearance, up to 0.15 mm, is considered suitable.[98] Further potential methods of reducing gas ingestion include the use of a semicircular groove for the thread and machining the thread in the stator.

The use of a pure viscoseal is usually a specialist application. However, the general principle is widely applied to improve the performance of other seal types. The best example is plastic lip seals as discussed in section 3.2.3. In these seals a helical thread form on the stationary seal lip provides a viscoseal action to improve dynamic sealing. As also discussed in section 3.2 the shaft machining lead can be a contributing factor to seal performance and some equipment designers use this phenomenon to provide improved leakage control. In these cases the lubricated contact of the seal creates a viscoseal with a clearance of one micron. They also, in the case of the plastic seals, demonstrate the problems of ingestion discussed here.

3.6.7 Circumferential seals

Circumferential seals are seals designed to provide a close fitting bush to limit leakage along the shaft. The most common form is a series of segments that are held together to form a bush, normally known as a segmented ring seal. Alternatively a one piece close fitting bush may be used for some high speed turbine applications.

3.6.7.1 Segmented seals

Segmented ring seals are produced in a variety of designs depending on the application. They are not a controlled clearance seal as they are not specifically designed with a prescribed clearance. They start as a lightly contacting seal that will bed into a small dynamic clearance and then operate with a low leakage.

They are used for both liquid and gas applications and are widely used on a range of equipment from large water turbines to high speed machinery such as fans, blowers and steam turbines. They are normally used in situations where some leakage is acceptable or there is some auxiliary means of collecting low pressure leakage for drain or vent. They provide a lower leakage seal than pure clearance types such as labyrinths but cannot provide either the low leakage or pressure capability that is achieved by a mechanical seal either on liquid or gas. They have the advantage of being relatively compact and have no rotating component apart from a shaft sleeve. This is a considerable advantage for both turbo machinery where shaft mass is a major consideration and also for large water turbines where space can be at a premium. However, to provide reasonable control over the leakage the pressure per stage will be quite limited, so to contain a reasonably high pressure a series of these seals will be required.

A particular advantage of these seals compared with a mechanical seal is that there is no restriction on axial movement of the shaft. They are therefore suitable for either high temperature machines where there may be appreciable relative movement between the shaft and casing, during warm-up and cool-down, and large machines such as water turbines or especially pump turbines where thrust reversals will occur in operation.

By far the most common material is carbon, although metal with suitable bearing properties, such as bronze, may occasionally be found on some oil lubricated applications.

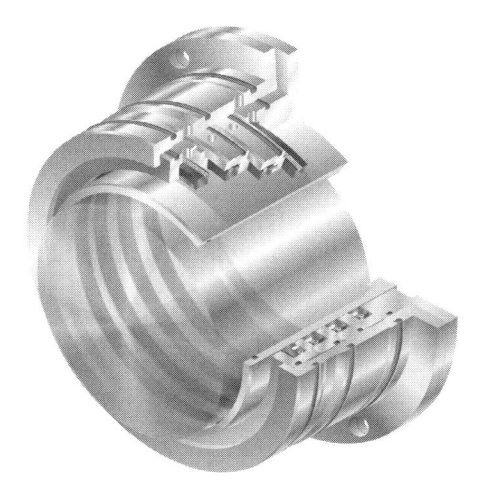

Figure 3.167 The basic arrangement of a segmented seal. (Source: Flowserve)

The basic design is for three or more segments that make up a bush to be a slight interference on the shaft, Figure 3.167. For large diameters, over one metre, up to 16 segments may be used. On the outside diameter is a groove, or some form of retainer that accepts a garter spring. The spring is in circumferential tension and holds the seal in contact at the butt joints and constrains the segments radially on the shaft. The joins will usually have some form of mating step to limit any leakage through the gap. Individual manufacturers have developed butt joints to suit the product areas in which they may specialize and many of these are similar to the range of butt joint designs that can be found on piston rings. The seal will be designed to have some clearance at the butt joints when initially fitted. The circumferential spring will hold the segments in contact with the shaft and the seal bore will wear to create the optimum contact profile so that the seal can run with minimum clearance. The seal is also designed to seal radially against the face of the housing to prevent excess leakage around the outside diameter of the seal. This may be achieved by having small axial springs to energize the seal against the face of the housing or by fitting the garter spring on a chamfered section of the outside diameter to create an axial force component. The seal segments as a whole are also free to float radially to maintain alignment with the shaft. Some form of pins or other mechanical stop is also necessary to prevent rotation of the seal in the housing, Figure 3.168.

The seals may have a plain bore and directly bed in to provide a close clearance bush. Such types will not be suitable for high pressure or speed as relatively rapid wear will occur. Various balanced designs are produced which generally have a relief diameter and step on the circumferential sealing surface. The sealed pressure on the inside diameter will then relieve some of the radially inward forces from the garter spring and seal pressure acting on the outside diameter of the seal, Figure 3.168. Some designs for high speed operation also has relief slots extending over a portion of the contact surface that act as hydrodynamic aids to lift the seal off during shaft rotation, Figure 3.169.

Depending on the application, pressure and degree of leakage retention required the seals may be used singly, but more usually in multiple sets, giving a progressive pressure breakdown. They may also be used in tandem and double arrangements where a barrier gas supply may be used. These seal types are normally used in conjunction with dry gas mechanical seals as discussed in section 3.4.6 to act as a barrier seal between the gas seal and bearing lubricants or to protect the mechanical seal from process liquids or contaminant.

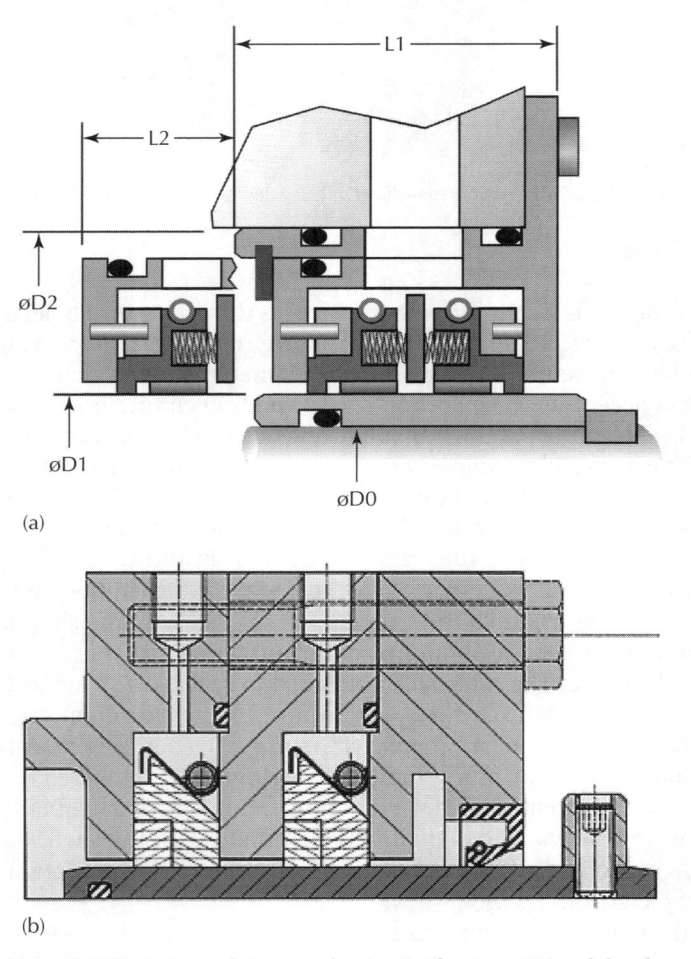

Figure 3.168 Spring and pins, etc. showing axial springs FSD and chamber.

3.6.8 Centrifugal or liquid ring seals

This type of seal depends on the centrifugal acceleration of the sealed liquid by a rotating disc in a confined chamber to generate sufficient pressure to prevent

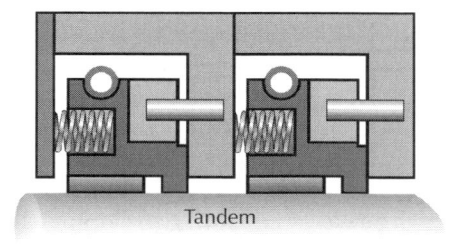

Figure 3.169 A high performance circumferential seal for gas compressor applications with hydrodynamic aids provided on the sealing bore to provide high speed lift. (Source: Flowserve)

leakage from the machine. It is also known as an expeller seal and in some literature a hydrodynamic seal.

In the simplest form it is a plain disc that is housed in a chamber containing liquid. This type of seal has been used for sealing gas,[99] where a reservoir of mercury around the disc is used to seal gas.

However, the most common form of the seal is to use a disc with vanes on the low pressure side which provide increased centrifugal acceleration to the liquid and hence a viable dynamic liquid seal. This type of seal is used in pumps on solids handling duties in the chemical, pulp and paper, and wastewater industries. Figure 3.170 shows an example of a pair of centrifugal seals fitted to a wastewater pump.

Figure 3.170 Cutaway of a centrifugal pump that is fitted with a pair of centrifugal liquid ring seals that work in series to provide the required pressure duty. (Source: Gould Pumps Inc, ITT Corp.)

Much of the original research for this type of seal was an investigation of the potential application to high speeds,[100, 101] but it is presumed that the power consumption involved combined with advances in conventional mechanical seals has reduced the interest in this area.

The pressure capability of the seal is basically dependent on the angular velocity squared and the ratio of the inside diameter of the seal housing throat to the outside diameter of the disc. It has been found that the pressure capability of the disc is not particularly sensitive to clearance although the best pressure ratio is obtained when the ratio of blade tip clearance to mean disc diameter is approximately 0.02; however, the height of the vanes can be important, Figure 3.171.[100]

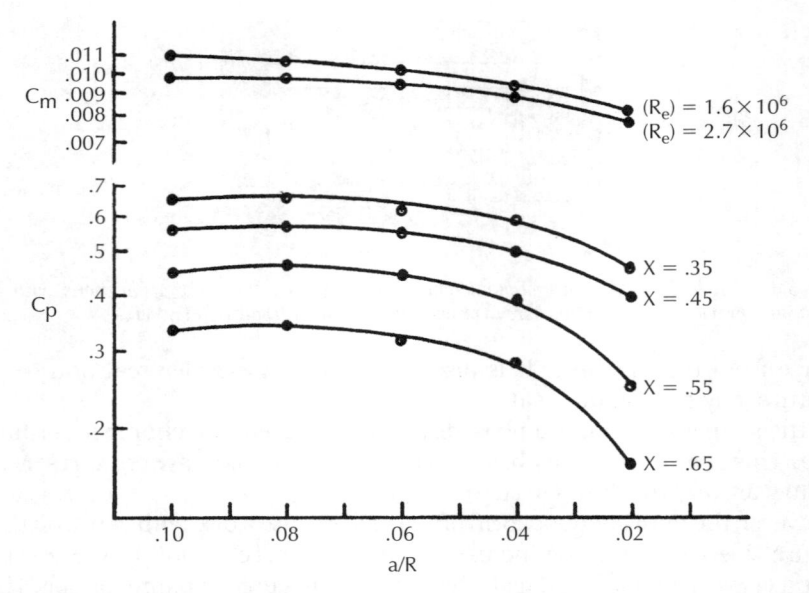

Figure 3.171 *The effect of vane height on the pressure capability of a centrifugal liquid ring seal. (Source: BHR Group)*

This type of seal has some attractive advantages but also some very significant disadvantages.

Advantages:

- A non-contacting seal so there is minimal wear and will hence last indefinitely in all but the most abrasive applications.
- Tolerant to both axial and radial relative movement without significant loss of performance and no long-term effects, provided contact does not occur. Thew and Saunders[100] suggest that up to 1.25 mm radial or axial movement can be accommodated by a small seal, with a 50 to 60 mm disc, and at least 2.5 mm by a seal in excess of 200 mm diameter.
- As the seal pressure capability is proportional to the square of speed a relatively small disc is required at high speeds.
- The seal design and application is relatively simple.

Disadvantages:

- The seal depends on shaft rotation so it is only a dynamic seal. It is necessary to provide an auxiliary seal to operate when the machine stops. Examples that have been used included mechanical seals with a centrifugal lift-off mechanism but in plant handling slurries such devices are not particularly reliable. Some pump manufacturers use a flushed lip seal arrangement, Figure 3.172.

- The generation of the sealing pressure consumes an appreciable amount of power. This is estimated at approximately 1 kW for a disc on a 50 mm shaft at 1750 rev/min.[102] The power consumption is also dependent on a factor that is proportional to the square of the speed. However, for a comparable pressure the disc diameter can be less as speed increases so each case will require individual evaluation.
- The power consumed in the disc cavity has to be dissipated. This generally requires a bleed flow of liquid. An example can be a flush flow provided from the pump suction, as shown in Figure 3.172, with a bleed into the pump through the neck bush between the seal and main pump impeller. The bleed flow has two deleterious effects; it reduces the pressure limit and also increases the power consumption. It is therefore preferable to keep it to the minimum consistent with reliable operation.
- If a pressure in excess of the limit of the seal occurs in operation it will either leak, or rely on the capability of the static seal to run at pressure. It may be necessary to renew the static seals after an overpressure event.
- The pressure limit for these seals is relatively low, especially if the pump will operate at low speeds. Suction pressures will generally be less than 2 bar and very unlikely to be above 5 bar.
- The overall reliability of the system is still reliant on a contact seal such as the lip seals or auxiliary packing that is used as a static seal.

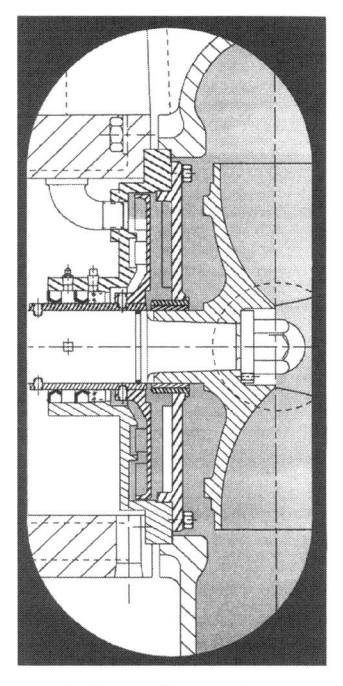

Figure 3.172 A centrifugal seal with lip seals for a stationary seal. (Source: Sealing Technology)

The development of mechanical seals capable of operating for long periods in abrasive slurries makes the disadvantages of these seals more significant in an up-to-date comparison of seal types than would have been the case when the research work detailed in references 100 and 101 was carried out.

3.7 Magnetic fluid seals

Magnetic fluid seals have been available since the early 1970s. The major application is as a rotary seal in vacuum applications and other specialist high integrity gas sealing.

A magnetic, or ferrofluid, seal takes advantage of the response of a magnetic fluid to an applied magnetic field. The basic seal components include ferrofluid, a permanent magnet, two pole pieces and a magnetically permeable shaft. The magnetic circuit, completed by the stationary pole pieces and the rotating shaft, concentrates magnetic flux in the radial gap under each pole piece, Figure 3.173. When fluid is applied to this gap it assumes the shape of a liquid O-ring and produces a hermetic seal. A rotary feed-through will comprise multiple rings of ferrofluid contained in stages formed by grooves machined into either the shaft or pole pieces, Figure 3.174. Typically a single stage can sustain a pressure differential of 200 mbar. The pressure capacity of the entire feed-through is approximately equal to the sum of the pressure capacities of the individual stages; a typical industrial seal is shown in Figure 3.175.

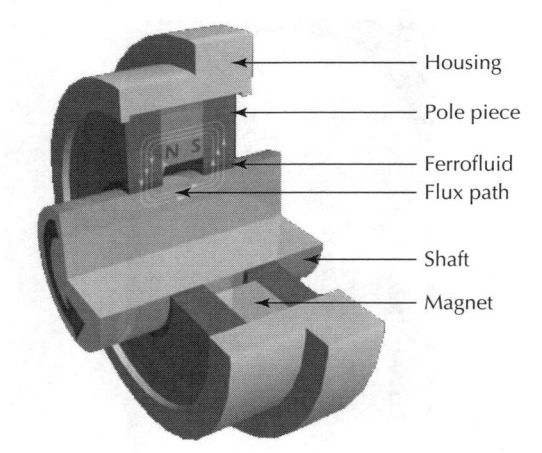

Figure 3.173 Basic concept of a ferrofluid magnetic seal. (Source: Sealing Technology)

3.7.1 The particular attributes of a magnetic seal are:

Sealing integrity
A ferrofluid seal provides a very low leakage seal with a leak rate in the region of 10^{-11} [He] mbar l/s. The performance is retained under both static and dynamic conditions against gas and vapour.

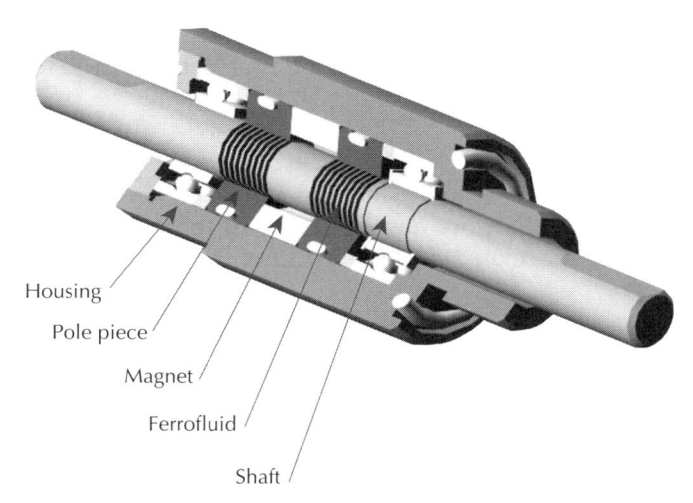

Figure 3.174 A typical magnetic seal arrangement with a series of magnetic ferrofluid rings and poles between a pair of bearings. (Source: Sealing Technology)

Figure 3.175 A multi-axial feed-through can provide a number of feeds through one shaft. (Source: Sealing Technology)

No wear
As the sealing medium is a liquid, there is a low friction between the rotating and stationary components so the seal does not wear. The fluids used are inert, stable and low vapour pressure. It is possible for the seals to operate for over 10 years without maintenance.

Non-contaminating
Since there are no wearing parts, no wear particles are produced which could contaminate the system. Low vapour pressure ferrofluid is resistant to outgassing in high vacuum. They can be used at vacuums down to better than 10^{-9} mbar.

Speed
Current technology has produced configurations that perform at dN values of 500 000 (where d = shaft diameter in mm and N = rotational speed in rev/min). This equates to a rotational speed of 20 000 rev/min on a 25 mm shaft.

Repeatable operation
The sealing performance remains consistent through static, start/stop and rotary operation.

Consistent torque
The viscous drag of the ferrofluid is independent of the pressure applied across the seal. The seal will therefore have a consistent torque without stick-slip or other speed or pressure related effects.

The typical operating limits that apply to ferrofluid seals include:

- Temperature range without cooling: 10 to 100°C. For higher temperatures a water cooled arrangement may be used. However, most seal units can be heated temporarily to 150°C for vacuum baking purposes.
- Vacuum 10^{-9} mbar, depending on the system design and pumping speed.
- Leakage rate of 10^{-11} mbar l/s.
- Standard seals use a hydrocarbon-based ferrofluid which is compatible with inert gas. For sealing in reactive environments a fluorocarbon-based ferrofluid may be used.
- Housing material is usually 300 series stainless steel.
- Shaft material is normally 400 series stainless steel.
- Maximum acceptable shaft runout is 0.076 mm.
- The application of ferrofluid seals is generally limited to gas and vacuum applications as the presence of liquid will contaminate the ferrofluid and potentially compromise the performance.

3.7.2 Ferrofluid types

The standard ferrofluid is based on a synthetic hydrocarbon with very low volatility and therefore low outgassing. It offers medium drag torque and a reasonable compromise of gas and temperature resistance. For reduced torque a synthetic ester-based fluid may be used. However, they have a higher volatility and so the temperature limit and potential life are reduced.

Fluorocarbon-based ferrofluids are used in applications involving the most reactive gases and highest temperatures. They have the lowest outgassing rates and offer the longest life. However, they have a higher viscosity which increases starting and running torque and can limit the maximum attainable speed due to heat generation caused by viscous shearing.

Bearing configurations
Bearing configurations of a sealed feed-through can be either simply supported or cantilevered.

Simply supported seals generally allow higher shaft loading, but this arrangement will require one bearing on the process side of the seal. This will normally be lubricated with a perfluorinated polyether (PFPE) grease. The use of a cantilevered bearing arrangement removes the need for a process side bearing, but may limit the radial and moment loads that can be applied because of the overhung load.

Water cooling
Most seals can be water cooled which allows operation at higher temperatures. This is usually achieved by passing a cooling liquid into the pole pieces through channels in the feed-through housing. For higher temperature applications, shaft cooling, where coolant is supplied to the rotating shaft through a rotary water union, is also available and can be used together with the housing cooling option.

3.7.3 Typical applications

Rotary gas unions
These can be employed for chemical vapour deposition (CVD), vacuum deposition systems and gas handling modules. A static gas feed runs into the rotating shaft which supports the component, such as a printed circuit wafer. This ensures that the gas outlet is in the centre of the wafer, resulting in uniform coating characteristics.

Multi-axial feed-through
These can be used for applications such as wafer handling and CVD wafer rotation with a stationary inner shaft. The feed-through can make it possible to achieve high repeatability with zero backlash. This arrangement provides a torsionally stiff system with a single shaft which enables high torque to be transmitted.

Reactive gas seal
Reactive gas seal designs can be used where aggressive cleaning agents are used. They will contain an inert fluorocarbon-based ferrofluid. A cantilevered seal design is used to protect the bearings from process gases. Special materials are used in the seal construction. Inert gas purge or protective plating can also be added to prevent corrosive attack of seal components.

In-line feed-throughs
Where there is a requirement to provide a rotary feed-through into a gas or vacuum chamber a magnetic seal is a potential option. The consistent torque capability is of particular potential interest where a servo controlled drive is used. An in-line motorized feed-through is also relatively compact compared to a shaft-coupled drive, Figure 3.175.

Hollow shaft feed-throughs
These can be used where a large diameter hollow shaft is required, in, for example, optical coatings for the manufacture of fibre optic filters. The subject component

can be observed or measured through the hollow shaft. 200 mm diameter shafts can be used at up to 1000 rev/min, Figure 3.176.

Figure 3.176 A hollow shaft feed-through provides a compact arrangement with access via the hollow shaft when required. (Source: Sealing Technology)

High precision spindles
Spindles can be designed to provide a high precision feed-through where very accurate alignment is required in precision electronics manufacture or detailed scientific manipulation. Shafts can be restricted to within 0.005 mm of runout.

Rotary-linear feed-throughs
If a combination of rotary and linear motion is required the magnetic fluid rotary seal can be combined with an edge-welded metal bellows linear seal. They use pre-loaded angular contact rotary bearings and sleeve or ball type linear bearings.

Gas seals
Industrial gas seals may be used to provide a hermetic seal for low pressure fans, blowers and vertical pumps in critical applications. This may include hydrocarbon processing where they can prevent the release of VOCs into the atmosphere, nuclear applications and other critical processes where it is essential to prevent either contaminant ingress or the emission of process gases.

Compact seal designs are used in aircraft for the protection of optical devices and sensitive electronic instrumentation.

The seals may also be used in equipment for the sealing of either pressurized or vacuum filled light bulbs during the manufacturing process. They can provide the benefit of avoiding lubricant and particulate contamination and the maintenance involved if a conventional lubricated seal is employed.

3.8 Rotary seal selection

3.8.1 Liquid sealing

Surface speed below 25 m/s		
High vacuum	**Mechanical seal:**	Low leak, long life potential
	Lip seal:	Elastomer or plastic depending on liquid. Finite life, some leak through plastic seal, suitable elastomer for vacuum required
Low vacuum	**Lip seal:**	Elastomer or plastic, as above
	Mechanical seal:	Optimum for long life and low leak
	Compression packing:	Some leakage
	Segmented rings:	Some leakage
Splash sealing, zero/very low pressure differential	**Lip seal:**	Elastomer or plastic, finite life, selection depends on liquid, temperature, environment
	Labyrinth:	No wear out but vapour leakage, high speed
	Labyrinth + vapour barrier seal:	No wear out
	Mechanical seal:	Compact design, e.g. magnetic or wavy spring
	Piston rings:	Potentially some leak
	Viscoseal:	Very high speed, but gas ingestion
	Segmented rings:	High speed, but some leak
Low pressure <5 bar	**Lip seal:**	Specially designed elastomer seal, limited life/speed. Provides a compact design
Lubricating liquid below vapour pressure	**Plastic designs:**	Potentially small leak, but compact design
	Compression packing:	If some leakage acceptable
	Mechanical seal:	Compact or elastomer bellows design depending on application, space/cost constraints

Low pressure <5 bar
Non-lubricating liquid or close to vapour pressure

- **Lip seal:** Plastic, provided some leak permissible
- **Elastomer:** Specific designs for application, limited speed or design for leakage may be constraints
- **Mechanical seal:** Selection of seal and system to suit liquid
- **Compression packing:** If some leakage permissible
- **Segmented rings:** Some leakage will occur, useful for large diameters

Medium pressure 5–20 bar

- **Elastomer:** Special designs if compact seal required
- **Plastic:** Lip or U section for compact design, some leakage may occur
- **Mechanical seal:** Very low leak and long life possible. Design and auxiliaries to suit application
- **Compression packing:** If some leakage acceptable
- **Segmented rings:** Large sizes possible, but some leakage will occur. Multiple rings in series required to break down pressure

Chemically aggressive

- **Plastic lip seal or U section:** Speed and/or pressure constraints. Compact but life may be limited
- **Mechanical seal:** Design, materials and auxiliaries to suit application
- **Compression packing:** If facilities for leakage

Slurry/abrasives

- **Pump design with expeller:** Effective as dynamic seal only at low suction pressures, static seal still required
- **Compression packing:** Lantern ring and flush required to keep packing clean
- **Mechanical seal:** Specialist slurry seal, dual seal or auxiliary system to keep seal clean required depending on type of application

Higher pressure >25 bar

- **Elastomer:** Compact but speed and/or life very limited for most designs
- **Plastic:** Lip, U or elastomer energized depending on fluid and application. Pressure × speed (PV) limit restricts surface speeds. Provides a compact seal, finite life
- **Mechanical seal:** Wide range of speed and pressure capabilities regardless of liquid. Best option for long life and very low leakage
- **Compression packing:** Speed limited at high pressure, some leakage must be permissible

High surface speed >25 m/s	**Plastic lip seal:**	Compact, limited pressure (1–2 bar), finite life, some leakage
	Mechanical seal:	High speed designs, including enhanced lubrication faces. Low leak, long life potential. Primary selection for pressure above 1–2 bar
	Segmented rings:	Compact, limited pressure capability, some leakage
	Viscoseal:	Dynamic seal only, low power loss, beware of gas ingestion. Limited pressure capability
	Labyrinth:	Very high speeds, high leakage, limited pressure drop capability on liquid

3.8.2 Gas sealing

High vacuum	**Ferrofluid magnetic seal:**	High integrity, temperature limited to 100°C
	Mechanical seal:	Dry running seal design
	Plastic lip seal:	Most compact but finite life, some leakage so will probably not be acceptable in many applications
Low vacuum	**Ferrofluid magnetic seal:**	High integrity, temperature limited to 100°C
	Mechanical seal:	Dry running seal design
	Plastic lip seal:	Most compact but finite life, some leakage may not be acceptable in all applications
	Compression packing:	Lubrication probably required so speed limited and application areas limited
Vapour barrier ±0.1 bar or less	**Ferrofluid seal:**	High cost, contamination with condensing vapour a potential problem. High sealing integrity
	Lip seal:	Plastic probably preferred if no lubrication. Finite wear life
	Labyrinth with vapour seal:	Long life, low power loss. Is dynamic vapour retention adequate?
	Mechanical seal:	Dry running design

Low pressure <5 bar

Lip seal, plastic: PV limits will restrict speed. Compact seal, finite wear life

Mechanical seal: High speed, minimal leakage, long life. Dry gas seal design. Lubricated dual possible depending on application

Segmented rings: High speeds possible, some leakage, multiple rings to seal other than very low pressure

Labyrinth: Very high speeds possible but high leakage. Rotor stability to be considered

Honeycomb: Alternative or complement to labyrinth

Brush seal: More compact and less leakage than labyrinth, no reverse rotation capability

Finger seals: Alternative to labyrinth under development, more stable and better pressure resistance than brush

Higher pressure >5 bar

Plastic seals: Compact design, speed very limited at high pressure

Mechanical seal: Dry gas seals up to high speeds, 200 m/s

Mechanical seal: Lubricated dual for specific applications such as heavily contaminated gas

Segmented rings: Lubricated multiple ring sets, problems of leakage and also lubricant contamination and loss

Labyrinth with leak off and vapour drains: High leakage relative to mechanical seals, wear of labyrinth increases leakage considerably. Only suitable for non-hazardous gas or vapour

3.9 References

1. D.E. Johnston, 'Design aspects of modern rotary shaft seals', *Proc. I.Mech.E.*, Vol. 213, Part J, 1999.
2. E.T. Jagger, 'Rotary shaft seals: the sealing mechanism of synthetic rubber lip seals running at atmospheric pressure', *Proc. I. Mech. E.*, Vol. 171, 1957.
3. R.F. Salant, 'Theory of lubrication of elastomeric rotary shaft seals', *Proc. I. Mech. E.*, Vol. 213, Part J, 1999.
4. R.F. Salant, Private communication, July 2005.
5. T. Kunstfeld and W. Haas, 'Shaft surface manufacturing methods for rotary shaft lip seals', *Sealing Technology*, July 2005.
6. L.A. Horve, 'A macroscopic view of the sealing phenomenon for radial lip oil seals', 11th Int. Conf. on Fluid Sealing, BHR Group, 1987.
7. Anon, *Simrit*, catalogue, 2005.
8. L.A. Horve, 'The correlation of rotary shaft radial lip seal service reliability and pumping ability to wear track roughness and microasperity formation', SAE910530, February 1991.
9. P. Embury and J. Armour, 'Influence of elastomer compound design on the performance of rotary shaft lip seals', *Sealing Technology*, August 2004.
10. B. Duhring and G. Iverson, 'The application of plastics in dynamic seals', *Proc. I. Mech. E*, Vol. 213, Part J, 1999.
11. R.F. Salant, 'Grease lubrication, bearing seals', *Seal Technology; Principles & Application*, Georgia Institute of Technology Short Course, 2004.
12. G. Medri, A. Strozzi, J. Bras and A. Gabelli, 'Mechanical behaviour of two elastomeric seals for bearing units', 10th Int. Conf. on Fluid Sealing, BHR Group, 1984.
13. A. Gabelli, 'Analytical modeling of load deflection characteristics of elastomeric lip seals', 11th Int. Conf. on Fluid Sealing, BHR Group, 1987.
14. Kim and Jun, 'Contact forces of lip seals for ball bearings', *Tribology International*, Vol. 27, pp. 393–400, 1994.
15. *SKF Bearing*, catalogue, www.skf.com, 2006.
16. *Rotary Seals*, catalogue, Busak + Shamban, 2004.
17. E. Bock, R. Vogt and P Schreiner, 'New radial shaft seal concepts for sealing hydraulic pumps and motors', *Sealing Technology*, November 2003. (First published in German at the 12th International Sealing Conference, VDMA, October 2002.)
18. 'Lip type stern tube seals', www.wartsila.com, 2006.
19. G.A. Fazekas, 'Helical O-ring seal', Invention Disclosure 6827, American Machine and Foundry Co., 1957.
20. D.F. Denny, 'The lubrication of fluid seals', Conference on Lubrication and Wear, I. Mech. E., London, 1957.
21. M.S. Kalsi, 'Development of a new high pressure rotary seal for abrasive environments', 12th Int. Conf. on Fluid Sealing, BHRA, 1989.
22. *Kalsi Seals Handbook*, Kalsi Engineering, www.kalsi.com/Rotary_Seal_Literature.htm, 2005.
23. F. Bauer and W. Haas, 'PTFE lip seals with spiral groove – the penetration behaviour, hydrodynamic flow and back pumping mechanisms', 18th Int. Conf. on Fluid Sealing, BHR Group, 2005.
24. 'Second generation PTFE engine oil seal offers space and cost saving', *Sealing Technology*, February 2005.
25. S. Watson and B.S. Nau, 'Analysis of a novel rotary seal', 11th Int. Conf. on Fluid Sealing, BHR Group, 1987.

26. A. Lebeck, '*Principles and Design of Mechanical Face Seals*', Wiley-Interscience, New York, 1991.
27. B.S. Nau, 'Hydrodynamic lubrication in face seals', Third Int. Conf. on Fluid Sealing, BHRA, 1967.
28. G.J. Field and R.K. Flitney, 'Diesel engine water pump seal investigations', BHR Group Report RR1293, 1975.
29. 'Water pump seal compensates for vibrations', *Sealing Technology*, October 2005.
30. R.K. Flitney and B.S. Nau, 'The effects of misalignment on mechanical seal performance', 15th Int. Conf. on Fluid Sealing, BHR Group, 1997.
31. N.D. Barnes, R.K. Flitney and B.S. Nau, 'Slurry seal investigation', BHR Group Report CR6376, 1994.
32. R.A. Smith, 'Introduction to sealing mixers in the pharmaceutical and biotech industries', 21st Int. Pump Users Symposium, Texas A&M, 2004 and *Sealing Technology*, September 2004.
33. *Goetze Mechanical Seals*, brochure, FTL Seals Technology, 2005.
34. J.S. Stahley, *Dry Gas Seals Handbook*, Pennwell, 2005.
35. E. Mayer, 'High duty mechanical seals for nuclear power station', Fifth Int. Conf. on Fluid Sealing, BHRA, 1971.
36. E.H. Iny, 'The design of hydrodynamically lubricated seals with predictable operating characteristics', Fifth Int. Conf. on Fluid Sealing, BHRA, 1971.
37. L. Young, E. Roosch and R. Hill, 'Enhanced mechanical face seal performance using modified face topography', 17th Int. Conf. on Fluid Sealing, BHR Group, 2003.
38. R.K. Flitney and B.S. Nau, 'Mechanical seal performance: procedure for the determination of temperature margin', BHRA Report TN3010, 1989.
39. J.L. Morton, J. Attard and J.G. Evans, 'Active lift seal technology impact on water injection services', 22nd Int. Pump Users Symposium, Texas A&M, 2005.
40. J.M. Heald, 'The seal's environment – its housing', Seventh Int. Conf. on Fluid Sealing, BHRA, 1975.
41. N.D. Barnes *et al.*, 'Mechanical seal housing optimisation study', BHR Group Report CR3005, 1988.
42. N.D. Barnes *et al.*, 'Mechanical seal housing optimisation study', BHR Group Report CR6125, 1992.
43. M. Williams and N.D. Barnes, 'Flow visualisation experiments on housing grade 2 (ISO 3069 Revision)', BHR Group Report CR6196, 1993.
44. M. Williams *et al.*, 'Seal housing design evaluation', BHR Group Report CR6378, 1995.
45. H. Azibert and R. Clark, 'Using CFD to improve performance and extend life of mechanical seals in slurry applications', 17th Int. Conf. on Fluid Sealing, BHR Group, 2003.
46. ISO 21049 and API 682, 'Pumps – Shaft sealing systems for centrifugal and rotary pumps'.
47. 'New pump design to improve seal life', *Sealing Technology*, July 2005.
48. P.E. Bowden and C. Fone, 'Containment sealing and the new API 682 standard', 17th Int. Conf. on Fluid Sealing, BHR Group, 2003.
49. D. Mathewson, 'Advanced mechanical seals for PWR reactor coolant pumps', *Advanced Topics and Technical Solutions in Dynamic Sealing*, Poitiers University, 2005.
50. T. Fuse, T. Shimzu and K. Nishiyama, 'Development of highly reliable mechanical seal for nuclear power plant', 17th Int. Conf. on Fluid Sealing, BHR Group, 2003.
51. 'Excellence in environmental protection awards from the FSA', *Sealing Technology*, January 2006.

52. T.R. Millar, 'The application of an angled face, large diameter, seal to a reversible pump-turbine shaft', Fifth Int. Conf. on Fluid Sealing, BHRA, 1971.
53. API 617, 'Axial and centrifugal compressors and expander-compressors for petroleum, chemical and gas industry services', 2003.
54. 'A guide to sealing hydrocarbon processing for low emissions', AESseal plc, 2005.
55. J.D. Summers-Smith, *Mechanical Seal Practice for Improved Performance*, Mechanical Engineering Publications Ltd, I.Mech.E., London, 2nd ed., 1992.
56. 'Design of mechanical seals for hygienic and aseptic applications', Document 25, European Hygienic Equipment Design Group, Belgium, 2002.
57. D.F. Denny, 'Tests on packed glands for rotary shafts', BHR Group Report RR487, 1953.
58. 'Supagraf control helps valves to meet latest regulations', *Sealing Technology*, February 2006.
59. Training video, James Walker, 2001.
60. R.M. Austin and M.J. Fisher, 'An investigation of methods of packing stuffing box seals for rotary shafts', BHR Group Report RR741, 1962.
61. F. Pugliese, 'Valve live loading using Belleville springs', Solon Manufacturing Company, 2001.
62. E. Staaf, 'A review of the properties of expanded graphite and selected forms of PTFE as an alternative in the sealing role', Eighth Int. Conf. on Fluid Sealing, BHR Group, 1978.
63. R.K. Flitney, 'Packing material alternatives', BHR Group Reports RR1771 and RR1772, 1981.
64. D. Harrison et al., 'Evaluation of graphite valve packings to reduce fugitive emissions on hydrocarbon duties', First European Conference on Controlling Fugitive Emissions from Valves, Pumps and Flanges, European Sealing Association, 1995.
65. S.E. Leefe and O.M. Davies, 'Laboratory based test for valve stem leakage correlation between test and working fluids', First European Conference on Controlling Fugitive Emissions from Valves, Pumps and Flanges, European Sealing Association, 1995.
66. 'Valve stem leak-tightness test methodologies', EOR 19772 EN, European Commission, 2000.
67. S.E. Leefe, 'SMT project on valve emission measurement methods', Second European Fugitive Emissions Conference, VDI Verlag, 1998.
68. S. Boyson, 'Control valve packing technology optimizing friction sealability and repeatability', *Valve World*, 2004, KCI World.
69. A. Jeffries and C. Edwin-Scott, 'Making a practical difference to site fugitive emissions', 18th Int. Conf. on Fluid Sealing, BHR Group Ltd, 2005.
70. R. Bartholdi, 'The development of a new packing geometry', *Valve World*, 2004, KCI World.
71. 'Valve packing tests leakage measurement equipment sensitivity', *Sealing Technology*, October 2003.
72. Y. Birembaut et al., 'Measurement of fugitive emissions: the different methods applied to industrial valves', 2nd European Fugitive Emissions Conference, VDI Verlag, 1998.
73. R. van der Velde, 'Industrial solutions to stop valve emissions', First European Conference on Controlling Fugitive Emissions from Valves, Pumps and Flanges, European Sealing Association, 1995.
74. 'Emission measurement creates animated discussion', *Sealing Technology*, January 2005.

75. J.A. Reynolds, 'Live loading – an economic solution', First European Conference on Controlling Fugitive Emissions from Valves, Pumps and Flanges, European Sealing Association, 1995.
76. B.S. Nau, 'Summary report: development of high reliability dynamic seals', BHR Group Report CR7599, 1991.
77. A. Roddis, 'Reducing moisture contamination in bearing lubrication', *Sealing Technology*, February 2006.
78. S. Hoeting, 'Protecting machine tool spindle bearings', *Manufacturing Engineering*, July 2001.
79. *Setco Airshield*, catalogue, Setco Sales Co., 2006.
80. K. Komotori and H. Mori, 'Leakage characteristics of labyrinth seals', Fifth Int. Conf. on Fluid Sealing, BHR Group, 1971.
81. A. Picardo and D. Childs, 'Rotordynamic coefficients for teeth on stator labyrinth seals at 70 bar supply pressure – measurements versus theory and comparison to honeycomb seal', ASME IGTI Conference, Vienna, 2004.
82. D.W. Childs, 'Recent advances in the rotordynamic behaviour of gas seals', Advanced Topics and Technical Solutions in Dynamic Sealing, EDF/LMS Workshop, Poitiers, October 2005.
83. P. Dowson, M.S. Walker and A.P. Watson, 'Development of abradable and rub-tolerant seal materials for application in centrifugal compressors and steam turbines', Proceedings of the 33rd Turbomachinery Symposium, Turbomachinery Laboratory, Texas A&M University, Texas, 2004. (Republished in *Sealing Technology*, December 2004.)
84. H. Benckert and J. Wachter, 'Flow induced spring constants of labyrinth seals', Second Int. Conf. Vibrations in Rotating Machinery, I. Mech. E., UK, 1980.
85. F. Romero, 'New seals solve vibration problem', *Sulzer Technical Review*, 2005.
86. W. Smarsly *et al.*, 'Advanced high temperature turbine seals materials and designs', Eighth International Conference on Mulitfunctional and Functionally Graded Materials, Lovain, Belgium, 2004.
87. 'Hybrid abradable labyrinth damper seal', Elliott Co., WO 2004/113771.
88. A. Delgardo *et al.*, 'Measurements of leakage, structural stiffness and energy dissipation parameters in a shoed brush seal', *Sealing Technology*, December 2005.
89. Patent Nos: WO 2005/103534, 103535 and 103536, Alstom Technology Ltd.
90. Patent No.: WO 2004/088180, Rolls Royce plc.
91. M.J. Braun *et al.*, 'A three dimensional analysis and simulation of flow, temperature and pressure patterns in a passive adaptive compliant finger seal', Advanced Topics and Technical Solutions in Dynamic Sealing, EDF/LMS Workshop, Poitiers, October 2005.
92. R.E. Chupp and P. Nelson, 'Evaluation of brush seals for limited life engines', AIAA/SAE/SAME/ASEE, 26 Joint Propulsion Conf., USA, 1990.
93. W.K. Stair and R.H. Hale, 'The turbulent viscoseal – theory and experiment', Third Int. Conf. on Fluid Sealing, BHRA, 1967.
94. J. Zuk and H.E. Reinkel, 'Numerical solutions for the flow and pressure fields in an idealized spiral groove pumping seal', Fourth Int. Conf. on Fluid Sealing, BHRA, 1969.
95. R.K. Flitney and B.S. Nau, 'Design and testing of an integrated viscoseal and mechanical seal', BHRA Report RR1203, 1973.
96. W. Yu-ming, 'Some experimental investigations about seal breakdown and gas ingestion in viscoseals', 10th Int. Conf. on Fluid Sealing, BHRA, 1984.

97. E.J. Bussemaker and G.G. Hirs, 'Viscoseals for free surface sodium pumps', Fifth Int. Conf. on Fluid Sealing, BHRA, 1971.
98. A.B. Crease, 'Windback seals – a simple theory and design method and the main practical limitations', Seventh Int. Conf. on Fluid Sealing, BHRA Fluid Engineering, 1975.
99. S.B. Rothberg, 'Centrifugal force keeps fluid in groove liquid shaft seals', *Production Engineering*, November 1963.
100. M.T. Thew and M.G. Saunders, 'The hydrodynamic disc seal', Third Int. Conf. on Fluid Sealing, BHRA, 1967.
101. M.T. Thew, 'Further experiments on the hydrodynamic disc seal', Fourth Int. Conf. on Fluid Sealing, BHRA, 1969.
102. N. Zaman, 'Dynamic seals lower life costs of wastewater pumps', *Sealing Technology*, August 2004.

CHAPTER 4

Reciprocating seals

4.1 Introduction

Reciprocating seals cover an extremely wide range of applications. Internal combustion engines have piston rings to seal between combustion gas and crankcase oil while diesel engines have high pressure plungers on injector pumps. Automobiles have seals on brake cylinders and shock absorbers that have very different duty cycles. Hydraulic cylinders are used on a wide range of mobile, earthmoving and agricultural equipment, industrial production machines such as presses, injection moulding and rolling mills, to very large cylinders in areas such as the marine industry. A huge range of industrial pumps and compressors use piston rings of various types, clearance seals, packing and elastomer or plastic seals depending on the fluid, application and duty requirement.

This chapter will discuss different seal types in turn covering elastomer and plastic seals for hydraulic cylinders, pneumatic seals, diaphragms and bellows, piston rings for engines and compressors, packing and clearance seals.

The selection of a particular seal type can depend not only on the duty – fluid, pressure, speed, leakage, life expectancy, etc. – but also on the production quantities and operating environment. Some seal types are readily available in small quantities but others may only be financially viable if large volume production is envisaged. For example, many plastic and elastomer seals are readily available in a wide range of standard or special sizes and can be obtained relatively economically. Packing can be purchased as lengths and cut to size. On the other hand the production of metal piston rings or manufacture to the very close clearances necessary to provide a sensible clearance seal are seldom viable unless a large production volume is contemplated. Seals involving metal-to-metal sliding contact are also extremely sensitive to lubrication and contamination.

4.2 Elastomer and plastic seals for hydraulic applications

Self-energizing elastomer or plastic seals are the sealing devices most widely used for both high and low pressure liquid and low pressure gas and pneumatic applications. They have the capability to provide very low dynamic leakage and positive static sealing. The design of the seal will be very dependent on the application. The operating conditions, environment, leakage integrity requirements and space envelope can all affect the selection of seal type.

4.2.1 Background

The behaviour of seals under pressure has been discussed in Chapter 2, sections 2.2.1 and 2.4.1. For reciprocating seals the pressure energization remains similar. A range of solid elastomer designs rely on compression in the groove to provide an initial sealing force, Figure 4.1. Many seals used are of a U configuration, Figure 4.2. In this case flexure of the material will provide the initial sealing force. A benefit of this style of seal for dynamic applications is that the flexure of the seal section will give more consistent interference force with eccentric operation and wider tolerances. It is also possible to optimize the static and dynamic geometry to ensure an optimum seal installation.

For plastic seals, such as PTFE, two energization methods are used, Figure 4.3. The metal spring energized U-ring, as discussed under static seals in section 2.4.1 can be used but for dynamic duty it will have a lighter spring design that provides

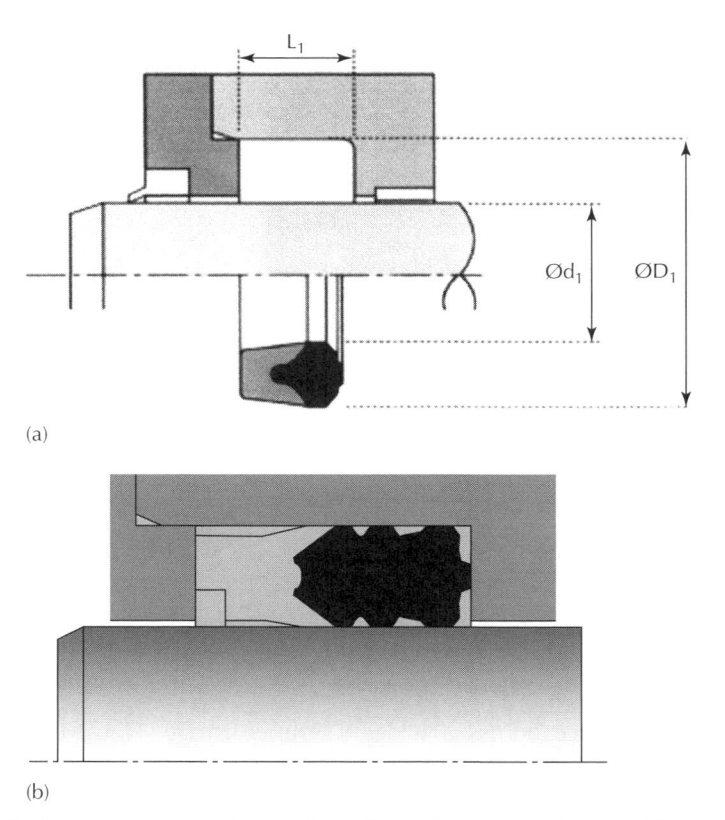

Figure 4.1 Basic squeeze seal types that may be used for reciprocating applications. (Source: (a) Hallite, (b) SKFI)

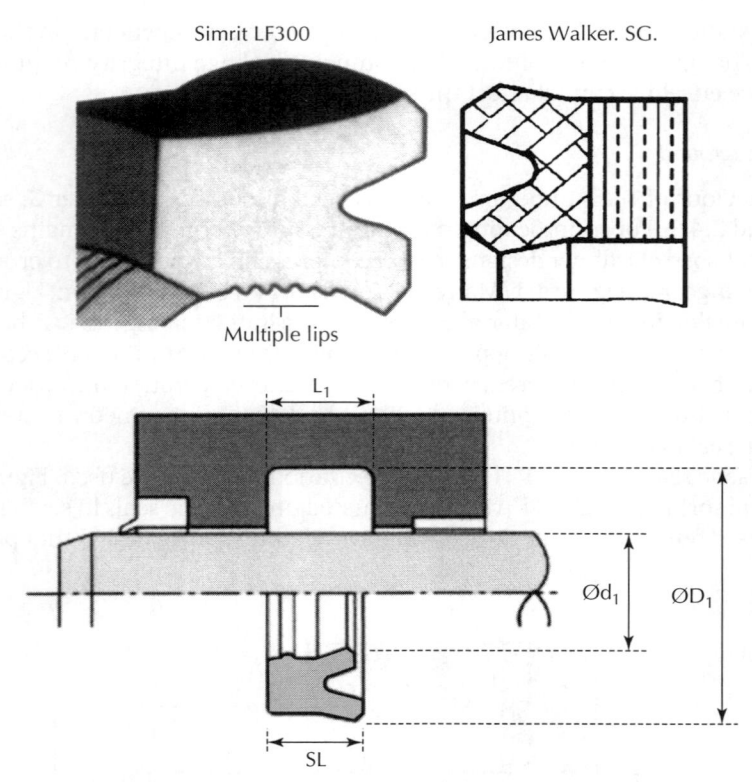

Figure 4.2 Typical U-seal configurations for reciprocating hydraulic cylinders. (Source: Freudenberg Simrit, Hallite, James Walker)

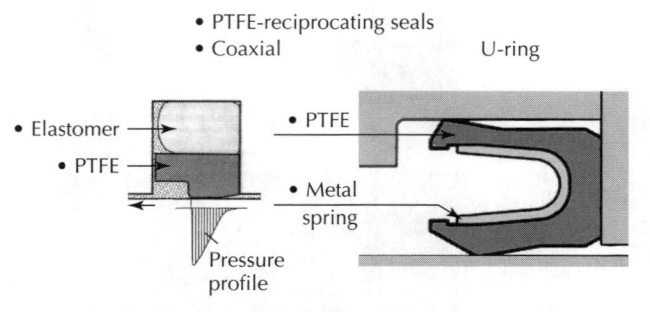

Figure 4.3 Typical plastic reciprocating seals, with metal U spring and elastomer energizers. (Source: BHR Group)

controlled interference across a wider tolerance range. However, the plastic material is more often used in cylinder applications for advantageous friction and wear properties. It is therefore viable to energize the plastic with an elastomer as the fluid resistant capabilities of the PTFE are not a prime concern. Plastic seals with a separate elastomer energizer are therefore popular for reciprocating duties, they can be more robust and easier to fit as it is possible to assemble them into a blind groove.

4.2.2 Seal stability

During operation a reciprocating seal is subject not only to the pressure cycles but also to the frictional forces from the relative motion of the rod in the housing, or piston in the cylinder bore. As the dynamic components are very unlikely to be precisely concentric the combination of interference stress and friction forces will probably vary around the circumference of the seal. This will lead to a considerable distortion force which can cause the seal to twist in the groove. A simple O-ring geometry has little resistance to this force and under all but very light duty reciprocating motion there is the possibility of spiral failure, Figure 4.4. For this reason O-rings were originally replaced by square or rectangular section seals, which are more resistant to spiral failure and can provide improved sealing. Spiral failure can lead to a sudden seal failure and hence this should be taken into account at the design and selection stage.

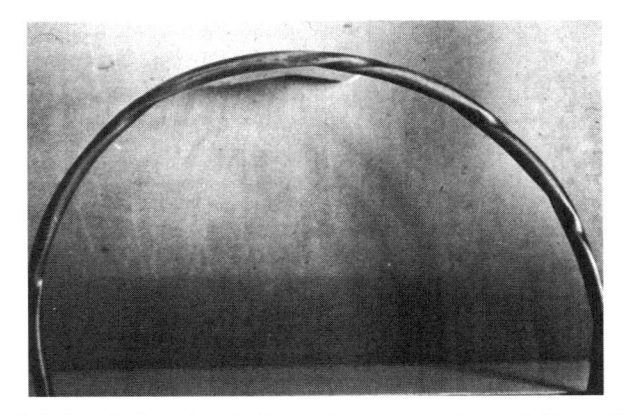

Figure 4.4 Spiral failure of an O-ring used as a reciprocating seal. (Source: Parker)

Seal designs have been developed that can be substituted for O-rings and provide improved resistance to rolling and other potential benefits. The X section seals, such as the familiar Quad ring, will resist a much higher friction force before rolling is likely to occur. Further rolling resistance and also backup rings are provided by T-seal designs. The inherent seal geometry aided by the backup-ring will inhibit rolling. Variants of the T-ring are widely used in aerospace and oilfield applications where a compact design of seal is required. The U-ring designs adopted for many commercial hydraulic systems are inherently very resistant to spiralling in the groove as the length to section ratio is selected to provide stability. For many applications, particularly single acting piston seals, it can be an advantage to use what is sometimes known as a 'distributor' seal geometry, Figure 4.5, where the seal has a more robust stationary sealing face that is also longer than the dynamic lip. This ensures that the seal remains stable due both to

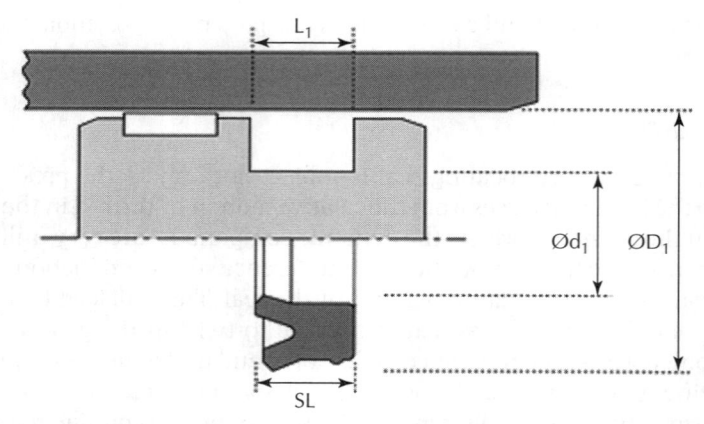

Figure 4.5 Distributor seal geometry with robust stationary sealing face. (Source: Hallite)

friction forces and in the event of reverse pressure, see section 4.2.4, the seal will not move and act as a pressure trap.

With O-ring energized plastic seals another sudden failure mode, blow by, can occur. If the plastic sealing element prevents access of the system pressure to the O-ring it will not be energized and create the sealing force necessary to hold the plastic element in place. This is shown diagrammatically in Figure 4.6(a). The pressure may then bypass between the O-ring and plastic, or across the surface of the plastic. This problem is overcome by providing either radial slots, Figure 4.6(b), or a series of small studs, Figure 4.6(c), in the end faces of the plastic to ensure that the pressure rapidly energizes the O-ring.

High dynamic leakage can also occur with the simplest form of O-ring energized plastic seal where a relatively slender plastic element is used, Figure 4.7. When the seal is pressurized the O-ring will be forced to the end of the groove and create an offset loading to the plastic which can tilt to create an entry profile that will give high dynamic leakage. The correct matching of plastic component and elastomer energizer is therefore essential to successful sealing. For this reason, the successful elastomer energized plastic seals have a contoured profile to the plastic, dependent on the application, which correctly distributes the energizing force.

4.2.3 Sealing mechanism

The mechanism of seals for reciprocating motion have been the subject of intensive research since the mid-1940s.[1] This early work demonstrated the problems of spiral failure with O-rings and initiated investigations into rectangular section seals that continued through until the mid-1970s. Detailed experimental work[2,3] provided valuable information on seal design. Measurement of the lubricant film between the seal and rod, Figure 4.8, showed that leakage was caused when the oil film was thicker as the rod extended, compared with that on the retraction stroke. This could vary with factors such as pressure but critically

Reciprocating seals 287

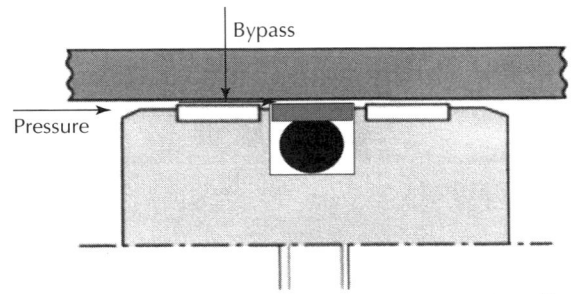

(a) If the plastic seal ring is not correctly designed to allow rapid energization of the O-ring then pressure may cause bypass leakage.

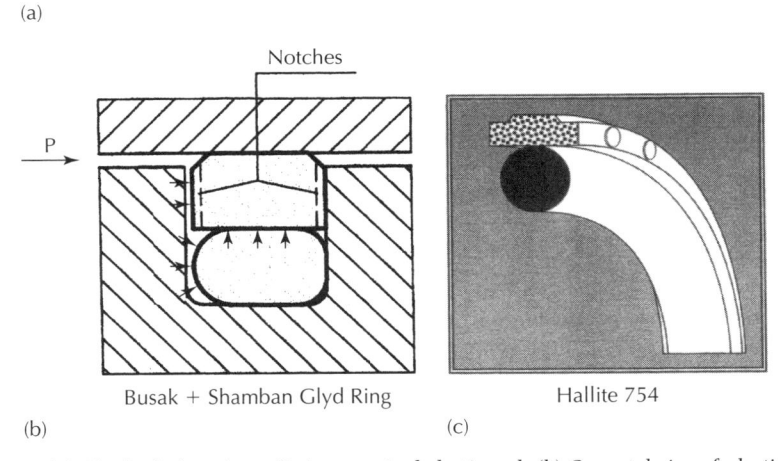

(b) Busak + Shamban Glyd Ring

(c) Hallite 754

Figure 4.6 (a) Blowby leakage in an O-ring energized plastic seal. (b) Correct design of plastic element with radial slots in the plastic. (c) Small studs on the radial face of a plastic coaxial sealing element to ensure pressurization of the elastomer. (Source: (a) Trelleborg Sealing Solutions, (b) Hallite)

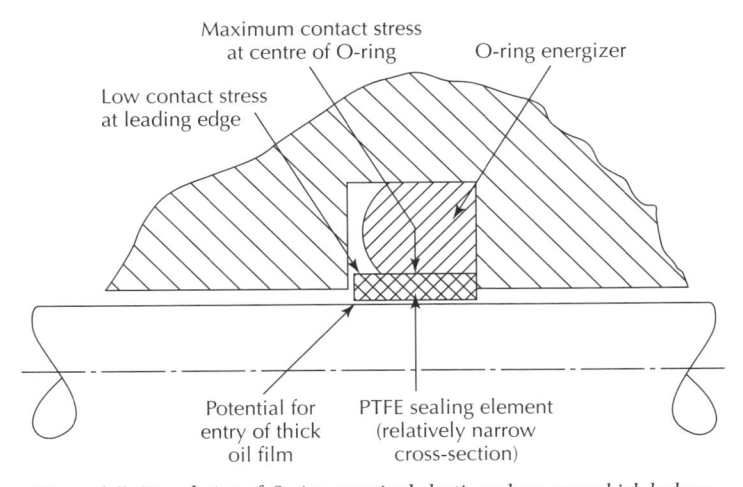

Figure 4.7 Poor design of O-ring energized plastic seal can cause high leakage.

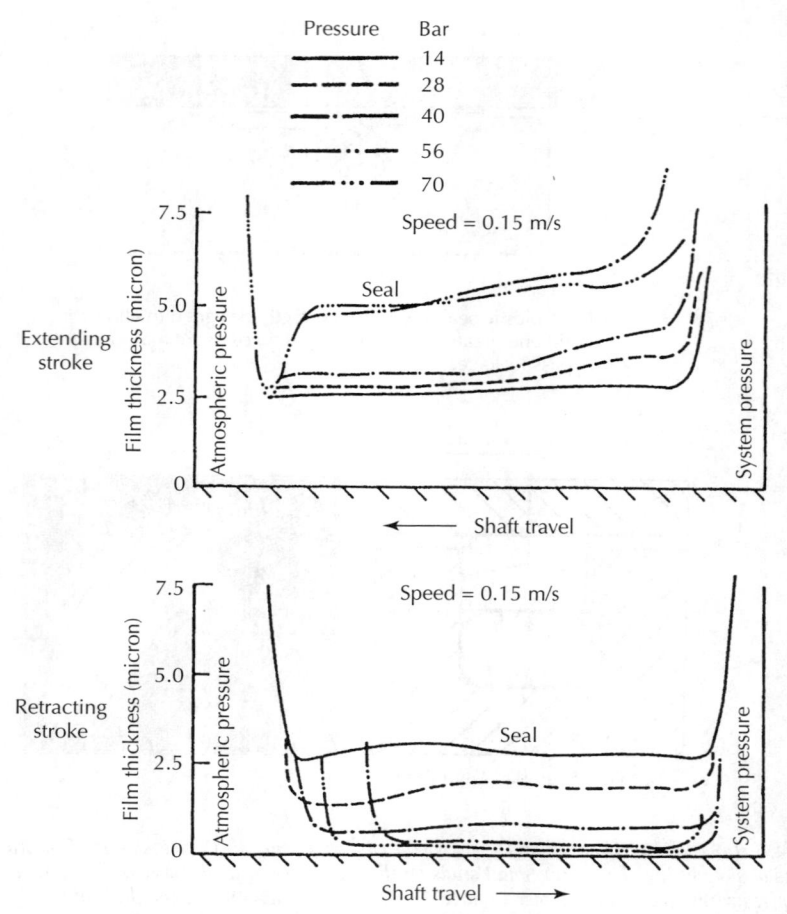

Figure 4.8 Oil film measurements under a rectangular section seal showing extending and retracting stroke and the effect of changing the pressure. (Source: BHR Group)

dependent on a number of factors. Changing speed could have a significant effect as increasing speed could cause entrainment of more lubricant film. However, the most significant factor governing the film of oil under the seal was the entry geometry of the seal rod interface exposed to the oil film. A sharp edge profile could cause sufficient contact stress to penetrate the oil film and prevent buildup of sufficient hydrodynamic pressure to lift the elastomer material and allow a thick lubricant film. This provides the ideal profile for the oil side of the seal to inhibit sufficient film to allow transport to the atmospheric side. Conversely it is necessary to encourage an oil film on the return stroke to promote return of any fluid that escaped past the seal. The sharp edge profile required is at the micron scale level as that is the maximum thickness of oil film under consideration.

It may be considered that this could be achieved by the use of a rectangular seal with one ground square edge and a radius on the atmospheric side edge. Apart from the complexity of ensuring that this is fitted correctly it is an approach that has proved to be unreliable. The presentation of a square edge is difficult with a rectangular seal. Figure 4.9 shows measurements looking along an oil film between shaft and seal for a ground square edge and it can be seen that streaks of oil film penetrate where there are inconsistencies in the finish of the edge.

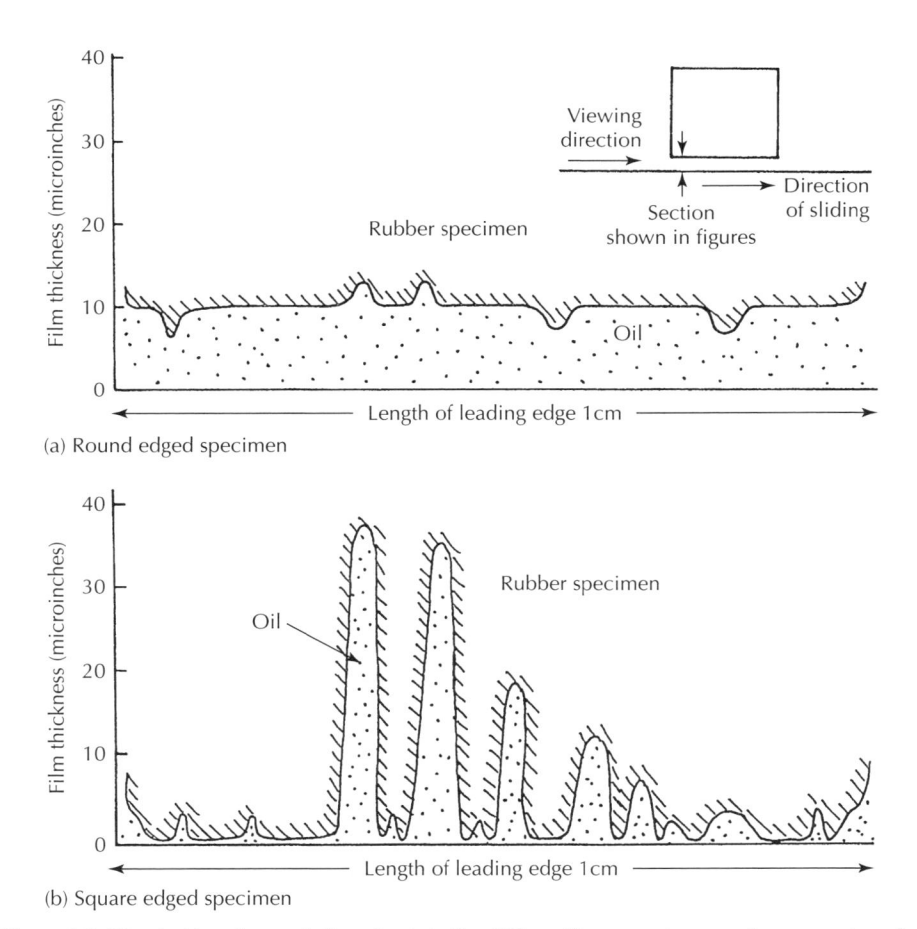

Figure 4.9 View looking along a shaft surface into the oil film with a ground square edge on a rectangular seal. (Source: BHR Group)

The atmospheric side of the seal introduces further complications. An initial assumption might be that a radius on the air side of the seal would encourage a thicker film and promote return of the oil film. Figure 4.10 shows that this is not always the case, under certain circumstances it is possible for a square edge seal to have thicker oil film than one with a radius. This is explained by the action of

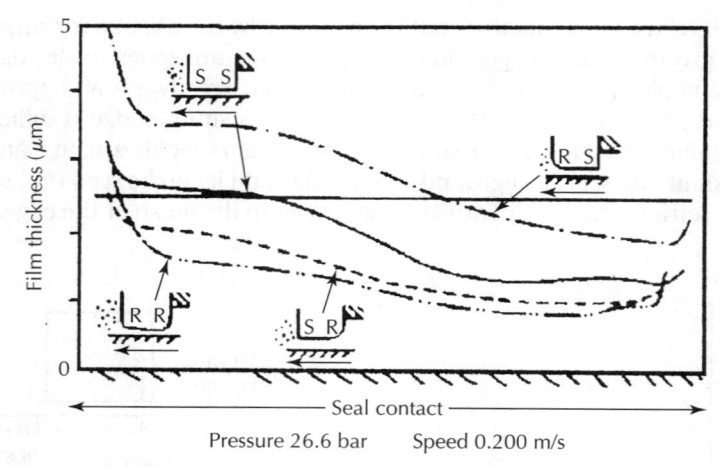

Figure 4.10 Oil film thickness measurements showing the retraction stroke with different edge geometries for rectangular section seals. (Source: BHR Group)

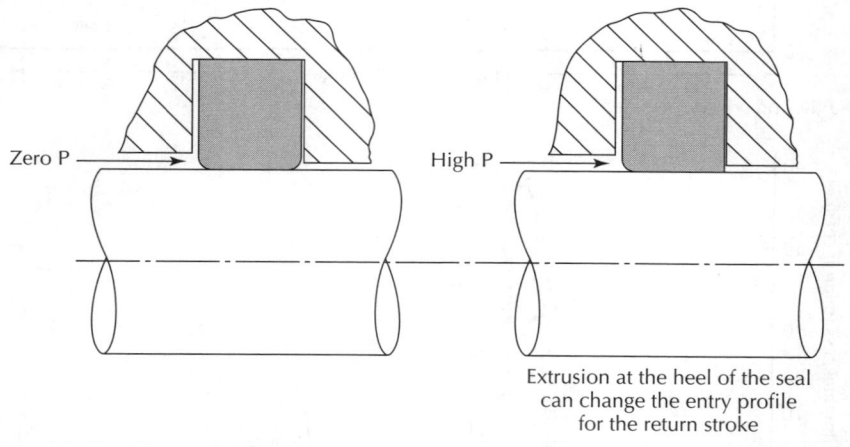

Figure 4.11 The effect of pressure on the distortion of a seal at the atmospheric edge of a rectangular section seal.

the pressure distorting the seal, Figure 4.11. The effect of pressure will tend to reduce the radius, and hence return film effect. Depending on a number of variables including pressure, elastomer hardness and extrusion gap the seal may distort to close to a square edge. Conversely a square edge may distort to provide an entry profile under certain conditions.

This work has demonstrated that while it may be possible to reduce leakage using some design features of a rectangular seal it will not be feasible to obtain a reliable seal under a wide range of operating conditions. It also shows why, in addition to problems of rolling an O-ring will not provide an optimum reciprocating seal. The geometry automatically encourages an oil film, and under the action of pressure the distortion of the O section will provide a geometry that is less favourable to the return oil film on the retraction stroke than the exiting film on the extending stroke. This work also indicates that a number of the O-ring substitutes, such as X-rings and T-rings, while providing an improved geometry compared with an O-ring do not offer the optimum geometry to control leakage. A T-ring in particular may give quite variable performance as the edge exposed for the liquid in action may be either the moulded radius of the elastomer or the machined edge of the backup ring.

For the majority of reciprocating applications it is therefore common to use seals specifically designed for the purpose. These have developed initially empirically from experience with more recent detail development benefiting from continued research. This research on the measurement of seal operation[4,5] has shown that the contact stress distribution is a critical factor governing the formation of a lubricant film under the seal. The steeper the pressure gradient of interference force on the entry to the seal contact zone, the higher the resistance to leakage. Research indicates that the gradient is far more significant than achieving a high level of interference force. A contact stress distribution which tapers away towards the atmospheric side of the seal will then also encourage the return of any residual film on the shaft during the return stroke. For this reason most modern reciprocating seals for rod applications have a profile that resembles that in Figure 4.12. The interference across the seal lips provides the necessary initial contact stress and the heel section on the atmospheric side has little or no contact with the shaft at atmospheric pressure. The trimmed profile of the seal lip is important to achieve the high contact stress gradient. Many early seals had a flat profile facing the pressure in the free condition. When installed the interference distorts the geometry of the seal to provide an entry profile for the oil film, Figure 4.13, and this can cause a leaky seal, hence the trimmed profile used on the majority of current seals.

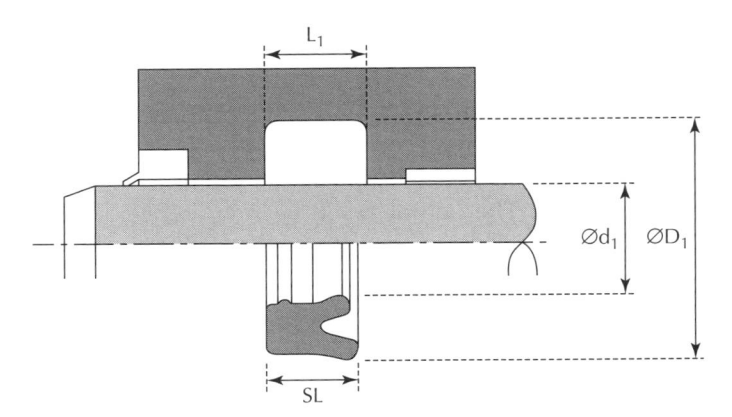

Figure 4.12 Typical reciprocating seal for rod applications. (Source: Hallite)

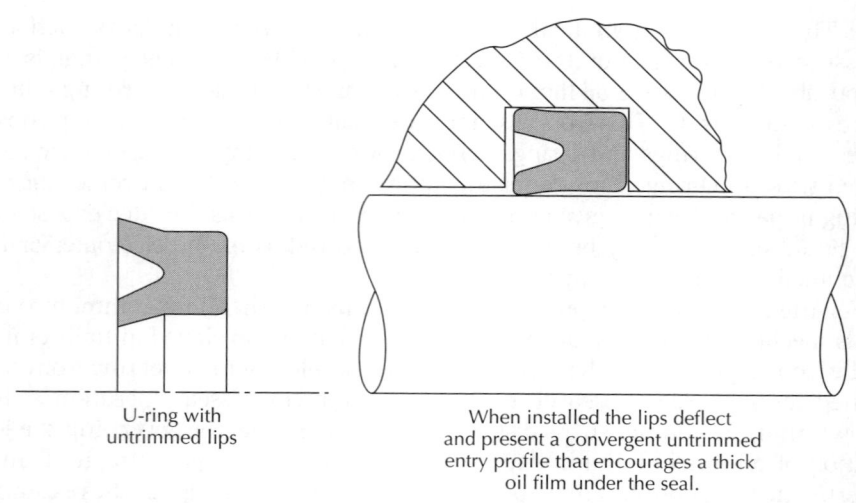

Figure 4.13 Early design of U-ring showing the effect of an untrimmed lip on installation.

U-ring with untrimmed lips

When installed the lips deflect and present a convergent untrimmed entry profile that encourages a thick oil film under the seal.

The basic profile of these rod seals has developed in a number of ways to serve different markets and individual preference. The seals may be manufactured from solid elastomer, usually nitrile, with a fabric and/or plastic anti-extrusion heel, as a U-ring of either thermosetting or thermoplastic elastomer, or designs energized by a resilient elastomer. A variety of the types available are shown in Figure 4.14.

We have seen that prevention of leakage from the seal is dependent on control of a sub-micron oil film. To ensure adequate lubrication of the seal, for minimum friction and wear, some oil film is necessary. A properly designed seal will encourage return of this oil film. However, to minimize leakage under all conditions it is now considered to be most reliably achieved by using some form of tandem seal arrangement. To achieve a high integrity under arduous operating conditions many seal configurations therefore use a seal design with more than one sealing lip or a multiple seal arrangement. These can also present problems of pressure buildup between the seals so must be used with extreme care, section 4.2.4.

To control the oil film reliably requires the correct design of seal, as discussed above. The actual performance will be very dependent on a number of operational factors. The seal interference and contact profile design will dictate the initial interference force. The actual working interference force will also be dependent on the pressure. This will have some direct effect on the interference. However, with elastomer materials this is not entirely straightforward. At a pressure, dependent on the hardness of the material, the elastomer will deform on the surface such that all the asperities on the surface have flattened into 100% real surface area contact with the metal counter-face. Increasing the pressure further cannot increase the real area of contact. The friction will therefore not increase with any further increases in pressure. Some typical friction

Examples: Hallite Seals:		Examples: Markel and Disogrin	
Type	Pressure (bar)	Type	Pressure (bar)
 16	300	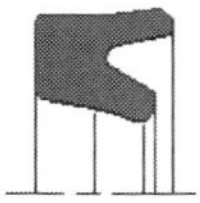 T20	400
 18	500	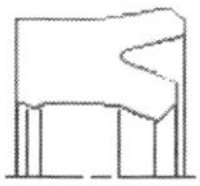 D-300	415
 21	400	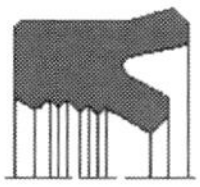 LF 300	325
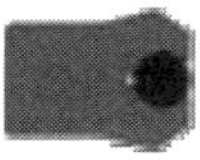 513	350	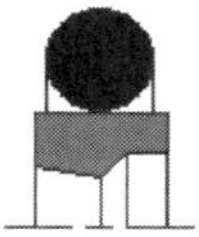 OMS-MR	400
601	400	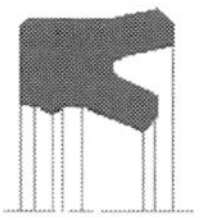 T22	400
605	400	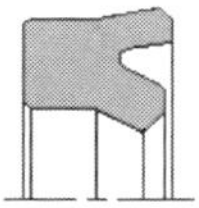 RO	345
 610	400	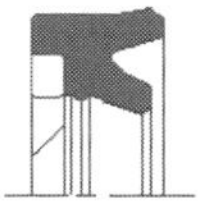 T23	500
 616	240	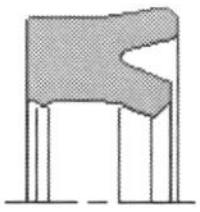 RB	345
 621	700	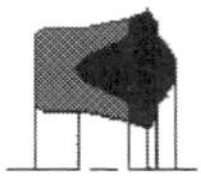 S 8	250
 652	700	 NI 150	100
 653	700		

Figure 4.14 Examples of the range of reciprocating rod seals available.

294   Seals and Sealing Handbook

(a)

(b)

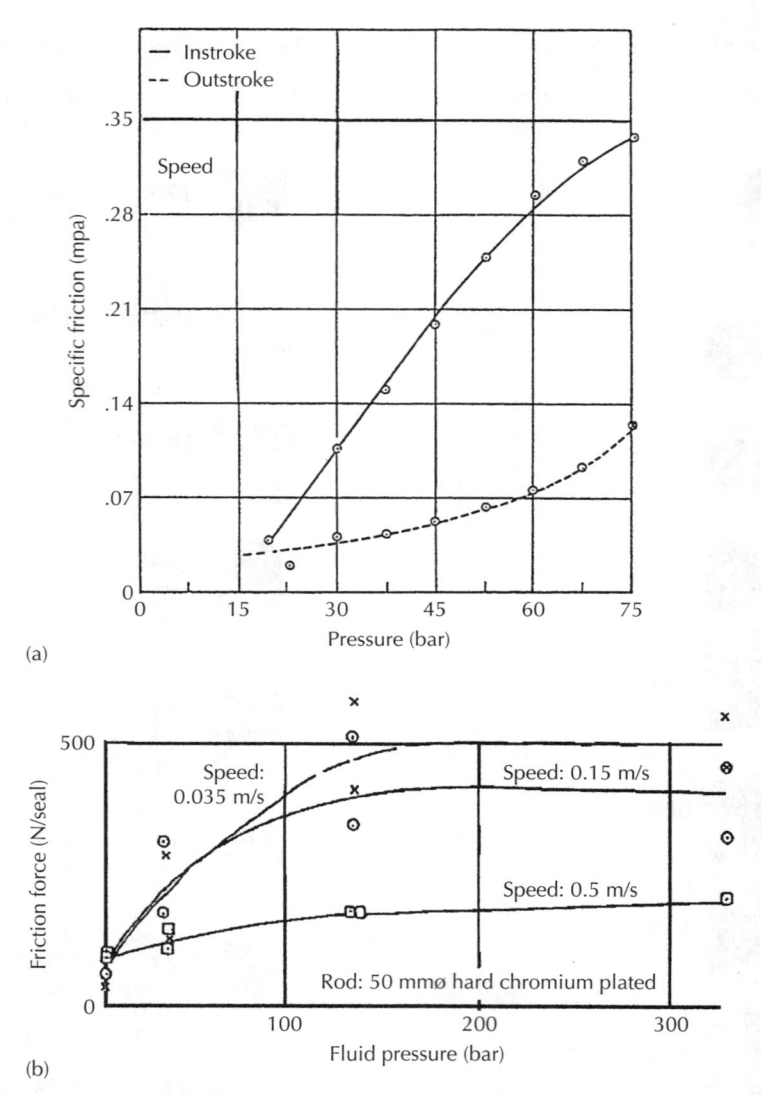

Figure 4.15 The effect of pressure on the friction force of an elastomer reciprocating seal. (Source: BHR Group)

curves for elastomer reciprocating seals are shown in Figure 4.15, the effect of the flattening out of the friction force curve can be clearly seen. For this reason, the use of a friction coefficient for elastomer seals is inappropriate, as it is continuously varying.

Plastic seals behave rather differently as the material is harder and will not flatten into full real area contact in the same way as elastomer. A sample curve for a plastic seal is shown in Figure 4.16. It can be seen that there is some deviation from a theoretical straight line as the material deforms on the surface, but not to the same extent as elastomer.

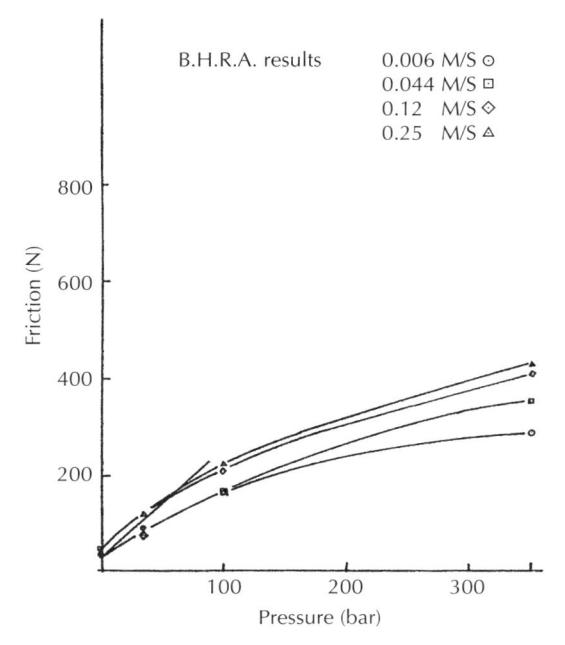

Figure 4.16 The effect of pressure on the friction force of a plastic reciprocating seal. (Source: BHR Group)

These friction curves are samples of representative seals, with an ISO 37 grade mineral hydraulic oil, showing the general trends that have been measured during research work. The actual friction values for individual seal types will be very dependent on the seal design and properties of the material. The rod material and surface texture can also affect the friction as it will have an effect on the lubricant film and actual area of contact.

Speed will affect the lubricant film. Elastomer materials can be deformed relatively easily by any hydrodynamic pressure in the lubricant film. At very low speeds, less than 0.1 m/s, the hydrodynamic effect is very small and the film extremely thin, so friction can be high. As speed increases the film thickness will also increase and the effect of this will be to reduce the friction. Such seals will tend to have a minimum friction force in the range 0.4 to 0.5 m/s. Once they are in a fully hydrodynamic regime the friction will again gradually increase proportionally with the speed. There may be some variation from this trend, if for instance a seal has a very efficient wiping action and runs effectively dry. This can cause a friction that is initially high and does not necessarily vary with speed in a predictable way. Figure 4.17 shows an example of the effect of speed on seal friction.

Viscosity can have a very significant effect, as it has a direct influence on the lubricant film under the seal, an example for a rectangular section seal is shown in Figure 4.18. Most seals are developed to operate with an optimum film when using hydraulic oil which has a viscosity within the range ISO VG 32 to 46. If a

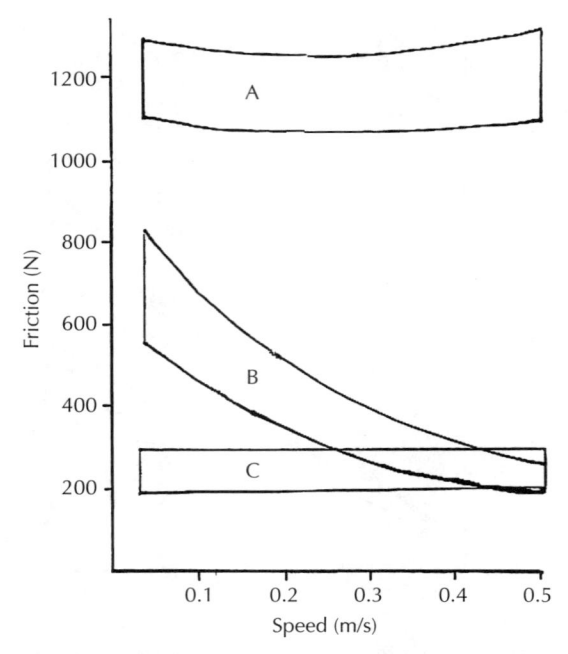

Figure 4.17 Effect of speed on the friction of different elastomer seals. Key: A: Extended content elastomer seal with efficient wiping action. B: Typical elastometer seals. C: PTFE seals

higher viscosity is used then a thicker oil film will result and the potential leakage can be high. As the viscosity reduces the lubricant film will become much thinner. The friction and also wear will increase.

With extremely low viscosities such as high water-based fluids or water itself then friction can be very high, but conversely leakage will be very low while the seal remains intact. In these conditions it is necessary to use seals especially designed for the conditions. If an elastomer seal is used then a design is required that creates a surface that is well lubricated, perhaps by having a fabric wear surface. It is now more usual to employ plastic faced seals using a material that provides good wear resistance in water. Examples are ultrahigh molecular weight polyethylene (UHMWPE) and thermoplastic polyester (TPE).

Surface texture has a direct effect on seal performance. Very often the only parameter quoted is a surface height parameter such as Ra, Rz or Rtm. Extensive work[6] has shown that while these height parameters are a dominant influence they are only a starting point for consideration of the surface. The overall subject is considered in more detail in section 7.3. It is normal for a dynamic seal to run on a surface that has been subject to a finishing operation such as polishing or honing. This style of surface will provide an increased 'bearing area' as shown in Figure 4.19. A grinding operation, no matter how fine will create a surface with sharp peaks and lead to high wear as these peaks will penetrate the lubricant film. Many seal catalogues recommend surface texture between 0.1 and 0.4 μm Ra or an equivalent value of Rz or Rtm. This has been shown to be quite inadequate, for the

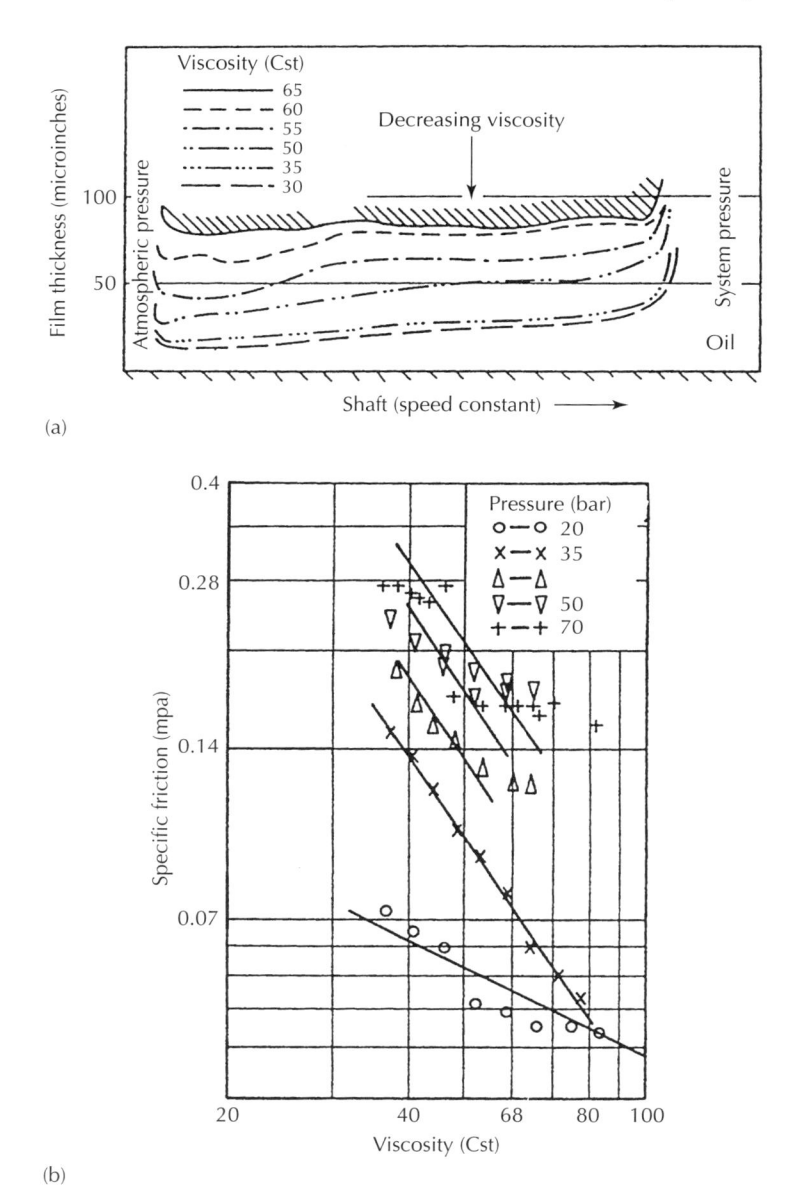

Figure 4.18 Effect of viscosity on the friction of a rectangular section seal. (Source: BHR Group)

reasons discussed above. Seals of plastic materials, such as PTFE or UHMWPE and also polyurethane seals, should not be used on surfaces with a roughness value above 0.25 μm Ra or equivalent. In addition it is necessary to specify a parameter or finishing operation that will provide an adequate bearing area. It has also been found that polyurethane and plastic seals will operate on surfaces smoother than

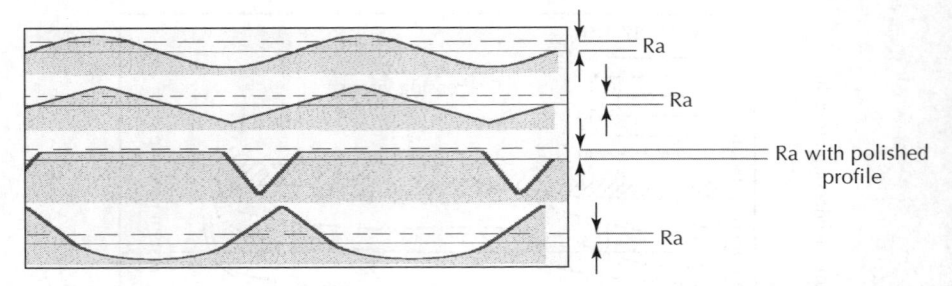

Figure 4.19 A polishing operation provides a surface that has a higher bearing area ratio and will provide lubrication to a seal surface without the high wear that can be caused by the sharp peaks of a finish ground surface. (Source: Taylor Hobson)

0.1 μm Ra. However, especially with plastic seals this must be considered carefully as friction can increase with time.

Traditionally for hydraulic cylinders it was common practice to use a hard chromed piston rod and a honed cylinder bore. Both of these provide a surface texture with a polished surface and lubricant reservoirs which then provide a good basis for seal operation. With the advent of alternative manufacturing techniques and increasing restrictions on the use of chrome, a number of alternatives are also now in use. Various types of drawn over mandrel (DOM) and also skived and roller burnished tube are used in cylinder manufacture. While these can provide nominally a similar surface finish the texture is very different. Most of the tube used in hydraulic cylinders is classed as 'smooth bore' which has a double drawing process and hence an apparently highly polished surface. This can be sufficiently fine to cause lubrication problems and seal design and material must be chosen to suit. A problem can occur with axial scoring, Figure 4.20,

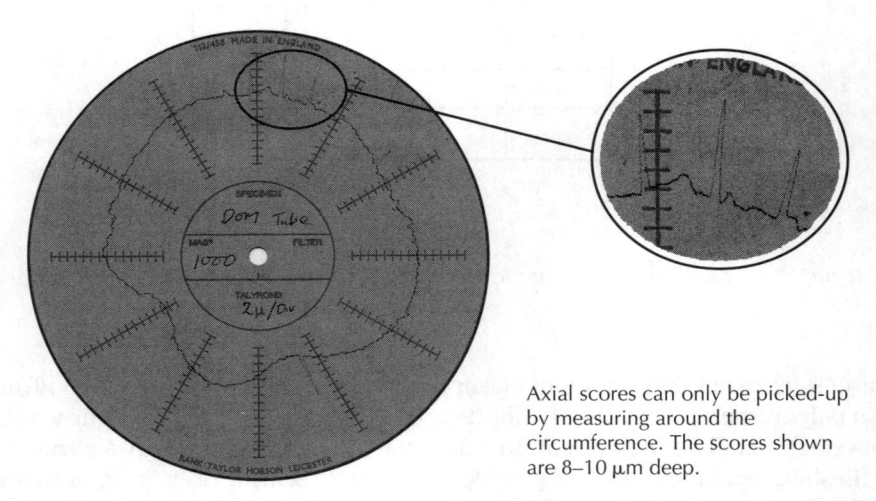

Axial scores can only be picked-up by measuring around the circumference. The scores shown are 8–10 μm deep.

Figure 4.20 Axial scoring in drawn tube can cause problems of leakage and high seal wear. (Source: Hallite)

and also in extreme cases traces of the weld line from the tube manufacture. This type of tube must therefore be inspected thoroughly. The manufacturing specifications of such tube permits some axial scoring[7] and these are likely to eventually increase the wear and erosion of the seals.

Skived and burnished tube is manufactured by skiving, effectively a controlled rough turning operation, and then burnishing with rollers to provide a smooth surface. There can be an apparent screw thread impression but on correctly manufactured tube this is purely an optical effect.

For piston rods alternatives such as stainless steel have traditionally been used where the corrosion resistance of chrome plate is insufficient. The finishing processes for both options have been similar. Alternative processes are now being used to replace chrome plating. For heavy duty and aerospace applications high velocity oxy-fuel (HVOF) coatings are being introduced. These involve the high temperature deposition of wear resistant material on the shaft surface. A number of different systems and a variety of coatings are used. Some examples are listed in Table 4.1. The characteristics of these materials are very different to those of chrome plate and as the table shows there are numerous options on the coating constituents. These can include tungsten or chromium carbide/cobalt, stainless steel, iron/molybdenum or high nickel alloy. Some promising coatings include a mixture of one of the carbides with a metal. The selection of optimum coatings, the appropriate finishing processes and seal design and material are still under development. The most recent data available should be checked before selecting seals for an HVOF surface treatment. It is also necessary to verify that any experience offered is for a similar coating type.

4.2.4 Tandem seal arrangements

The potential for pressure buildup between a pair of seals is something that must be taken seriously and considered at the design and specification stage. Section 2.2.1.3 discusses some further details of temperature related problems with trapped volumes of liquid. This section discusses some of the more usual dynamic multiple seal arrangements. Tandem seal arrangements have been widely used for many years but they require caution. They are found as a pair of back-to-back seals on a piston and as a pressure seal and backup on rod glands.

We have seen in section 4.2.3 that in operation a sub-micron film of oil exists under the seal. It is possible for the fluid film to transfer oil past the seals such that the volume between the seals can become flooded with oil. If this situation continues the interspace can become pressurized[8] and the interseal pressure can even exceed the system pressure. Figure 4.21 shows an example of this from a cylinder seal test rig. It is not feasible to predict whether this interseal pressure may build up because it is dependent on many factors including seal design and operating conditions. Even if the space becomes flooded but not pressurized when operating it is still an undesirable situation as discussed in section 2.2.1.3. It is therefore necessary to ensure that the seal arrangement is designed to permit venting of any pressure buildup both static and dynamic.

Table 4.1(a) Typical powder materials used for hydraulic cylinders, plungers and other pump applications, etc.

Sulzer Metco material designation	Chemistry	TS process typically used	Properties	Some typical seal applications
Amdry 5843; Woka 365; SM 5847	WC-10Co-4Cr	HVOF	Coatings with excellent resistance to abrasion and erosion and very good corrosion resistance in diverse environments; maximum operating temperature 450°C (840°F)	High wear mining and pump applications (piston rods, plungers); landing gears as alternative to chrome plating
Diamalloy 2005	WC-17Co	HVOF	Used to protect against sliding wear, hammer wear, abrasion and fretting; do not use in corrosive environment; maximum operating temperature 500°C (930°F)	Approved for landing gears as alternative to chrome plating; pump seals
Metco 136; Metco 106	Cr_2O_3-SiO_2-TiO_2; Cr_2O_3	HVOF	High wear and excellent corrosion resistance; good low friction characteristics; use up to 540°C (1000°F)	Pump seals, piston rods; compressor applications
Diamalloy 4006	Ni-20Cr-10W-9Mo-4Cu-1C-1B-1Fe	HVOF or APS	Coatings offer sliding wear, medium abrasion resistance and excellent corrosion protection; coatings resist scuffing and galling	Piston rods in corrosive environments with medium wear
Diamalloy 3007; SM 5241; Woka 710x	Cr_3C_2-NiCr	HVOF	Generates hard dense coatings that provide good protection from abrasion and erosion; good fretting resistance; very good oxidation and corrosion resistance; used as alternative to hard chrome plating; maximum operating temperature 815°C (1500°F)	Piston rods, e.g. for earthmoving machines; hydraulic rods,
Diamalloy 1004	Cu-9.5Al-1Fe	HVOF	Aluminium bronze material with moderate oxidation, wear and fretting resistance at low temperatures; very good emergency dry running properties	Pumps; soft bearing surfaces such as piston guides; compressor air seals

(Source: Sulzer Metco)

Table 4.1(b) Proprietary coatings offered from SM Coating Service

Designation	Composition	TS process	Max operating temp.	Properties	Application
SUME®2700	WC/CrC/Ni	HVOF	500°C (930°F)	Excellent corrosion resistance in diverse environments. Good wear and tribological characteristics	Reciprocating pump plungers. Carriers, sleeves and seats for mechanical seals. Bearing, gland and seal sleeves for centrifugal pumps
SUME®3652	Chromium oxide/ silicon oxide/ titanium oxide	APS	540°C (1000°F)	Excellent corrosion resistance in diverse environments. Good wear characteristics. Excellent tribological characteristics	Reciprocating pump plungers and rams. Bearing, gland and seal sleeves for centrifugal pumps
SUME®2651	WC/Co/Cr	HVOF	450°C (840°F)	Good corrosion resistance in diverse environments. Excellent wear and tribological characteristics	Piston rods for reciprocating compressors. Reciprocating pump plungers
SUME®3650	Cr_2O_3	APS	540°C (1000°F)	Excellent corrosion resistance in diverse environments. Good wear characteristics. Excellent tribological characteristics	Reciprocating pump plungers and rams. Sleeves and faces for mechanical seals

(Source: Sulzer Metco)

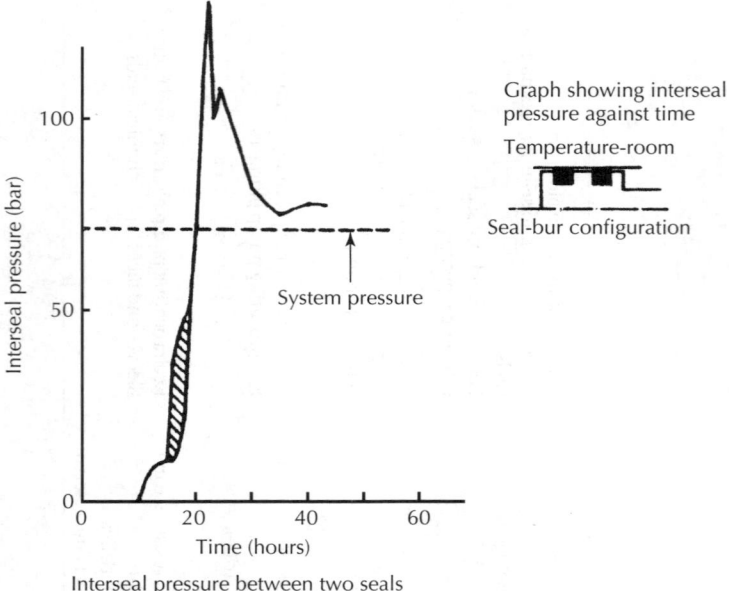

Figure 4.21 Example of interseal pressure using rectangular section piston seals. A pressure of 125 bar has built up between the seals in a 70 bar system. In this situation it is necessary to ensure that the seal design allows the interseal pressure to vent. (Source: BHR Group)

4.2.4.1 Piston back-to-back seals

A common arrangement for both elastomer and plastic seals is to use a pair of U section seals back to back on a piston, Figure 4.22. In the event of pressure buildup it may look as if this arrangement can easily vent by deflection of the seal lips. However, it is possible for the interspace pressure to trap the seal lips in the clearance and hence seal the possible venting route.

It is important to design the arrangement so that this potential trapping is avoided. This can be achieved by using a non-asymmetric seal design with the stationary seal portion longer than the dynamic lip, Figure 4.23(a). This is sometimes known as a 'distributor' seal. This will position the seal in the groove so that the dynamic lip cannot be trapped in the low pressure clearance. An alternative if it is not practicable to fit a seal of this design is to use a spacer ring to hold the seal in position, Figure 4.23(b). This style of spacer has been successfully applied to both elastomer and plastic U-ring seals where reverse pressure problems have occurred.

4.2.4.2 Tandem rod seals

A variety of rod seal systems are used where the seals are used in series. The usual philosophy is for the first or pressure side seal to form the primary pressure seal. Most or virtually the entire pressure drop will be expected to occur across

Reciprocating seals 303

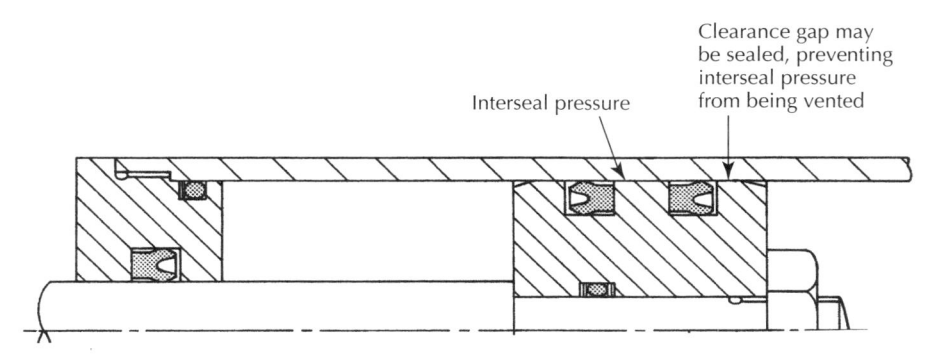

Figure 4.22 A piston with symmetrical back-to-back U-rings. Care is required to ensure that pressure buildup does not occur that could be trapped between the seals.

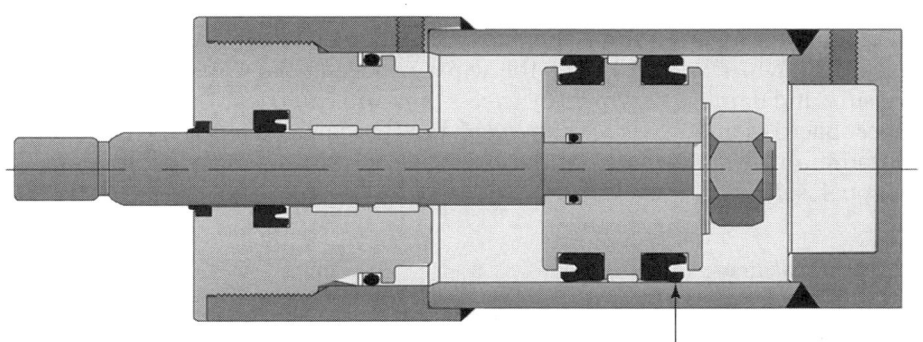

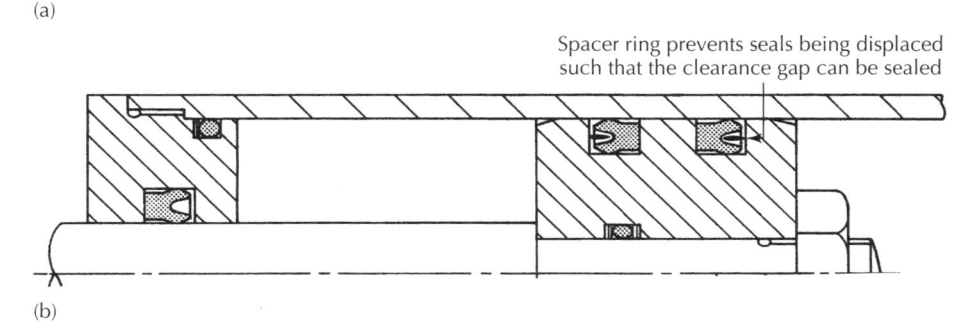

Figure 4.23 Pistons with back-to-back seal arrangements that are designed to ensure the venting of any interseal pressure: (a) non-symmetric seal geometry locates the seal in the groove where it cannot be easily displaced to seal of the clearance gap; (b) a spacer ring locates the seal to prevent displacement.

this seal. The second seal will act as a low pressure wiper to reduce the oil film that is transferred out of the cylinder to the minimum. These systems may be made up in several ways:

- A single pressure seal and a double acting wiper.
- A pair of single acting plastic rod seals.
- A heavy duty elastomer pressure seal with a light duty seal as wiper.
- Plastic pressure seal with an elastomer U-seal as the wiper seal.
- One of the above pairs of seals with the addition of a double acting wiper to provide a particularly dry rod.

In each case it is important to ensure that some form of venting can occur. If the pressure seal is a single acting design then provided the seal lip cannot become trapped as described in section 4.2.4.1 then some venting can occur. The pressure at which it occurs will be dependent on the design of seal. The use of a rectangular section or solid single acting seal on the pressure side will potentially lead to unacceptable pressure buildup.

Additional care is required where a double acting wiper is used. The construction of the wiper and design of the wiper groove are not intended for a high pressure and damage to wipers by excess pressure can occur. A vent hole to the wiper space is sometimes recommended, but this can lead to the introduction of contaminant. An alternative is the use of a self-venting wiper, as discussed in section 4.2.7.

4.2.5 Rod seal arrangements

Table 4.2 provides a summary of potential seal types that may be considered for different application conditions. This table is intended as a starting point to guide the reader into the correct general area for a given application. Classifications are very general and depend on the environment and quality of manufacture of the equipment. In general light duty would be considered up to 160 bar, medium duty to 210 bar and heavy duty over 300 bar. Most hydraulic systems are within the range up to 400 bar, but some specialized systems work up to 700 bar.

4.2.6 Piston seal arrangements

Table 4.3 provides an equivalent selection guide for piston seals. Double acting cylinders may have either a pair of single acting seals or one double acting seal. A pair of double acting seals should not be used. This would not only be wasteful but there would be a very definite risk of creating very high interseal pressure buildup.

Single acting, displacement type cylinders would normally be fitted with a single acting piston seal, or if good leakage control is required a tandem piston arrangement equivalent to that discussed in the section on rod seals. A double acting piston seal would be expected to cause higher leakage in this case and should be avoided.

Table 4.2 Hydraulic system rod seal application areas

Duty	Seal type	Advantages	Disadvantages	Standards
Light duty compact	O-ring	Lowest initial cost compact dimensions	Spiral failure, limited offset Poor dynamic leakage control	BS 1806, MIL-G 5514 BS 4518, SMS 1588
Light duty compact	X-ring	Improved resistance to spiralling compared to O-ring	Will have dynamic leakage Limited offset	BS 1806, MIL-G 5514 BS 4518, SMS 1588
Light duty compact	T-ring	Improved resistance to spiralling Extrusion resistance	Liable to have dynamic leakage Limited offset	BS 1806, MIL-G 5514 BS 4518, SMS 1588
Light duty compact	L section + similar	Reduced leakage Extrusion resistance	Average leakage control Limited offset	BS 1806, MIL-G 5514 BS 4518, SMS 1588
Compact low friction	O-ring/PTFE, Slipper, Delta	Low friction Compact dimensions	Leakage Stability poor	BS 1806, MIL-G 5514 BS 4518, SMS 1588
Compact low friction	Plastic U-ring spring energized	Low friction Low leakage	Higher cost Difficult fitting	BS 1806, MIL-G 5514 BS 4518, SMS 1588
Compact low friction	Elastomer energized Single acting design	Low leakage Low friction	Medium cost Care fitting	BS 1806, MIL-G 5514 BS 4518, SMS 1588
General purpose, medium duty 160–210 bar	U-ring, elastomer or polyurethane	Improved leakage control Improved offset performance More stable design	Increased space	ISO 5597
Reduced leakage	As above + double acting wiper	Reduced leakage	Interspace pressure	ISO 5597 + ISO 6195 Type C
Heavy duty Fluctuating pressures Higher offset 300–400 bar	Heavy duty U-ring polyurethane	Improved resistance to pressure fluctuations and extrusion	Higher friction	ISO 5597

(*Continued*)

Table 4.2 (Continued)

Duty	Seal type	Advantages	Disadvantages	Standards
Lower leakage	As above + double acting wiper	Reduced leakage	Interspace pressure	ISO 5597 + 6195 Type C ISO 7425 Part 2
Heavy duty Minimized leakage	Tandem seals + double acting wiper	Minimized leakage	Additional space Interspace pressure	+ ISO 6195 Type C
Medium duty Minimized friction	Elastomer energized plastic	Minimum friction	Possible leakage Increased care and handling	ISO 7425 Part 2
Medium duty Consistent low friction Reduced leakage	Tandem elastomer energized plastic	Minimum friction reduced leakage	Increased care and handling Increased space requirement Interspace pressure	ISO 7425 Part 2
Heavy duty Minimum leakage and friction	Tandem seals elastomer energized plastic and polyurethane U-ring + double acting wiper	Reasonable friction dry rod performance offset capability	Space requirement Interspace pressure	ISO: 7425 +6195
Heavy duty, high offset vibration and contaminants	Solid elastomer	Robust seal	Higher friction, especially breakout. Leakage control less than optimums above	ISO 5597
Large diameter heavy duty environment	V-ring set	Robust, multiple redundancy	High friction Space requirement Axial compression of seal	ISO 5597 long housing
Heavy duty reduced low pressure leakage	V-ring set with elastomer seal as header	Robust, multiple redundancy header gives low pressure seal No axial compression improved low/zero pressure leakage control		ISO 5597 long housing

Table 4.3 Hydraulic system piston seal application areas

Duty	Seal type	Advantages	Disadvantages	Standards
Light duty compact double acting	O-ring Double acting	Lowest initial cost compact dimensions	Spiral failure, limited offset Poor dynamic leakage control	BS 1806, MIL.-G 5514 BS 4518, SMS 1588
Light duty compact double acting	X-ring Double acting	Improved resistance to spiralling compared to O-ring	Will have dynamic leakage Limited offset	BS 1806, MIL.-G 5514 BS 4518, SMS 1588
Light duty compact double acting	T-ring Double acting	Improved resistance to spiralling Extrusion resistance	Liable to have dynamic leakage Limited offset	BS 1806, MIL.-G 5514 BS 4518, SMS 1588
Light duty compact single acting	L section + similar Single acting	Reduced leakage Extrusion resistance	Average leakage control Limited offset	BS 1806, MIL.-G 5514 BS 4518, SMS 1588
Compact low friction double acting	O-ring/PTFE, Slipper, Delta Double acting	Low friction Compact dimensions	Leakage Stability poor	BS 1806, MIL.-G 5514 BS 4518, SMS 1588
Compact low friction single acting	Plastic U-ring spring energized Single acting	Low friction Low leakage	Higher cost Difficult fitting, can need split piston	BS 1806, MIL.-G 5514 BS 4518, SMS 1588
Compact low friction single acting	Elastomer energized Single acting	Low leakage Low friction	Medium cost Care fitting	BS 1806, MIL.-G 5514 BS 4518, SMS 1588
General purpose, medium duty, single acting	U-ring, elastomer or polyurethane Single acting	Improved leakage control Improved offset performance More stable design	Increased space	ISO 5597 L1 or L2

(Continued)

Table 4.3 (Continued)

Duty	Seal type	Advantages	Disadvantages	Standards
Double acting medium duty	Energized polyester or polyurethane Double acting	Higher pressure, compact, wear resistant, low lubricity fluids	Increased radial space	ISO 7425 Part 1
Double acting minimum friction	Elastomer energized PTFE	Low, consistent friction	Some leakage across piston	ISO 7425 Part 1
Single acting minimum friction	Elastomer energized PTFE	Low, consistent friction	Some leakage across piston	ISO 7425 Part 1
Double acting robust seal integral bearings good static seal	Elastomer piston seal with backup rings and bearings	Good leakage control High offset Zero static leakage	High friction especially breakout Less resistance to side loading than design with wider spaced bearings	ISO 6457
Double acting robust seal good static seal	Elastomer piston seal with backup rings or fabric reinforcing	Good leakage control High offset Zero static leakage	High friction especially breakout Improved side loading if remote bearings are used Split piston required for fabric designs	ISO 5597
Double acting robust seal good static seal improved friction	Elastomer energized polyurethane or polyester	Good leakage control High offset Zero static leakage Improved wear and less friction	Higher friction at breakout than plastic seals	ISO 5597

Wherever feasible it is recommended that seals complying with a popular standard are used. This will ensure availability from numerous reputable suppliers. The ISO hydraulic seal system standards provide details of seal housings that comply with the various ISO hydraulic cylinder standards. This ensures that a wide range of seal types and sizes are available, including the excluders as discussed in section 4.2.7.

The most compact seals in Tables 4.2 and 4.3 are all widely available to fit in the popular O-ring standard groove sizes. The imperial grooves are covered by BS 1806 in the UK and by MIL–G-5514 in the USA and are also commonly used worldwide by the aerospace industry. The most widely used metric groove sizes are those provided by BS 4518 and the Swedish SMS 1588. At the time of writing there is no usable ISO O-ring standards covering both seals and groove sizes.[11] It is hoped that this situation will be resolved in the relatively near future.

Seals designed specifically for hydraulic cylinder seals are covered by three standards, each of which covers one of the popular categories of seal configuration discussed above.

ISO 5597:1987 Hydraulic fluid power – Cylinders – Housings for piston and rod seals in reciprocating applications – Dimensions and tolerances

This standard covers seal grooves for single acting seal designs typified by elastomer seals with fabric backing and most of the U-ring designs. Grooves for single acting seals used on both pistons and rods are covered. In the piston application the two seal grooves are shown for back-to-back installation of single acting seals for a double acting cylinder. Grooves are specified for cylinder bores in the range 16 to 500 mm diameter, and for rods in the range 6 to 360 mm diameter. Within the size range the document includes the rod and cylinder bore sizes of the ISO hydraulic cylinder ranges. For each diameter ISO 5597 provides a choice of short, medium or long seal housing lengths. The short and medium length housings are suitable for compact seals, whereas the long housings are for multi-lip seals such as vee packs and the extended contact unit seals that are used for some earthmoving. A smaller series of rod and piston seal grooves with reduced cross-sections is given for ISO 6020-2 160 bar cylinders, with bores in the range 25 to 200 mm and rods 12 to 140 mm.

ISO 6547:1981 Hydraulic fluid power – Cylinders – Piston seal housings incorporating bearing rings – Dimensions and tolerances

This standard specifies T-shaped grooves for piston seals, incorporating bearing elements, Figure 4.14. It covers cylinder bores from 25 to 500 mm. In the range 25 to 160 mm bore size, there is a choice of two seal housing sections.

ISO 7425 Hydraulic fluid power – Housings for elastomer-energized, plastic-faced seals – Dimensions and tolerances

This standard provides groove dimensions for seals such as profiled plastic elements with an O-ring energizer and the variants that may have rectangular elastomer energizers. There are two complementary parts to the standard:

- Part 1: Piston seal housings: This standard gives groove dimensions for piston seals for bores in the range 16 to 500 mm, complementing the

cylinder bores covered by both ISO 5597 and ISO 6547. The grooves are specifically designed to suit coaxial plastic faced seals with an O-ring that is used to energize the plastic dynamic seal. Most of these seals employ a standard imperial size O-ring (ISO 16031, BS 1806, AS568)
- Part 2: Rod seal housings: This rod seal standard complements ISO 7425-1 above for rod diameters in the range 6 to 360 mm. The rod seal standard features the potential use of a pair of seals in tandem that is often preferred for this style of rod seal. For applications where plastic seals may not provide a sufficiently dry rod a number of manufacturers also supply a U-seal design, typically of polyurethane material, that will fit the ISO 7425-2 groove sizes.

4.2.7 Excluders

The exclusion device is an important component in the overall rod sealing system. The primary duty of the excluder is to remove contaminant from the rod surface to prevent it either damaging the rod seal or passing under the seal in the lubricant film to contaminate the hydraulic system. The construction and materials can vary widely depending on the environment. A very broad and general classification adopted by the author is to consider them either as:

- Wipers, generally hard elastomer or softer plastics that will remove water, dust and wet mud, etc.
  Or:
- Scrapers, which are either made from harder plastic or a softer non-damaging metal such as brass. These will be expected to remove ice, dry mud or other solidified debris from the rod.

The excluder is therefore designed with an outward facing wiper profile to remove a film from the rod. If it is considered that a sub-micron film of hydraulic fluid will have passed out through the excluder on an outstroke, an effective exclusion design can be expected to remove that film on the return stroke. It has been shown that fitting excluders will cause higher seal leakage than a system that comprises a seal only.[9]

Figure 4.24 demonstrates some leakage results for one seal design under constant operating conditions with several wiper designs and without a wiper at all. It can be seen that the selection of the wiper in conjunction with the seal is an important consideration to overall system performance.

The first important consideration is the operating conditions. If it is necessary to remove ice or solidified contaminant then a scraper of hard plastic or brass will be required. However, a harder material will be less effective at removing very fine dust and condensation or water. For the most aggressive conditions a metallic scraper is likely to be necessary, Figure 4.25. This design has an elastomer secondary lip to remove fine dust particles and any potential liquid contaminant. If the removal of very hard contaminant is not required then one of the

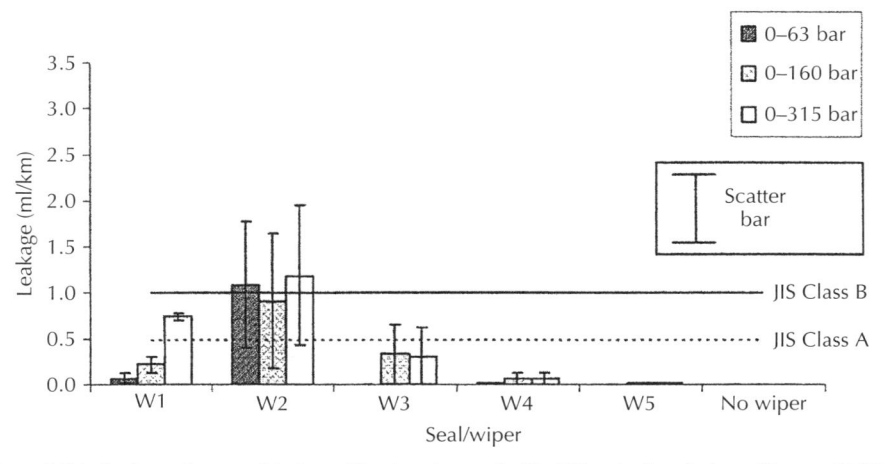

Figure 4.24 Leakage of one seal design without a wiper and with different wiper designs. (Source: Hallite)

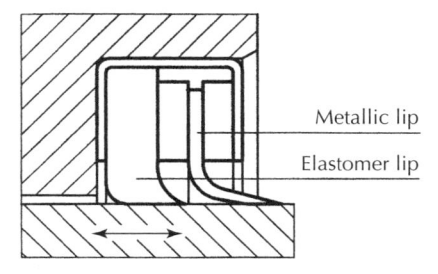

Figure 4.25 Metal scraper with elastomer wiper lip to remove fine dust and liquid contaminant. (Source: Trelleborg Sealing Solutions)

wide range of plastic and hard elastomer wipers or scrapers can be considered. For removal of mud or ice in normal agricultural or mobile hydraulic applications a range of wipers manufactured from materials such as filled PTFE and polyester are available. These may depend on direct interference of the wiper to provide the scraping action, or can be energized by an elastomeric element such as an O-ring. In applications where there is less likely to be ice, or dry mud, an elastomer wiper can be considered.

To provide the minimum leakage conditions a double acting wiper can be considered, Figure 4.26. But, these must be used with caution as there is the possibility of pressure buildup between the seal and wiper, see sections 4.2.3 and 4.2.4. Excess pressure can damage the wiper or even blow it out. It is important that a seal design that facilitates inward pumping of the oil is used. Some seals are specifically designed for this purpose, Figure 4.27. However, under certain operating conditions, such as high speed extending strokes, it may still be possible for pressure buildup to occur. Some designs of wiper[9] and seal,[10] are designed to vent if excess interspace pressure occurs, Figure 4.28.

312   Seals and Sealing Handbook

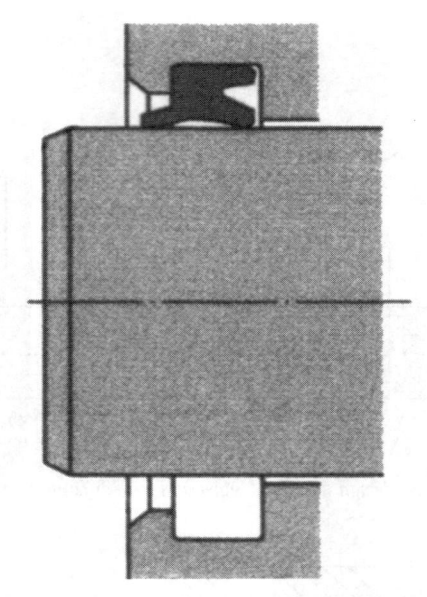

Figure 4.26   A double acting wiper acts as a low pressure seal to retain any small amounts of leakage from the pressure seal and also as an excluder for contaminant. (Source: Hallite)

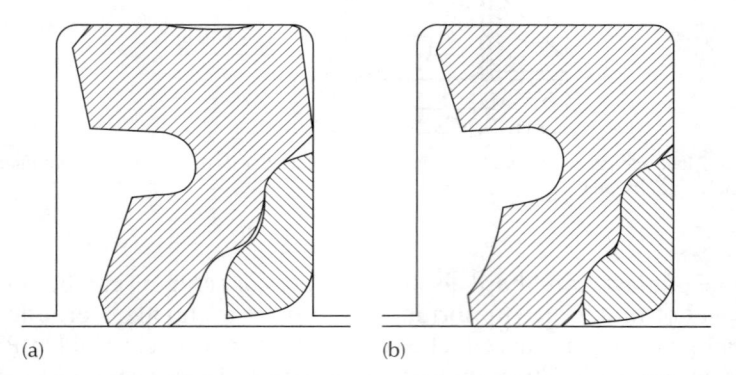

(a)   (b)

Figure 4.27   Rod seal specifically designed to facilitate inward pumping of external oil film that is intended for use in conjunction with double acting wipers. (Source: Patent WO 2004/088135)

There are a series of housing designs that accommodate each of the excluder designs, and these are included in ISO 6195:2002 Fluid power systems and components – Cylinder-rod wiper-ring housings in reciprocating applications – Dimensions and tolerances.

This standard details grooves that cover the requirements of four different types of wiper, Types A, B, C and D all of which are shown in Figure 4.29. Type A are single lip wipers that clip into a groove in the cylinder housing. This housing is used for elastomer and plastic wipers such as those in Figure 4.30. Type B

Reciprocating seals 313

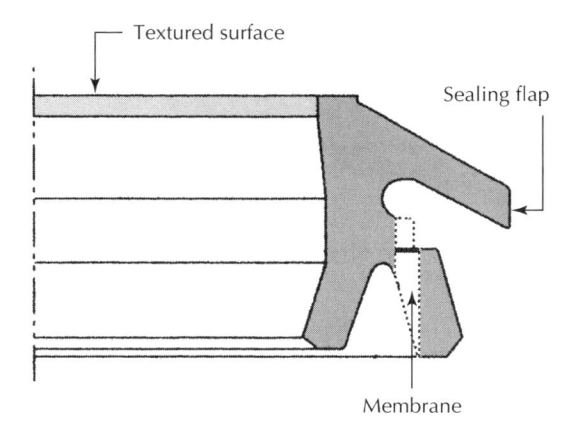

Figure 4.28 Self-venting double lip wiper designed to relieve pressure buildup that may occur between seal and wiper. (Source: Hallite)

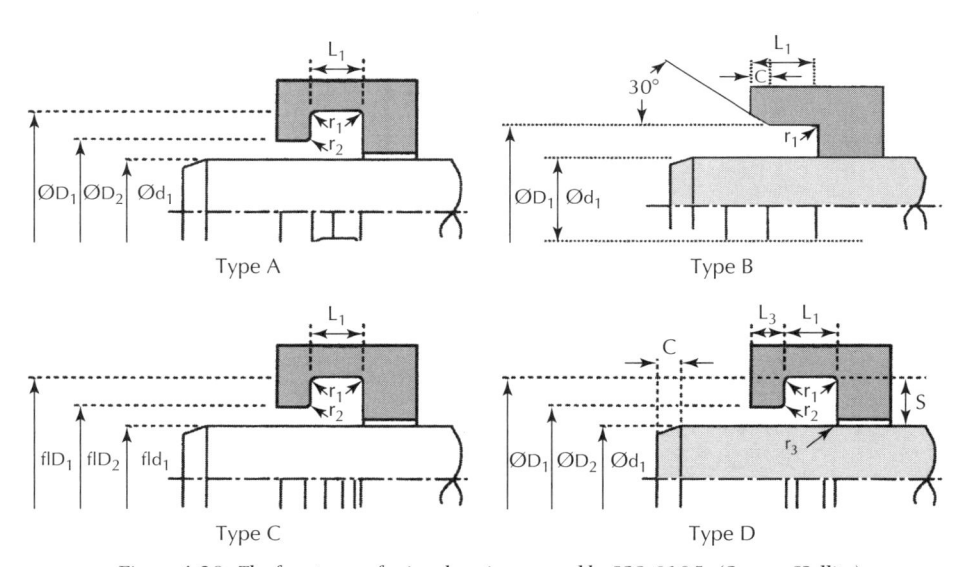

Figure 4.29 The four types of wiper housing covered by ISO 6195. (Source: Hallite)

wipers are also single lip, but are designed for a press fit. This will be used for metallic scrapers, Figure 4.25, and also plastic or elastomer wipers that have a moulded-in metal reinforcement, Figure 4.31. Double lip wipers can be used with a Type C groove. This design has a complete groove to allow for the pressure buildup that may occur between the seal and wiper with this type of wiper arrangement, see sections 4.2.3 and 4.2.4. The Type D grooves are for elastomer-energized, plastic-faced wipers, Figure 4.32.

314 *Seals and Sealing Handbook*

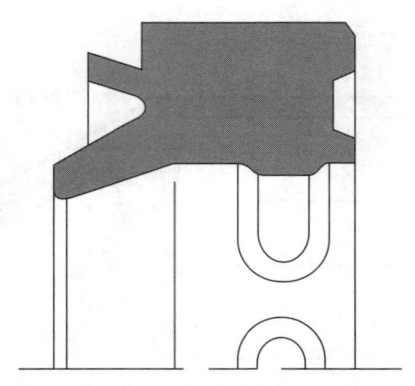

Figure 4.30 Plastic wiper to be used with an ISO 6195 Type A clip-in housing. (Source: Freudenberg Simrit)

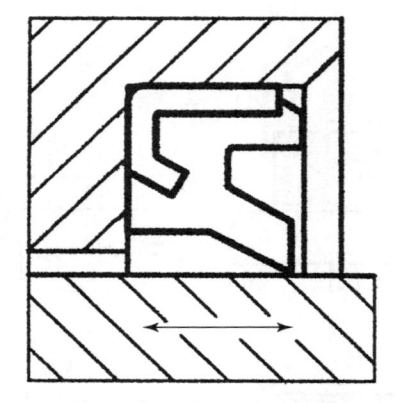

Figure 4.31 Press fit plastic wiper with metal reinforcing ring for fitting to an ISO 6195 Type B housing. (Source: Trelleborg Sealing Solutions)

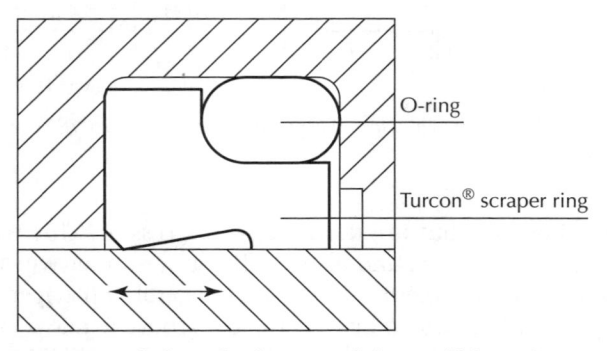

Figure 4.32 Elastomer energized plastic faced wiper seal that would be used in an ISO 6195 Type D housing. (Source: Trelleborg Sealing Solutions)

4.3 Pneumatic cylinder seals

Historically it was common to use light duty hydraulic cylinder seals for pneumatic applications. It is now normal practice to use seals specifically designed for pneumatic applications, for several reasons:

- Pneumatic systems now have to work with minimal lubrication or totally dry which places very different criteria on seal selection.
- Pressures are much lower, the maximum for pneumatic normally being 10 bar, which means that much lighter duty seal designs can be used.
- The lubricants used can be very different to those in contact with hydraulic seals. It is typical to use grease during installation, with the type of grease being dependent on the application.
- The seal material must be resistant to the oxidizing effects of compressed air which may also be at higher temperatures than those experienced by hydraulic systems.
- A small leakage of compressed air is not a major problem, and may be preferable to high friction that will cause premature wear.
- The lower pressure with consequent lower stresses and high manufacturing volumes provide the potential for alternative materials of construction and manufacturing processes.
- The seal needs to retain the lubricant in the contact zone rather than exclude it. This is in direct contrast to the requirements for an oil hydraulic seal.

For these reasons pneumatic seals have now evolved as a very distinct class of seals with very different characteristics to hydraulic cylinder seals.

The lubricant available for a seal in pneumatic systems will consist of:

- Very small amounts of carryover from the system compressor.
- Minute amounts from a system lubricator.
- Grease pre-applied to the seals when assembled.

The increasing restrictions on the exhaust of any form of oil mist aerosol mean that in most modern systems the only available lubricant is that applied to the seals during assembly. This is especially so in applications concerned with food, pharmaceuticals and other products where cleanliness is paramount. In such applications the air is rigorously dried and cleaned before it is used and exhausted to atmosphere. It is therefore important that any lubricant applied during assembly is retained in the area of the seal and encouraged to lubricate the seal lip and other potential wearing parts of the seal.

The basic contact zone design for pneumatic seals is therefore quite different to a hydraulic seal. The seals are designed to have a much lower contact force and so the lip design will be a more flexible and reduced section of material. In addition to the lighter loading the lip design is intended to encourage any lubricant present under the seal. The pressure side of the seal will have either a radius or a shallow chamfer to encourage entrainment. An example is shown in Figure 4.33.

Rod seal

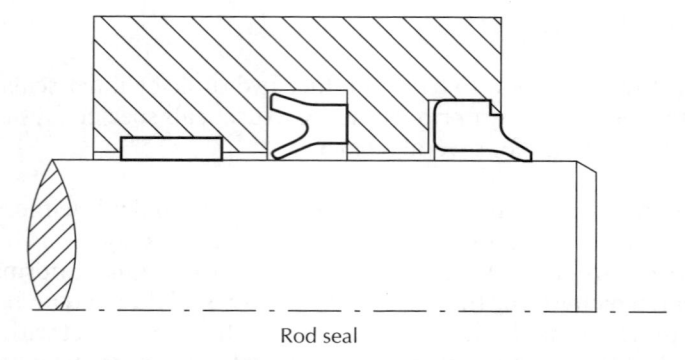

Figure 4.33 Examples of pneumatic cylinder seal, showing the low contact pressure design and entry profile of the contact zone.

4.3.1 Pneumatic rod seals

These are typically a light construction of U-seal. As the seal does not have to withstand a very high pressure loading it is also quite common to integrate the wiper seal with the pressure seal. In this case it is important to ensure that the wiper does not exclude any lubricant on the return stroke. The seal and the wiper will therefore both have a chamfered or radiused profile to ensure retention of lubricant under the seal, Figure 4.34.

The major difference between individual designs is the method of fitting, which will be dependent on the design of the cylinder. There are three basic types, a snap-in housing, press fit and those with a retaining bead that mates with a groove in the housing for retention.

The snap-in housing is similar to those used for some hydraulic cylinder excluders. The seal can be flexed to fit into the housing. An example is shown in Figure 4.35(a). A variation that may be used to facilitate installation has a location step on the atmospheric side of the static part of the seal that mates with the housing recess, Figure 4.35(b). This is normally used at smaller diameters where the load on the seal and hence retention forces will be lower.

Press fit seals rely on the interference with the housing, these will normally include a metal reinforcing element, Figure 4.36, to provide the necessary interference fit. This style of seal will require a press or other assembly machine to facilitate correct fitting.

Lubrication film

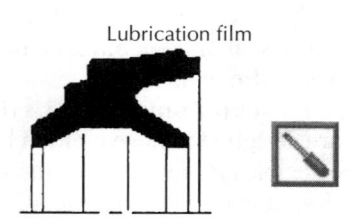

Figure 4.34 A typical rod seal for a pneumatic cylinder, incorporating a wiper lip. Both lips have a profile to encourage lubricant retention under the seal. (Source: Freudenberg Simrit)

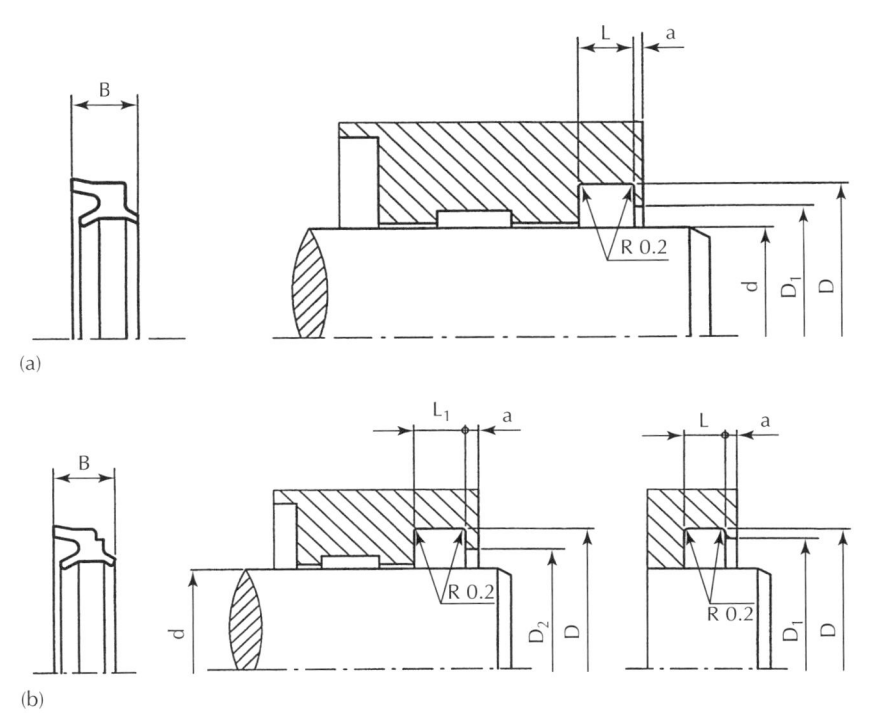

Figure 4.35 Pneumatic cylinder rod seals that are designed to snap fit into the seal housing. A standard design (a) and a recessed design (b) for additional retention in a shallow recess housing. (Source: Trelleborg Sealing Solutions)

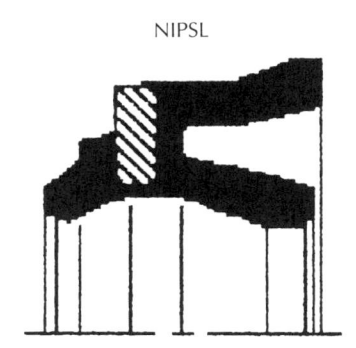

Figure 4.36 A press fit pneumatic cylinder rod seal with metal reinforcing within the seal. (Source: Freudenberg Simrit)

One style of seal has an external circumferential bead, Figure 4.37, which matches a recess in the seal housing. The seal can be a press fit in the housing and retention is by the bead snapping into the groove. As the bead and external interference provide the retention of the seal it is necessary to ensure a dry fit on the outside diameter of the seal. The advantage of this type of seal is that it can

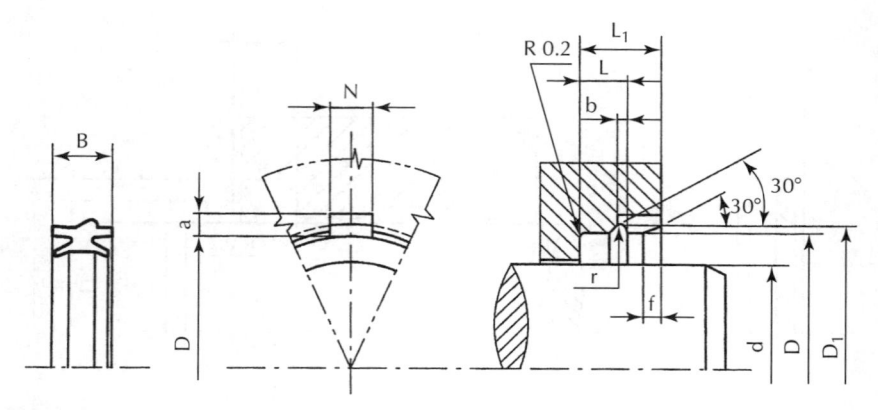

Figure 4.37 Pneumatic rod seal with retaining bead, this design has the advantage of replacement without disassembly of the cylinder. (Source: Trelleborg Sealing Solutions)

be fitted without separating the cylinder and rod components, making seal replacement considerably quicker and easier than for most other rod types.

4.3.2 Excluders

The relatively light construction of pneumatic cylinders means that in most cases a one piece seal and wiper can be used. Separate wipers are also available. In both cases these are of different design and construction to hydraulic excluders. They are designed for low friction and also with a rounded or chamfered profile to encourage return of any lubricant. They will not provide the same degree of exclusion performance that may be expected from a hydraulic cylinder excluder, but there is likely to be a very low level of compressed air leakage during operation that will assist with removal of fine contaminant from the seal contact zone.

4.3.3 Piston seals

The simplest design of piston seal is an outside diameter contacting version of the rod seals, Figure 4.38. This again has a lip that is designed for a low contact load and with either a rounded or chamfered entry profile to encourage lubrication of the lip.

However, the lightweight construction of pneumatic cylinders combined with the relatively high manufacturing volumes and requirements for compact designs has prompted the introduction of a number of alternative arrangements. Some types still use a seal in a groove configuration. To provide a low friction compact seal they will be relatively short but with adequate radial depth to provide control of the interference forces. They may also have a compliant radial construction to ensure a low contact force across the tolerance range. One of the first examples of this type of seal was the Airzet, Figure 4.39. This is a double acting, compact seal with grooves to the face side for pressure activation.

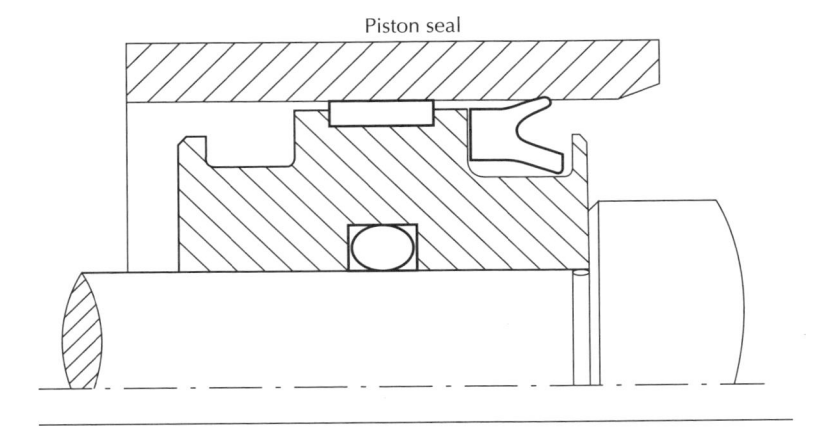

Figure 4.38 A U-seal design for pneumatic cylinder pistons.

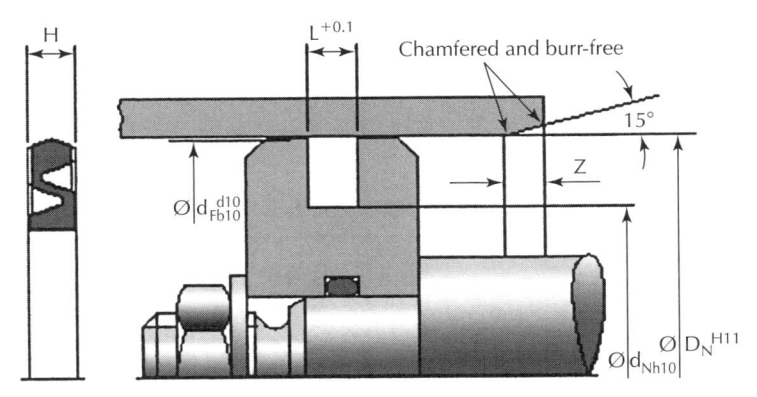

Figure 4.39 The Airzet double acting piston seal for pneumatic applications with a low interference force radially compliant design. (Source: Freudenberg Simrit)

The compact design permits short piston designs (e.g. for short-stroke cylinders). The rounded-off sealing profile and the flexible construction provide low friction and maintain the lubrication film.

A common construction method is to use an integral piston and seal arrangement, Figure 4.40. A double acting lip seal provides the dynamic sealing. The interseal space is designed to act both as a lubricant reservoir and provide bearing support to the piston. The elastomer on the radial faces of the piston can also be designed to act as an end-stop buffer to cushion the ends of the stroke. This can be either a thin layer of elastomer, or a more comprehensive buffer design, if required. In either case it will be necessary to ensure that the buffer does not prevent access of pressurizing air to the full piston area and so some radial grooves will be provided in the buffer stop. This type of construction provides a compact and economical design of pneumatic piston. However, if the seals require replacement it is necessary to replace the entire piston.

NADUOP

SIMRIT complete pneumatic piston seal
Compact piston with steel body bonded to buffers and sealing lips with special pneumatic sealing edges. Ready to install, double acting, complete piston with integral guide. Bonded buffers for end-position damping of piston in the cylinder. Radial air relief passages ensure pressure distribution at the end of stroke.

Figure 4.40 Pneumatic cylinder piston with integral seal and end stop buffers. (Source: Freudenberg Simirt)

An alternative that uses a similar design of seal can be piston seals that clip over the piston, Figure 4.41. This type of seal is also used if it is required to include a magnet on the piston for functions such as position sensing.

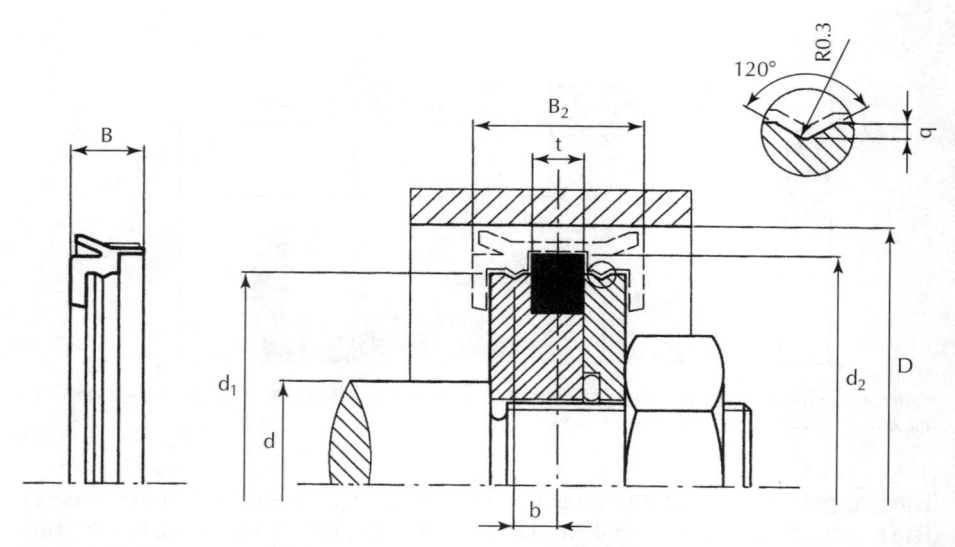

Figure 4.41 Pneumatic cylinder piston with clip over piston seal that encloses the magnet fitted to the piston. (Source: Trelleborg Sealing Solutions)

4.3.4 Material considerations

The seals in pneumatic cylinders are expected to work with minimal lubrication and low friction while achieving a long life, for example 2×10^6 cycles. The materials are therefore usually specifically formulated for pneumatic applications. Nitrile elastomers are still used regularly but there is also increasing use of polyurethane and polyester material. Pneumatic systems are also often used in high temperature and hazardous areas where it may not be considered safe to use hydraulic systems. It may then be necessary to use a higher temperature resistant material such as a fluorocarbon. If the conditions are particularly aggressive then plastic seals such as PTFE may be used. In such cases a metal spring energized design will be most suitable to provide the fluid and temperature resistance together with a low lip load, Figure 4.42.

More details on the materials can be found in Chapter 5.

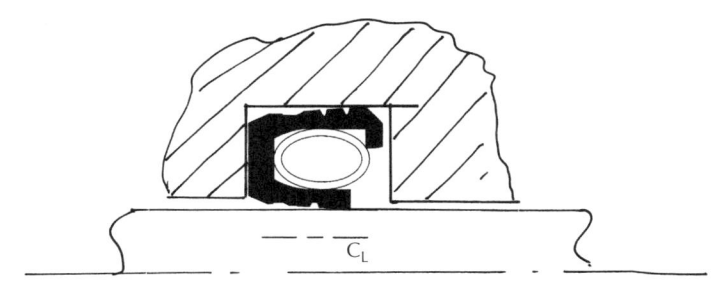

Figure 4.42 Spring energized PTFE for use on a high temperature pneumatic application where resistance to aggressive fluids may also be required. (Source: Balseal)

4.4 Piston rings

4.4.1 Introduction

Piston rings are widely used. The most common use is in internal combustion engines, but they are also used in reciprocating gas compressors and in hydraulic cylinders. Gas compressors can also be subdivided into two classes, lubricated and dry. In a lubricated compressor the piston rings provide a similar function to those in an internal combustion engine, but dry compressors normally use filled plastic rings, these have a somewhat different sealing action and are dealt with separately.

Piston rings are generally used in a high speed and very dynamic pressure situation. They are a split component with a butt joint, and as such do not provide a high integrity static seal as we may expect with something like an elastomer or a static metal compression joint. However, as modern engines demonstrate, when correctly specified they perform a quite remarkable job in limiting oil consumption. But, it is important to consider this distinction when assessing the potential use of piston rings for other duties where alternative seal types may be used, such as in hydraulic cylinders.

4.4.2 Internal combustion engines

Piston rings in an internal combustion engine are required to carry out two distinct sealing functions. The first is to seal the gas in the combustion chamber and the second to prevent the oil that is used to lubricate the cylinder walls and piston rings from being passed into the combustion chamber while permitting sufficient oil film for their own lubrication and that of the piston skirt.

The combustion pressure is continuously varying during the engine cycle, from vacuum on the induction stroke through to full combustion pressure on the power stroke. Unsatisfactory piston rings will create power loss through leakage on the suction and pressure strokes. The leakage (blow-by) of gas, particularly combustion gas into the engine crankcase, will cause rapid degradation of the lubricant, and create problems with the engine breather system.

The piston rings are lubricated by oil sprayed onto the cylinder bore. A major function of the piston rings is to control this lubricant film such that there is sufficient oil film to lubricate the rings, and assist with the gas sealing by providing a film between the bore and ring. At the same time this oil film must be strictly controlled to prevent excess oil passing into the combustion chamber. This will cause oil consumption, increased emissions and also problems with carbon deposits in the engine and other components such as the catalytic converter.

In addition to the sealing function the piston rings of an internal combustion engine are an important component in the transfer of heat away from the piston to the cylinder block. They are therefore also required to provide an adequate heat transfer path.

In a typical automotive engine the piston ring friction accounts for 20–30% of the total engine losses[12] as shown in Figure 4.43. The friction losses are also proportionately much higher in the low load condition, which is the regime that predominates for the average car. The task for the engine designer and piston ring supplier is to achieve the correct balance between satisfactory engine performance, economy, emissions, friction, oil consumption and life. The rings are required to seal in a bore that can vary due to changes in temperature and pressure, causing distortion and the oscillating side load during rotation of the crankshaft.

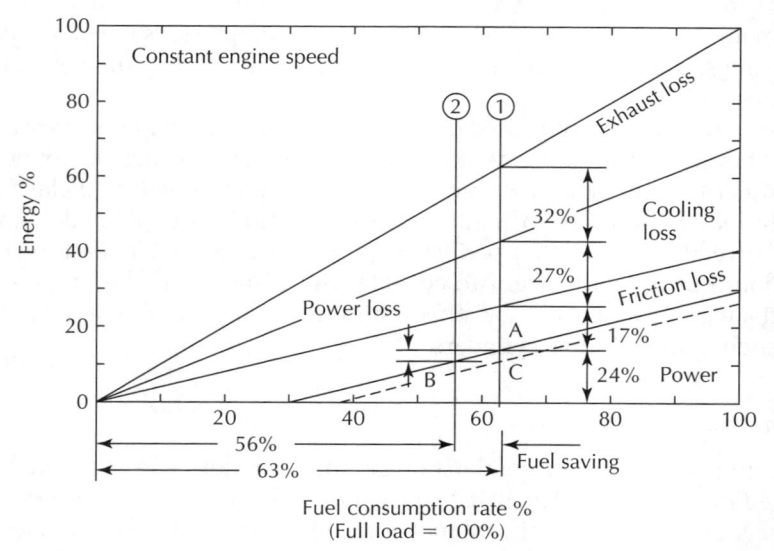

Figure 4.43 A summary of the various losses in an internal combustion engine under different load conditions. (Source: Tat seishi)

The most common arrangement is to use three piston rings, Figure 4.44. This achieves the most favourable compromise between satisfactory sealing and friction. Two rings are to be found on light duty engines and also some high performance engines. It is to be expected that development will see the trend of two ring engines continue in an effort to improve fuel economy. By contrast large marine diesel engines where the cylinder bore diameter may be 250 to 1000 mm may use four or five piston rings, Figure 4.45. Such large engines are very different from the average car engine. They are designed to burn heavy oil, which creates significant problems for the piston rings and their lubrication. It is normal to provide lubrication directly to the piston rings with a specific grade of oil that is different to that used in the crankcase. Side load is less of an issue as such engines are of cross-head design.

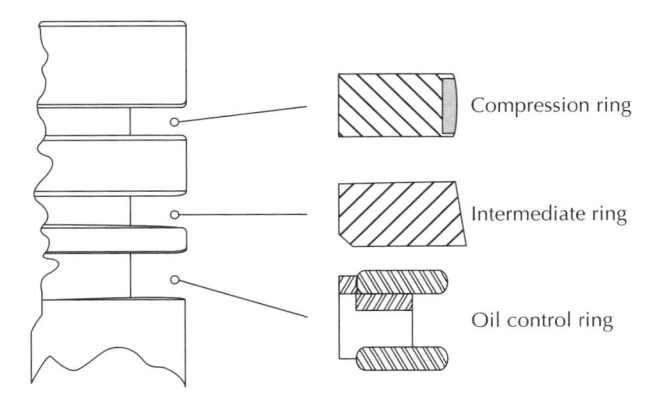

Figure 4.44 A typical automotive piston assembly with three piston rings.

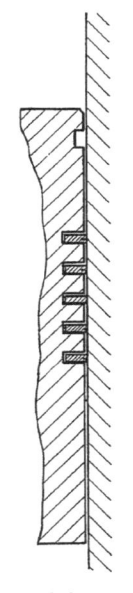

Figure 4.45 A large marine diesel engine piston with five piston rings. (Source: Patent WO 2004/090389)

4.4.2.1 Construction

Piston rings are machined from either a high quality cast iron or increasingly from alloy steels. The rings are designed to spring outwards onto the cylinder bore so they are accurately machined over size and then split. When assembled in the bore the butt joint will be a small but definite gap. This will typically be 0.07 mm per 25 mm of bore diameter. As the ring will be one of the highest temperature components in the engine allowance is included for thermal expansion so that the operating gap is minimized. The stress in the ring is carefully calculated to optimize the contact stress for sealing and lubrication. In modern engines the top ring may be only 1.0 or 1.2 mm wide as designers strive to reduce overall piston height and mass.

The rubbing face of the ring may be coated to improve the tribological properties, Figure 4.46. This has traditionally been chrome plating but increasingly it will be molybdenum disulphide treatment, plasma nitriding or a number of the metal matrix composites including carbides, etc. Particularly in large engines the lower surface of the ring and the groove may also have a wear resistant coating to resist wear as the highly loaded ring oscillates in the groove, Figure 4.47. Recent developments in chromium plating techniques have enabled such coatings to be applied to the side faces of truck engines to reduce groove wear. The butt joint of the ring may vary from a simple strait cut, which is widely used, to 45° cuts, to a stepped butt and a number of complex geometries, Figure 4.48, designed to reduce leakage at the joint and also be pressure balanced. In many cases the simple butt joint is selected because the complexity of the alternative geometries does not necessarily provide sufficient benefit to justify the additional cost.

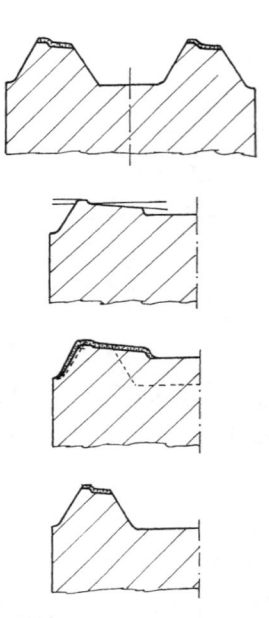

Figure 4.46 Oil control piston ring with a wear resistant insert. (Source: Patent WO 01/52737)

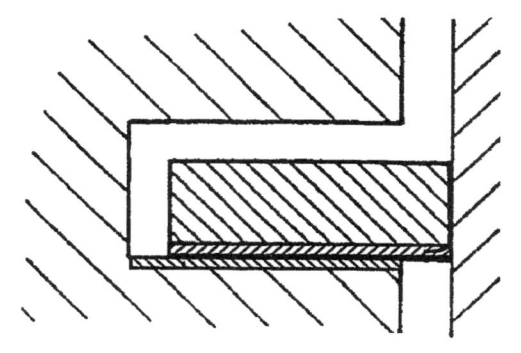

Figure 4.47 Piston ring lower surface and groove with wear resistant coating. (Source: Patent WO 2004/090389)

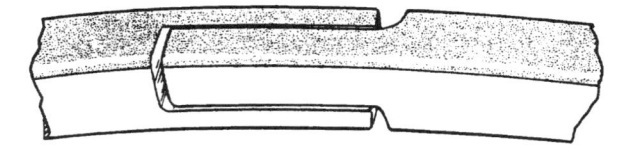

Figure 4.48 A complex piston ring butt joint. (Source: Patent WO 01/52737)

4.4.2.2 Ring design

A typical automotive internal combustion engine three ring arrangement comprises a compression ring, intermediate ring and oil control ring. Each ring has an individual but complementary function. The basic designs of piston rings are extensively covered by a series of standards for engine sizes up to 200 mm bore. These are published as ISO standards that are dual numbered with national standards. The common piston ring profiles are all designated an abbreviation code which may be one or up to three letters. The ISO standards are summarized in section 7.2.

The compression ring

As the name implies this ring primarily seals the gas in the combustion chamber. However, this must be achieved with an adequate but not excessive oil film, and without scraping oil into the combustion chamber on the compression stroke. The selection of compression ring will depend on the engine type and performance requirements. The simplest form is the rectangular ring, type R, which is found on many gasoline engines. However, in practice as the ring is tipped by the pressure and piston movement it can become worn to a tapered or offset barrel faced geometry, Figure 4.49. This geometry has some benefit as it encourages oil film under the piston ring on the compression stroke, and assists with oil control on the downstrokes of the piston. The rings may be symmetrically barrelled, asymmetrically barrelled or tapered depending on oil consumption and blow-by requirements. Approximately one third of European passenger car gasoline engines are designed with either an asymmetrical barrel or a taper to improve oil consumption.[21]

Figure 4.49 A barrel face piston ring.

Both the material and the wear resistant coating are important for the top ring as it has to operate at the highest temperature and is energized with the full gas pressure at close to top dead centre of the engine while it is virtually stationary. It is necessary to avoid scuffing and erosion damage both to the piston ring and the cylinder bore and provide an effective heat transfer path. In gasoline engines the top ring is predominately of nitrided steel.

In diesel engines a keystone piston ring may be used, particularly as the bore diameter increases. These have a tapered radial profile, Figure 4.50. The standard taper may be at an included angle of 6 or 15°. Alternatively a half keystone ring has a taper only on the top face of the ring. The purpose of the keystone ring is to avoid the sticking of the ring that can be caused by the deposition of oil coking deposits at the high temperatures and pressures in a diesel engine. As the ring flexes in the groove due to the motion of the piston on alternate strokes the ring will move on the taper and break up the deposits.

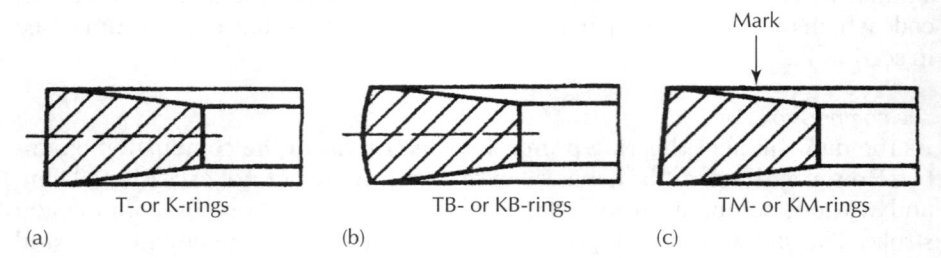

(a) T- or K-rings   (b) TB- or KB-rings   (c) TM- or KM-rings

Figure 4.50 Examples of keystone piston rings.

Keystone rings may have a straight face on the outside diameter, known as a type T, or may be either barrel faced, TB, or taper faced, TM. The barrelling will typically be of the order of 0.005 to 0.02 mm on a 2.5 mm width piston ring. The taper on a taper face ring may be in one of five ranges between 10′ and 120′. The molybdenum coating will usually be a minimum of 0.2 mm thick.

An alternative design of compression ring is the L-shaped ring, also known as a Dykes ring, Figure 4.51. At high engine speeds the inertia of the piston ring can

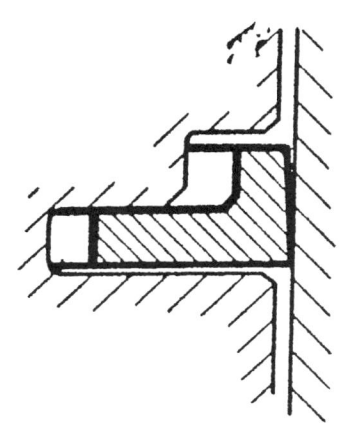

Figure 4.51 Dykes or L-shaped piston ring.

cause it to be loaded sufficiently against the top of the ring groove at the start of the power stroke to prevent gas pressure passing around behind the piston ring to energize it outwards. This can lead to what is known as ring flutter, where the piston ring loses sealing interference contact and blow by leakage can occur. The Dykes ring ensures that there is a path for the gas pressure to energize the back of the ring. The ring itself is not only more complex, a particular disadvantage is that a more complex piston groove is required in the area of the most critically stressed region of the piston. Hence alternative arrangements are often used to overcome this problem. A popular method is to design the ring to have a higher spring load in the area of the butt joint to provide increased control of the ring in this area. This may be achieved by oversizing the ring and cutting a larger butt joint that is closed up on assembly, but more sophisticated methods include a variation in ring geometry around the circumference and varying the profile of the ring circumferentially. This modifies the bending characteristics close to the butt joint to achieve improved conformability. The conformability of a one-piece compression ring with a constant wall thickness is governed primarily by the free gap. This defines the closure and opening stress. The conformability of one-piece rings is not uniform around the circumference but is at its greatest opposite the gap. The lack of bending moment at the ring ends reduces the local conformability at the gap to zero. A recent development has been to introduce variations in the radial wall thickness near the gap. This has the effect of improving the local flexibility and adaptability of the piston ring to irregular bore deformations. Test results demonstrate potential for oil consumption reduction and the first ring was introduced in series production during 2005.[21]

It is sometimes necessary to ensure that there is not excessive pressure drop across a piston ring, one example being very large diesel engines, where there are more piston rings. To ensure that there is a suitable pressure distribution across several rings it may be necessary to ensure that there is some controlled gas passage past an individual ring. To achieve this, the rings may have axial slots or holes to act as gas pressure balancing passages, Figure 4.52. The use of radial slots or holes is yet another alternative to alleviate potential ring flutter discussed above.

The intermediate ring
This ring is also considered as a compression ring, but a major function is assisting with oil control. The intermediate ring will most likely be either a taper face rectangular ring or a stepped design known as a Napier ring, Figure 4.53. The shape of these rings is designed so that they will assist the compression ring with gas retention, but also assist with oil control. The tapered face, and the step, provide a sharp edge facing towards the lower ring, and so will tend to scrape oil downwards during a down piston stroke. The tapered face will also discourage upward pumping of the oil film during an upwards piston stroke.

The profile of the inner diameter is also important to ring performance. The piston ring will tend to twist in the groove, depending on the pressure load and frictional forces. This twisting can be beneficial, such as assisting oil scraping, but

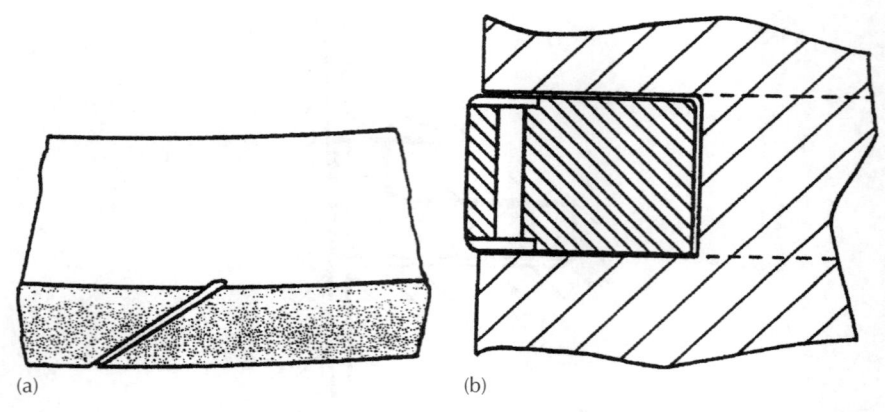

Figure 4.52 Pressure equalizing passages across a piston ring. (Source: Patent WO 97/11294)

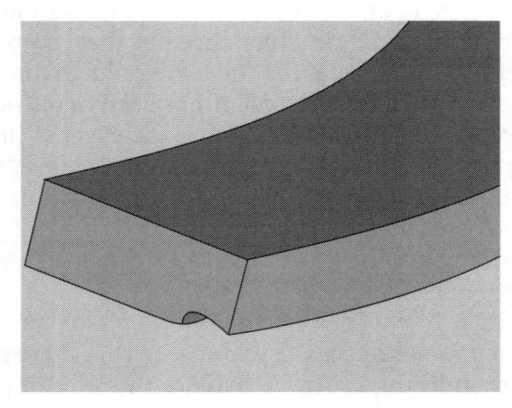

Figure 4.53 Napier stepped rings.

Figure 4.54 Intermediate piston rings with a chamfer at the inside diameter edge and tapered o.d. for oil control.

may allow gas blow-by around the back of the ring. Excessive tipping may also cause wear of the ring groove. To overcome such problems the rings may be chamfered on the inside diameter edge, Figure 4.54. This may be at the top or bottom of the ring depending on the particular situation. A compromise has to be achieved between the conflicting demands of achieving minimum oil consumption and also minimum wear and friction. Selection of the appropriate ring design will enable the engine designer to achieve the correct compromise for the intended application. The design shown in Figure 4.54 creates a contact stress at the outside bottom edge of the groove and inner top edge which prevents oil access to the rear of the ring on a downward stroke and ensures that the gas pressure loads the ring into this preferential orientation.

A traditional material for uncoated intermediate rings is fine-flaked cast iron in a heat-treated condition. These have good wear resistant properties from the special carbides formed by chromium, vanadium, manganese and tungsten in the martensitic matrix structure.

The oil control ring
A single function of the oil control ring is to remove oil from the cylinder walls and return it to the crankcase. The oil may be the splash from the bearings and piston undersides and also that scraped down the cylinder walls by the intermediate ring. The oil control ring does not rely on gas pressure for internal energization. The sealing contact force is generated by the inherent spring generated by making the ring slightly oversize and springing into place or additionally by using a spring inside the piston ring. This may be a coil spring that is compressed circumferentially to provide an outspring force, or a variety of flexible metal rings to expand the ring into place.

Most oil control rings have a pair of scraper edges, typically with circumferential slots in the centre portion of the ring, to drain away the oil from the upper ring, Figure 4.55. To ensure that the piston has a high local contact stress to scrape the oil film the contact zone is often contoured to a small axial length. The two scraper portions of the ring may be square, tapered or profiled. Some examples of individual designs are shown in Figure 4.56. This type of ring will generally be manufactured from cast iron.

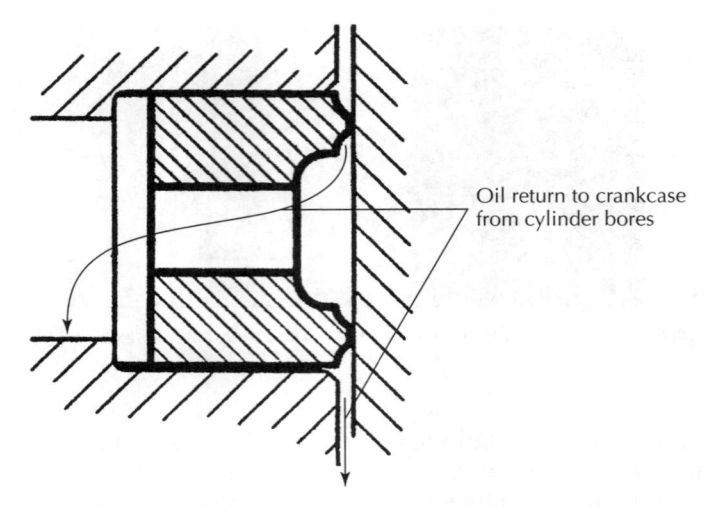

Figure 4.55 Typical oil control ring.

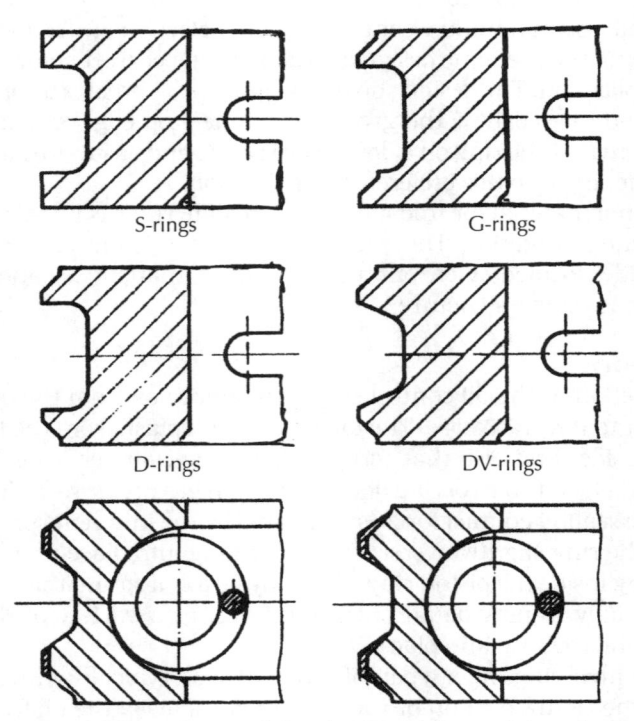

Figure 4.56 Examples of different designs of slotted cast iron oil control rings showing four profiles of ring without spring and two different styles of coating on a sprung ring.

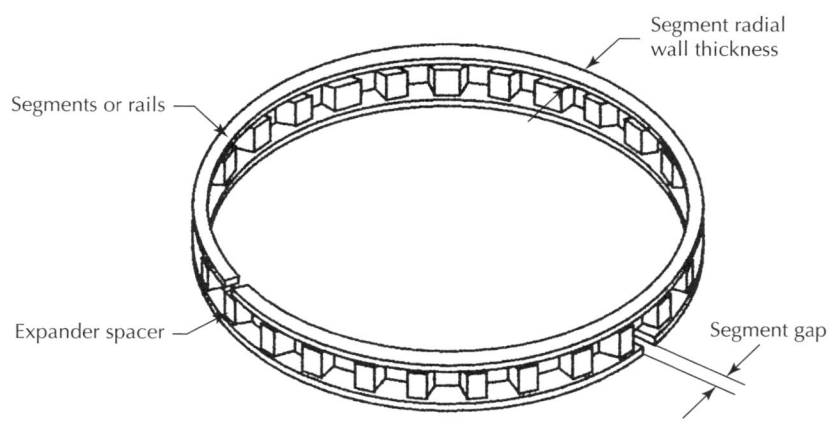

Figure 4.57   Multi-part steel rail oil control piston ring.

An alternative design is a multi-part steel rail oil control ring, Figure 4.57. These are typically a three-piece ring and use a stainless steel expander between two chrome-faced steel rails. This configuration provides a high degree of conformability within the cylinder bore, a narrow contact zone for high interference force at a minimized friction to scrape the oil and a compact piston ring that helps to minimize the overall piston size and weight.

4.4.2.3 Piston ring coatings

A major factor in the advance of piston ring performance has been the development of the coatings. This is particularly true of the top compression ring.[21]

Standard hard chromium coatings are now primarily used on intermediate and oil control rings. To achieve a higher thermal capacity and improved wear performance chromium ceramic coatings were developed and have been widely used in diesel engines. A more recent development is a hard chrome matrix reinforced with minute diamond particles that are firmly embedded in the extremely fine crack network, which has a special sub-structure. A material of this type is claimed to have the lowest wear of any known coating on the market. It also has the benefit of being produced with a sharp bottom running face edge for low oil consumption.

Plasma spray coatings are widely used and can provide a high ceramic content, making them highly suitable for reducing scuffing. Plasma spraying, however, is not suitable for depositing hard metal-like structures, which further improve the wear characteristics. The high velocity oxygen fuel (HVOF) spraying is used to deposit CrC, WC and metallic Ni-Cr-Mo alloys. These can improve wear resistance by a factor of three or four compared to plasma coatings.

Nitriding is used to treat high chromium alloy martensitic steels. A further advantage with this treatment is the additional side wear protection provided by the formation of an all-over nitride case.

A variety of metal carbides and metal matrix combinations using deposition techniques such as HVOF are also being investigated to improve the life of piston rings.[19, 20]

The ongoing development of high performance diesel engines has brought about a need for piston rings that are designed to cope with the increasing temperatures, stresses, longer service intervals and consistently good sealing to help reduce emissions and fuel consumption. Ceramic coatings have been used to increase performance over the conventional chromium coatings. A more recent development from one supplier is a diamond coating which coats the surface of the rings with nano-sized diamond particles to offer increased resistance to scuffing between the piston and the cylinder with its resultant benefits. Piston rings with this coating have regularly offered more than four times better wear resistance than plain chromium coated rings and more than twice that of the ceramic coated rings. This has been achieved with a cylinder wear rate that is comparable to that of other hard coated rings.[21]

The most recent generation of piston ring coatings is the physical vapour deposition (PVD) process. These coating systems are predominantly chromium nitride (CrN) based, have high hardness and low friction. These coatings also offer low wear and high thermal capability.

The manufacturing of the cylinder bore has also been the subject of considerable research. This has included a study of the optimum surface texture, and methods of measuring it repeatably in practice to ensure that a reliable bedding-in and engine life are obtained. The objective is to provide a surface that requires an absolute minimum of bedding-in by presenting the optimum texture to the piston rings from the machining process. Hence, the texture analysis involves not just the Ra of the surface, but an analysis of the bearing area ratio and peak angles. This is described further in section 7.3 where surface texture is discussed.

4.4.3 Compressor piston rings

4.4.3.1 Lubricated pistons

Piston rings for lubricated compressors work under relatively similar conditions to internal combustion engine rings. The one obvious difference is the lack of combustion and hence there is less cyclic temperature. However, they may generally be required to operate continuously and the continuous compression of gas creates elevated temperatures. The requirements to minimize blow-by and also inhibit oil consumption remain. As the presence of oil in compressed air is now a major health concern the design and operation of these rings is as critical as those in an internal combustion engine. Butt joints designed to reduce leakage may be used, Figure 4.58.

4.4.3.2 Dry compressors

In many applications it is necessary to run the compressor with a dry gas and zero lubrication. At one time these would all have been manufactured from carbon, but filled plastic piston rings have been developed for many of these duties. The applications include oxygen service, where oil must be excluded to prevent an explosion hazard, food and pharmaceutical plants, oil-free compressed air services and chemical plant compressors where lubricant in the product stream

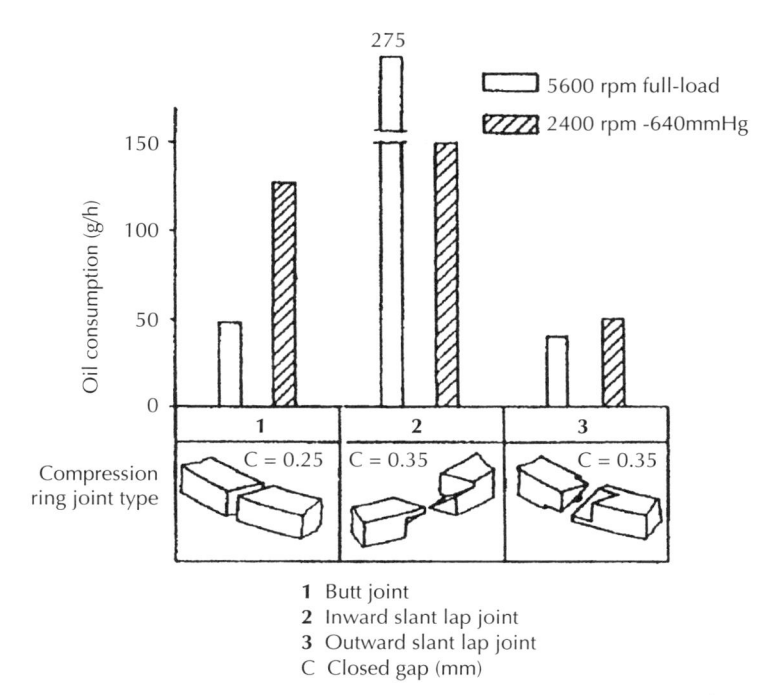

Figure 4.58 The effect of piston ring joint design on oil consumption form[2].

1 Butt joint
2 Inward slant lap joint
3 Outward slant lap joint
C Closed gap (mm)

may be an unacceptable contaminant or a safety hazard. A typical dry compressor will still have a lubricated crankcase, but use a crosshead and sealed piston rod to prevent access of lubricant to the piston area. In critical circumstances such as oxygen service additional precautions may be taken to ensure that no oil is present around the piston.

There are some additional design factors to consider when plastic piston rings are used. The guidance and support of the piston cannot be accommodated by the cylinder wall as there is no lubrication. It is therefore normal to design the piston with a larger clearance than a lubricated metal piston ring design and use one or more bearing rings, also known as rider rings. The number of piston rings will then be selected to suit the pressure. Figure 4.59 shows a typical dry compressor piston, with two bearing rings and six piston rings, and Table 4.4 the number of piston rings recommended for different pressures.[13]

The design of the bearing rings is important to the performance of the seal assembly. They may be solid rings or split. Split rings are easier to fit, and will be necessary if the diameter to thickness ratio is such that excessive stretching would be required. They have the benefit that they are less susceptible to pressure acting behind the ring and causing increased wear. Solid rings have the potential disadvantage that it will usually not be practicable to remove them without damage. This will normally be in the form of axial notches in the surface to allow

334 *Seals and Sealing Handbook*

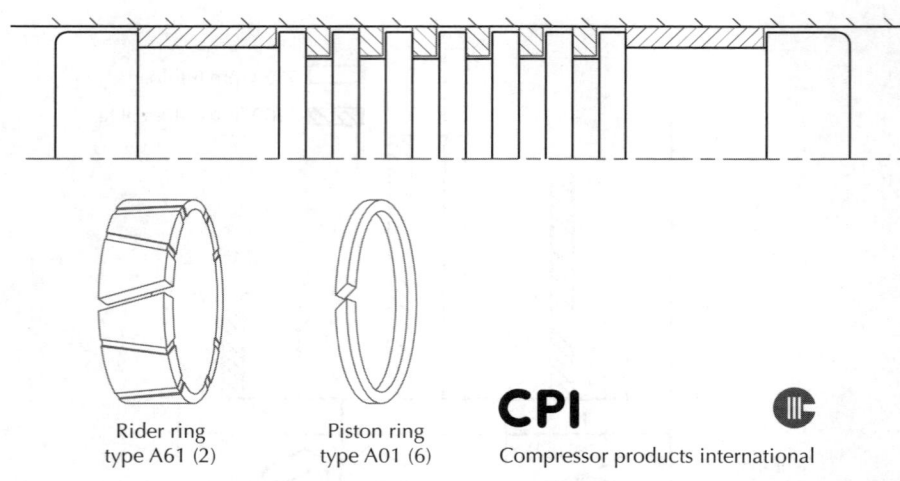

Figure 4.59 A typical dry compressor piston showing piston rings and rider rings. (Source: CPI)

Table 4.4 Number of piston rings for different pressures

Typical differential pressure ranges bar		Number of rings per piston				
		2	3	4	5	6
Up to 20 bar		X				
20 to 60			X			
60 to 100	Plain Face			X		
	Balanced Face				X	
100 to 210	Plain Face				X	
	Balanced Face					X
Over *210*		Minimum of 6 balanced face				

Based on normal piston speeds and temperatures
(Source: Koppers)

passage of the gas, Figure 4.60, or a defined gap in the assembled condition with a split ring.

Plastic piston rings are energized by the system pressure acting behind the ring in the same way as metal piston rings, Figure 4.61. Depending on the plastic polymer compound used the material may retain some inherent outward energization when not pressurized.

Early applications of plastic to piston rings used a metal expander ring to overcome any plastic flow effects and load the piston ring onto the cylinder bore. Current practice is not to use expander springs under piston rings. This is because:

- The gas pressure loads the piston ring onto the cylinder bore from underneath, and does not need the assistance of an expander spring, irrespective

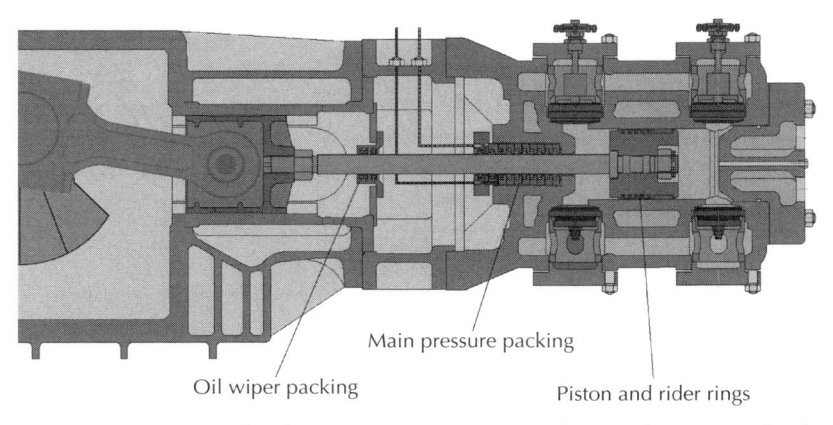

Figure 4.60 A cross-section of a dry reciprocating compressor showing the piston and rod sealing arrangements. (Source: HBRP)

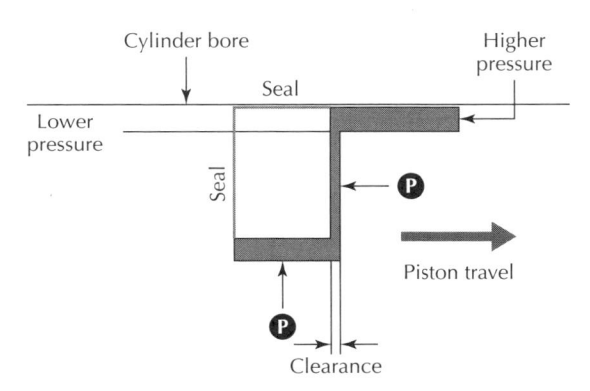

Figure 4.61 The pressurizing action of a dry piston ring. (Source: CPI)

of the physical properties of the piston ring material (i.e. even if it is naturally 'floppy', such as would be the case with most filled PTFE materials).
- Metallic expander springs always cause damage to the sides of the piston ring groove.
- Expander springs have no real benefit, other than when they are used with step-cut piston rings (the expander seals the leakage paths formed by the gaps of each step). However, other designs without metallic expander springs are available which achieve this improved seal when required.

A variety of butt joints are used for plastic piston rings. The most common are a straight cut or a 45° scarf joint. A step joint is used but is not a preferred arrangement, Figure 4.62. The bending load on the step is a crucial factor with a plastic and some suppliers use an offset step to provide increased strength to the step portion that takes the pressure load. This in turn means that the ring must be installed correctly.

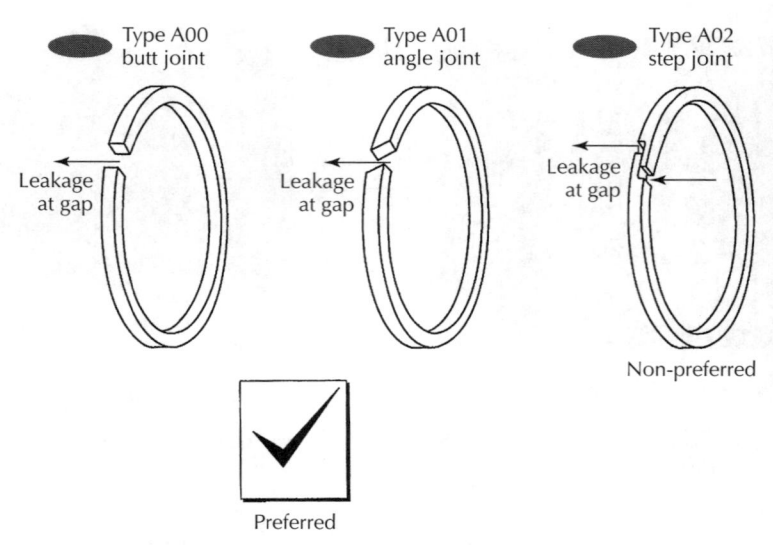

Figure 4.62 Dry piston ring joint configurations. (Source: CPI)

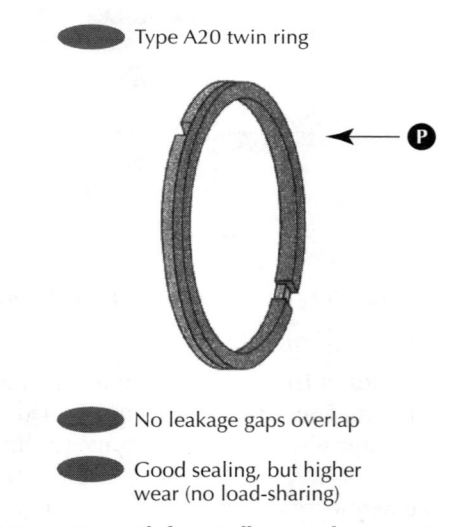

Figure 4.63 Two part piston rings with diametrally opposed gaps to minimize leakage. (Source: CPI)

Two part rings can also be used, Figure 4.63. The joints are displaced 180° to each other. This provides a benefit of lower leakage as there is no overlap of the leakage gaps but as there is no load sharing between the rings the wear is higher. A compromise must therefore be achieved between leakage requirements and life.

For high pressure applications it is possible to reduce the contact stress by providing a hydrostatic feature to relieve the pressure load, Figure 4.64. This will

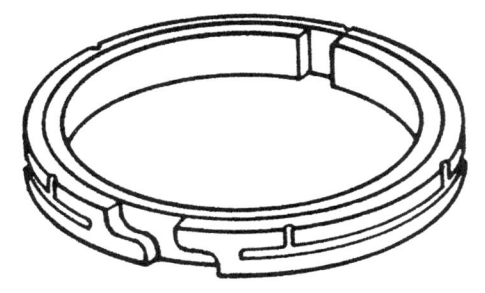

Figure 4.64 Hydrostatically balanced piston ring for high pressure application. (Source: Koppers)

reduce contact stress and hence friction, temperature and wear with some potential increase in leakage across a single ring.

4.4.3.3 Materials for dry gas compressor rings

With a dry gas application the deposition and integrity of the transfer film is critical to success. The surface finish and texture of the mating counter-face is therefore an important factor in the success of an installation. The interaction of the polymer, counter-face and compressed product is extremely complex. It will be dependent not only on the materials but also on factors such as the gas, whether it is reactive or inert, and any contaminants. The presence of water and whether it is above or below the dew point can have a dramatic effect on the formation and retention of the transfer layer. This has been extensively investigated for dry polymer bearing.[14,15,16] Further work has been reported for compressors.[17,18,22] It has been shown that there are very significant differences in transfer film characteristics between conventionally dry gas and what is termed 'bone dry'. In a gas that is formed by boil-off from a cryogenic liquid or from a chemical process the water content may be much less than one part per million. The behaviour is very different to conventional 'dry air'. Hence a combination that works satisfactorily on air may be quite unsuitable for nitrogen or any oxygen-free environment. Similarly very dry inert gases, nitrogen, argon, helium, etc., will require separate consideration. At the other extreme oxygen service requires special consideration as it is necessary to avoid the possibility of ignition due to unsuitable tribological conditions.

The majority of the materials used have traditionally been filled PTFE compounds. However, for higher duty compressors where high temperature, high pressure, aggressive gas or long life requirements are important factors alternatives such as compounds termed polymer alloys may be used.

The traditional PTFE compounds have been based on materials filled with glass, carbon/graphite or bronze. The choice of filler will depend on the application, counter-face and supplier user preferences.

Common fillers include carbon, graphite, glass fibre, carbon fibre, molybdenum disulphide, bronze powder, ceramic powders, etc. Fillers are often used in combination, particularly carbon and graphite. For dry gas usage, at low and cryogenic dew points, it is useful to add specific polymer fillers. Fibre fillers are added to make the material stronger and more resistant to creep; wear resistance can also be improved due to the increased hardness.

Glass filled materials provide good chemical resistance. As the glass is abrasive they can provide rapid bedding-in for a dry application. However, the choice of counter-face material will require careful consideration. It is also possible for scoring and high wear to occur with these materials.

Carbon/graphite filled materials are the most commonly used of the PTFE compounds. The additional lubrication provided by the graphite, together with a higher thermal conductivity to disperse frictional heat makes it the preferred PTFE compound for many dry applications. The counter-face texture is still an important factor.

Bronze filled PTFE may also be used. It is a higher strength material than carbon/graphite filled with relatively good conductivity although it can be subject to attack by contaminants and may cause higher counter-face wear than carbon/graphite.

Many filler combinations involve a mixture of harder carbon, cokes and softer graphitic carbons tailored to give the desired properties. These fillers reduce friction, increase hardness and heat transfer and play an important part in the tribochemistry of the transfer film. The carbon also provides a hard phase in the soft PTFE matrix and this helps to support the normal contact load. The surface reactions and bonding which occur with carbon/graphite require the presence of a small, finite amount of water vapour in the contact zone. In dry or cryogenic gases the important water molecules are missing.

More recently major manufacturers have developed what are termed special polymer alloys incorporating chosen polymers (such as PTFE and other high temperature engineering polymers) and fillers in a variety of combinations, specially processed to achieve the necessary wear resistance and mechanical properties for self-lubricating piston and rod seals in oil-free compressors. The development of these materials is based on wear testing in different gas conditions to provide a sufficient understanding of the behaviour of a range of materials in different gas conditions, so as to ensure the proper selection of material to be made for the real compressor application.

Table 4.5 provides an example of the range of materials provided by one manufacturer to operate in different gas conditions. It will be essential to consult with the supplier to establish the preferred material for individual applications as the transfer film characteristics are quite unique depending on the gas, operating conditions and counter-face. Individual manufacturers maintain confidentiality on the formulation of the materials for individual applications but some details have been published by Radcliffe.[22]

Dry gas requirements
If the gas has a low dew point it is necessary to include fillers such as molybdenum disulphide which can provide an active surface layer without needing water vapour. Some sulphur containing polymers such as polyphenyline sulphide (PPS) work in a similar way as fillers for PTFE.

Table 4.5 Piston ring materials for dry compressors

Material code	Specific gravity	Tensile strength MPa	Elongation %	Hardness Shore D	Mean temp limit* deg C	Application Non-lube	Lube
CPI 100	2.0	13	50	60–65	125	Atmospheric air	All gases
CPI 111	3.0	13	75	70–75	130	Atmospheric air	N/A
CPI 113	2.3	12	125	60–65	120	Dry air, dry oxygen	N/A
CPI 114	2.4	20	200	60–65	120	Dry air, dry oxygen	N/A
CPI 124	3.8	15	120	65–70	130	Atmospheric air	N/A
CPI 184	1.9	11	5	65–70	125	Dry gases (H_2, HC, CO_2, NH_3); wet gases	N/A
CPI 188	2.0	9	4	65–70	120	Dry nitrogen, argon, helium	N/A
CPI 192	1.5	35	2	80–85	175	Dry gases (H_2, HC, CO_2, NH_3); wet gases	All gases
CPI 193	1.5	70	3	85–90	175	Dry gases (H_2, HC, CO_2, NH_3); wet gases	All gases
CPI 196	1.7	30	2	80–85	175	Dry nitrogen, argon, helium	N/A

*Mean temp is defined as (inlet gas temp + discharge gas temp)/2   Suggested temp limits given are for non-lube operation
Source: Compressor Product's International.

Oxidizing gases

Gases such as oxygen and air and even those containing a small percentage of oxygen can exhibit considerable oxidizing potential at high pressures and temperatures. At these conditions carbon filled PTFE materials wear rapidly as the carbon fillers will suffer oxidative wear. It is common to use bronze as a filler to provide a hard phase for supporting the load and providing heat transfer and MoS_2 to reduce friction and help with the transfer film.

Inert gases

Inert gases like nitrogen have different tribochemical reactions because the gas plays no part in the system. Without any oxygen present the breakdown reactions of the PTFE and fillers are different and alternative types and quantities of fillers are needed. Often nitrogen gas is also very dry and so additional solid lubricants have to be added.

Hydrogen and reducing gases

The situation in hydrogen is similar to that in nitrogen, but a further effect is that metal oxide films are both reduced by the hydrogen and worn away by the sliding contact. It is particularly important to use an active filler to replace the lost metal oxide layer.

The material for the cylinder liner or piston rod is also crucial to success. Cylinder liners will typically be manufactured from grey cast iron, but a minimum hardness should be 200 Hb. In corrosive conditions Ni-resist or austenitic (flake graphite) cast iron 436 type 2b may be used, again with a minimum hardness as above.

Special surface treatments and hard coatings must be used with caution. They can be successful, but can also cause very high wear. The detail surface texture, bedding-in, surface chemistry and transfer film are all important to success. The preparation of the surfaces and integrity of coating deposition are also critical.

The liner surface finish should be in the range 0.4–0.6 μm Ra. This texture range is important for bedding-in and transfer film formation, so if the bores are worn smooth, they should be retreated to re-establish a satisfactory texture.[18]

4.4.4 Piston rings for hydraulic cylinders

The use of metal piston rings in hydraulic cylinders was a common practice in heavy duty industrial cylinders until the 1980s. They had the benefit of long wear life compared with many other seals available at the time. With the introduction of improved polymer materials for seals and bearings this advantage no longer exists. A typical hydraulic cylinder piston ring is a plain rectangular ring with a step butt joint. As there are several rings on the piston there is usually no additional bearings fitted.

The comparative benefits of metal piston rings compared to alternatives such as polyurethane, polyester or filled PTFE seals are:

Advantages:

- Durability.
- High pressure limit.
- Fluid resistance.

Disadvantages:

- Friction, at low pressure they can cause stick slip, and it can be more than polymer seals at high pressure.
- Leakage across the piston, typically 15 to 50 ml/min.
- Risk of corrosion with high water content fluids.
- Risk of damage to bore with low lubricity fluids.
- Higher cost.
- Not suitable for high side loads.

For these reasons they are not often specified in Europe, but are still offered as a standard option on some industrial cylinders in the USA.

4.5 Compression packing

Compression packings can be used for a wide range of reciprocating applications. The basic arrangement is similar to that described in section 3.5. They are of particular benefit in applications that have poor lubricity or where there are solids or a crystallizing liquid. They can be found in water duties, where either high pressure or a high suction lift is required, chemical pumps and other duties, where a reciprocating plunger pump is operating with a difficult liquid. Water duties may vary from potable water to wastewater and marine duties. Hence the packings used may be potable water grade or with lubricants to suit contaminated water. In poorly lubricated conditions such as water and chemical solutions packing can provide one of the few potential solutions at high pressure. By the nature of their operation the packing will have a small amount of leakage, although depending on the operating conditions this may be as vapour. If leakage local to the machine is not acceptable for environmental or health and safety reasons it will be necessary to provide a low pressure collection and drain system to route leakage for disposal.

Packings are widely used to seal the plungers of high pressure water pumps at pressures in the range 150 to 700 bar. In these applications a limited number of packing rings are used in conjunction with spring loading to maintain the initial sealing force on the packing as pump conditions vary and the packing wears. An example of such an arrangement is shown in Figure 4.65. It is important to avoid using more packing rings than necessary. Additional packing does not provide increased security or redundancy, it increases friction, causes overheating and shortens packing life. For high pressure it is necessary to pay particular attention to the clearance at the gland plate and the neck bore of the stuffing box. At pressures above 200 bar it is normally necessary to use anti-extrusion rings of filled PTFE or a higher duty plastic such as filled PEEK.

Section 4.6.1 provides examples of a typical range of packings and examples of the application areas for reciprocating duties. When selecting packings for general plant maintenance it is also important to remember that many varieties are suitable for both reciprocating and rotary shafts as well as valves. This can

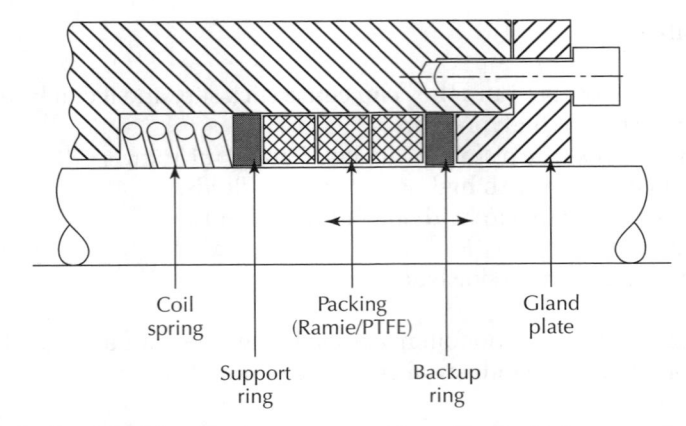

Figure 4.65 Spring loaded stuffing box arrangement for packing in a high pressure plunger pump.

provide simplification of spares stockholding, but care must be exercised if the optimum packing for a critical duty is required.

4.5.1 Packing types and application areas for reciprocating duties

Cotton
Cotton packing is used particularly for cold water applications, but is also useful for other liquids such as oil. The main applications are the water and marine industries. It is also easy to cut, fit and handle compared with some synthetic fibres. Water quality grades are available depending on lubricant and it is particularly useful for pumps that have soft shaft material.

Application conditions:

Media in the range:	pH 6–8
Operating temperature:	−40 to +90°C
Maximum rod speed:	1.0 m/s
Maximum pressure:	50 bar

Flax
Flax can provide a dense but flexible packing and is used in the marine and water industries particularly for seawater, wastewater and other applications involving water with suspended solids. Lubricants can be mineral-based grease and oil, tallow and impregnants such as mica or PTFE dispersion depending on the application. The flexibility of the natural fibres can allow a more controlled response to gland adjustments than many synthetic fibres.

Application conditions will vary depending on the lubricant additives but are within the range:

Media in the range:	pH 5–10
Operating temperature:	−40 to +95°C
Maximum rod speed:	1.0 m/s
Maximum pressure:	100 bar

Ramie

Ramie is a tropical nettle plant that produces an extremely durable and rot resistant fibre with significantly greater strength than the other natural fibres. It has very good extrusion and abrasion resistance with low friction and does not cause excessive shaft wear. It is normally lubricated with a specialized PTFE dispersion. Ramie-based packings have been used with considerable success in the mining and quarrying industries on reciprocating pumps working at 300 bar with water containing highly abrasive particles and for water-based hydraulic systems. For many high pressure plunger pumps a ramie-based packing is the standard method of sealing. It can also be applied successfully to process fluids such as cellulose slurry, brine circulation, cooling water systems and with fluids that crystallize or contain suspended solids. The limitations are primarily the chemical and temperature range when compared to synthetic fibres.

Application conditions:

Media in the range:	pH 4–11
Operating temperature:	−30 to +120°C
Maximum rod speed:	2.0 m/s
Maximum pressure:	200 bar standard, 700 bar with special backup rings

PTFE

A variety of lubricated PTFE packing types may be used for reciprocating shafts. Many packings are combined with graphite to provide both lubrication and heat dispersion benefits. These may be manufactured from either expanded PTFE, such as Gore GFO, or a solid PTFE fibre depending on the application. PTFE dispersion may also be used to provide a denser packing to resist leakage, but with the potential disadvantage of being more sensitive to gland adjustment. An elastomeric core may also be used which enables this packing to absorb misalignment.

A range of lubricants and additives may be incorporated with the packing during manufacture to provide compatibility with chemicals, food and potable water. These packings can be used with aggressive chemicals and at temperatures beyond the range of the natural fibres.

Application conditions:

Media in the range:	pH 0–14 but not strong oxidizing agents and molten alkali metals
Operating temperature:	−100 to +260°C
Maximum rod speed:	2 m/s
Maximum pressure:	80 bar

Aromatic polymer fibre

The most familiar fibre used for these packings is the yellow Aramid fibre, but alternatives are also manufactured. These are extremely strong and abrasion resistant fibres. They are therefore used in applications on reciprocating pumps that handle highly abrasive slurries or aggressive chemical solutions in the mineral, pulp and paper, wastewater and chemical processing industries. They may also be used on liquids that are liable to solidify such as tar and bitumen. The strength and abrasion resistance of these fibres means that they can cause excessive shaft wear and be excessively sensitive to gland adjustment. The manufacture of the packing is therefore very important. Variants may include special lubricant treatments, incorporation of a flexible elastomer core and a combination of Aramid fibres with expanded PTFE/graphite fibre. This latter improves both the adjustability and the lubrication while reducing shaft wear. It should be noted that hardened shafts, coating or sleeve, are essential with the majority of Aramid packings. Some more recent fibre developments provide improved performance over the more familiar yellow Aramid fibre.

The actual application conditions will be very dependent on the packing construction and lubricants used. An elastomer core for instance may limit the fluid and temperature resistance:

Media in the range:	pH 2–13
Operating temperature:	−50 to +250°C (280° for special fibres)
Maximum rod speed:	2.0 m/s
Maximum pressure:	150 bar

Glass

General purpose glass fibre packings are manufactured with a combination of other fibres which aids the manufacture and lubrication of the packing. They may be woven from a reinforced glass insert with a spun cover of polyolefin fibre or from yarn that is spun from a blend of glass and other fibres. The application of these packings is limited by the additional fibres used. They offer a wide chemical compatibility but potentially limited temperature range. The major benefit is as a general purpose replacement packing. Dependent on the fibre mix they can offer low shaft wear and also be potable water compatible.

Application conditions:

Media in the range:	pH 0–14 (dependent on fibre mix)
Operating temperature:	−50 to +130 or 280°C (dependent on fibre mix)
Maximum rod speed:	2 m/s
Maximum pressure:	100 bar (dependent on fibre mix)

Specialist glass packings will use a glass fibre yarn with high temperature lubricants and graphite impregnation. This will primarily be used if high temperature resistance is required. It may be used up to 350°C but dynamic properties are more limited.

Graphite filament

These are typically manufactured from graphite yarn, impregnated with PTFE dispersion and graphite powder. The benefits are a high temperature and speed capability together with wide chemical resistance. The packing is expensive and also very sensitive to gland adjustment so can be difficult to use. However, it can be used in a wide variety of applications so has the possibility to be a standard packing across a site. It would only be considered directly as a reciprocating packing if the application involved particularly high temperature or speed.

Application conditions:

Media in the range:	pH 0–14 (excluding strong oxidizing agents, molten alkali metals, fluorine gas and fluorine compounds)
Operating temperature:	−50 to +400°C
Maximum rod speed:	4.0 m/s
Maximum pressure:	50 bar

Metallic foil

Metallic foil packings are manufactured using a foil, usually aluminium or lead, which is coated with lubricant such as oil and graphite. The foil is usually crinkled, then twisted and folded over a core of lubricated yarn to provide flexibility. A glass yarn is employed to provide the temperature resistance. Packings made entirely from foil without a fibre core will have poor flexibility. This will make adjustment difficult and limit their ability to cope with shaft misalignment, etc. These packings can be used on pumps and compressors handling oils, solvents and refrigerants. It is worth noting that with some modern refrigerants it can be difficult to find a compatible elastomer or plastic.

Application conditions:

Media in the range:	pH 6–8 (aluminium) excluding steam and corrosive agents
	pH 4–10 (lead)
Operating temperature:	−70 to +540°C (aluminium), +260° (lead)
Maximum rod speed:	1.0 m/s
Maximum pressure:	70 bar

4.6 Clearance seals

Clearance seals are widely used as a reciprocating seal, but in quite specific applications. These are generally at high pressure and relatively high speeds or where seal friction may be a problem. The most obvious disadvantage of a clearance seal is that there will be a consistent and definite leak. This means that they can only be used where there is a facility to collect the leakage. Examples of applications where a clearance is used as the primary high pressure seal are:

- Diesel fuel distributor pumps.
- Hydraulic fluid power piston pumps and motors.
- Some designs of high pressure plunger type water pumps.
- Hydraulic fluid power spool valves.

In each of these examples the high pressure leakage is contained within the equipment body and there is a low pressure atmospheric seal, often on the pump input shaft, for example.

To achieve a successful clearance seal it is necessary to meet two very stringent criteria:

- An extremely small and consistent clearance between the plunger and bore.
- Satisfactory tribological conditions to preserve the bore clearance at this low value for the working life of the equipment.

The basic equation defining leakage through a small concentric clearance as shown in Figure 4.66 is:

$$Q = \frac{\pi D \cdot h^3 \cdot P}{12 \mu L} \text{ ml/hr}$$

Where:
viscosity, μ	$N \cdot s/mm^2$ (1 cP water is $10^{-9} N \cdot x/mm^2$)
clearance, h	mm
length, L	mm
diameter, D	mm
pressure difference, P	N/mm^2
leak rate, Q	ml/hr

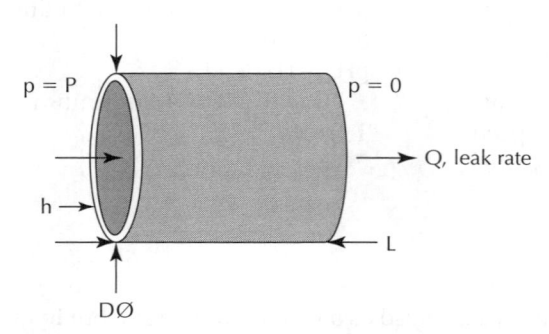

Figure 4.66 Calculation of leakage through a clearance.

To take account of potential eccentricity the equation must be modified to:

$$Q \times (1 + 1.5e^2)$$

Where:
e = the ratio of the offset to the clearance
Hence when concentric e will = 1
And at full eccentricity e will = 2.5

It can be seen that for a given liquid viscosity and mechanical set of dimensions the major governing factor is the clearance. A doubling of what may be a very small clearance will cause an increase in leakage by a factor of 8.

The viscosity is also an important consideration. Low viscosity liquid such as water will leak much more than say an oil that may be 10 to 100 times more viscous. With hydrocarbon-based liquids there is an additional benefit that the viscosity will increase at high pressure whereas this does not occur with water.

The materials for the clearance seal plunger and bore must be chosen very carefully to be compatible with the sealed liquid, and this selection is quite individual to the application. At high pressures and loads this is absolutely essential as otherwise terminal damage will occur. There are numerous examples of damage to hydraulic pumps by selection of a hydraulic oil with alternative additive package and damage to diesel pumps by use of either the wrong or poor fuel.

4.7 Diaphragms and bellows

Diaphragms and bellows have a number of potential advantages compared to seals for reciprocating applications. They are widely used for short stroke applications at lower pressures. They also have a number of other applications usually involving a low pressure differential. The most common forms are moulded elastomer but for higher temperatures and pressures welded metal bellows are also used. The advantages include:

- Effectively zero leakage.
- Do not require lubrication.
- No friction.
- With correct design there is a low hysteresis loss.
- There is no breakout force on startup.
- For low speed operation there is no stick-slip effect.
- As the contact surfaces should only experience rolling movement there is a minimum of wear.
- There are much lower requirements on tolerances and surface finish of the mating components which can provide potential savings during manufacture.

They do have a number of disadvantages which can limit the areas to which they are applied:

- Stroke is limited, although rolling diaphragms do permit a fairly wide range.
- Pressure is limited, reinforcing the diaphragm to withstand a higher pressure limits the movement and reduces the benefits.
- When failure occurs it can be sudden and catastrophic, compared with the gradual decay of seal performance as wear occurs.

4.7.1 Flat and dished diaphragms

The simplest diaphragms are a simple flat disc, Figure 4.67. These have a relatively limited stroke but to increase it a dished design can be used, Figure 4.68. The application limits of the diaphragm are dependent on a number of design factors:

- A larger unsupported area allows a longer stroke but limits the pressure.
- Thicker or reinforced material allows a higher pressure but will create increased resistance and potentially limit the stroke.

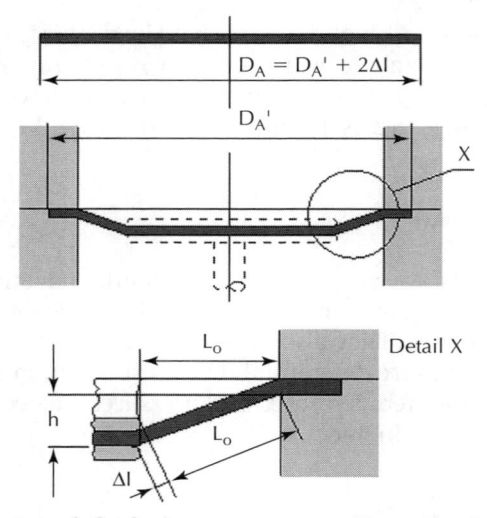

Figure 4.67 A simple flat diaphragm arrangement. (Source: Freudenberg Simrit)

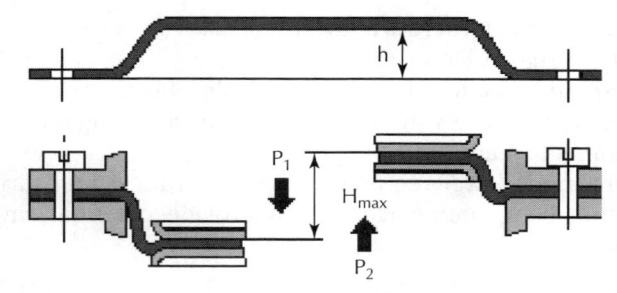

Figure 4.68 A dished diaphragm that allows a longer stroke than a simple flat design. (Source: Freudenberg Simrit)

Although diaphragms do not have any friction there is inherent resistance caused by the deflection of the diaphragm. If required this can be determined by the measurement of the pressure that is necessary to overcome the resistance. An example is shown in Figure 4.69. This curve is caused by tensile stresses in the diaphragm surface on elongation during the rolling movement, and by bending stresses on the deflection of the diaphragm from the production position.

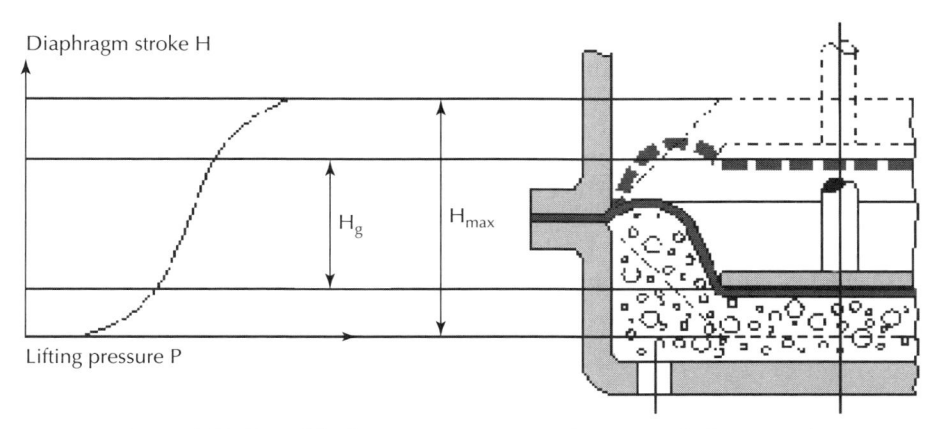

Figure 4.69 Typical diaphragm resistance curve. (Source: Freudenberg Simrit)

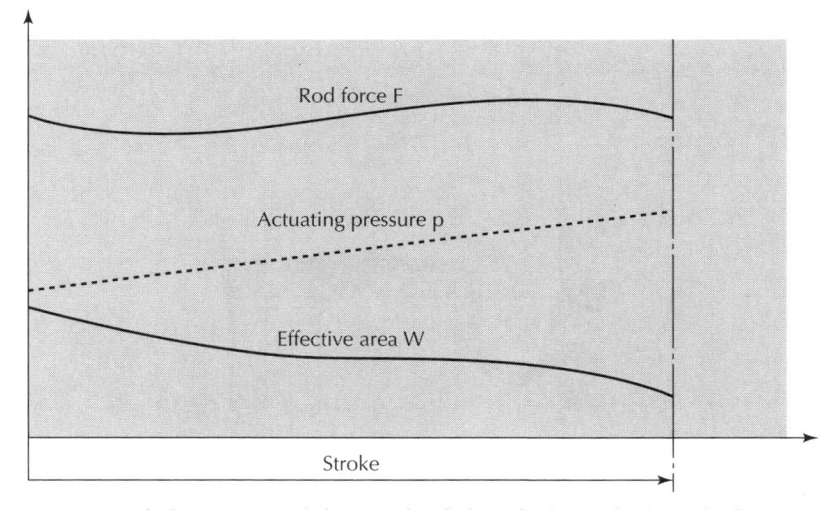

Figure 4.70 Typical characteristic of force produced through the stroke for a diaphragm. (Source: Freudenberg Simrit)

The actual force produced will depend on the applied pressure, the effective area and the above resistance. Figure 4.70 shows an example of a characteristic curve. As the force depends on actuating pressure and effective area or effective diameter, the characteristic curve is essentially determined by the dependency of the effective diameter on the stroke. The choice of the basic shape for the diaphragm plays a decisive role here. In the figure the principal relationship can be seen. As the change on the characteristic curve increases at the ends of the stroke, the characteristic curve can be optimized by the selection of the working area. By increasing the height of the diaphragm these outer areas can be truncated and a better overall characteristic curve is obtained with a lower stroke effect. With a dished diaphragm the effective diameter is very dependent on the stroke.

350 Seals and Sealing Handbook

If the correct material is selected for the application the primary limitation on diaphragm life is fatigue. The localized stresses at the clamps are the areas of highest bending and stress. The design of the attachments to the cylinder body and the actuator rod are therefore extremely important. It is necessary to ensure that all the clamp components have a satisfactory radius to avoid high stresses in the diaphragm as it bends about the clamp during the stroke, Figure 4.71. The radius required will vary with the type of diaphragm, stroke and clearances but a typical range is between 0.8 and 3 mm for devices with diameters between 10 and 600 mm.

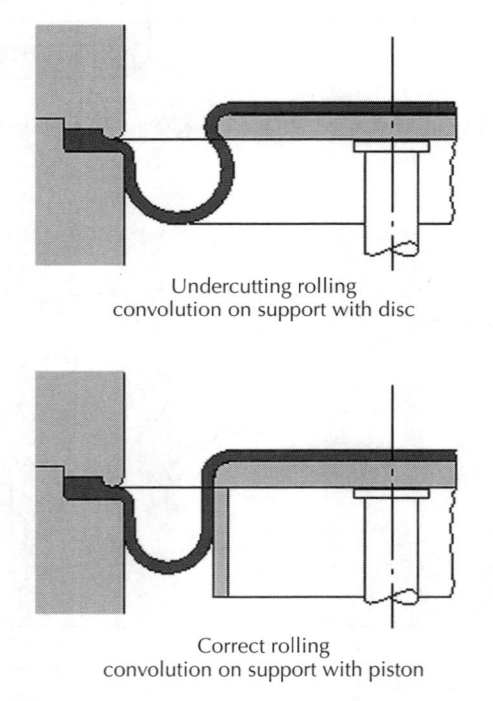

Undercutting rolling
convolution on support with disc

Correct rolling
convolution on support with piston

Figure 4.71 Example of radii on clamps for diaphragm. (Source: Freudenberg Simrit)

4.7.2 Rolling diaphragms

The long stroke rolling diaphragm can be regarded as a special form of dished diaphragm. The possible stroke is limited by the manufacturing property of the diaphragm. Rolling diaphragms must be supported on the outside by the housing and on the inside by the piston. This results in a virtually stroke-independent, constant effective diameter. As the diaphragm rolls rather than stretches this provides a long stroke with a relatively constant characteristic.

These were originally supplied only by Bellofram, now Marsh Bellofram, but are now also available from other suppliers. They are formed in the shape of a truncated cone, or top hat, the diaphragm is turned in on itself when installed so that, during the stroke, it rolls and unrolls alternately on the piston skirt and cylinder wall. The rolling action is smooth and eliminates any sliding contact. Figure 4.72 demonstrates the principle of the rolling diaphragm.

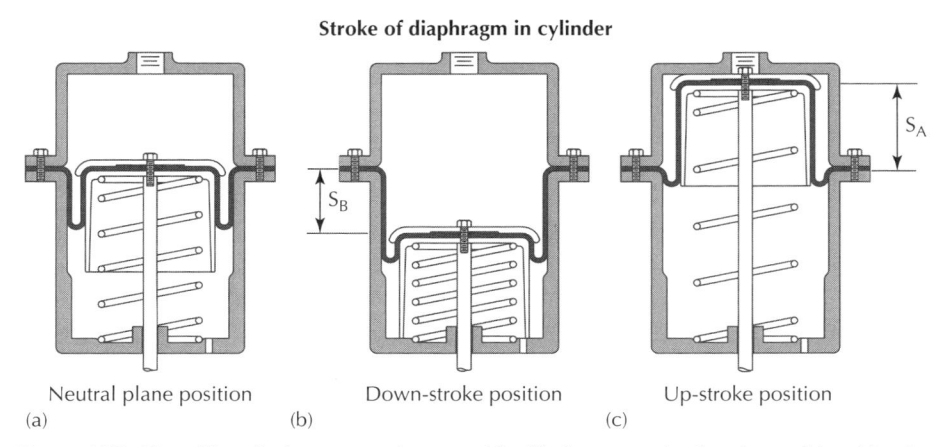

Figure 4.72 The rolling diaphragm permits a considerably longer stroke than is possible with other designs. (Source: Marsh Bellofram)

The rolling diaphragm is constructed from a layer of specially woven fabric, impregnated with a thin layer of elastomer. The total thickness is usually 0.4 and 1.2 mm. The fabric, which is typically a polyester material, provides the high tensile strength to the diaphragm and is designed to permit free circumferential elongation, allowing free rolling action, while preventing axial distortion. This eliminates stretching or ballooning during the stroke. Although this allows a considerably longer stroke than the flat and dished diaphragms the overall stroke length is still limited by the constraints of the manufacturing process and rolling action. A practical limit that is generally considered feasible is for the displacement from the rest position to be equal to the diameter. So the maximum limit for the overall stroke allowing for displacement in both directions is twice the diameter.

The flex life of the rolling diaphragm depends on operating pressures, amount of axial and circumferential stress applied during the stroke and the materials which form it. A correctly designed and installed arrangement will provide a life of several million cycles.

The pressure rating of the diaphragm is dependent on the strength of the fabric and the clearance gap in which the diaphragm rolls, between the piston head and the cylinder bore. This is known as the convolution width, Figure 4.73. Examples of the relationship of pressure and convolution width are given in Table 4.6.

4.7.3 Diaphragm materials

Diaphragms are manufactured either from elastomer or a woven fabric such as polyester proofed with elastomer. The selection will depend on application but for higher pressure differentials or applications where there is a higher stress on the diaphragm, due to a wide convolution gap for instance, then it would be expected to use a fabric reinforced material.

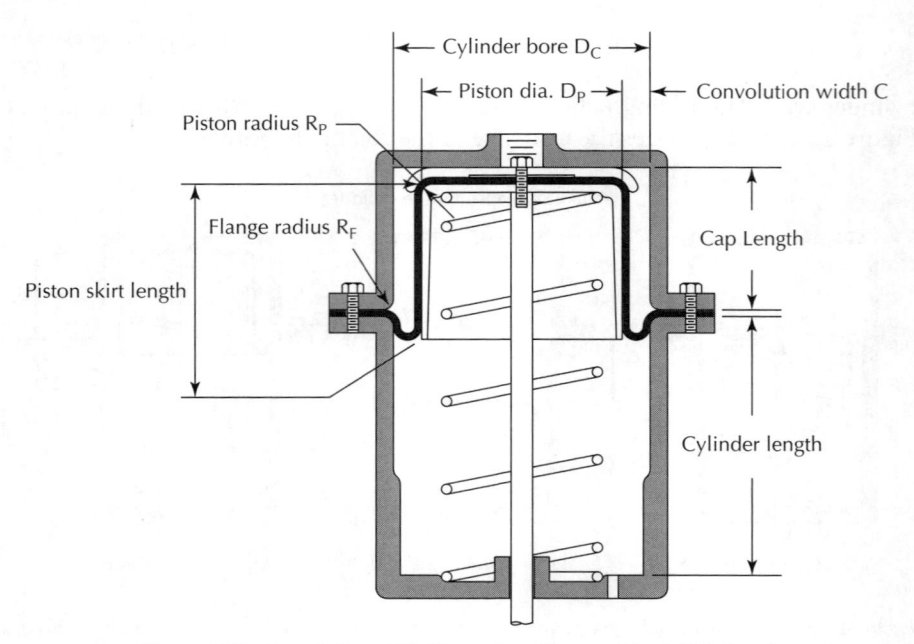

Figure 4.73 Convolution width illustration. (Source: Marsh Bellofram)

Table 4.6 Pressure rating of rolling diaphragms at different convolution widths

(a) Working pressure (bar) of diaphragms at four different convolution widths

Fabric code	Convolution width (mm)			
	1.6	2.4	4.0	6.4
A	19	12.5	8.5	4.7
B	32	21	13	8
C	53	35.5	21	13
V	26	18	11	6.6
L	50	34	20	13

(b) Comparison of fabric types used for rolling diaphragms

Fabric code	Type	Strength factor	Duty
A	Polyester	17	General purpose
B	Polyester	28	General purpose
C	Polyester	47	Heavy duty
V	Nomex	23	High temperature
L	Polyester	44	Heavy duty

(c) Typical cylinder bore sizes and the convolution widths used

Cylinder bore Dc (mm)	Convolution width C (mm)
9.4–25	1.6
25.4–63.5	2.4
63.6–101.5	4.0
102–203	6.4

(Source: Marsh Bellofram)

The general range of elastomer materials is available for manufacture of diaphragms and selection will depend on the application requirements much the same as it would for seals. Factors to consider include temperature, working fluid, strength and permeation. As a diaphragm has a relatively large exposed area and is a thin sheet of elastomer the permeation losses can be expected to be significantly higher than for a seal on the equivalent application. This should be considered carefully if permeation is an issue, and if necessary a low permeation material used. Properties of individual elastomers are discussed in Chapter 5.

To provide increased resistance to chemical attack, simple diaphragm designs can be produced with a PTFE coating to one side. This provides a wide chemical resistance combined with the flexibility of the elastomer.

4.7.4 Diaphragm applications

Applications include a very wide variety of functions where a short reciprocating movement is required, usually at low pressure or a low pressure differential. Examples include pneumatic controllers and brake systems, solenoid valves, pressure switches, many pumps, pressure controllers and valve actuators. Although the operating pressure differential is usually relatively low the ambient pressure may be quite high, such as for instance in hydraulic system accumulators. Many of the applications include frequent short strokes, an area which is difficult for reciprocating seals.

4.7.5 Polymer bellows

These are used almost exclusively to provide some form of exclusion protection. They can be used to protect seals, or other components from solid and liquid debris and also to retain lubricant around components. The shaft may be reciprocating or even rotary with some additional plunging motion, such as for instance vehicle drive shafts. A particular benefit of a bellows is that it can accommodate lateral movement and vibration in addition to the reciprocating movement. They are available in a wide variety of configurations depending on the application from very small actuators to in excess of one metre for accommodating axial movement, thermal expansion and vibration in large industrial machinery from vibratory hoppers to gas turbine exhausts.

A particular problem that can occur with a bellows is the displacement of the internal volume as reciprocating movement takes place. This can be readily accommodated if the stroke is short relative to the volume, such as the plunging of a drive shaft. However, if a bellows is used to protect the exterior of a hydraulic cylinder then a large change in internal volume may be involved. This can often only be accommodated by having some form of vent. This in effect removes some of the purpose of the bellows and must be taken into consideration in the design of system excluders and corrosion resistance.

On short stroke bellows, particularly in automotive applications the designs have been amended to improve the location and sealing at the ends of the bellows. This helps to provide improved contaminant exclusion and better retention of enclosed lubricant, Figure 4.74. If a bellows such as this is well sealed it must also be designed to accommodate any pressure variations caused by temperature changes.

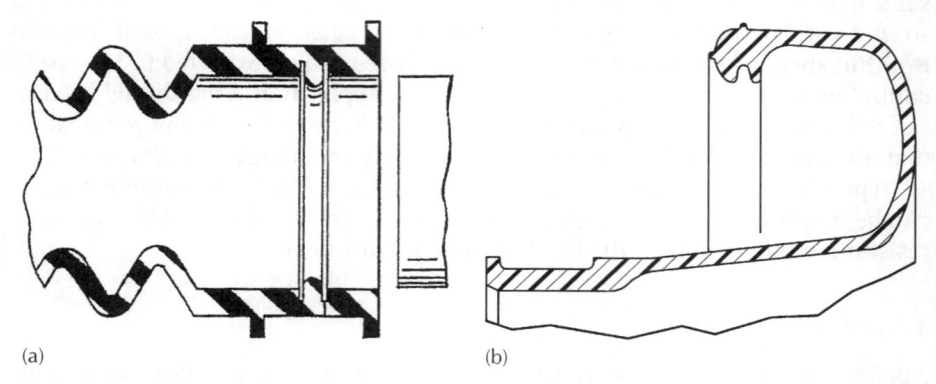

(a)   (b)

Figure 4.74 Examples of bellows designed with specific sealing arrangements to improve exclusion of debris and retention of lubricant.

Bellows may be manufactured from moulded elastomers, fabric reinforced elastomers such as polyester or Aramid, and can also be sewn directly from fabric to make highly conformable bellows for specialist applications. The majority of bellows will be used with very low pressure differential but reinforced materials can be used up to 7 bar with appropriate designs such as those shown in Figure 4.75.

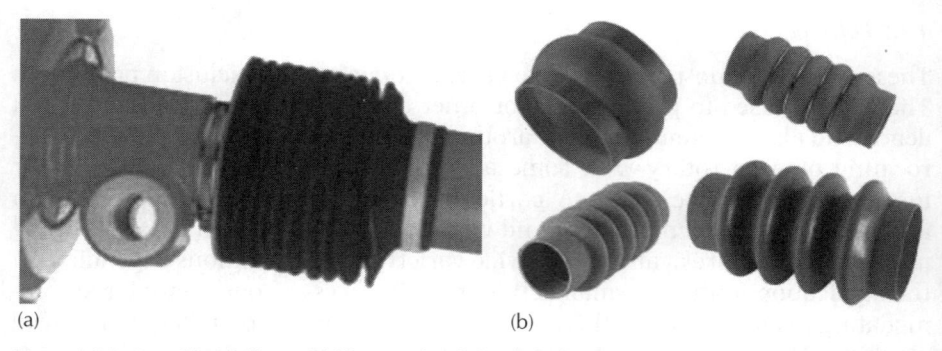

(a)   (b)

Figure 4.75 Sewn fabric bellows and fabric reinforced elastomer for pressures up to 7 bar. (Source: Beakbane)

4.7.6 Metal bellows

Metal bellows may also be used as a reciprocating seal. Two regular applications are for sealing the valve stems of process valves and for high duty accumulators.

4.7.6.1 Valve stems

Metal bellows are used for the sealing of valve stems where it is considered necessary to remove the risk of emissions due to valve packing. They are used for high temperature or corrosive and toxic services, for high vacuum and also where a high degree of cleanliness is required such as electronic wafer manufacture.

Many of these valves use a hydroform type of bellows, Figure 4.76. Valves with metal bellows sealed stems are available for applications ranging from high vacuum to in excess of 70 bar. They can be used at temperatures from cryogenic to over 500°C. To achieve a high pressure rating and sufficient flexibility a multiple ply bellows may be used.

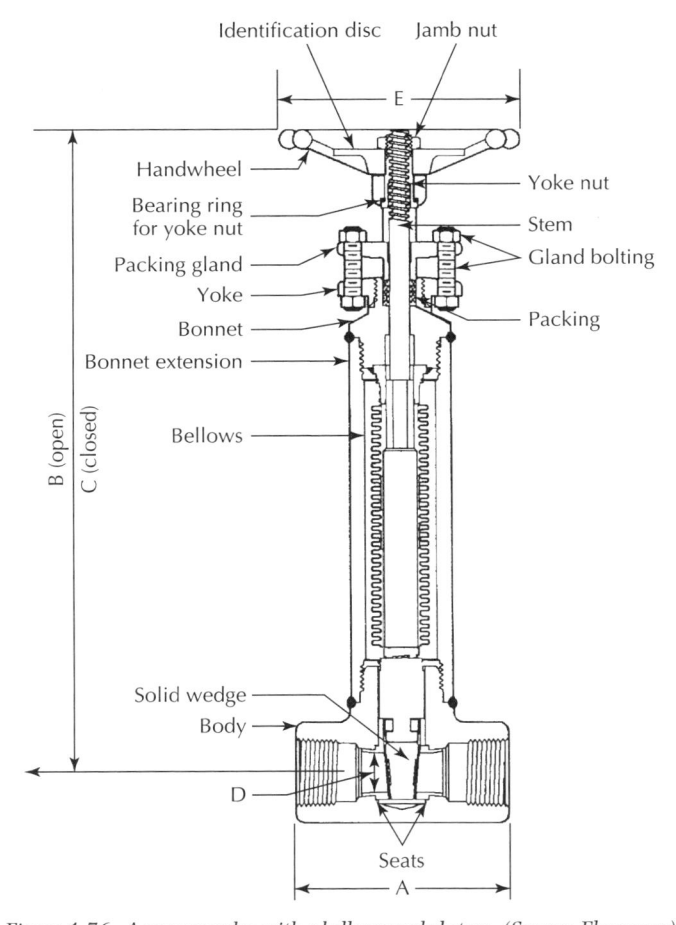

Figure 4.76 *A process valve with a bellows sealed stem. (Source: Flowserve)*

The use of a hydroform bellows provides economic manufacture and avoids the narrow convolutions that could trap debris compared with a welded metal bellows. A backup seal, usually packing, is fitted to prevent excess leakage in the event of bellows failure.

The disadvantages of bellows for valve sealing include:

- Application is restricted to parallel rising stem valve designs as the bellows will not accept any stem rotation.

- Increased height of valve body and longer stem due to the length required by the bellows to allow for the axial movement of the stem.
- Increased cost due to manufacture of the bellows and increased stem height.
- Pressure limited to capability of the bellows.
- The latest specialist low emission valve packings have proved capable of approaching the emission levels expected of bellows sealed valves.
- Unsuitable for applications such as clean in place due to the convolutions that are well away from the flow and will not flush easily.

4.7.6.2 Accumulator bellows

For high duty applications a metal bellows may be used for hydraulic and other fluid system accumulators, Figure 4.77. They are widely used in aircraft hydraulic systems. The advantages compared with conventional accumulators include:

- Removes problems of leakage and breakout associated with piston accumulators.
- Avoids the gas porosity that will occur over time with a bladder accumulator.
- Suitable for wide temperature extremes beyond those of elastomers and plastics.
- Removes the fluid compatibility problems associated with elastomers.

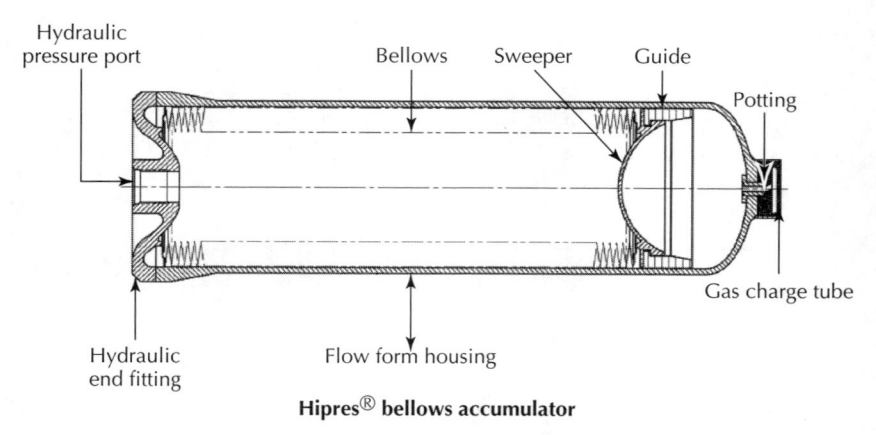

Hipres® bellows accumulator

Figure 4.77 Hydraulic accumulator design with welded metal bellows. (Source: Senior Aerospace)

These factors make bellows suitable for accumulators and pressure surge devices on critical systems particularly if the temperature range is at the limit of the capability of elastomers or where corrosive or toxic fluids are concerned. In addition to aircraft hydraulics and fuel systems they are used in weapon and other rocket systems, spacecraft and nuclear power fluid circuits. The reduction of leakage or porosity also reduces the maintenance requirements for gas charging.

The primary disadvantage compared with other accumulator designs is high cost. As it is necessary to provide a maximum of bellows travel in the minimum envelope they are generally welded bellows.

4.8 References

1. C.M. White and D.F. Denny, 'The sealing mechanism of flexible packings', Ministry of Supply Technical Memorandum No. 3/47, HMSO, 1948. Republished by BHRA, 1972.
2. G.J. Field and B.S. Nau, 'Film thickness and friction measurements of a rectangular section rubber seal ring', Sixth Int. Conf. on Fluid Sealing, BHR Group, 1973.
3. G.J. Field, R.K. Flitney and B.S. Nau, 'The lubrication of reciprocating rubber seals: U-seal tests', BHRA Report RR1315, 1975.
4. R.M. Austin, R.K. Flitney and B.S. Nau, 'Research into factors affecting reciprocating rubber seal performance', BHRA Report 1449, 1977.
5. P.W. Wernecke, 'Analysis of the reciprocating sealing process', 11th Int. Conf. on Fluid Sealing, BHR Group, 1987.
6. R.K. Flitney, I. Hansford and B.S. Nau, 'The effect of surface texture on reciprocating seal performance', BHR Group Report CR 3069, 1989.
7. N.A. Peppiatt, 'The influence of the cylinder tube surface finish on reciprocating seal performance', 13th Int. Sealing Conference, VDMA, 2004.
8. B.S. Nau and G.J. Field, 'Interseal pressure of piston seals in thin-walled jacks', BHRA Report RR1088, 1971.
9. N.A. Peppiatt, 'The influence of the rod wiper on the leakage from a hydraulic cylinder gland', *Sealing Technology*, December 2003.
10. H. Jordan, 'PTFE rod seal with high pressure relief technology', 13th Int. Sealing Conference, VDMA, 2004.
11. R.K. Flitney, 'International O-ring standards: where are they?', *Sealing Technology*, October 2003.
12. Y. Tatseishi, 'Tribological issues in reducing piston ring friction issues', *Tribology International*, 1994.
13. *Engineers Handbook of Piston Rings*, Koppers Inc.
14. J.K. Lancaster, 'Accelerated wear testing of PTFE composite bearing materials', *Journal of Lubrication Tribology*, April 1979.
15. J.K. Lancaster, 'Abrasive wear of polymers', *Wear*, October 1969.
16. R. Schubert, 'The influence of a gas atmosphere and its moisture content on sliding wear in PTFE compositions', *Lubrication Technology*, April 1971.
17. R.S. Wilson, 'Developments in piston and rod sealing materials for dry gas compressors', Fluid Machinery Congress, I.Mech.E., 1990.
18. R.S. Wilson, 'Advances in piston and packing ring materials for oil-free compressors', Fourth Workshop on Piston Compressors, Kötter Consulting Engineers, October 2000.
19. 'Thermally applied coating for piston rings, consisting of mechanically alloyed powders', Federal-Mogul Burscheid, Patent No. WO 2002/024970.
20. 'Wear protection layer for piston rings, containing wolfram carbide and chromium carbide', Federal-Mogul Burscheid, Patent No. WO 2002/048422.
21. J. Esser, S. Hoppe, R. Linde and F. Münchow, 'Compression piston rings in gasoline and diesel engines – current engineering practice and outlook', *MTZ*, July 2005.
22. C. Radcliffe, 'Sealing material developments for reciprocating gas compressors', 18th Int. Conf. on Fluid Sealing, BHR Group, 2005 and *Sealing Technology*, November 2005.

CHAPTER 5

Materials

The material used to manufacture seals covers the entire range of engineering materials. Metals, plastics, elastomers and ceramics are all widely used and will be discussed in turn. In some cases materials, such as some specialist elastomers, have been primarily developed to cater with the demands of sealing and related industries. Similarly some of the specialist fibrous materials have been developed to provide the properties required for packing and gasket materials.

This chapter describes the basic properties of the major material classes that are used for seals. Some aspects of the application of the materials is very specific to the type of seal. For this reason the description of materials specifically used for packing and gaskets is primarily included in the sections discussing these seal types.

5.1 Elastomers

Elastomers are very widely used and potential applications cover the full range of static and dynamic seals. They are also a key component of other seal types. For instance the majority of mechanical seals contain several elastomer seals and these are often one of the limiting components in the fluid resistance and temperature range of the seal. There is quite often a severe lack of understanding of elastomer materials even among those who use them regularly. The properties and capabilities of the individual material types vary enormously, as we shall see in the following sections. Even within an individual type of elastomer the potential range of properties is extremely wide and very dependent on the formulation of the ingredients and the manufacturing process. An appreciation of the wide differences between both the different material types and the grades within a single type is made more difficult because a very large proportion of the products are black. There is very little obvious difference, even to a knowledgeable observer, in the visual appearance of a practically worthless O-ring made from a small proportion of low temperature elastomer compound and a ring made from high temperature and extremely fluid resistant material costing five orders of magnitude more.

5.1.1 Why do we use elastomers?

The short answer is that the materials have some quite unique properties that make them particularly suitable for use as a seal, and many seal geometries depend on the properties of the material to provide the sealing action. The method by which elastomers are able to provide a sealing action is described in detail in section 2.2.1. The majority of successful elastomer sealing applications use these material characteristics to provide an efficient and reliable seal, usually much more economically than can be achieved by materials used for other sealing methods.

The properties of the material that enable it to be so effective for sealing include:

- Large deflections only require low stress: Elastomers have a very low modulus of elasticity (E) of typically 5 to 20 MPa compared to 50 to 200 GPa for typical engineering metals. This is combined with an elongation to break that is normally well in excess of 100%. It is therefore possible to use relatively large strains both during installation and to provide seal interference. Elastomer seals will typically be designed to have an installed strain between 10 and 30%. This large value of elastic strain will be achieved with a relatively low stress. Strains of 50–100% may be used during installation without adverse effects on most elastomer materials. (Note that some especially hard elastomers may have an elongation to break of less than 100%.) The seals can accommodate a relatively wide range of tolerances and misalignment without high contact stresses and will conform to the mating groove surface texture to provide a high integrity seal without plastic flow of either the seal or the counter-face.

    Figure 5.1 provides examples of stress/strain properties of elastomers compared with both plastics and metals.
- It is a resilient material with low hysteresis: When used within the appropriate temperature range for the material grade elastomer materials provide a high degree of resilience with low hysteresis. This provides the ability to respond rapidly to pressure changes, vibration plus conditions such as runout in dynamic applications.
- They have relatively low creep: The tensile strength of the materials is relatively low, compared to most plastics and obviously metals, but when strained within the range of material properties they do not suffer excessive creep. This is important for maintaining sealing stress for long periods of time and through high temperature cycles. This can be a particular advantage of elastomers when compared with plastic seals. Figure 5.1 shows that plastics will generally creep when strained beyond 5–30%, depending on the type of plastic and filler content.
- Elastomers have a high Poisson's ratio, very close to 0.5: This means that the materials are almost incompressible. The high Poisson's ratio combined with the very low elastic and shear modulus combine to make a material that is both easily deformed and incompressible. The combination of the two properties means that the material behaves in a similar

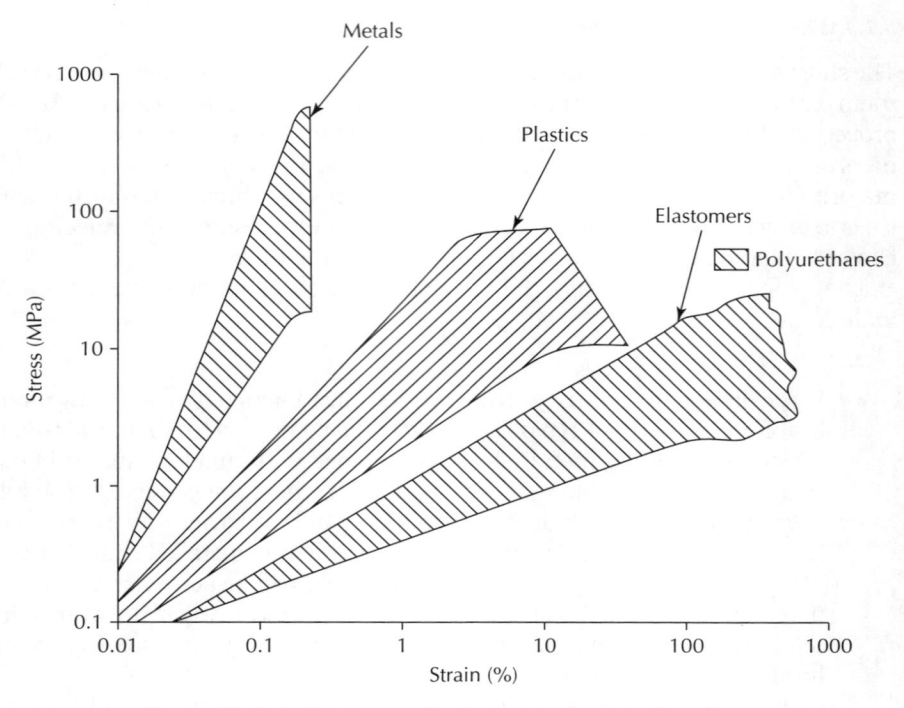

Figure 5.1 Stress/strain range for elastomers, plastics and metals.

manner to a liquid and will transmit pressure evenly through the bulk of the seal. This provides the pressure energizing effect discussed in section 2.2.1. The inherent resilience of the material ensures that when the distorting force, such as pressure or groove constraint, is removed the seal will return to the original geometry.

In practice the degree of resilience and ability to restore completely to the original geometry will be dependent on the individual characteristics of the material. However, there are also some disadvantages associated with these materials that introduce design constraints and limit the applications to which they may be applied.

- They have a limited temperature range: The individual temperature limits of material types are discussed in section 5.1.4, but the basic temperature ranges are limited to −30 to +100°C for general purpose materials, −50 to +200°C for more specialist elastomers and a maximum of 300°C for very exotic elastomers.
- Chemical resistance is very dependent on both elastomer type and individual grades within a material type. This is an area that can require very careful evaluation.
- Materials are soft and easily torn, requiring careful design and handling to avoid damage.

- High and erratic friction: The potential use in dry sliding is very limited. The lubrication mechanisms are still not fully understood and friction can be high and erratic compared to other materials. With a few notable exceptions the application of elastomers to dynamic seals is limited to liquids that provide adequate lubrication such as lubricants and hydraulic oils. Water seals are generally limited to relatively restricted operating conditions or life expectancy.
- Non-linear behaviour and hysteresis: Elastomers do not exhibit linear stress/strain properties. Samples of typical stress/strain curves are shown

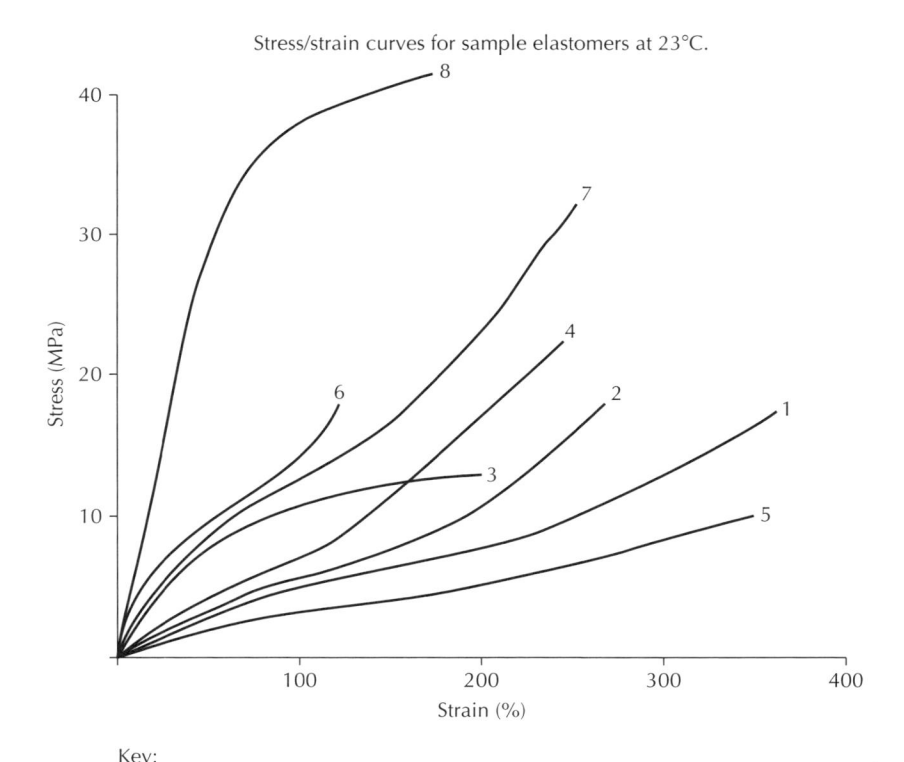

Key:

No.	Material type	Hardness (IRHD)
1	Nitrile	65
2	Nitrile	70
3	Fluorocarbon type A	75
4	Fluorocarbon, terpolymer, peroxide cure	70
5	Fluorosilicone	60
6	Fluorocarbon, type B	90
7	Hydrogenated nitrile	90
8	Polyurethane	95

Figure 5.2 Typical elastomer stress/strain curves.

in Figure 5.2. The actual curve will be dependent on the type of elastomer, type of cross-links and the proportion of filler. Materials such as unfilled natural rubber are highly non-linear but highly filled seal materials have been found to have essentially linear characteristics over a large proportion of the extension range.[1] They are also subject to hysteresis effects which inhibit recovery. This can limit the ability to follow vibration and runout. These properties are also temperature sensitive.

5.1.2 Elastomer material basics

The original elastomer material was natural rubber and it is interesting to note that it probably still provides the highest elasticity and resilience of any material. However, the temperature range and ageing resistance are not satisfactory for most industrial sealing applications so it is seldom found as a sealing material, although it is widely used for many other engineering applications where abrasion resistance and resilience are required. The first synthetic elastomers were developed in the 1930s with the early development of polychloroprene material. Since that time many material types have been developed, each of which has individual properties. In this section the basic makeup of an elastomer is discussed. Subsequent sections discuss relevant material properties, the attributes of individual material types and where they may be used.

The primary constituent of an elastomer is the organic polymer. The basis of the polymer is a very long chain of organic molecules which contain a repeating monomer unit. In hydrocarbon polymers these molecules are primarily hydrogen and carbon, as for instance the acrylonitrile polymer, Figure 5.3. The chemistry of the bulk polymer unit dictates the basic properties of the elastomer.

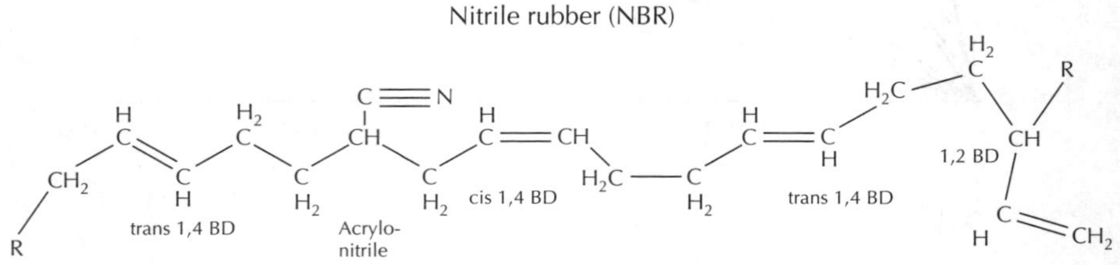

Figure 5.3 Acrylonitrile polymer.

And so, for instance, a fluorocarbon polymer which has very different chemical properties, as discussed in section 5.1.4.13, will have a molecule that contains a large proportion of fluorine atoms, Figure 5.4.

The basic polymer alone is not very useful. There is no chemical attachment between the separate polymer chains and the material will behave like a soft plastic, easily distorted with no recovery. To provide a usable elastomer the material requires what is usually known as curing. This involves the creation of

Copolymer fluoroelastomer:

$$-(CF_2-CF)-(CH_2-CF_2)- \\ | \\ CF_3$$

Terpolymer fluoroelastomer:

$$-(CF_2-CF)-(CH_2-CF_2)-(CF_2-CF_2)- \\ | \\ CF_3$$

Improved low temperature fluoroelastomer terpolymer:

$$-(CF_2-CF)-(CH_2-CF_2)-(CF_2-CF_2)- \\ | \\ O-CF_3$$

Non-VF$_2$ fluoroelastomer terpolymer:

$$-(CF_2-CF)-(CH_2-CH_2)-(CF_2-CF_2)- \\ | \\ O-CF_3$$

Figure 5.4 Fluorocarbon polymer molecules.

cross-links between the polymer molecules. To achieve this, a small proportion of curing agent is mixed with the polymer and the mixture subjected to heat and pressure to create a chemical bond between the polymer molecules and the cross-links. The curing agent may variously be sulphur, zinc oxide, peroxide compounds or for fluorocarbons bisphenol.

Diagrammatically the effect of the curing process is shown in Figure 5.5. The molecules of the elastomer are linked together. A further constituent that is usually incorporated into the mixture is filler. These are micron size particles that are mixed in between the polymer molecules and contribute to the physical properties of the material. This will create something similar to the elastomer, shown diagrammatically in Figure 5.6, which will provide the properties that distinguish it from other material classes.

The polymer provides a series of long chain contorted molecules, which at the working temperature of the material are in a thermodynamically free state. When a stress is applied to the material it will be distorted. The molecules may become more aligned as a stress is applied, but the cross-links will maintain the attachment between the molecules. When the stress is released the energy within the material will return it to the free state. The energy required to strain the material, the degree of resilience and any residual strain that may remain are very dependent on a number of factors. These include the polymer type, formulation of the elastomer compound, proportion and type of fillers and also the temperature. The ability of the polymer molecules to retain their active properties is fundamental to the temperature range of the materials and one of the significant limitations to elastomer applications, as discussed further in the individual material properties, section 5.1.4.

(a)   (b)

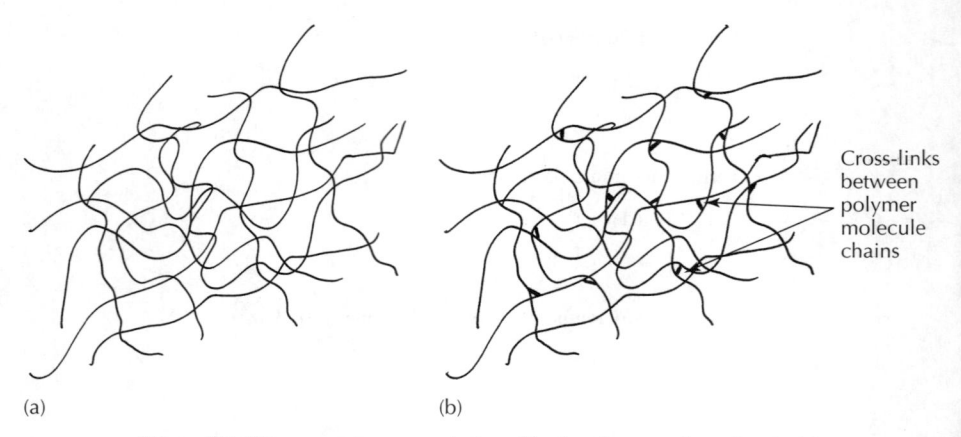

Figure 5.5 Diagrammatic representation of basic polymer and cured material.

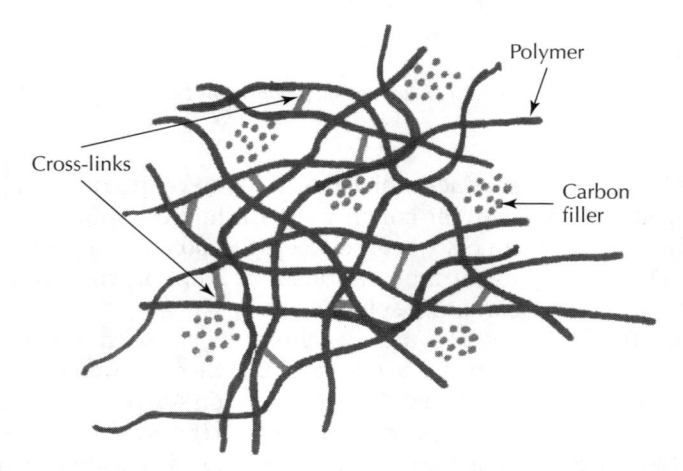

Figure 5.6 Diagrammatic representation of a cured and filled elastomer showing the interaction of the filler to provide additional strength.

The relatively simple diagrams in Figures 5.5 and 5.6 do not portray the complexities of elastomer formulation and manufacture. The selection and production of the polymer will include a number of factors:

- Basic seal requirements, e.g. oil resistance may suggest a nitrile compound.
- Temperature range may affect the choice of low or high acrylonitrile content.
- Curing system, generally a sulphur-based cure may give higher strength but a peroxide cure a wider temperature range.
- Filler size and type, carbon is the primary reinforcing filler but this needs to be very fine and well-mixed particles. Alternative non-reinforcing fillers provide bulk but do not necessarily contribute to the integrity of the material.

Materials   365

- Consistency of the mixing and curing process.
- Type of moulding process, which will be very dependent both on the volumes required and the seal geometry.
- Seal type to be manufactured. Oilfield seals may require a very high strength whereas for a rotary shaft lip seal a material that will generate the correct surface texture for hydrodynamic lubrication is required.

Hence seal manufacturers have a wide variety of material grades available to meet the large range of applications to which seals are subjected.

As an example of the considerations that may be applied Table 5.1 summarizes the properties that will contribute to the decision to select a nitrile polymer with a low or high acrylonitrile content. Further considerations then include the cure system to give the physical and temperature resistance required, filler type and amount to provide hardness plus further processing and property aids.

Table 5.1 The effect of varying the relative amounts of acrylonitrile and butadiene in a 'nitrile' elastomer compound

NBR with lower acrylonitrile content		NBR with higher acrylonitrile content
	Processability	⟶
	Cure rate w/sulphur cure system	⟶
	Oil/fuel resistance	⟶
	Compatibility w/polar polymers	⟶
	Air/gas impermeability	⟶
	Tensile strength	⟶
	Abrasion resistance	⟶
	Heat-ageing	⟶
⟵	Cure rate w/peroxide cure system	
⟵	Compression set	
⟵	Resilience	
⟵	Hysteresis	
⟵	Low temperature flexibility	

Additional constituents may include:

- Accelerators to provide a shorter curing time.
- Anti-oxidants to improve the ageing resistance of the material.
- Additives to improve the flow in the mould.
- Additives to improve the release of the cured seal from the mould.
- Various compounds known as plasticizers or extenders to improve the low temperature flexibility of the material.
- Occasionally lubricant additives such as a PTFE or MoS_2 compound may be incorporated for seals in marginally lubricated applications.

It can be seen from the above description that elastomer compounds are extremely complex and that selection should be undertaken with some care. It is very common to see a seal material specification as a material type and hardness, such as for instance 70 hard nitrile. This is about as helpful as specifying the metalwork as steel.

To help specify the steel material there are a wide range of international and national standard alloy specifications to allow the engineer to select the material properties required for the application. There are very few similar widely accepted standards to assist in a similar way with specifying an elastomer. There are, however, some aerospace material specifications, section 7.2, and suggested minimum property values are provided in BS 7714.

To demonstrate the potential problems that can occur, Table 5.2 illustrates how such a very simple inadequate specification can cause considerable problems. Table 5.2(a) provides two potential formulations of compounds that are aimed to provide something that will produce a nitrile material of 65IRHD. It can be seen that compound A contains the base polymer, curing agents and a fine carbon black filler. Compound B starts with the same polymer and cure system, but has more sulphur and accelerator, to speed the curing, very little of the costly carbon black but considerable quantities of low cost fillers that will provide volume, plus some oil extender to preserve some flexibility.

It can be seen from the final PHR (parts per hundred parts of rubber) that it is possible to produce 2.5 times more seals using less costly ingredients.

Table 5.2 NBR formulation examples

	(a) U-T-A-Q	(b) D-T-A-P
34% ACN NBR	100	100
Zinc oxide	5	5
Stearic acid	1	1
Sulphur	1	2
Accelerators	1	4
Ester plasticizer	7	–
AO2/AO3	2	–
Carbon black N550	70	20
Oil extender	–	50
Chalk dust	–	200
Talcum powder	–	50
Plain flour	–	50
Candle wax	–	5
Total: parts/hundred rubber	190	487

(Source: James Walker)

Table 5.2(b) shows the relative properties of these two materials. Both have a nominal hardness of 65IRHD ± 5°. They therefore meet the required specification. However, the remainder of Table 5.2(b) shows that the potential performance of the two materials will be very different. It can be seen that Material B has only approximately 10% of the tensile strength of A, less than a quarter of the tear strength and double the compression set. The reality is that Material B, which has been compounded to meet a price specification, will look visually quite similar to A but will provide very poor performance. Materials such as Compound B are on the market and they can fail even during assembly due to the poor strength. If they are the lowest cost 65IRHD seals available and that is the only specified parameter then the buyer can claim to have done a good job. The problem is caused by the specification provided to the buyer. Production lines have been halted and major warranty problems created by inadequate elastomer specifications of the types described above. It can be seen that a material specification should include at least as a minimum some mechanical properties, such as tensile strength. It will also be normal to specify some fluid resistance parameters, but this is discussed further in section 5.1.3.1. This is necessary for any application where some degree of sealing integrity is required. For more extreme applications, such as high pressure gas, then further validation testing of individual compounds is undertaken.

Guidance on suitable material properties has not until recently been very widely considered. There are a number of aerospace material specifications and these are one potential source of a guide to minimum sensible material properties, some of these are listed in section 7.2. Some outline guidance on minimum material properties is also provided in 'Seal material/fluid compatibility for fluid power applications'[2] and BS 7714 which is also referenced in section 7.2.

5.1.3 Factors to consider when selecting an elastomer

When selecting an elastomer the first consideration is generally the fluid resistance. It is quite normal to approach this subject rather simplistically along the lines of; we are sealing oil so we need an oil resistant seal. While this is obviously true, it is also only the starting point as will be illustrated by the two examples in Figure 5.7.

A lip seal on an automotive application, Figure 5.7(a), may be used to retain hot lubricant in the engine or transmission. Resistance to the lubricants is a prime concern, but the other fluids must also be considered. Most seals are also exposed to air, and so air, or particularly oxygen, resistance is a major concern. If the seal may also be exposed to spray from the road, and salt in winter, plus cleaning fluids that may be used for the vehicle then these should be considered.

In the steam seal example, Figure 5.7(b), it is common practice to select EPDM material which has particularly good resistance to steam and hot water. The long-term operation of seals in this application can be limited by the air ageing resistance of the material rather than that of the steam.[3,4]

These examples demonstrate that it is necessary to consider all the environmental aspects of the application, not just the fluid being contained. A list of the type of factors that should be considered when preparing an elastomer selection specification is given in Table 5.3, some of them are discussed in more detail in this section.

5.1.3.1 Fluid resistance

Having confirmed that the material has suitable mechanical properties as discussed in section 5.1.2 the next task is generally to confirm the fluid resistance. This is normally achieved using a 'soak test' in which the elastomer is soaked in a

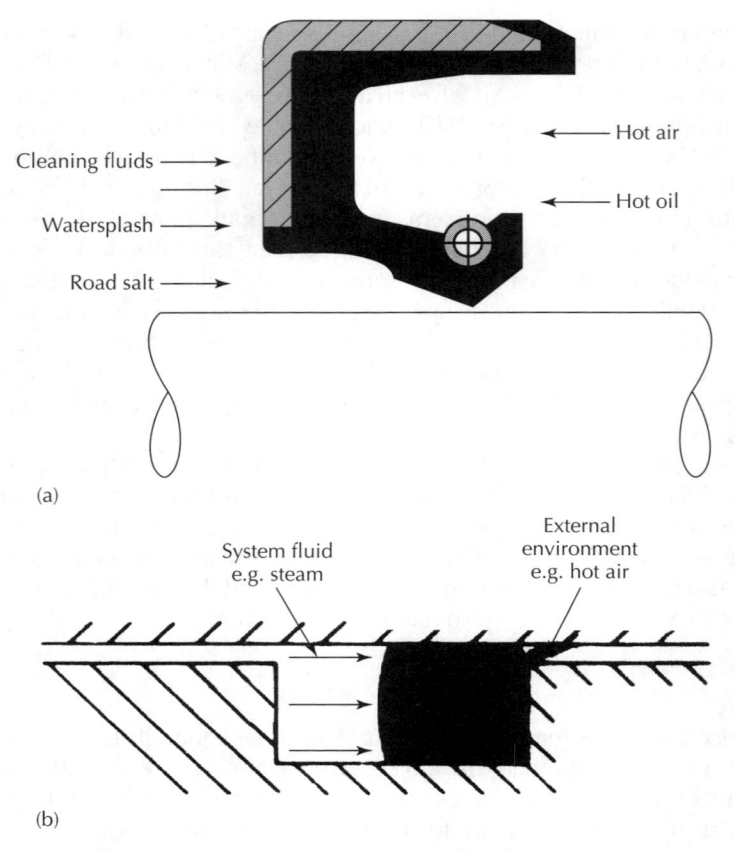

Figure 5.7 Examples of the range of fluids to which a seal may be exposed when in service.

Table 5.3 Application factors to consider when selecting an elastomer material

- How and where will the seal be used? How will it be stored and transported?
- What is the environment in which the seal will operate, including liquids, gases, contaminants, pressures, temperatures, etc.?
- What life and duty cycle is required?
- Is the potential price compatible with market expectations of price and performance?
  Factors to be considered:
- The primary fluid(s) to which the elastomer will be exposed.
- Secondary fluids to which the elastomer will be exposed, such as assembly lubricants, cleaning fluids, sterilizing cycles, etc.
- The temperature extremes, both hot and cold.
- The presence of abrasive external contaminants.
- Presence of ozone from natural and artificial sources, such as electric motors.
- Exposure to ultraviolet light and sunlight.
- The potential for outgassing in vacuum applications.
- Requirements for contact with food, pharmaceuticals or drinking water.
- Any industry specific approvals such as aerospace.
- Any assembly requirements, such as coatings for dry assembly or very high stretch for small one piece components.
- Is colour coding required?

vessel of the liquid concerned. For optimum results this will be carried out at a range of temperatures including some close to the maximum rating of the liquid elastomer combination. This has traditionally been carried out for relatively short periods such as 48, 72 or 168 hours. While this will readily demonstrate totally unsuitable combinations, where very high swell occurs, it will not demonstrate potentially harmful long-term effects. There is now a generally gathering acceptance that longer-term testing is necessary and this is reflected in some of the more recent standards, such as ISO 6072.

Two examples demonstrate how short-term tests can be misleading.

Figure 5.8(a) shows some long-term swell tests of four different EPDM compounds in hot water. This is generally considered to be a material that has good compatibility with water and low swell. These results show that in some compounds swell may continue for long periods with little sign of saturation after in excess of 1200 hours' exposure. This has proved to be true of a number of EPDM compounds and it was not common knowledge in the industry as previously such long-term tests had not been carried out. As in this case the swell probably counteracts compression set, it may not be of any concern, but it demonstrates that there is often an incomplete knowledge of material behaviour.

The graph in Figure 5.8(b) demonstrates a potentially serious problem that will not be disclosed by short-term testing. The specification concerned called for a soak test of seven days and a material was classed as having passed if the swell was less than 20%. Longer-term testing demonstrated that after the seven days the seal material was quite liable to start losing volume, probably due to the loss of extenders used to provide low temperature performance. In some cases it was possible for an eventual shrinkage and consequent loss of seal interference to occur. This oil and seal combination had been in use for a number of years, but again the problem had not been sufficiently investigated to demonstrate one of the causes.

As a general guide, soak testing results may be considered acceptable if the material swell is less than 10%. Higher swell, up to 20%, may be considered acceptable for static seals, provided there is no concern at the reduced mechanical properties that this will cause. The effect on groove fill must also be considered, as excessive swell may cause extrusion damage to the seal. Shrinkage should be avoided as this will cause loss of seal interference and probably also a reduction in seal flexibility and resilience, especially at low temperature.

To ensure that the material is not adversely affected by the fluid in other ways it is also necessary to carry out comparative measurements of physical properties after exposure to the fluid. Excessive swell, or other ageing effects, may cause an unacceptable loss of either strength or elongation of the material. Guidelines for acceptable changes are given in fluid acceptance and testing documents such as ISO 6072.

5.1.3.2 Compression set

Compression set is a term used in the elastomer industry to measure the recovery of a material after it has been compressed. It is a very useful parameter to help with the selection of materials but an appreciation of the methods of measurement is necessary to ensure that the correct interpretation is used.

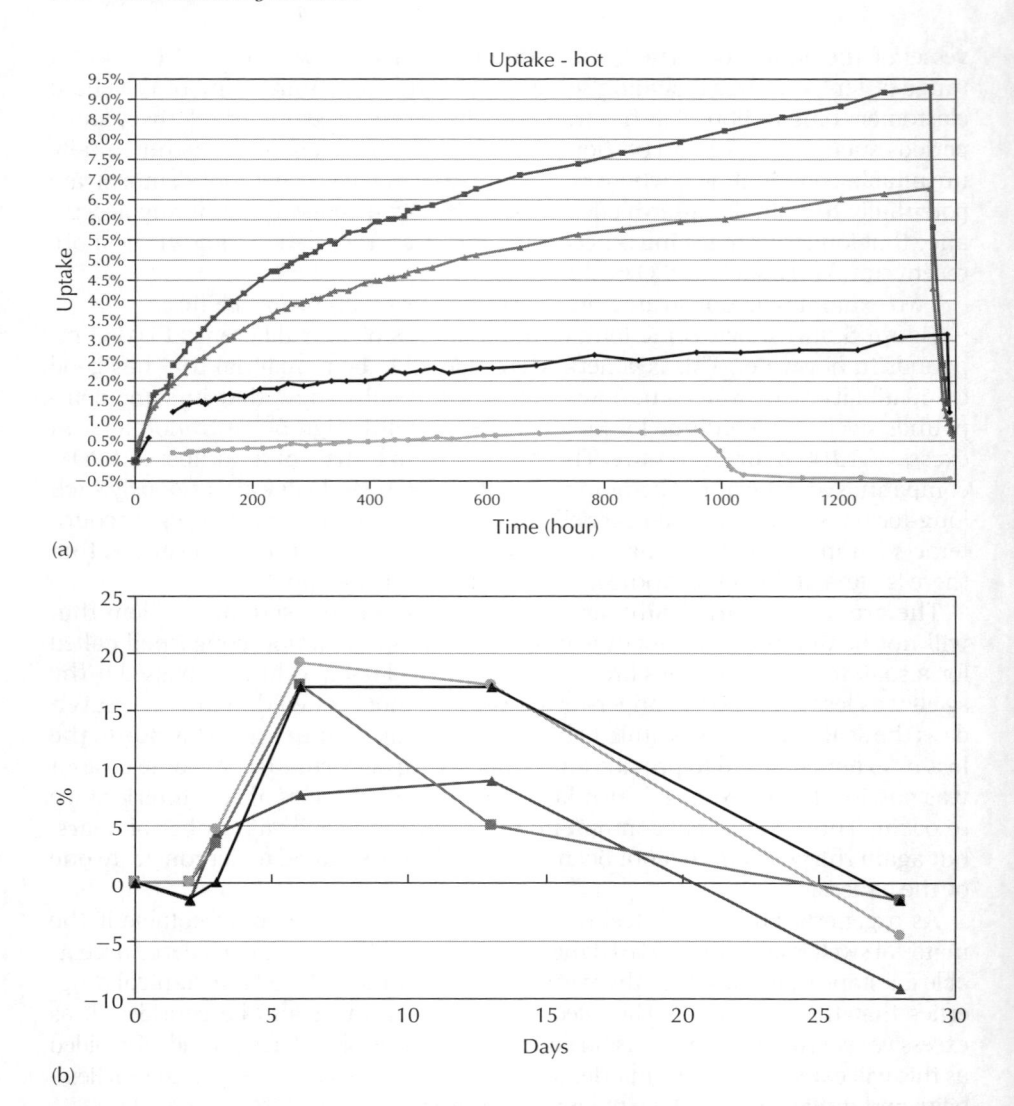

Figure 5.8 Examples of elastomer fluid compatibility testing by swell testing over extended periods showing that fluid interaction may continue for long periods. (Source: BHR Group)

Compression set is quoted as a percentage and is obtained from the expression:

$$\text{Compression set} = \frac{X - R}{X} \times 100$$

where X = the amount of compression in mm
R = the amount of recovery in mm

Using the example of the O-ring in a groove in Figure 5.9, assume that the compression when assembled is 5 mm. If when the seal is removed it recovers

4 mm, then the compression set will be $5 - 4/5 = 0.2$, or 20%. Thus 0% set means that the seal has totally recovered to the original dimension and 100% set means that there is no recovery and the seal remains at the compressed dimensions.

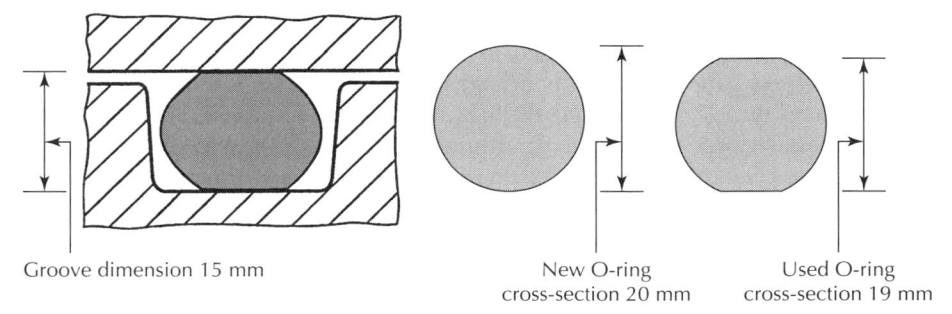

Figure 5.9 The measurement of compression set on a seal.

There is a standard for the measurement of compression set, ISO 815, and several equivalent national standards. This should be used with great caution for selection of materials for sealing as it was originally developed by the elastomer manufacturers for manufacturing quality control purposes. It measures compression set in air over a limited period and releases the material while it is still hot. This does not necessarily provide optimum information for seal material selection.

It is more useful for assessment of sealing performance to measure compression set in more detail and by a method that will discriminate between the two potential components that set can comprise. These are:

- Recoverable, also known as physical set, and
- Permanent, also known as chemical set.

Physical set is attributable to the inability of the material to recover at the measured conditions even though there is no chemical change that prevents the recovery. It is usually due to the material being at a temperature, which can be room temperature, at which the material is not able to readily recover. This can be a problem peculiar to high temperature fluid resistant materials. If the material is reheated to the working temperature it will recover.

Chemical set is caused by changes in the chemical structure of the elastomer that prevent it from recovering. The most common cause with many materials is that while the seal is in operation in a compressed state further cross-links are created by a residual curing agent in the mix. When the seal is released some cross-links have formed which compete to retain the deformed geometry. This is a permanent change to the material and cannot be reversed.

It is possible to discriminate between these two types of set by using an alternative procedure for measurement to that provided in the general material measurement standards. This involves allowing the seal to cool in the compressed state before removing for measurement. The unconstrained seal is then reheated to the test temperature for 24 hours to permit recovery.[5]

5.1.3.3 Stress relaxation

Stress relaxation, when applied to seals, is a measure in the change of the seal interference with time. It is applied in standard material tests by measuring change in stress with time at constant strain on tensile specimens. To apply the measurement directly to seals a technique was devised, Figure 5.10, to measure the compressive force required to maintain an O-ring compressed in a groove.[6]

There have been many arguments on the relative merits of the measurement of stress relaxation and compression set. Stress relaxation has the merit of providing a potentially continuous measurement of the interference force of the seal. However, it requires specialist measuring equipment to be dedicated long term to each individual seal, and is hence expensive. Compression set, if used wisely, employing the measurement method described by Ho[5] will provide spot measurement of considerable value on a large number of samples. It is also important to use stress relaxation in a procedure that will provide all the necessary information. Many tests are performed at one elevated temperature, Figure 5.11(a). This appears to show that sealing stress is adequate for long periods of time, but does not consider what happens if the seal is cooled, as shown in Figure 5.11(b). There are also situations where a seal material will harden considerably with time. This will show as a high retained sealing force, even when cooled down. However, the fact that the seal will have little recovery to be able to seal during cyclic conditions will not be demonstrated.

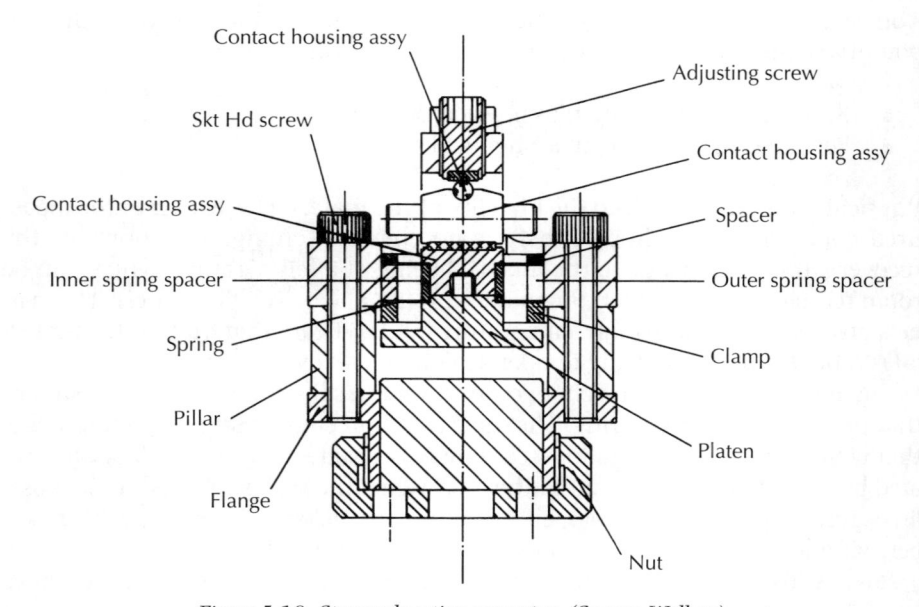

Figure 5.10 Stress relaxation apparatus. (Source: Wallace)

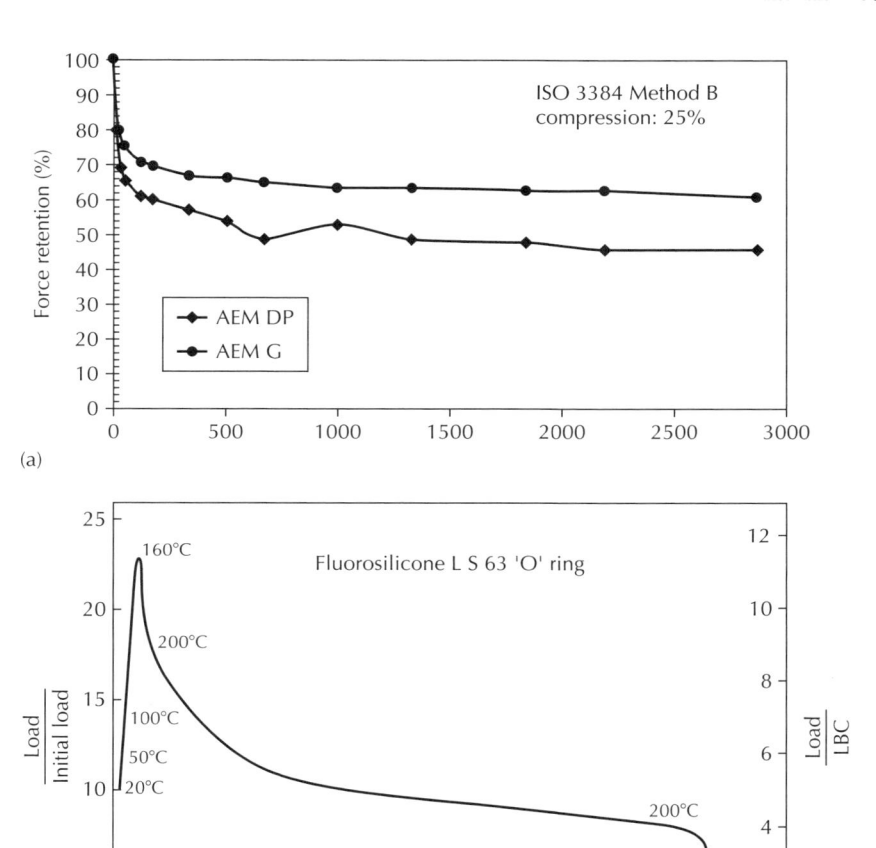

Figure 5.11 Typical stress relaxation curves: (a) constant temperature; (b) cyclic temperature.

This test procedure is widely used in both the aerospace and automotive industries to test potential seal performance. It is, however, necessary to ensure that tests are carried out over the entire temperature cycle as otherwise poor low temperature performance is easily overlooked.

5.1.3.4 Elevated temperature properties

Seals are continuously exposed to chemical action that can degrade their properties and hence sealing performance. As discussed in section 5.1.3, this can include air which contains 20% oxygen. Both liquids and gases will permeate into the material and this effect is accelerated with increasing temperature. As described in section 5.1.3.2 on compression set, it is possible for the curing to continue and increase the number of cross-links. The presence of other chemical species in the elastomer may initiate further chemical activity. This may increase or decrease the rate of cross-linking, attack the cross-links or directly attack the elastomer chain, or polymer backbone. It may also be possible for the filler to suffer chemical attack if it has not been correctly selected. Although very often chemical degradation is associated with the seal becoming harder it is also possible for the material to soften if cross-links or some of the polymer chain has been broken. This is often referred to as chain scission.

The processes that affect the changes in elastomer properties are very temperature dependent. This includes the physical effects such as permeation but more particularly the chemical changes in the material. These changes will follow an Arrhenius expression so the degradation is proportional to a power of the temperature rise. Hence as temperature increases the potential useful life of a seal may be significantly reduced. There have been many attempts to use elevated temperatures to predict long-term seal life,[7, 8, 9] but at present there is no overall consensus on an effective method for reliable prediction.

The actual temperature limits for elastomer materials are extremely dependent on the environment and operating fluids. Those quoted, whether by suppliers or by independent sources, must therefore be used with considerable caution. In some cases the useful life may be limited to a few hundred hours at the quoted temperature. Aggressive constituents in a fluid can significantly reduce the long-term maximum temperature. Conversely it has been found that in an inert environment, where air is excluded, that the ageing of seals can be reduced and they may operate reliably well above the normally recommended maximum.

A further factor to be considered is that the physical properties of elastomer materials are extremely temperature dependent. They are also not related to the chemical stability of the material. Consequently the tensile strength at 100°C may be less than 50% of the value at room temperature. It should be noted that the more chemically resistant materials such as fluoroelastomers will often reduce in tensile strength much more than materials such as nitrile, Figure 5.12. Also, the silicone-based elastomers, which are not noted for their strength at room temperature, are less temperature dependent than other material types. The potential for physical damage by extrusion, abrasion or decompression damage is therefore considerably increased as temperature rises. It should be noted that many fluoroelastomers become extremely weak at temperatures well below the chemical temperature limits of these materials. It is often necessary to consider anti-extrusion rings for these materials at relatively modest pressures if the temperature is approaching 100°C. The physical properties of elastomers at elevated temperatures are not widely published, partly because of the difficulty of accurate measurement, but also because, apart from some notable exceptions,[10] they do not often provide data that the suppliers would wish to advertise.

The majority of elastomers discussed are what may be termed thermo-set elastomers. Once they have been cured then the basic form of the material will remain and any gross changes outside the elastic range will only occur if there is physical damage to the material such as extrusion slicing sections of the material away.

There are materials with elastomeric properties that are classed as thermoplastic elastomers. The two commonly used for seals are polyurethane and polyester. These materials can be compounded to provide an elastomer, but at high temperature and load they will undergo plastic deformation which will not be recoverable. For this reason the temperature limits of these materials should not be exceeded even for short periods of time as irreversible deformation may occur.

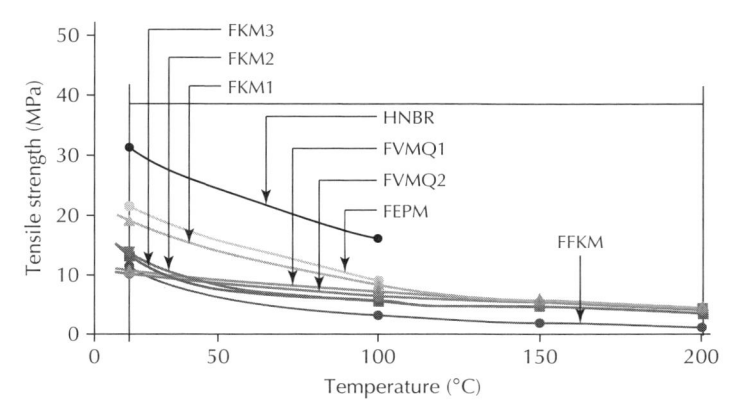

Figure 5.12 Tensile strength of a range of elastomer types across the working temperature range.

It is therefore necessary to ensure that a considerable proportion of testing is carried out at the maximum application temperatures, otherwise the potential performance of the seals may be considerably less than expected.

5.1.3.5 Low temperature properties

The potential low temperature performance is often equally important, but very often overlooked. The chemical degradation of the elastomer is of little concern at low temperature, the chemical processes being extremely slow. Similarly diffusion of gas or liquid will be slow, so the material will not degrade further and any changes to the material at low temperature are generally reversible as temperature increases again. What is of concern is the change in properties and characteristics of the elastomer material. As the temperature is decreased the thermal energy of the material is reduced and the polymer molecules have reduced resilience. The material will gradually become stiffer and resilience will reduce. Eventually as temperature is reduced it will reach the glass transition temperature of the material. The material will then have lost all of the elastomeric resilient properties. Many conventional elastomers such as nitrile may also be just like glass and extremely brittle. Other materials such as fluoropolymers may be in a plastic-like state with very little recovery. While it is possible to use some materials at or even below their glass transition temperature without damage to the seal it can be seen that it will not behave as would be expected of an elastomer. It is therefore very likely to leak as there will be no seal energization. As the thermal expansion coefficient of elastomers is also an order of magnitude higher than metals the interference will already be reduced.

The onset of reduced resilience will vary both with generic elastomer types but also with individual grades. However, for many materials this phenomenon will have at least some effect from room temperature downwards. There are two accepted methods of measuring relative material properties at low temperature. The original method was low temperature retraction. In this test the material is elongated and then cooled to a very low temperature, well below the glass transition. It is then released and the temperature gradually increased. The recovery of the material, as it retracts back to the original length, is then plotted against temperature. This temperature retraction curve will demonstrate the ability of that material to recover at a given temperature. The results of tension retraction tests are quoted as TRn, where n is the percentage retraction at that temperature. The most commonly quoted results are for the TR10, which indicate that the material has retracted 10% from the elongated position. This value is sometimes used to judge the low temperature operation of a material, but at this temperature 90% of the recovery potential of the material is inhibited. Figure 5.13 is an example of the retraction curves for two low temperature materials. Both of these materials were rated for use at −45°C but it can be seen that their behaviour between room temperature and −45°C is very different with one material having very little resilience below −10°C. This can be a severe problem if the application requires the ability to compensate for some relative movement of the metal components at low temperature. It is therefore preferable to consider a curve of TR covering the range TR70 down to TR10 to assess whether the material has adequate recovery over a broad low temperature range.

Many high temperature fluid resistant elastomers have quite poor low temperature flexibility and resilience. Some are really only suitable for cyclic duties above room temperature. The TR10 for a number of FEPM compounds is quoted as +2°C.

An alternative is to use a dynamic stiffness test called dynamic modulus analysis (DMA) which will demonstrate the change in properties of the elastomer as temperature is reduced. Figure 5.14 shows an example of measurements on several materials. A user or manufacturer can select a limiting value of stiffness above which it can be considered that the elastomer material is too stiff to behave in a suitable elastomeric manner. A typical limiting figure used is 30 MPa.

This can also be compared with the loss modulus, tan δ. The peak in tan δ occurs when the material has the lowest recovery. As it continues to become

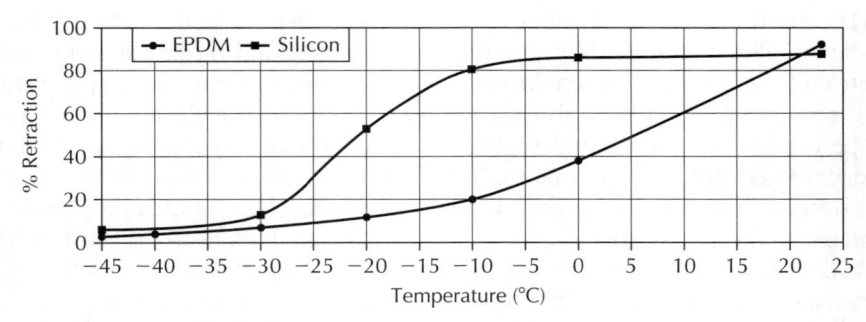

Figure 5.13 The low temperature retraction of two materials specified for a low temperature application. (Source: BHR Group)

stiffer the recovery increases as the material becomes harder and more glass like, Figure 5.16. As the material approaches a low tan δ value again it is becoming very hard and will not perform as an elastomer. If such values are available it would be sensible to take the peak of the tan δ curve as the lowest sensible temperature for dynamic seal operation. In this context dynamic includes relative thermal or pressure induced movement of the sealed components.

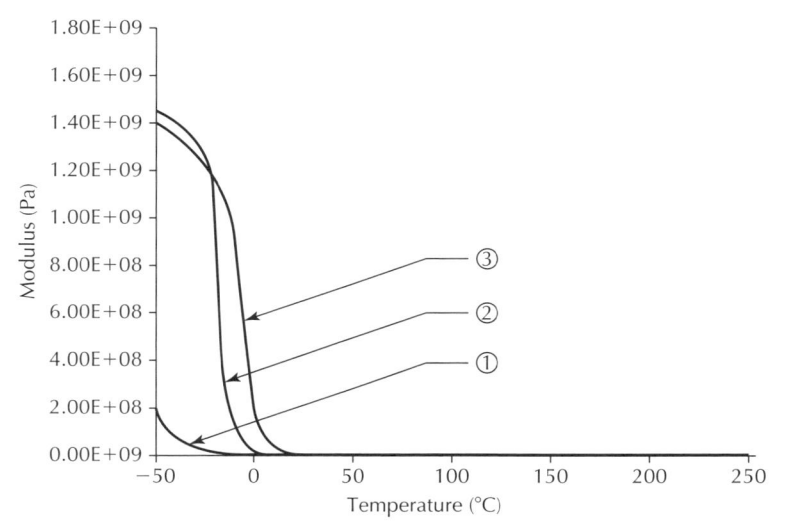

Figure 5.14 Low temperature stiffness measurements of two fluorocarbons and one fluorosilicone material. (Source: Sealing Technology, Dow Corning)

Figure 5.15 High pressure gas permeation measurement apparatus at BHR Group. (Source: BHR Group)

It is important to remember that although seals may leak at low temperature, as mentioned initially in this section, the elastomeric properties will recover as temperature increases and so leakage may only be transient during a low temperature startup. For continuous low temperatures these properties must be considered seriously.

5.1.3.6 Mechanical properties

A number of mechanical properties can be considered but a meaningful property that is regularly measured by manufacturers and is hence readily available is the tensile strength and elongation. These may be reported in a number of different ways but it is normal to quote the stress at a range of strains. These values are known in the elastomer industry as a modulus, and so the 100% modulus is in fact the elastomer stress at 100% strain. Useful numbers to compare to assess the relative properties of materials for a sealing application can be the 50%, 100% and ultimate tensile stress in association with the elongation at break. To achieve a material that will provide a satisfactory seal it is necessary to achieve a compromise between sufficient strength to resist pressure and wear, etc. while also retaining the flexibility required to allow fitting, conform to the groove and react to pressure changes and dynamic movement. To assess these numbers it should be considered that most regular engineering elastomers will provide at least 150% elongation to break, with some providing over 300%. However, some highly filled hard materials may break at below 100%. This may be necessary for some arduous applications but can limit the fitting and application of these materials. Some general guidance on typical values that may be expected for the regular seal elastomers are provided in the BFPA Guidance document.[2] More detailed data on material properties for specific elastomer types is provided in a series of elastomer specifications for the aerospace industry listed in section 7.3 on standards. These standards also provide a useful reference for typical properties of a reliable seal elastomer for other applications where guidance may not be available.

Tear strength is also a useful indicator as it will provide comparative information on the potential resistance to extrusion, damage on assembly and wear. There are a number of standard test methods in the standards.

Abrasion testing may also be used particularly for dynamic seals. Most such testing is based around user specific comparisons of material removal against time.

5.1.3.7 Permeation

Permeability is an important parameter for all sealing applications. It has already been discussed in simple terms as it controls the swell discussed in section 5.3.1.1. Both liquids and gases will permeate into an elastomer but the mechanisms are quite different. However, in both cases there are two important aspects of permeation to consider; the rate of diffusion into the material and the solubility. The diffusion rate will govern how rapidly the fluid permeates into the seal and the solubility will dictate how much fluid remains dissolved in the elastomer.

This can be readily appreciated in the case of liquids. The diffusion rate will affect how rapidly a material will swell and the solubility will govern the eventual extent of the swelling. The values of these two parameters can be obtained by carrying out a detailed swell test to obtain the rate of swell and the saturation value, as demonstrated in Figure 5.8. In most applications this is not particularly necessary although it is important to ensure that sufficient data is obtained to ensure that the saturation value of swell is known as discussed in section 5.3.1.1. Swelling in liquids is generally accelerated by increasing temperature but inhibited by increasing pressure.

The effect of liquid permeability on leakage is not normally a consideration. However, in the case of volatile liquids such as gasoline the liquid that permeates into the seal can evaporate out on the atmospheric side. This will then become a volatile organic (VOC) emission and is a major concern. The permeability of volatile compounds through elastomers is now a major consideration in the selection of elastomers for vehicle fuel systems and will potentially gradually affect their selection in other applications. In such cases the level of emission will be dependent on the pressure and temperature of the application, the solubility of the liquid in the material, the vapour pressure of the liquid and the area of seal exposed to the atmosphere.

Gaseous permeation is very different to that of liquids. It is dependent on the size of gas molecule, the chemistry and also whether the gas may be considered as a pure gas, i.e. well above the critical point. This means that CO_2 behaves very differently to gases such as nitrogen and methane. The overall rate of diffusion will vary considerably for different elastomer types but in general small molecule gases, such as helium and hydrogen, will permeate very rapidly compared with others. In general increasing pressure will accelerate diffusion, but it has been shown that at high pressure compaction of the elastomer material can inhibit diffusion.[11] Swell is not normally considered with gases but it can occur[12] and will affect the material properties. The actual values can be difficult to measure.

To measure permeation of gases some specialist equipment is required but it is an important consideration in some applications. One is high pressure gas sealing, where it can be a contributory factor in decompression damage, discussed in section 6.1.1.8. Another area is high integrity sealing where permeation of gas or vapour through the seal can be crucial. Examples include electronic casings, where permeation of water vapour into the casing can be a problem, and delicate apparatus that is required to be kept under an inert gas blanket. Permeation of gases is governed by the gas laws so the important pressure is the partial pressure of the gas. Hence high pressure nitrogen will not actually prevent permeation inwards of atmospheric oxygen or water vapour, it will only reduce the partial pressure. In such applications a detailed analysis of the seal materials and seal geometry is necessary to limit permeation.[13]

Standard methods of permeation measurement are covered in the general material measurement standards. General data are available from some reference sources, but actual values are very dependent on the compound formulation and processing so in critical applications it is important to make actual measurements. Measurement at high pressure requires the ability to handle high pressure gas at accurately controlled pressure and temperature and delicate instrumentation to measure the permeated gas, Figure 5.15.[14]

(a)

(b)

(c)

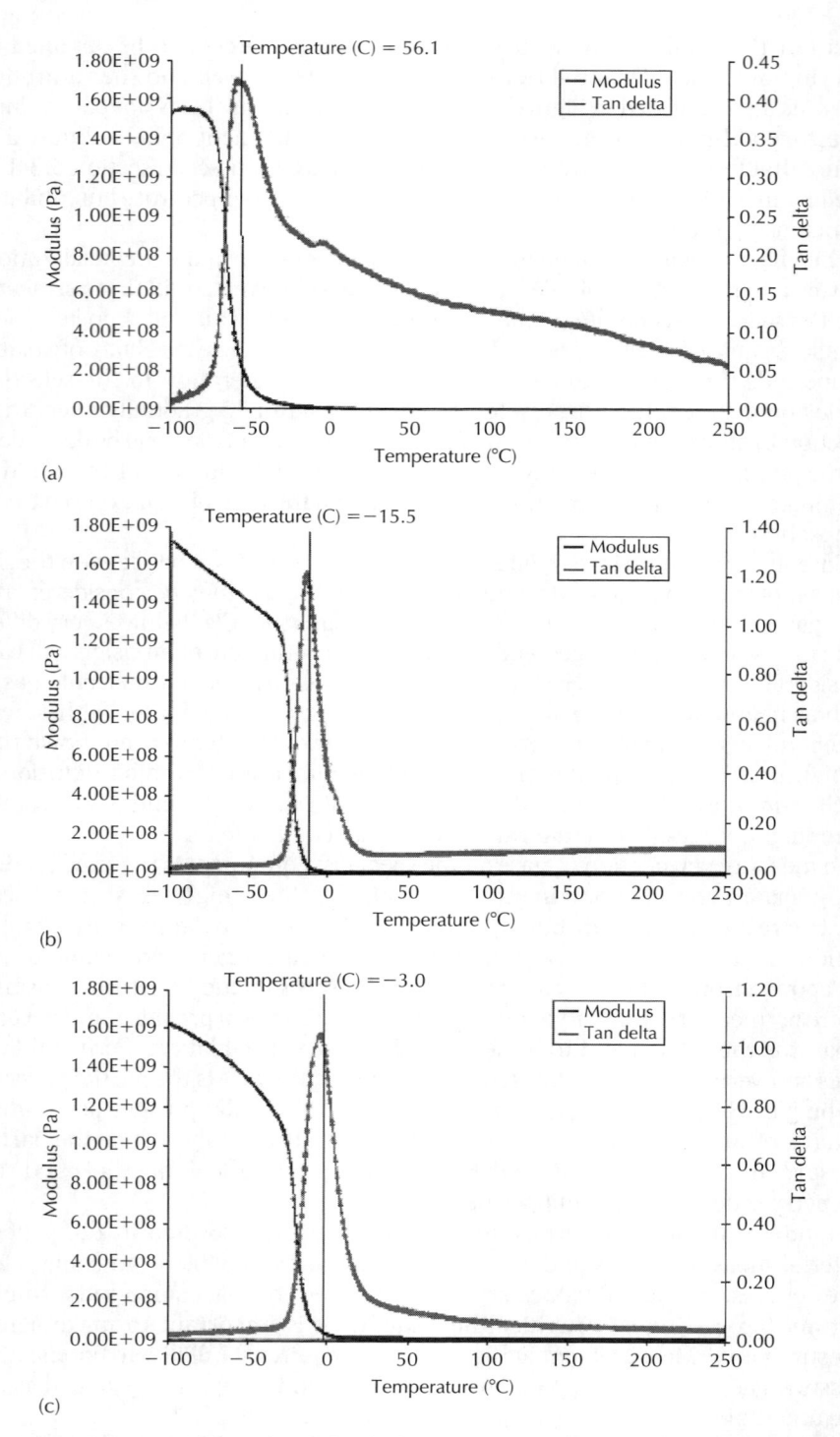

Figure 5.16 Dynamic modulus and tan δ plots for the two FKM materials and FVMQ shown in Figure 5.14. (Source: Sealing Technology and Dow Corning)

Conversion of data measured on flat sheet samples to an actual seal configuration also requires careful analysis of the geometry. The permeation both into and out of the seal will be dependent on the exposed areas to the seal and emission on the downstream side. Software is available to calculate actual gas permeation[14] and provide data on potential material behaviour.

5.1.3.8 Resilience

Resilience in an elastomer provides a measure of the speed of recovery when a stress is released. It therefore gives an indication of the ability of the material to respond to dynamic movements or if for instance a pressure is applied or released. At one extreme, natural rubber has a very high resilience which can be seen in elastic bands and the highly bouncy balls that can be obtained. Often an engineering elastomer can have a much lower resilience and only respond relatively slowly. As previously discussed in section 5.1.3.5 the materials also become much more sluggish as the temperature is reduced.

It is possible to measure resilience by assessing the bounce of a hard ball on the surface of the material. However, the most common method in industry is to use the dynamic modulus analysis (DMA) method discussed in section 5.1.3.5. This provides a graph of the stiffness of the material and also a hysteresis plot, tan δ, which is a measure of the recovery, Figure 5.16. As this value increases the material is reducing in resilience and will not respond so rapidly to changing conditions. These curves are characterized by a peak, after which the material appears to become more resilient again. It will also be seen that this corresponds to the material becoming much stiffer. It has now reached the glass transition point and has become hard. It is resilient, but can only work at very small strains so behaves much more like any other engineering material. They may be liable to either brittle fracture or plastic strain depending on the characteristics of the individual materials. In general traditional elastomers such as nitrile will become brittle and fracture. Highly fluorinated elastomers will behave more as a plastic.

The use of much of this data will be quite subjective and based on previous experience. In general, for a dynamic seal it is necessary to have good resilience, so a low tan δ will be required. As the values begin to increase at lower temperatures this indicates the area where dynamic performance will decrease. This is an especially important consideration where the seal may operate continuously at low temperature such as in refrigeration equipment.

5.1.4 Elastomer types and their applications

There is a very wide range of elastomer types which possess a very wide range of overall properties. This section will summarize the basic types and the typical applications in which they may be used. It is again important to remember that within each elastomer type any one seal manufacturer may have 15 or 20 different grades of material to meet the requirements both of general applications and also those developed to operate in specific environments. The materials are discussed in generally ascending order of temperature range.

Elastomer types are designated by a series of standard abbreviations that are internationally recognized by ISO Standard 1629:1995. The materials of most interest for sealing applications are listed in Table 5.4. The abbreviations in ISO 1629 are aligned with the chemical group of the polymer chain on which the elastomer is based. These are known as the symbol groups. These groups are:

- M Group: Elastomers with a saturated chain of polymethylene.
- O Group: Material having carbon and oxygen in the polymer chain.
- Q Group: Elastomers with silicone and oxygen in the backbone of the polymer.
- R Group: Materials with a diene in the polymer chain.
- T Group: Elastomers having carbon, oxygen and sulphur in the polymer chain. Usually known as polysulphide rubbers.
- U Group: Elastomers with carbon, oxygen and nitrogen in the polymer chain. The familiar materials are the polyurethanes.

Most of the elastomer base polymers are produced by the major chemical manufacturers and these are then formulated and processed into the engineering compound suitable for seals by the seal manufacturer, or professional contract compounding company. Where elastomers may be familiarly known by a trade name, the common trade names have been included.

Table 5.4 Standard elastomer abbreviations according to ISO 1629

Elastomer type	Abbreviation	Symbol group
Butyl	IIR	R
Chloroprene	CR	R
Ethylene acrylic	AEM	M
Ethylene propylene	EPM	M
Ethylene propylene diene	EPDM	M
Fluorocarbon	FKM	M
Fluorosilicone	FMQ, FVMQ (*FSR*)	Q
Nitrile (acrylonitrile butadiene)	NBR	R
Nitrile, carboxylated	XNBR	R
Nitrile, hydrogenated	HNBR	R
Perfluoroelastomer	FFKM	M
Polyacrylate	ACM	M
Polyester elastomer (thermoplastic)	(*TPE*)	
Polyurethane (polyester urethane)	AU	U
Polyurethane (polyether urethane)	EU	U
Polysulphide	OT, EOT	T
Silicone (methyl phenyl)	MFQ	Q
Silicone (methyl phenyl vinyl)	MPVQ	Q
Silicone (methyl vinyl)	MVQ	Q
Tetrafluoroethylene propylene co-polymer	FEPM (*TFE/P*)	M

Note: Abbreviations in italics are commonly used non-standard abbreviations.

5.1.4.1 Butyl (IIR)

Butyl is noted for low permeability to gases and may be used in applications where this is an advantage. It is not noted for resilience and may therefore be used as a low permeability liner.

Temperature range: -40 to $+100°C$.

Resistant to: Hot water and steam, glycol-based brake fluids, polar solvents, polyglycol (HFC) and phosphate ester (HFD-R) fire resistant hydraulic fluids and silicone oils, also ozone and weather resistant.

Not suitable for: Mineral hydrocarbon liquids, oils, greases or fuels.

5.1.4.2 Chloroprene (CR)

Chloroprene was the first synthetic elastomer material, developed by DuPont around 1930, and marketed by that company under the trade name Neoprene. It has good air and ozone resistance, reasonable all-round chemical resistance and good mechanical properties which it retains reasonably over a wide temperature range. The good ozone resistance leads to applications in the civil engineering field and is also found as a seal material in compressed air systems. Also one of the few elastomers with good resistance to refrigerant R12 and was used extensively for this. In many general sealing applications it is outperformed by nitrile materials.

Typical temperature range: -40 to $+100°C$.

Resistant to: Paraffinic mineral oils, silicone oils, water, refrigerants, weather and ozone.

Not suitable for: Aromatic hydrocarbons such as benzene and high octane gasoline, chlorinated hydrocarbons, polar solvents, with limited use in naphthene-based hydrocarbons, and LPG.

5.1.4.3 Nitrile (NBR)

Nitrile compounds are very widely used general purpose sealing materials with the benefit of good mechanical properties, reasonable resilience, good wear properties and resistance to most mineral-based oils and greases. Also suitable for use with water, fire resistant hydraulic fluids (types: HFA, HFB and HFC), many dilute acids, LPG, diesel fuel and other fuel oils. It therefore finds very wide application across many sectors of industry for both static and dynamic seals. The properties vary quite widely depending on the acrylonitrile content, Table 5.1. The grade of material will have to be selected to achieve the correct compromise of properties. It is for instance difficult to achieve both good resistance to fuels and low temperature flexibility. Most general purpose compounds are therefore based on a medium nitrile content which provides a reasonable compromise across the properties required. The low temperature properties are also quite compound specific. The cure system will also impact on the temperature range. Peroxide cured material has a higher temperature range but generally poorer mechanical properties making it unsuitable for dynamic applications or where good extrusion resistance is required.

Typical temperature range: Minimum temperature -20 to $-45°C$ depending on compound. Maximum general temperature $+100°C$, up to $120°C$ for short periods. A major factor for temperature limitations is air ageing and it has been found that in an oxygen-free environment, such as down hole in oil wells, the temperature range can be extended to perhaps 120 or 130°C in some circumstances.

Resistant to: Mineral oils and greases, LPG, fuel oils, water, dilute acids, water and glycol-based hydraulic fluids, vegetable oils.

Not suitable for: Aromatic hydrocarbons (benzene, high octane gasoline, etc.), polar solvents, glycol-based brake fluid, ozone and weather ageing, strong acids, high temperatures.

5.1.4.4 Carboxylated nitrile (XNBR)

The addition of carboxylic acid groups to the NBR polymer backbone produces a polymer matrix with increased strength, which provides a material with improved tensile and tear properties plus good abrasion resistance. The negative effects include reduction in compression set and water resistance, resilience and some low temperature properties. The additional strength and abrasion resistance are valuable for dynamic seals and carboxylated nitrile is widely used for high pressure reciprocating seals.

Temperature range: -20 to $+100°C$.

Resistant to: Mineral oils and greases, LPG, fuel oils, water and glycol-based hydraulic fluids, vegetable oils.

Not suitable for: Aromatic hydrocarbons (benzene, high octane gasoline, etc.), polar solvents, glycol-based brake fluid, ozone and weather ageing, strong acids, high temperatures. Not recommended for hot water.

5.1.4.5 Polyurethane (AU and EU)

Polyurethane is a thermoplastic elastomer and so as discussed in section 5.1.3 the material properties are rather different. It has tensile strength and wear resistance that is typically two or three times higher than comparable elastomers such as nitrile and also good resilience. It has good resistance to mineral oils and together with the excellent mechanical properties this makes it a popular material for hydraulic cylinder seals. The seals will maintain an effective dynamic sealing geometry at high pressure for extended periods. It is widely used for both hydraulic and pneumatic reciprocating seals and it may also be used for O-rings. The high strength means that most polyurethane seals will not require backup rings at pressures below 300 bar. The potential for plastic flow at elevated temperature also means that it is not suitable for a typical rotary seal as the high underlip temperature will cause plastic flow. The elastomer surface texture is also very different to other elastomers. As these are thermoplastic materials they will only have elastomeric type properties within a limited temperature range. At low temperature they will have very limited flexibility while at high temperature they will be subject to plastic flow. For this reason they should not be used at excessive temperature even for limited periods of time.

There are two types of polyurethane, one based on polyester (AU) and the other on polyether (EU). While polyurethane has good resistance to mineral oil it suffers from hydrolysis in hot water and this can limit potential applications. It is not suitable for water-based fluids unless the temperature is well below 50°C. The polyester and polyether urethanes are used either individually or in combination to optimize the seal properties. In general the polyester materials provide the best mineral oil resistance, wear and dynamic properties. Polyether provides better hydrolysis resistance where this is a factor. Combinations of the two materials may be used for applications involving water-based fluids and to provide hydrolysis resistance for applications where it may be required. Some grades are also suitable for synthetic and rapeseed-based biodegradable oils. The limited fluid resistance to many fluids restricts this range of materials primarily to the fluid power industry.

Temperature range: −35 to 80°C (depending on grade of material the range may be extended to −45 and up to 100 or even 110°C). The temperature limit is also very dependent on the sealed fluid. Depending on grade the limits in water-based and environmentally acceptable fluids are between 40 and 60°C.

Resistant to: Mineral oils and greases, synthetic hydrocarbons, water-based and environmentally acceptable hydraulic fluids, plus hydrocarbon fuels, at restricted temperatures, water up to approximately 50°C depending on grade.

Not suitable for: Phosphate ester hydraulic fluids, solvents, alcohol, glycol, brake fluids, hot water, steam, alkalis and acids.

5.1.4.6 Thermoplastic polyester elastomer (TPC-ET)

The most common trade name is Hytrel manufactured by DuPont.

As the name implies this is a further thermoplastic elastomer material. It is particularly suitable for injection moulding and is widely used for very high volume integral seals such as those on bottle caps and also automotive moulded covers. It also provides a sealing function in application to flexible boots such as those on constant velocity joints. It can be made as a very flexible material or relatively hard depending on the grade of material. The flexible grades also demonstrate high resilience at normal working temperatures and this helps to make it a good potential as a seal material. It is used as the sealing element in some coaxial seals in place of PTFE. It offers the benefit in this application of flexibility and resilience which permits easy fitting compared with the care and fitting aids required for PTFE. It is also used occasionally as a backup ring as the resilience and flexibility provide good gap filling and response to pressure changes although the extrusion resistance will be limited compared with harder materials. Both the temperature range and fluid resistance are very dependent on the grade of material. Grades designed for low temperature flexibility have a more restricted fluid resistance and upper temperature limit.

Temperature range: −40 to +120°C for typical sealing material grades. (Limitations for many fluids similar to polyurethanes.)

Resistant to: Mineral oils and greases, hydrocarbon fuels, water-based hydraulic fluids HFA and HFB, environmentally acceptable hydraulic fluids, water, alcohol, glycol, dilute acids and bases.

Not suitable for: Water glycol (HFC) and phosphate ester (HFD) hydraulic fluids, glycol brake fluids, solvents, concentrated chemicals.

5.1.4.7 Polysulphide (OT/EOT)

This material has particularly good resistance to solvents and gasoline as well as air and ozone. However, it has poor mechanical properties and is not suitable for directly manufacturing seals. It is used as a liner for stronger elastomers, such as in fuel hoses, and also as a sealant where it has improved fluid resistance compared with those based on silicone. A major application is as a sealant in the aerospace industry for aircraft fuel tanks.

Temperature range: −50 to +120°C.

Resistant to: Gasoline fuels, petroleum solvents, ketones, ethers, oxygen, ozone.

Not suitable for: Amines, direct manufacture of seals.

5.1.4.8 Hydrogenated nitrile (HNBR)

Hydrogenated nitrile is a development of conventional nitrile polymer in which the polymer chain is subjected to a hydrogenation process which has the effect of improving the resistance of the polymer chain to chemical attack. This provides a material with improved chemical and temperature resistance and also good mechanical properties. A well-developed HNBR will have both a high strength and high elongation to break making it a very useful material. However, it will have a higher cost than standard NBR materials. The combination of a higher temperature capability, better fluid resistance and high strength make this material popular for seal applications in the oil and gas industry, chemical industry, diesel engines and similar high reliability aggressive environments. The good ozone resistance will also make this material useful for high temperature, high pressure compressed air which causes rapid ageing of conventional nitrile. It also finds use in other applications where high strength at elevated temperature is an advantage such as timing belts.

Temperature range: −20 to +150°C. Special compounds can provide flexibility to −40°C usually at the expense of other properties. Intermittent use up to 175°C may be possible in some circumstances.

Resistant to: Mineral oils and greases, LPG, fuel oils, water, dilute acids, water- and glycol-based hydraulic fluids, vegetable oils, ozone and air ageing.

Not suitable for: Polar solvents, glycol-based brake fluid, strong acids.

5.1.4.9 Polyacrylic (ACM)

Trade name.   Hytemp, Zeon Chemicals

This material was developed as an oil resistant elastomer that could be used with automotive engine and transmission lubricants at a much lower cost than fluorocarbons. It has good resistance to mineral oils including those with high additive packages such as transmission lubricants, including automatic transmission fluids (ATF). The high temperature capabilities also give better air and ozone resistance than NBR. The material has seen only limited areas of application as it has relatively low resilience and poor low temperature properties. It is used for some rotary shaft lip seal duties for gearboxes but the main use is for static seals and non-seal applications such as hoses in automotive applications.

Temperature range: −15 to +150°C. (Intermittent use to 175°C.)

Resistant to: Mineral oils including transmission lubricant additives, ATF, air and ozone.

Not suitable for: Hot water, steam, chemicals, low temperature. Limited use for dynamic seals.

5.1.4.10 Ethylene acrylic (AEM)

Trade name   Vamac, DuPont

This material was designed to have improved low temperature flexibility compared with polyacrylic material (ACM). The low temperature properties are quite good but the oil resistance is inferior to materials such as polyacrylic and HNBR. The resilience is also generally inferior to HNBR. The majority of uses are again in static seals for automotive applications. It may for instance be used as the outer static elastomer on a rotary shaft lip seal casing where fluorocarbon is used for the dynamic lip.

Temperature range: −30 to +150°C. (Intermittent use to 175°C.)

Resistant to: Air ageing and ozone, paraffinic lubricating oil, water and engine coolants.

Not suitable for: Transmission oils, ATF, aromatic mineral oils, acids, esters.

5.1.4.11 Ethylene propylene (EPM, EPDM)

The ethylene propylene-based materials are extremely useful elastomers that have a wide temperature range and are usefully resistant to some fluids that cause problems with most other elastomers. The major limitation is the complete lack of resistance to hydrocarbon liquids which means that they are easily swollen and damaged if they come into contact with mineral oil or grease. They therefore have to be handled with some care in the average workshop environment.

Ethylene propylene is produced from a co-polymer of these two polymers. EPDM includes a third monomer, diene, and is the material that provides the best resistance to hot water, steam and phosphate ester hydraulic fluids such as Skydrol. The material is widely used for seals in steam and hot water systems, aircraft hydraulic systems that use Skydrol, other hydraulic systems using phosphate ester and vehicle brake systems that use glycol-based fluid. It can also be used with synthetic ester lubricants that are used for low temperature applications such as refrigerant compressors. The major limitation to performance in high temperature applications is the air ageing resistance. Seals in an inert total steam environment can be used to higher temperatures than those sealing between steam and air.

Temperature range: −45 to +150°C (some grades may be usable up to 180°C in steam provided there is no access for air).

Resistant to: Water, steam, phosphate ester (HFD-R and Skydrol), silicone oil and grease, some polar solvents, some acids and alkalis, ozone and air ageing.

Not suitable for: Mineral hydrocarbon oils, greases and fuels, air ageing and ozone at high temperature (130°C+).

5.1.4.12 Silicone (MVQ, MPQ, MPVQ)

Trade names   Silastic, Dow Corning; Silopren and Silplus, GE Silicones; Elastosil, Wacker

A large group of compounds comprise the silicone elastomers. The most common is MVQ. The main attribute of silicone elastomers is the wide temperature range. They can be used from below −50°C to in excess of 200°C. The major problem is the relatively poor tensile strength and tear resistance which in turn makes them susceptible to wear and also more prone to damage on assembly. Although the physical properties are relatively poor at room temperature compared to other elastomers used for seals they have the benefit that the properties are less affected by temperature. Hence they retain some viable strength at high temperature and remain flexible to very low temperatures. They are generally inert with respect to the human body which makes the material popular for food and medical applications. Silicones are also widely used as sealants, also known as liquid silicone rubbers (LSR). These are either a single part, incorporating the curing agent, or a two part, which is mixed immediately prior to application. The curing process is triggered by atmospheric moisture and occurs at room temperature (hence these compounds are also known as RTV) so some introduction of moisture to the area may be necessary to achieve a sensible cure rate in conditions of low humidity. Major applications of silicone are medical, food, automotive and aerospace. Care is often required due to the limited mechanical properties which can be further reduced by a higher swell than some other elastomers. The poor resistance to steam limits the food applications if high temperature sterilization is required. Silicone has limited use as a dynamic seal but is used for some automotive rotary shaft lip seals because of the superior low temperature flexibility.

Temperature range: −60 to +200°C. (Some grades retain some flexibility to even lower temperatures.)

Resistant to: Mineral oils, brake fluids, water, ozone and air ageing.

Not suitable for: Hot water and steam (120°C+), aromatic hydrocarbons and solvents, acids and alkalis.

5.1.4.13 Fluorocarbon (FKM)

Trade names   Viton, DuPont; Technoflon, Solvay Solexis; Dyneon fluoroelastomer, Dyneon (previously 3M); Dai-el, Daikin

Fluorocarbons are now a complex range of polymers.[15] They were first developed in the 1950s to provide a high temperature oil resistant elastomer. They are now an important class of elastomers and capable of providing a very wide chemical compatibility at temperatures up to 200°C. There are several different fluorocarbon groups which have quite different properties and are often not interchangeable. The basis of the materials is the incorporation of fluorine into the polymer chain. This forms a very strong bond within the chain and prevents other fluids attacking the polymer. The original development was a co-polymer, also known as a dipolymer, of hexafluoropropylene (HFP) and vinylidene fluoride (VDF). These comprise the Viton A and GA polymers from DuPont and the FL21xx grades from Dyneon. These are typically 65% fluorinated and form the basis of many of the general purpose fluoroelastomers. They have good resistance to hydrocarbons and good compression set resistance.

Terpolymers were developed using tetrafluoroethylene (TFE) in addition to the HFP and VDF which increase fluorine content to 68% providing improved fluid resistance but inferior compression set and low temperature performance. These materials are not usually serviceable below about −10°C. Typical materials are Viton B, FL2320 and FL2350 from Dyneon and Technoflon T838K.

What are sometimes known as tetrapolymers are further developed from terpolymers using fluorinated cure site monomers to improve curing and provide more stable cross-linking with increased fluorine content to 69 or 70%. These are, for example, the Viton GF and Dyneon FLS2650 grades. Special polymers may also include fluorinated ethers to provide premium grades with improved low temperature performance, such as Viton GFLT. These grades have a much wider chemical resistance and can be used with methanol and a number of other polar solvents. A problem with the development of these materials has been that increasing the fluorine content has usually provided improved fluid resistance but this has been associated with a reduction in the physical properties so that the materials are more prone to compression set and extrusion problems.

A summary of the relative low temperature, compression set and general fluid compatibility is provided by the comparison of Viton grades in Table 5.5.

Temperature range: −15 to +200°C (−20 to −30°C with GL and GFLT grades).

Resistant to: Mineral oils and greases, some phosphate esters (HFD), silicone oils and grease, chlorinated solvents, air and ozone, fuels. This is very general and applies to most fluorocarbon grades; however, it is necessary to evaluate grades individually for specific fluids. Some guidance is provided in Table 5.6.

Not suitable for: Steam and hot water above 100°C, phosphate esters (Skydrol), polar solvents, fuels containg methanol, gear lubricants with EP additives, engine oils with amine additives, amines, alkalis, organic acids, brake fluids. (Speciality terpolymers overcome many of these limitations.)

Table 5.5 Comparison of general properties of Viton fluoroelastomer grades

Standard and speciality types of Viton elastomer	Principal applications	Polymer composition	Weight (%) fluorine
A, AL, GAL, GLT-S	General purpose sealing: automotive, aerospace fuels and lubricants	VF_2/HFP	~64–67%
B, BL, GBL-S, GBLT-S	Chemical processing, power, utilities. Seals and gaskets	VF_2/HFP/TFE	~68%
F, GF-S, GFLT-S	Oxygenated automotive fuels, concentrated aqueous inorganic acids, water and steam	VF_2/HFP/TFE	~69–70%
TBR-S	Automotive and off-road (high pH requirements)	TFE/Propylene	~60% TBR-S
ETP-S	Oil exploration/production. Special sealing requirements – harsh environments	Ethylene/TFE/PMVE	~66% ETP-S

(Source: DuPont Performance Elastomers)

390 *Seals and Sealing Handbook*

Table 5.6 Fluorocarbon fluid resistance summary

Standard and speciality types	Low temp. flexibility TR-10(°C)	Compression set, O-rings 70 h/ 200°C (%)	Fluid resistance % volume increase after 168 h					Base resistance % loss of elongation at break	
			Toluene at 40°C	Fuel C/ MeOH (85/15) at 23°C	Methanol at 23°C	Concentrated H_2SO_4 at 70°C	33% solution of KOH 336 h/40°C	ASTM Ref. oil 105 500 h/ 150°C
AL	−19	20–25	20–25	35–40	85–95	12–15	−50	−80
A	−17	12–17	20–25	35–40	85–95	10–12	−45	−80
BL, GBL-S	−15	25–40	12–15	20–25	25–35	3–5	−25	−65
B	−13	25–30	12–15	18–23	25–35	8–10	−25	−70
ETP-S	−11	45–50	6–8	8–10	1–2	4–6	0	−10
GFLT-S	−24	35–40	8–12	10–15	3–5	3–5	−30	−40
GBLT-S	−26	35–40	12–15	27–32	25–35	3–5	−40	−50
F	−8	30–45	8–12	5–10	3–5	7–9	−50	−55
GF-S	−8	30–45	8–12	5–10	3–5	7–9	−45	−50
TBR-S	+3	45–50	60–65	80–90	–	3–5	−10	−20

(*Source: DuPont Performance Elastomers*)

5.1.4.14 Fluorosilicone (FVMQ/FMQ)

Trade names   Silastic LS or Silastic FSR, Dow Corning; Silplus, GE Silicones

Fluorosilicone also uses the introduction of fluorine to provide improved chemical resistance. These materials have very similar properties to silicone, but the latest grades have improved mechanical properties.[10] They offer the very wide temperature capabilities with reasonable fluid resistance but not the wide chemical compatibility of the specialist terpolymer fluorocarbons. For automotive and aerospace applications the low temperature performance can be a primary factor in the selection of a fluorosilicone material. A comparison of the relative flexibility of examples of fluorosilicone and fluorocarbon elastomers is shown in Figure 5.16. For automotive applications a range of silicone/fluorosilicone polymers are also used to provide a combination of fluid resistance, moulding and economy.[16] The primary uses are for static seals in hydraulic and fuel systems.

Temperature range: −60 to +220°C.

Resistant to: Mineral oils and greases, silicone oils, synthetic lubricants, ATF, brake fluids, water, alcohol. Gasoline and phosphate esters may be suitable depending on grade, swell can be up to 20%.

Not suitable for: Aromatic solvents, amines, phosphate esters (Skydrol), concentrated acids.

5.1.4.15 Tetrafluoroethylene propylene copolymer (FEPM) (TFE/P)

Trade names   Aflas, Ashahi Glass; Fluoraz; Greene, Tweed; Viton (BRE), DuPont

This material was first introduced by Ashahi Glass in 1975 and this company remains as the primary polymer supplier. The fluorine contributes to the good fluid resistance and temperature capability, but the ethylene base causes higher swell in hydrocarbon liquids than experienced with fluorocarbons. The particular benefits of this material are the resistance to amines, solvents and hot water for which most other fluoroelastomers are unsuitable. It has also proved to have good resistance to mixtures of fluids which has led to widespread use in oilfield applications. With the increasing use of amines and other aggressive additives in automotive lubricants FEPM is also being considered for vehicle applications. A major limitation of these materials has been the poor low temperature performance and low resilience. Although they may not be brittle and actually do not break until quite low temperatures they have very poor elastomeric properties over a considerable range. Many grades are not considered suitable for use below 0°C.

Temperature range: 0 to 200°C.

Resistant to: Water, steam, methanol, amine corrosion inhibitors, brake fluids, ethylene glycol, many acids, mineral-based lubricating and hydraulic oils, some synthetic lubricants, air, ozone.

Not suitable for: Lighter, aromatic, hydrocarbon liquids such as gasolines, solvents such as benzene, ethers and ketones, chlorinated solvents, refrigerants. Performance in hydrocarbons such as kerosene can be marginal.

5.1.4.16 Perfluoroether

Trade name   Sifel, Shin-Etsu Chemical Co.

This material is relatively new but has some potentially useful looking properties as a seal elastomer. It is a perfluoroether polymer backbone combined with silicone for cross-linking. The perfluorinated polymer provides a good fluid resistance and the silicone cross-linking ensures that the material has good flexibility. It is therefore available as a liquid or paste as well as for moulding. The fluid resistance appears to compare well with both FEPM and FKM elastomers and the low temperature flexibility is comparable with fluorosilicone, Figure 5.17. It therefore has

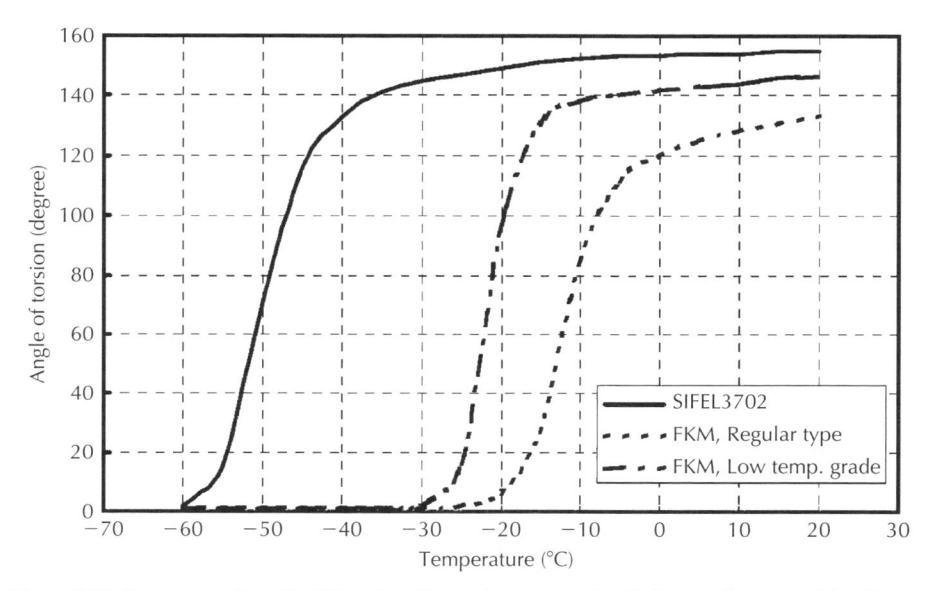

Figure 5.17 Low temperature flexibility of perfluoroether compared with fluorocarbon materials. (Source: Sealing Technology)

the possibility to provide a high degree of fluid resistance to temperatures well below −20°C which has not been previously available from fluoroelastomers. A potential disadvantage is the low mechanical properties at room temperature. However, as with other silicone-based elastomers the mechanical properties are less affected by high temperature than other material classes, Figure 5.18. It is currently used for some automotive applications.

Temperature range: −50 to 200°C.

Resistant to: Hydrocarbon oils, fuels and solvents, methanol and ketones, synthetic oils, acids and other chemicals.

Not suitable for: Ozone environments.

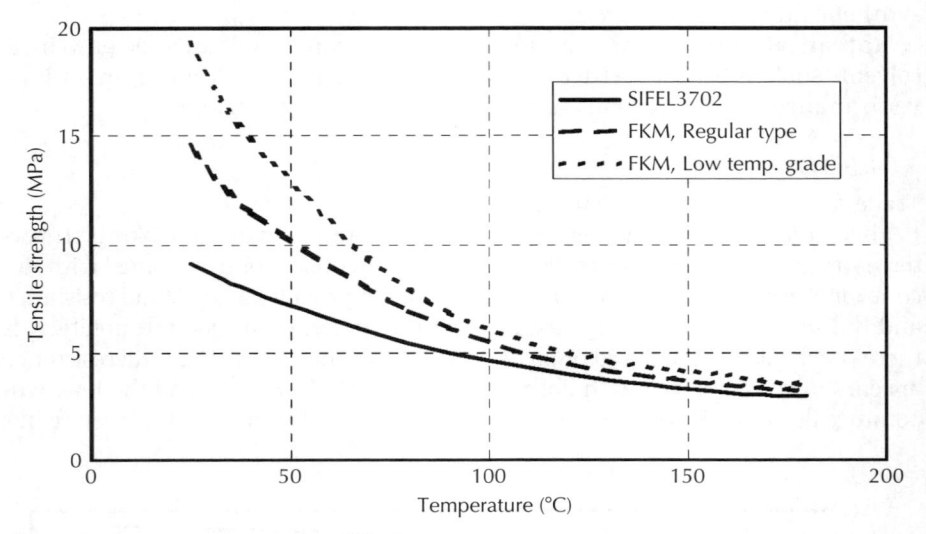

Figure 5.18 Comparison of mechanical properties at room and elevated temperature for perfluoroether and fluorocarbon materials. (Source: Sealing Technology)

5.1.4.17 Perfluoroelastomer (FFKM)

Trade names   Kalrez, DuPont; Chemraz, Greene, Tweed; Simriz, Freudenberg Simrit; Isolast, Trelleborg Sealing Solutions; Parofluor, Parker Hannifin; Perlast, Precision Polymer Engineering; Technoflon, Solvay Solexis

Perfluoroelastomers are a development of fluoroelastomers where the polymer chain is fully fluorinated, with no hydrogen present, which in turn provides a very high degree of chemical resistance to the polymer and high temperature capability. The first materials available were produced by DuPont in the 1970s. Due to the high value and difficulties in manufacture they were originally only available as finished parts. Progressively other producers have introduced perfluoroelastomer production and materials are now available from a number of manufacturers as indicated above by the list of trade names. However, the raw materials for the production of these polymers are very expensive so the materials remain as high cost specialist compounds.

The materials can have a very wide chemical resistance with the notable exceptions of molten and gaseous alkali metals such as sodium and potassium, and fluorinated liquids such as some solvents and refrigerants. There are also some indications of disintegration in steam over 150°C although this appears to be quite compound specific. With such a chemically resistant polymer the filler and cross-links can be the cause of material degradation. Chemical compatibility can therefore be quite compound specific and the recommendations for individual grades should be checked carefully before specification.

The compounding of these polymers with high chemical resistance does impose some disadvantages on the mechanical properties. They tend to have relatively poor resilience and in turn this causes physical relaxation which becomes apparent as high compression set at room temperature. The materials can therefore suffer from leakage if there are wide temperature cycles.[3,17] Some specialist grades have been compounded to minimize this problem.

As the materials can be used to higher temperatures than other elastomers, and suffer considerable loss of mechanical properties at elevated temperatures, thermal expansion can become important. It is generally recommended that the interference of perfluoroelastomer seals is restricted to a maximum of 18% unless it is required to seal at low temperature. Applications involving cycling between low, below 0°C, and high, above 150°C, temperatures can require special design considerations. For reliable sealing at low temperatures, below 20°C, after exposure to high temperature it may be advisable to consider some form of seal energization.

There are many applications exposed to chemicals and solvents where a perfluoroelastomer is the only elastomer that will be suitable. They therefore find wide application in the oil and gas, chemical processing, pharmaceutical and semiconductor processing industry. The cost of the seals can be high, but this is often the cost-effective solution as an overall design that avoids the use of elastomers may be much more costly. They are for instance widely used as the secondary seals of mechanical seals as an elastomer is required to provide the sealing integrity necessary to provide low emission sealing. The PTFE seals that were often used for high temperature and chemical resistance do not provide sufficient integrity under the conditions of temperature and vibration experienced. For semiconductor and pharmaceutical applications specialist high purity grades are manufactured.

Temperature range: −15 to 310°C (very dependent on grade and application).

Resistant to: Hydrocarbon liquids and gases, water and steam, solvents, amines, brake fluids, many acids and alkalis, wide range of chemicals, air, ozone.

Not suitable for: Liquid sodium and potassium, fluorinated solvents and refrigerants, some chlorine compounds, steam over 150°C depending on grade.

Individual grades vary widely depending on the cure system and fillers so specialist selection is necessary for new applications.

5.2 Plastics

Plastic materials are widely used for seals, either directly as the sealing component or as part of the overall assembly such as anti-extrusion rings or bearings.

Elastomer seals may be moulded directly onto plastic components, or different types of plastic may be moulded into one component with an integral seal. With the increasing use of materials such as thermoplastic elastomers the division between plastics and elastomers becomes increasingly blurred. However, to be classed as an elastomer a material must demonstrate the capability to be substantially strained (e.g. up to 100%) with a low stress and exhibit more or less complete recovery. Most plastic materials have an elastic recovery region of less than 10% strain. Although they may be strained to large values this is normally plastic flow which cannot recover.

Plastics materials are also based on long chain polymers, like elastomers. The chemical format of the plastic may be more varied than that of elastomers. They may be a polymer that is sintered but does not have any chemical cross-links. Alternatively they can be a cross-linked material which is below the glass transition point, so it does not behave as an elastomer. The properties of plastics can change quite substantially at transition points. If for instance they pass through the glass transition temperature the stress/strain characteristics may alter significantly in a relatively small temperature variation. The potential range of properties available from plastics is even wider than that demonstrated by elastomers in section 5.1. In this section the plastic materials that are most commonly used for fluid sealing are summarized and the key properties of each discussed in turn.

5.2.1 Benefits of plastics

The key potential attributes that can be obtained from plastics are summarized below. However, some of these benefits are only available from a limited number of plastic types:

- Low friction and dry running potential: Many plastics will permit dry running which is an advantage for dry or marginally lubricated seals. They can also have good boundary lubrication properties which benefit low speed or poorly lubricated applications. Many plastics offer this benefit, within the temperature and fluid limitations of the material.
- Good wear resistance possible: When compounded with appropriate filler many plastics can provide good wear resistance.
- High strength possible: Some plastics offer comparatively high strength and when assessed on a strength to weight basis can often compete with metals, which can make them attractive for structural components. This high strength provides valuable attributes for sealing such as good extrusion resistance.
- Wider temperature range possible than when using elastomers: Plastics are available that can be used at both higher and more particularly lower temperatures than elastomers.
- Wide chemical resistance possible: Some plastics widely used for seals, such as PTFE, have a very wide chemical resistance. As with elastomers this is very dependent on the individual material and chemical compatibility must be checked carefully.

5.2.2 Potential disadvantages of plastics

- They do not provide inherent energization like an elastomer: The higher modulus and low resilience mean that a plastic seal will generally require some form of energization such as an elastomer or a metal spring to provide the initial sealing force. The lack of resilience also inhibits the ability to follow any movements or vibration.
- Low limit of elasticity: Plastic seals cannot be stretched during fitting as is often possible with elastomer seals. This can limit design options and cause fitting problems. It also means that a plastic seal can often only provide a high integrity by plastic flow into the counter-face texture. This is practicable for a static seal but will cause high wear for a dynamic seal.
- Creep: Plastics will undergo plastic flow, or creep. The amount of creep will depend on the plastic and types of filler. It causes loss of sealing stress and in particular makes the seals less able to operate in cyclic situations such as alternate high and low temperature. Creep can also be very temperature dependent. If a temperature rise passes through a material transition point the creep may increase significantly.
- Fillers can inhibit useful sealing properties: Fillers are used to impart strength and resistance to creep and wear. While this is obviously beneficial they will also reduce the elastic limit and elongation to break. In most general purpose plastics these parameters are well optimized but some filled high performance plastics can have a high strength but very low elongation to break, well below 10%. They cannot be treated as a material with the inherent flexibility associated with elastomers and regular plastics.

5.2.3 Mechanical properties of plastics

The mechanical properties of plastic materials vary over a very wide range as they can be considered as spanning almost the entire gap between elastomers and metals. Some materials have a relatively low strength and can have a substantial elongation or other deformation without breakage, although as it will be plastic flow it is usually not recoverable. Other materials can have exceptional strength but be almost brittle in character. The actual properties will be very dependent on the base polymer, fillers and method of processing.

Table 5.7 provides some outline comparative data of sample plastic materials from a range of suppliers. This table is based on data obtained from various supplier data sheets and public reference sources. The properties of individual materials can vary widely within any one group depending on the material grade, manufacturing method, type of filler and proprietary processes used by individual suppliers. This data can only be taken as a general guide and is primarily intended to demonstrate the very wide range of properties that are available from the plastic materials in general use as sealing materials. Where a large range of values is included this may be for a variety of reasons. In some cases, particularly reinforced materials the properties may be very axis dependent. This can affect both mechanical and thermal properties by an order of magnitude.

Table 5.7 Basic mechanical properties of examples of plastic materials regularly used for sealing applications

Material	Service temperatures Min (°C)	Service temperatures Max (°C)	Flex temp (°C) at 1.80 MPa	Tensile (MPa)	Tensile at temp (MPa)	Elongation at break (%)	Flex modulus (MPa)	Thermal exp. coeff ($\times 10^{-6}$)
Polyamide 66	−30	110	60	50		30	2000	100
Polyamide 66/MoS$_2$	−30	100	90	85		25	3200	70
Polyamide 33% glass fibre	−50	140	200	170		2.5	8400	20–100
Acetal	−50	100	95	70		50–70	2600	10–120
Acetal/PTFE	−50	100	110	55		15	3000	90
Acetal (30% glass)	−50	120	160	140		15	6000	30
Polyester	−40	120	80	70		5	2500	81
UHMWPE	−250	80	110–120	37–47		300	900	
PPS		220	120	70–120		1–5%	2500–16 500	15–180
PPS 30% carbon fibre		260	260	150–200		0.5–2%	2000–2500	10–16
PTFE (virgin)	−250	175	50	20–35		300	500	100
PTFE glass filled	−200	250	200	30–80		10–100	1400	100
PTFE polymer reinforced	−200	250		14		180	1000	50
PTFE carbon filled	−250	250		30		20	500	30
PTFE bronze filled	−250	250		14		90	1400	60
PEEK	−70	275	175	90		40	4000	25–75
PEEK carbon and graphite filled	−70	275	320	145		2	9500	8–17
PEK				115		14	4500	
Polyimide (thermoset)		370	not applicable	50–80	45 @ 260	2 to 8	2500–4000	25
Polyimide (thermoplastic)		350	300–360	175	30–40 @ 260	2 to 10	4000	14
Polyimide (carbon filled TP)		290–320	280–330	220	46 @ 200	2	11 000	3–900
Polyamide – imide	−200	250	280	190	65 @ 230	15	5000	30
PAI 30% carbon fibre	−200	250	280	200	100 @ 230	6	16 000	9
Polybenzimidazole	−15	400	420	140		3	6500	13
Polybenzimidazole/PEEK	−15	300	310	95		1	13 500	25

A particular feature of most plastics is the softening as temperature is increased. Where suitable data have been located these are included in Table 5.7. This softening will severely limit the operating parameters at higher temperature. The wear rate and extrusion resistance will decrease significantly well below the maximum temperature.

Further general aspects of the various seal materials that are most commonly found as seal materials are dealt with in section 5.2.4.

5.2.4 Plastic material types and their applications

5.2.4.1 Polyamide (PA)

Polyamide is also known familiarly as Nylon. This was the first synthetic plastic material to be developed, by DuPont, and first appeared in 1938. It is now manufactured by a number of major petrochemical companies.

The most regularly used material for engineering applications is Nylon 66.

The material is tough and abrasion resistant which makes it useful for wear resistance particularly in dry or marginally lubricated conditions. The temperature range over which the material may be used is relatively limited. It may be reinforced by fillers such as glass to provide higher strength and grades filled with MoS_2 are also used to provide improved lubrication for dry running. A major limitation is the high water absorption which leads to geometric instability. The mechanical properties will also change after exposure to air with a high relative humidity. This is a particular problem for the manufacture of seal components. The major applications are for bearings and also backup rings in equipment which is not exposed to temperature extremes or a wide range of fluids.

Temperature range: -30 to $100°C$. (Fillers may extend this range.)
Resistant to: Air, mineral oils and greases.
Not suitable for: Water or high relative humidity, solvents, acids.

5.2.4.2 Acetal (POM)

Trade names   Delrin, DuPont; Ertacetal, Quadrant; Ultraform, BASF

Acetal, or polyacetal, is available in many forms and is a widely used material for bearings and also structural components. It is available as both a co-polymer and homopolymer. It has better stability, lower water absorption and wider temperature capabilities than polyamide and is hence much more widely used. It is available with a very wide range of fillers. Strength may be improved by many types of filler from glass to stainless steel, and lubrication improved by the incorporation of PTFE, MoS_2 or carbon. It can be seen in Table 5.7 that the incorporation of fillers has very significant effects on the mechanical properties. The better stability combined with good resistance to mineral oils make this material a popular choice for backup rings and integral bearing rings for applications such as hydraulic cylinder seals.

Temperature range: $-40°C$ to $100°C$.
Resistant to: Air, mineral oils and fuels.
Not suitable for: Wide range of chemicals.

5.2.4.3 Polyester

Polyester material is used as a thermoset resin to make an effective bearing material that is widely used in hydraulic cylinders and similar applications. A core of woven polyester fibre is impregnated with molten resin which is cooled as it passes through a mould to make a bearing strip. This material has proved to be dimensionally stable and possesses good wear resistance.

In applications such as hydraulic cylinders it will usually outperform metal bearing materials such as phosphor bronze without creating the side effects of wear debris and polishing of the cylinder bore or rod.[18, 19]

Temperature range: −40 to 120°C.

Resistant to: Mineral oils, greases and fuels, fire resistant fluids from water emulsions through to phosphate esters HFA, HFB, HFC and HFD.

Not suitable for: Brake fluids, water above 80°C.

5.2.4.4 Polyethylene (ultrahigh molecular weight) (UHMWPE)

This material is noted for wear resistant properties and particularly resistance to aqueous media. It has been found to give superior performance to PTFE in water applications and is widely used for dynamic plastic seals in this area. Another major application is as the conformable bearing component in artificial joints such as hip joints. A further application where it has been found to provide good service is in cryogenic seals where it is used for liquefied natural gas loading swivels that are typically operating at −150°C.

Temperature range: −250 to 80°C.

Resistant to: Water, aqueous solutions, human implant requirements, hydrocarbons including aromatics.

Not suitable for: Strong acids and alkalis, esters, chlorine and chlorinated hydrocarbons, refrigerants, amines.

5.2.4.5 Polyphenylene sulphide (PPS)

Trade names   Ryton, Chevron Phillips; Techtron, Quadrant; Fortron, Ticona Polymers (Celanese)

PPS is a high temperature, high strength plastic with a useful range of fluid resistance. Although temperature and fluid resistance capabilities are not so comprehensive as PTFE it has a higher strength and provides a useful material as an economic alternative for many applications, especially where the high strength is an advantage. It is used for backup rings and also a V-ring material in oilfield applications and widely used as a high temperature structural plastic for automotive components where the high temperature strength is an advantage. Some 50% of production is for automotive components including brake, coolant and fuel systems and in the power train and transmission. It is used for components such as plastic pistons, valve bodies, and the housing of integral seals and gaskets. As with other fluid resistant plastics and elastomers the fluid resistance can be limited by the filler and so it is important to assess individual grades for actual compatibility.

Temperature range: Up to 220°C (no reliable data for low temperatures).
Resistant to: Water, mineral and synthetic oils, automotive coolants and brake fluids, non-oxidizing aqueous acids, bases and salts, non-oxidizing organics, refrigerants at low temperature.
Not suitable for: Oxidizing chemicals, strong acids, chlorine, chlorinated solvents and refrigerants at elevated temperature.

5.2.4.6 Polytetrafluoroethylene (PTFE)

Other fluoroplastics in the same family may include ETFE, FEP and PFA.

PTFE and derivatives are probably the most widely used plastic class in sealing technology. The material has virtually universal chemical resistance and a very low coefficient of friction with excellent dry running properties. In unfilled (virgin) form it is also relatively soft and so will conform to the surface texture to provide a low leakage seal. It is found in both dynamic and static seals. For dynamic seals it is found in a variety of rotary and reciprocating seal designs and as a key material for packings. As a static seal it is widely used as a gasket material where fluid resistance and also low flange loading are required. Various forms of energized seal such as spring energized U-rings and elastomer filled O-rings are used as static seals where either the temperature or fluid conditions are beyond normal elastomers. They are also particularly useful where intermittent motion is required such as valves and rotary joints.

PTFE itself is found in both unsintered and sintered form. The unsintered material is relatively soft and easily deformed but this makes it useful for packings and low stress gasket materials. Seals will usually be made from sintered PTFE prior to machining. Materials such as ETFE and FEP have been developed to provide materials that can be injection moulded and are more likely to be found as high volume components and as for instance the FEP coating around an O-ring. A further type of the material is expanded (ePTFE). This is manufactured by a process in which microscopic voids are created in the material, the primary supplier being Gore. This provides a pliable and conformable material that is especially suitable for low stress gaskets and packing.

To provide strength, wear resistance and extrusion resistance a range of fillers are used. Glass is very popular as it provides strength without compromising the fluid resistance in many circumstances. The highest chemical resistance is generally required in gaskets and the fillers used for these are discussed in section 2.8.8.1. Glass is less successful as a filler for dynamic seals as it is highly abrasive. A range of fillers are used and a general guide is provided in Table 5.8. However, the type of fillers used and their individual effectiveness varies very widely from one supplier to another. The size and geometry of the filler together with the manufacturing process have a significant effect on the performance of the seal. The material, hardness and surface texture of the counter-face are also important factors in dynamic seal performance. The dry sliding capability is provided by a transfer film of the PTFE to the counter-face so that PTFE runs against itself. The surface texture is important in this process. For lubricated sliding it is necessary for the PTFE and filler matrix to bed-in satisfactorily on the counter-face to provide a long lasting seal with minimal wear.

Table 5.8 Typical applications of PTFE with different fillers

Filler material	Temperature range (°C)	Typical applications	Counter-face
None, virgin PTFE	−200 to 200	Static Low pressure, light duties Easily extrudes at temperature Widest fluid resistance Rotary, reciprocating at very light duties	Steel, chromed steel, cast iron, stainless Nickel alloys
Glass	−200 to 260	Static (very abrasive for dynamic) Good fluid resistance	Steel, chromed steel, cast iron, stainless Hardened if any oscillation of seal in groove
Glass fibre/MoS_2	−200 to 250	Reciprocating Mineral oils, synthetics, phosphate esters, aqueous fluids, chemicals Rotary mineral oils, synthetics, good fluid resistance	Steel, hardened Chromed steel
Graphite	−200 to 230	Reciprocating light duty Aqueous and chemicals Dry gas	Steel, chromed steel, cast iron
Graphite fibre	−200 to 250	Rotary Medium duty, high wear resistance	Steel, hardened
Carbon/graphite	−200 to 260	Rotary, reciprocating Mineral oils, aqueous fluids, steam, dry gas, pneumatics	Steel, chromed steel, stainless
Carbon fibre	−200 to 260	Rotary, reciprocating Mineral oils, aqueous fluids, synthetic fluids and phosphate esters Suitable for soft counter-faces Good extrusion resistance	Steel, chromed steel, cast iron, stainless, aluminium, bronze alloys
Bronze	−150 to 260	Reciprocating Mineral and synthetic oils Phosphate esters High pressures High extrusion resistance	Steel, chromed steel, cast iron
Polymer (usually Ekonol)	−200 to 260	Rotary Mineral, synthetic and aqueous lubricants Dry gas Relatively low abrasion	Steel, chromed steel, cast iron, stainless

The applications for PTFE are found across all the many sectors of industry where sealing products are used. It is very widely used in the process industry where chemical resistance is required for gaskets and packings, in food, pharmaceutical and medical applications for inertness and across a wide range of general industry for dynamic seals where low or consistent friction characteristics are required. It is also finding increasing use as a rotary shaft seal for automotive engines and transmissions as the increasing demands of higher temperatures and lubricants with new additives are beyond the capabilities of conventional elastomers.

Temperature range: −250 to 250°C.
Resistant to: Most media.
Not suitable for: Molten alkali metals, fluorine and other halogens, strong oxidizing agents. (With filled PTFE the fluid resistance will be dictated by the filler material.)

5.2.4.7 Polyetheretherketone (PEEK)

Trade names   Victrex, Victrex; Arlon, Greene, Tweed; Ketron, Quadrant

PEEK compounds are notable for a usable temperature range similar to that of PTFE but they have a much higher strength and also retain their mechanical properties to high temperatures. The flex temperature of filled materials can be well above the continuous use temperature of 260°C, Table 5.7. The chemical resistance, although not as comprehensive as PTFE, is still very wide including good hydrolysis resistance. A wide variety of grades are available with either reinforcing fillers to increase strength or lubricant additives to provide improved tribological properties.

PEEK is extensively used as a backup ring material for arduous conditions such as in the oil and gas industry and high pressure, high temperature chemical process applications. It is also used as a seal material in spring energized U-rings and V-ring sets for high pressure and temperature applications. It can also be used for piston rings, valve plates and similar applications where high temperature stability is an advantage.

Temperature range: −70 to 260°C.
Resistant to: Hydrocarbons including lubricants, fuels and aromatics, phosphate ester fluids, water and steam, refrigerants and halogenated solvents, many inorganic chemicals, ketones, esters, bases and many acids.
Not suitable for: Some acids including hydrochloric, nitric and sulphuric, some phenols and sulphur compounds. Fluid resistance with some media is reduced above 200°C.

5.2.4.8 Polyamide-imide (PAI)

Trade name   Torlon, Solvay

Torlon provides the overall highest performance capability of any melt processable thermoplastic. It has good resistance to elevated temperatures and is capable of performing under severe stress conditions at continuous temperatures up to 260°C. Machined parts provide greater compressive strength and higher impact resistance than most other engineering plastics. It also has a low coefficient of thermal expansion which together with the high creep resistance provides good dimensional stability in dry conditions. Torlon is an amorphous material with a glass transition temperature of 280°C which helps to provide the stable material characteristics over the operating range. The high strength and low elongation to break also mean that it is only suitable for seal designs where limited deflections occur. It can be used as a backup ring material but only if care is taken to ensure that only limited deflection is required. The major area of application for seals is as mechanical seal faces, piston rings and similar applications for low lubricity or weight sensitive applications. It can also be used for thrust bearings and dry or partially lubricated journal bearings. A developing potential application is for rub tolerant labyrinth seals and is also used as the ball in non-return valves.

A particular limitation is the high water uptake, Figure 5.19, which will occur in humid conditions or exposure to hot water. This is sufficient to affect dimensional stability and mechanical properties.

Temperature range: −200 to 260°C.

Resistant to: Hydrocarbon oils, fuels and solvents including aromatics, some acids including sulphuric, chlorinated organics, some alcohols and ethylene glycol.

Not suitable for: Hot water or hot humid conditions, methanol, caustic.

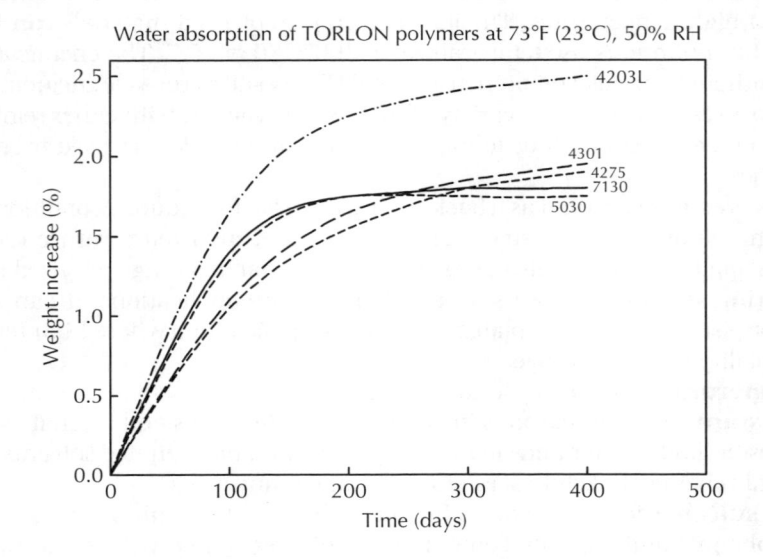

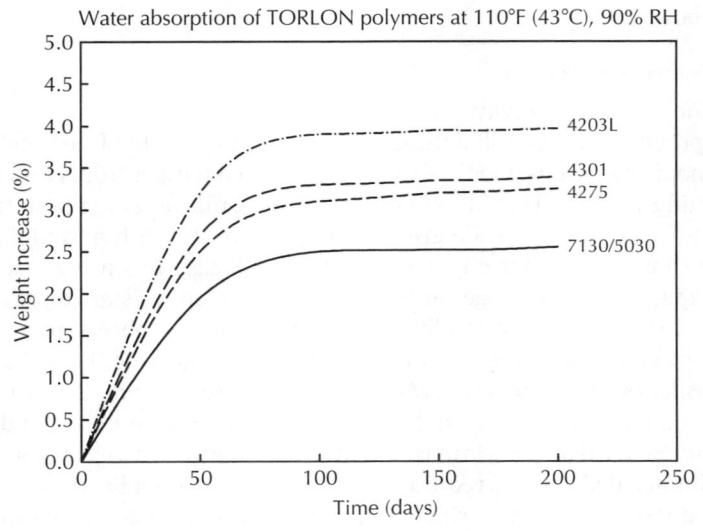

Figure 5.19 Water uptake of polyamide-imide. (Source: Solvay)

5.2.4.9 Polyimide (PI)

Trade names   Vespel, DuPont; Meldin, Solvay; Aurum, Mitsui

Polyimide is available as both a thermoplastic and also a thermoset material. The thermoset offers the benefits of higher strength and thermal stability. It is generally provided as finish machined parts or pressed components for higher volume applications. The thermoplastic material is suitable for injection moulding so is used on high volume components. A variety of grades are available with graphite, carbon or PTFE fillers to provide enhanced lubrication properties. For high temperature, high stress applications the thermoset grades have the benefit that they do not have a softening point. The major applications are again for bearing and thrust rings, mechanical seal faces, piston rings for dry running or in applications such as automatic transmissions. The material is very susceptible to hydrolysis and suffers a loss of properties at low temperatures in water. Above 100°C the material will have less than 50% of the original mechanical properties and cracking may occur after extended exposure.

Temperature range: Up to 280°C. (Excursions to 350 or 400°C possible for thermosets. No reliable data on low temperature properties.)

Resistant to: Mineral oils and fuels, organic solvents, some acids including nitric and dilute hydrochloric and acetic, air and oxygen.

Not suitable for: Hot water and steam, aqueous media above 100°C, bases, some oxidizing agents, chlorinated solvents and refrigerants.

Note: Some grades of Vespel from DuPont are not based on polyimide, check the grade carefully.

5.3 Carbon

Pure carbon graphite is strictly speaking a ceramic material, but in the form that it is used for seals it will usually be impregnated with other materials that are provided to variously improve the sealing, by reducing porosity, improve the wear performance or increase strength. It is widely used, being one of the face pair on the vast majority of mechanical seals, but also found as the most popular material for segmented circumferential seals and for piston rings in dry or marginally lubricated conditions.

It has the advantage of being a conformable material with good dry running properties, wide temperature range and extensive chemical resistance. Compared with the high duty plastics, such as PEEK and polyimide it will have a lower load capacity, but is superior with respect to temperature, chemical compatibility range and recovery from transient upsets, with the ability to re-establish a carbon transfer film.

5.3.1 Manufacture of carbon

Carbon and graphite materials are manufactured from raw materials such as petroleum cokes, pitch cokes, carbon black and graphite.

Important parameters for the manufacturer to control include grain size distribution of the raw material and the mixing rate and temperature profile. The mixed ingredients are formed into what are often known as green shapes by a process such as die moulding, isostatic moulding or extrusion depending on the material type and the manufacturing quantities involved. This may be carried out at a range of pressures and temperatures depending on the individual process.

The green carbon preforms are then baked. Depending on the type of material, dimensions and the required material characteristics, the baking process materials are mixed with a thermoplastic binder at elevated temperature. The binder may be coal tar, petroleum-based pitch or synthetic resin.

The baking process will decompose the binder into volatile components and carbon, the resulting binder coke ensuring the integrity of the finished ring. In this form the material is known as carbon/graphite or hard carbon. It is relatively brittle and has good mechanical strength and hardness. These properties help to make it suitable for applications such as seal rings and bearings.

For many applications the material can then be graphitized, often called electrographite. During this process recrystallization occurs, creating more graphite and more orientation of the structure. The material properties of the graphitized carbon are defined by the structural properties of graphite. These materials generally possess improved sliding properties, low electrical resistance, high thermal conductivity and an improved corrosion resistance. They can be used for applications where improved tribological properties, resistance to chemical attack or temperature cycling are required.

The majority of seal carbons are then impregnated to improve their properties. It is very common to impregnate the seals with synthetic resins to convert the porous structure into a low permeability material. Higher duty materials can be impregnated with metal which can provide an increase in hardness and strength together with improved conductivity. A popular metal impregnant for seal carbons is antimony. It is important for the metal to also have good bearing properties. The temperature limit of the finished carbon component is usually dictated by the manufacturing process and impregnants that may be used. This will be for instance the limit of the resin for resin bonded or impregnated materials and the 'wiping' temperature of a bearing metal. For non-impregnated grades the short-term limit of carbon graphite materials is up to 350°C and up to 500°C for graphitized material. The long-term limit will be lower due to progressive oxygen attack.

5.3.2 Tribology of carbon

The friction and wear of carbon has been the subject of extensive research for a number of applications in addition to mechanical seals, brushes of electrical machines and carbon aircraft brakes. The carbon creates a complex boundary lubrication mechanism. Initial wear of the carbon face creates very fine particles of carbon debris. As very fine particles they are extremely reactive and can link to form a protective carbon particle layer. A similar but much thinner layer may also form on the counter-face.

However, it has been found that the formation of this layer is dependent on the local environment; it will not occur in a vacuum, but happens much more readily in the presence of moisture, air and hydrocarbons.[23] Research into mechanical seal face properties[20, 21] has shown that while a seal is operating this boundary layer film will gradually build up and then break down. This work was initiated following unexplained seal failures that were shown to be due to changes in the supply of carbon.[22] The distinguishing property of a good seal face material is the ability to re-establish the carbon boundary layer. In normal lubricated operation, with design face loading, this buildup and breakdown phenomena will occur very gradually. During upset conditions or transient dry running the breakdown may occur rapidly. The quality of the carbon will control whether it will recover from the upset. Tests have shown[20] that good quality seal carbons will survive dry running and upset conditions and re-establish a stable film. Materials with a less optimum carbon structure or degree of impregnation will wear rapidly during an upset and may not recover to a stable boundary layer film condition.

Research also suggests that the carbon boundary layer formed in this way will break down in the region of 180°C.[24] Hence transient dry running can precipitate an increase in wear due to the high surface temperatures, and high temperature seals will require carbons that can operate at high temperatures such as those of electrographite. Electrographite also has a high thermal conductivity which will help to disperse frictional heat and avoid excessive temperatures during transient conditions. Some typical properties of materials used for seal applications are summarized in Table 5.9.

5.4 Silicon carbide

Silicon carbide has progressed from a position as an exotic high duty mechanical seal face material to become a standard seal face material for many application areas. The combination of good tribological properties with high modulus and thermal conductivity combine to make it an excellent seal face material and a major contributor to the improved reliability of mechanical face seals over the last two decades.

The particular attributes of silicon carbide for a seal material are high abrasion resistance, high elastic modulus and high thermal conductivity combined with a very wide chemical resistance. It also has a low density which is particularly useful where rotating mass is a concern.

The combination of high elastic modulus, approximately 400 GPa, and high thermal conductivity, 100–150 W/mK, are major factors in producing reduced face temperatures and low distortion seal faces which in turn provide stable seal operation. The thermal expansion coefficient is approximately 4×10^{-6} depending on grade and manufacturing process. This compares with typical carbon mating faces that may be in the range 2 to 6×10^{-6} depending on grade.

A number of types of material are available and these will be selected according to the application. Examples of mechanical properties are given in Table 5.10.

5.4.1 Reaction bonded

The earliest use of silicon carbide involved reaction bonded material. The blank shapes of material are formed by mixing the ground silicon carbide particles with

Table 5.9 Typical properties of carbon materials used for dynamic seals

Supplier + grade	Density (gm/cc)	Compressive strength (MPa)	Young's modulus (GPa)	Porosity (Vol. %)	Thermal expansion $\times 10^{-6}$	Thermal conductivity (W/mK)	Temperature limit (°C) [oxidizing service]	Typical applications
Morganite								
CY10 carbon graphite	1.6	138	18	15	3.6	21	250	Piston rings
CY10C carbon graphite resin impregnated	1.8	207	23	2	4	21	n/a	General purpose seal material
MY10K carbon graphite antimony impregnated	2.5	276	33	2	4.7	23	n/a	General purpose seal material
P-658RC carbon graphite resin impregnated	1.8	234	24	3	4.9	9	260	High duty
CTI-1812 graphite resin impregnated	1.86	186	18	1.5	4.3	62	450	Corrosive applications
P-03 graphite	1.82	98	122	8	4.5	71	540	Segmented rings
Schunk								
FH27S carbon graphite	1.55	90	9	15	3.5	10	350	Segmented seals, large diameter
FH82Z(H)2 carbon graphite resin impregnated	1.8	250	24	2	4.7	8	260	General purpose seal material
FH82A carbon graphite antimony impregnated	2.15	350	26	n/a	4.5	9	350	High pressure and high temperature seals
FE45Z(H)2 graphite resin impregnated	1.75	150	12	3	4	50	260	Seals for dry running and aggressive media
FE45Y2 graphite	1.7	100	12	8	3.6	65	500	Dry and wet seals, high temperatures and corrosives
FE45A graphite antimony impregnated	1.5	170	16	1.5	3.5	65	500	High temperature seals

Table 5.10 Typical properties of silicon carbides used for mechanical seal faces

Manufacturer	Morganite	Coors	Morganite	Coors	Morganite	Coors
Material grade	Purebide R	SC-RB (SC-2)	PGS 100	SC-DS (SC-30)	Purebide G	SC-DSG (SC-35)
Material type	Reaction bonded	Reaction bonded	Sintered	Sintered	RB with graphite	Sintered with graphite
Property						
Density (gm/cc)	3.08	3.1	2.9	3.15	2.8	2.8
Flexural strength (MPa)	n/a	462	140	480	n/a	220
Elastic modulus (GPa)	345	393	210	410	140	310
Compressive strength (MPa)	2500	2700	690	3500	550	675
Thermal conductivity (W/mK)	150	125	155	150	155	125
Thermal expansion (10^{-6}/°C)	4.5	4.3	4	4.4	4.1	4.4
Maximum temperature (°C)	1400	1400	540	1750	540	1750

carbon. The preformed material is reacted in a furnace with an inert atmosphere and silicon metal. This creates further silicon carbide that binds the original particles together leaving the free spaces between the particles filled with silicon metal. The typical silicon content of a correctly manufactured material is between 8 and 12%. This does not adversely affect the mechanical and thermal properties and the material will function very well as a seal face material, in fact it is generally considered to be less brittle than the sintered variety. However, it limits the fluid resistance as any chemical that will attack the silicon metal will gradually dissolve it out of the silicon carbide matrix. This will eventually weaken the material and adversely affect the mechanical properties, or lead to porosity. Reaction bonded silicon carbide is therefore not suitable for use with caustic, high pH chemicals and many acids and is not recommended for applications with a pH range outside 4–11.

Reaction bonded material is also considered to have a slightly lower coefficient of friction when run against carbon in certain conditions.

5.4.2 Self-sintered silicon carbide

Self-sintered silicon carbide is also known as direct sintered or alpha sintered. In this process the silicon carbide particles are directly sintered together with a sintering aid in an inert environment at over 2000°C. This produces a material that is almost 100% silicon carbide with no free metal or other component that will adversely affect the fluid resistance. It has for most practical purposes an almost universal chemical resistance and can therefore be used in the majority of mechanical seal applications where silicon carbide is selected.

As the range of sintered materials has become more readily available there is a tendency to standardize on these materials for general purpose seals as this avoids the problem of the free silicon.

5.4.3 Graphite loaded sintered silicon carbide

Self-sintered silicon carbide is also available with a graphite loading. This provides free graphite particles dispersed in the structure of the silicon carbide. This material is marketed as offering the low frictional benefits of graphite within the mechanical strength of silicon carbide. This may well be true in marginal lubrication and startup conditions.

An alternative reason for the effectiveness of this material is that the relatively soft graphite particles will cause depressions in the surface of the carbide and hence promote lubrication.[27]

5.5 Tungsten carbide

Tungsten carbide is cermet, a ceramic crystalline material that is bound together with a metallic binder. In the case of seal rings this is either cobalt or nickel. Tungsten carbide was the first hard, high strength and high conductivity material to be used for seal faces and was quite widely used in the 1960s and 1970s. It has been replaced by silicon carbide for most applications but is still used where particular strength may be required. The lack of corrosion resistance of the metallic binder and the high density are major disadvantages compared with silicon carbide.

The metallic binder makes it more resistant to fracture than silicon carbide, so it may be preferred where shock or high centrifugal forces are involved. It is for instance used as a standard material as the rotary face on some high speed seals where the design does not allow for a containment of the rotating seal ring. However, the significantly higher density can also be a disadvantage in this situation especially for high speed multistage machines such as compressors and many designs include a shroud around the rotary ring so that an alternative such as silicon carbide may be used. For conventional speed pump seals the higher density has also been found to be a benefit in some high runout situations.[25]

Tungsten carbide is likely to be specified where the additional toughness is an advantage. The chemical compatibility is usually limited by the metal binder, and so for higher corrosion resistance a nickel bound material will be used.

5.6 Silicon nitride

Hot pressed silicon nitride may also be used as a seal face for special applications. It offers high toughness and wear resistance with better fracture resistance than most of the alternative carbide materials and may be used where these characteristics are required.

It is used for a number of high load and high temperature applications such as valve seats, ceramic rolling element bearings and cam followers, etc. but does not demonstrate sufficient benefit over silicon carbide to be used other than where the high fracture toughness is required.

5.7 Alumina ceramic

Alumina ceramic seal faces are produced by sintering of aluminium oxide powder, alumina. This material has high hardness, high stiffness and good strength together with a more or less universal chemical resistance. It was widely used as a face material for chemical applications and also water duties. However, a major limitation is a poor resistance to thermal shock which will cause face fracture. This may readily occur if there is any tendency to dry running or a sudden change in temperature.

It is to be found in low duty mechanical seals, such as light duty water pumps where thermal shock is unlikely to occur. The major sealing use currently is as the face pair for the seal plates of quarter-turn water taps.

5.8 Hard/hard mechanical seal face combinations

The vast majority of mechanical seals use carbon as one of the dynamic seal faces. However, in certain circumstances a hard versus hard face combination may be required. Typical examples[28] are:

- Slurries or other products with abrasive particles.
- High viscosity fluid that may shear a carbon face.
- Where crystallization of the fluid may occur causing abrasive particles.
- Products that may polymerize, these can adhere to the carbon and shear particles from the surface.
- The presence of high vibration or shock which may chip a carbon face.

The primary materials selected for hard/hard face combinations are tungsten carbide and silicon carbide. There is a general tribological philosophy that like materials should not be run against each other and for this reason a common selection is to run tungsten carbide against silicon carbide. This provides good service in oil and can also be used in aqueous abrasive duties.

Other combinations are also used:

- Tungsten carbide versus tungsten carbide is used in heavy oils, tars and asphalt duties. It is only used at low pressure and speed combinations and should not be used in water.
- For corrosive duties sintered silicon carbide running against itself is used, and is a regular choice for chemical plant hard/hard face combinations. It does not have any boundary lubrication properties and precautions must be taken to ensure that it does not run dry. It is not recommended if the lubrication conditions are marginal.
- In non-corrosive duties such as hydrocarbon processing reaction bonded silicon carbide versus itself provides good performance.

5.9 Metals

A wide range of metals are used for the manufacture of seals. They can be the seal material itself, as in a metal seal, a major proportion of the seal as in a mechanical seal, or a less obvious part such as the energizing spring of a polymer seal or the metal components of a semi-metallic gasket. In this section an overview is provided of the main metal types that may be encountered.

There are a number of different standards, designations and trade names that are used when referring to metals. The AISI designations for stainless steels are very familiar, but the wide range of more corrosion resistant alloys are often referred to by a trade name that can be very non-specific. Trade names such as Hastelloy and Inconel now embrace a range of alloys, and in some cases it is possible to purchase the same alloy with either of these two trade names. This section has therefore used the Unified Numbering System for Metals and Alloys (UNS)

The UNS number is a systematic scheme in which each metal is designated by a letter followed by five numbers. It is a composition-based system of commercial materials and does not guarantee any performance specifications or exact composition with impurity limits. Other nomenclature systems have been incorporated into the UNS numbering system to minimize confusion. For example, Aluminium 6061 (AA6061) becomes UNS A96061 and AISI 316 becomes UNS S31600.[26]

5.9.1 Stainless steel

Stainless steels are the general purpose materials for a wide range of sealing products from the springs of plastic U section seals, the metallic components of gaskets to the metal components of mechanical seals. These are primarily the austenitic '300' series alloys.

S30400 (AISI 304)
This is the standard material for many gasket and static seal components such as energizing springs. As a high chrome alloy it has good corrosion resistance. It is also ductile which facilitates a number of manufacturing processes. The low carbon content is beneficial for welding and it is widely used for food and medical applications.

S31600 (AISI 316)
This has improved corrosion resistance and better creep strength at high temperatures than S30400. It is the standard material for the majority of mechanical seal manufacturers and for improved performance in other seal types where it can be used with a wide range of chemicals. It is easily welded and can be used for metal bellows and springs. 316 stainless steel or one of the variants described below are the general default materials for most mechanical seal components for general duties specified in ISO 21049.

Neither S30400 nor S31600 is suitable for applications where there is a high chloride content and they are susceptible to pitting corrosion in seawater environments. This material is used almost universally and can be found in many chemical, food and pharmaceutical applications.

S31603 (AISI 316L)
This material is a derivation of S31600 which has superior resistance to intergranular corrosion following welding or stress relieving. It may therefore be selected in preference to other grades for certain manufacturing processes.

S31635 (316Ti)
This is a further derivation of S31600 which is stabilized by the addition of titanium. It can be used to reduce intergranular corrosion and ensure good mechanical properties at room and elevated temperature.

S31803
This is a duplex stainless steel which has up to twice the yield strength of S31600 and much improved corrosion resistance. It is used as an alternative to the 316

series for mechanical seal components where the improved corrosion resistance is required. The higher strength may also be an advantage in some cases.

AISI 321

This does not appear in the UNS series. It is a development of AISI 304 with some Ti content which helps prevent chromium carbide precipitation resulting from welding or elevated temperatures. It also has good resistance to scaling and vibration fatigue. It is therefore a useful material for high temperature process equipment and gas turbines and is used for metal gasket components.

S32750 and S32760

These are super duplex stainless steel with 25% Cr, a high resistance to pitting corrosion and high strength. The high resistance to stress corrosion cracking and crevice corrosion makes it a suitable material for mechanical seals and other metallic seal components in oil exploration, refining, seawater and geothermal applications. The high strength can also facilitate weight savings compared with conventional stainless steels such as the 304 or 316 series materials.

S34700 (AISI 347)

An alternative option for metal gaskets selected for good heat resistance.

S35000 (AISI 350)

A stainless alloy that includes molybdenum, it has good high temperature properties, and is easily welded. This made it popular for welded metal bellows but the lack of high temperature corrosion resistance means that it can only be used where there is a low corrosion risk.

5.9.2 Nickel alloys

A variety of high nickel alloys are used for high corrosion resistance and high temperature capabilities. These are often referred to by the trade name attributable to the original supplier of the alloy. However, as explained above this can be misleading as with the progress of time alternative suppliers become available and the trade name expands to encompass a family of alloys from one manufacturer. A good example is N10276 often referred to as Alloy 276 or Hastelloy. This material is actually available under both the Hastelloy and Inconel trade names. There are at least 14 different Hastelloy alloys and 20 with the Inconel trade name. A specific trade name reference is generally only valid when the alloy is still within patent protection.

N10276 (Alloy 276)

One of the most widely used nickel alloys, and is likely to be referred to as 'Hastelloy'. It is suitable for use in seawater, brine and many acids. It is readily welded and retains good properties in the welding zone. Widely used for the metal parts of mechanical seals in high temperature or corrosive applications. It can be used as a spring material, it is the default material for the springs of multi-spring seal designs within ISO 21049/API 682, and springs for PTFE seals. This also makes it suitable for welded metal bellows and it is the default material for Type B seals in ISO 21049. This alloy is also selected for metal gasket components in high temperature or high corrosion applications.

N07718 (Alloy 718)
This material is also known sometimes as Haynes 718. It is a high nickel, chrome alloy with some iron content that has excellent corrosion resistance and high temperature properties that is readily heat treated to obtain the required characteristics. It is used for high temperature aerospace seals and as the material for welded metal bellows in hot hydrocarbon service. It is the default material for Type C welded metal bellows in ISO 21049. The spring characteristics also make it suitable for other types of seal such as static metal E section seals.

N06600 (Alloy 600)
This material is familiarly known as Inconel 600. It is a nickel-chromium alloy with good oxidation resistance at high temperatures and resistance to chloride-ion, stress-corrosion cracking, corrosion by high purity water, and caustic corrosion. It is used for the manufacture of metal O-rings.

N07750 (Alloy X750)
This is a nickel-chromium alloy similar to N06600 also known as Inconel X750. The addition of aluminium and titanium make it precipitation hardenable. It has good resistance to corrosion and oxidation with high tensile and creep-rupture properties up to 700°C. Its excellent relaxation resistance is useful for high temperature springs and bolts and it is used for the manufacture of high temperature C section metal seals.

N07080 (Nimonic 80A)
This is another high nickel content nickel-chromium alloy that is also capable of precipitation hardening because of some aluminium and titanium content. The alloy has good corrosion and oxidation resistance and high tensile and creep-rupture properties up to 815°C. It is widely used for high temperature applications such as gas turbine components and exhaust valves in internal combustion engines. It is used for high temperature C section seals for aerospace applications.

N04400 (Alloy 400)
This material is also often known familiarly as Monel, but again it is one of several alloys available under that trade name. It is a copper-nickel alloy that has particularly good corrosion resistance against sulphuric and hydrochloric acid, alkalis and seawater. It is also not liable to stress corrosion cracking and is often the material of choice for seawater applications. It is also used for mechanical seal components in hydrofluoric acid.[27]

N05500 (Alloy K-500)
A further material of the Monel family which has higher strength and is used where this may be required in a seal manufactured from N04400, such as set screws and fasteners.

5.9.3 Low thermal expansion alloys

The range of low expansion alloys, of which many are known as Invar, are either high nickel or nickel-chrome steels. They are useful for mechanical seals that have a shrunk fit seal face as the thermal expansion of the metal component can be matched more closely to that of the silicon carbide or carbon seal face. For example, the majority of the stainless steel alloys will have a coefficient of thermal expansion between 10 and $16 \times 10^{-6}/°C$ compared with 4 to $7 \times 10^{-6}/°C$ for the seal faces.

The low expansion alloys will generally have a very low thermal expansion up to approximately 250°C.

K93601 (Carpenter 36)

This material is also known as Invar 36, a registered trademark of Carpenter Technology Corporation.

It has a low coefficient of expansion from cryogenic temperatures to about 260°C. The alloy also retains good strength and toughness at cryogenic temperatures. The thermal expansion coefficient is 1.3×10^{-6} at room temperature.

There are a number of low expansion alloys which due to their specialized applications do not appear with a UNS designation.

Other materials in this series include Carpenter Super Invar 32-5, which has an extremely low thermal expansion coefficient around room temperature, $0.2 \times 10^{-6}/°C$, and Carpenter Thermo-span. Thermo-span is a higher strength material with consistent thermal expansion coefficient close to $8 \times 10^{-6}/°C$ over a wider temperature range and improved corrosion resistance compared with many of the low expansion materials. This can make it particularly suitable for a mechanical seal material.

The primary use of these alloys is for metal bellows mechanical seals for high temperature service. A problem can be the reduced corrosion resistance compared with the high nickel alloys such as N10276 or N07718. In these circumstances the seal manufacturer will have to select a compromise solution optimizing the thermal expansion with corrosion resistance.

5.10 Soft metal overlay

Metal static seals may use a soft metal overlay on the sealing surface. The purpose of this overlay is to flow into the surface texture and improve the integrity of the seal. The major deciding factors for selection of the overlay material are temperature, fluid resistance and cost. Such overlays are primarily used on metal O- or C-rings and similar configurations. However, they may also be used on other metal-to-metal seal designs to improve sealing integrity.

The most common materials are aluminium and silver. Aluminium will have limitations on fluid resistance and temperature but will provide a lower cost. The coating or overlay does not form part of the structure of the seal and is not taken into account in the groove interference calculations. The coating thickness will be in the range 0.010 to 0.08 mm depending on the surface conditions

Table 5.11 Soft metal overlay for metal static seals

Coating	Applications
Aluminium	Economical coating used for non-corrosive gas and vacuum applications, temperature limit 200°C.
Silver	The most common metal coating. It has a wide temperature range and corrosion resistance, and can be used up to 800°C.
Indium	Very soft and can be used for low temperatures only. Also requires a good surface finish.
Lead	Soft, with a maximum temperature limit of 150°C. Also requires good surface finish on the flanges.
Copper	Can be used on similar flanges to aluminium.
Nickel	Wide corrosion resistance and high temperature, it can be used up to 1200°C. A better surface finish than that required for silver may be necessary.
Gold	Very expensive coating. Used where a soft coating with wide corrosion resistance is required.

of the flanges and sealing requirements. The coating will flow when the seal is assembled and not contribute directly to the sealing stress. The most commonly used range of metal overlay materials are shown in Table 5.11.

5.11 References

1. K. Edmond, C. Newland and K.J. Monaghan, 'MODES Final material test data report', BHR Group Report CR 7773, 2005.
2. 'Seal material/fluid compatibility for fluid power applications', BFPA Document P/81, British Fluid Power Association, Bicester, UK, 1998.
3. R.K. Flitney and B.S. Nau, 'Elastomer seals in a steam buffered process valve', Ninth Int. Conf. on Fluid Sealing, BHR Group, 1981.
4. A.F. George, 'Guidance for the selection, assurance, procurement, storage and installation of elastomer seals for nuclear plant applications', British Energy Engineering Specification TGN 005, British Energy, 2005.
5. E. Ho, 'Summary for seal life test procedures', BHR Group, CR 6819, June 1998.
6. M.W. Aston, W. Fletcher and S.H. Morrell, 'Sealing force of rubber seals and its measurement', Fourth Int. Conf. on Fluid Sealing, BHRA, 1969.
7. E. Ho and B.S. Nau, 'Elastomer seal life prediction', 14th Int. Conf. on Fluid Sealing, BHR Group, 1994.
8. Cameron Elastomer Technology Inc., 'Improved life prediction techniques for offshore marine elastomeric sealing elements', Paper OTC006130, Offshore Technology Conference, USA, 1989.
9. J. Duarte and M. Achenbach, 'A finite element methodology to predict age-related mechanical properties and performance changes in elastomeric seals', 12th Int. Sealing Conf., VDMA, 2002.
10. R.K. Flitney, 'Extending the application of fluorosilicone elastomers', *Sealing Technology*, February 2005.
11. R.P. Campion and G.J. Morgan, 'High pressure permeation and diffusion of gases in polymers of different structures', *Journal for Plastics Rubber and Composites: Processing and Applications*, 17, No. 1, 1992.

12. K.J. Monaghan, C. Newlands and E. Ho, 'Specification of elastomeric materials for rapid gas decompression applications', Oilfield Engineering with Polymers 2006, RAPRA, MERL, March 2006.
13. E. Ho, 'Sealing electronic equipment casings', *Sealing Technology*, September 2003.
14. E. Ho, K. Edmond and D. Peacock, 'Effect of temperature and pressure on permeation, ageing and emissions of elastomers', *Sealing Technology*, October 2002.
15. S. Jagels and S. Arrigoni, 'The role of base resistant FKM technology in oilfield seals', Oilfield Engineering with Polymers 2006, RAPRA, MERL, March 2006.
16. 'FSRs in extreme applications', Dow Corning Corp., 2004.
17. 'Kalrez seal design guide', DuPont Performance Elastomers, www.dupontelastomers.com.
18. R.M. Hobson, R.K. Flitney and B.S. Nau, 'Effects of operational factors on reciprocating polymeric seals for fluid power', BHR Group Report CR 6221, 1994.
19. R.K. Flitney and B.S. Nau, 'The effect of surface texture on reciprocating seal performance,' BHR Group Reports CR 3068 and CR 3069, 1989.
20. B.S. Nau, 'Mechanical seal material performance', BHR Group Report CR 3009, 1988.
21. B.S. Nau, 'Mechanical seal face materials', *Proc. I. Mech. E.*, *211*, 165–183, 1997.
22. R.K. Flitney and B.S. Nau, 'A study of factors affecting mechanical seal performance', *Proc. I. Mech. E.*, 1986, Vol. 200, No. 107.
23. W.T. Clarke and J.K. Lancaster, 'Breakdown and surface fatigue of carbon during repeated sliding', *Wear*, 6, 1963.
24. J.K. Lancaster, 'Transition in the friction and wear of carbons and graphite's sliding against themselves', *ASLE Trans.*, 18, 1975.
25. R.K. Flitney and B.S. Nau, 'The effects of misalignment on mechanical seal performance', 15th Int. Conf. on Fluid Sealing, BHR Group, 1997.
26. www.matweb.com.
27. M. Heubner, 'Material selection for mechanical seals', Proc. 22nd Int. Pump Users Symposium, Texas A&M, 2005.
28. ISO 21049/API 682, 2004 'Pumps – Shaft sealing systems for centrifugal and rotary pumps'.

CHAPTER 6

Failure guide

6.1 Introduction

This chapter is designed to provide a first step to the solution of seal failures. Seal failure analysis should not be undertaken casually. A simple look at the seal in isolation is unlikely to provide a reliable diagnosis. This chapter should be complemented by both the relevant technical chapter earlier in the book and also the references for the seal type being investigated, as they will potentially provide further in-depth information. It is important to have as much evidence as possible for the investigation. Not only the seal but the equipment hardware and samples of fluid are also necessary together with good information on the history of the equipment.

6.2 Static seal failure guide

This section is divided into a section for seals, as covered in sections 2.2 to 2.7 and a further section for bolted joints and gaskets as discussed in section 2.8. The section on seals is further subdivided to cover elastomer, plastic and metal seals.

6.2.1 Elastomer seals

The majority of the description in this section, and also other literature,[6] relates to O-rings, but the general principles will apply to most types of elastomer seal, including the static areas of dynamic seals.

6.2.1.1 Surface cracking of an otherwise unused or good condition seal, Figure 6.1

Ageing of the material has been caused most probably by exposure to UV radiation or excessive ozone. The storage and handling of the seals should be reviewed. If the application is the cause then an alternative material grade with improved resistance to UV and ozone will be required. Review storage in line with recommendations in relevant standards and guides, section 7.3.

418 *Seals and Sealing Handbook*

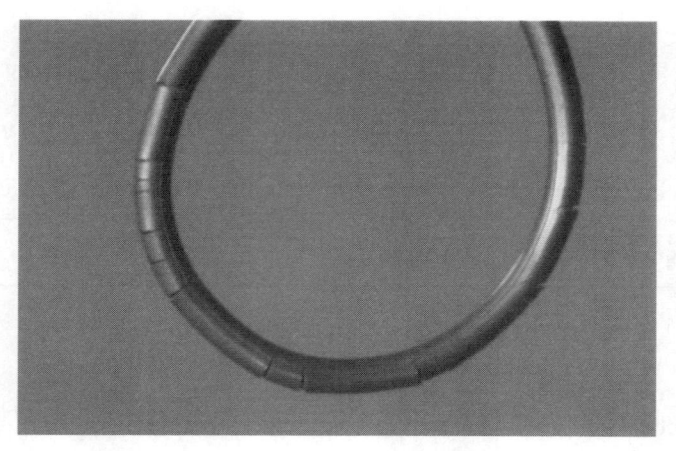

Figure 6.1 Surface cracking due to ozone attack. (Source: Freudenberg Simrit)

6.2.1.2 Flash on atmospheric side of seal

Extrusion damage has been caused to the seal due to excess pressure for the extrusion gap, Figure 6.2(a). With cyclic pressure it is also possible to have a nibbled appearance, Figure 6.2(b). This is caused when the extrusion gap opens under pressure allowing elastomer into the gap. When pressure is reduced the cylinder wall can contract cutting off a portion of the seal.

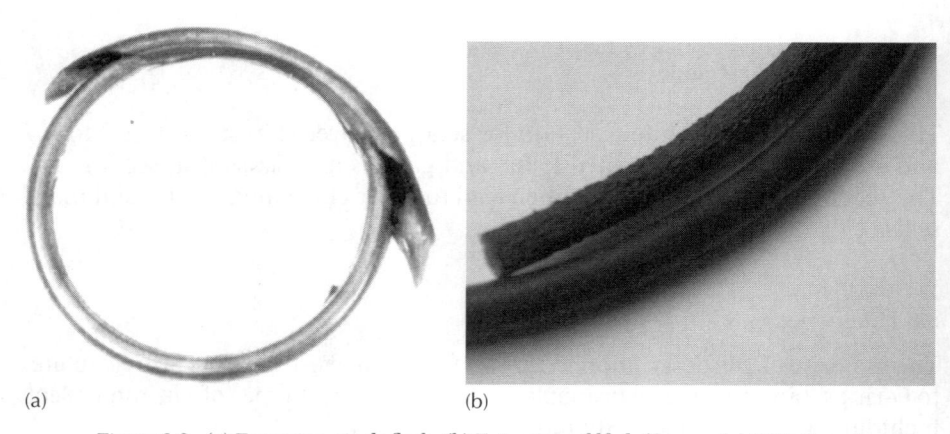

(a)   (b)

Figure 6.2 (a) Extrusion, single flash. (b) Extrusion nibbled. (Source: (a) BHR Group)

Solutions can include:

- Reduce the extrusion gap with closer tolerance metalwork. This is liable to involve higher cost machining and may also be ineffective if metal movement or expansion is occurring due to pressure or temperature.
- Use a harder seal material. This may not always be possible due to other considerations.

- Find an alternative more extrusion resistant grade of material. Many highly fluorinated grades of material have poor mechanical properties and extrude relatively easily, section 5.1.4. If it is possible to use an alternative grade with improved physical properties this can reduce extrusion.
- Backup rings will generally provide a satisfactory solution. The selection of material will depend on pressure, temperature and fluid resistance required. Backup rings require additional axial space in the groove and so it is necessary to extend the groove length by the axial width of the backup ring, section 2.2.1.
- An alternative seal design may be substituted for the original. A number of potential options are discussed in sections 2.3.3 to 2.3.6. Space and application requirements, such as single or double acting, will influence the selection of either a backup ring or alternative seal design.
- Alternative seal material such as plastic may be considered. This would also need to include an assessment of sealing criteria such as leakage retention which may be inferior if a material other than elastomer is used.
- Redesign of equipment to reduce potential for extrusion. Moving the seal position from, for instance, an axial sealing configuration to a flange seal where extrusion gap can be eliminated.

6.2.1.3 Excessive swell of seal

Excessive swell may be evident from a variety of symptoms. The free seal dimensions when removed will be larger than when fitted. There may be extrusion damage to both low and high pressure sides of the seal. Other components may be damaged, it is possible for swell to bend or break other components, Figure 6.3. There may be difficulty disassembling the equipment.

An alternative elastomer suitable for the sealed fluid is required. This problem is often caused because the sealed product has not been fully specified or assessed. Figure 6.3 was caused by a failure to include within the seal specification a small percentage of solvent that was present in the sealed fluid.

Figure 6.3 Failure of a mechanical seal because O-ring swell has broken the stationary seal face. (Source: BHR Group)

6.2.1.4 Leakage from new

This can be attributed to either design or fitting problems:

- If there is no damage to the seal then there is probably insufficient seal interference allowing leakage even with a new seal. A review of all dimensions and tolerances is required, including tolerance stackup with maximum potential eccentricity. If all dimensions are satisfactory check seals are of the correct specification. If a high temperature seal material is fitted, such as some FEPM or FFKM materials the properties may be such that they will not seal at temperatures below 15–20°C. It may be necessary to warm the equipment before starting.
- Damage to seals caused by faulty fitting or lack of correct provision of tapers, etc. Ensure that all potentially damaging edges have been considered including cross drillings, etc.
- Poor fitting or where assembly may be difficult such as a large size or where a long press fit into place is required may cause spiralling of an O-ring, Figure 6.4. This may require a review of procedures, special design to facilitate fitting or use of an alternative seal geometry that will not roll in the groove on assembly. This type of failure is a particular problem with O-rings used for reciprocating applications, section 6.3.

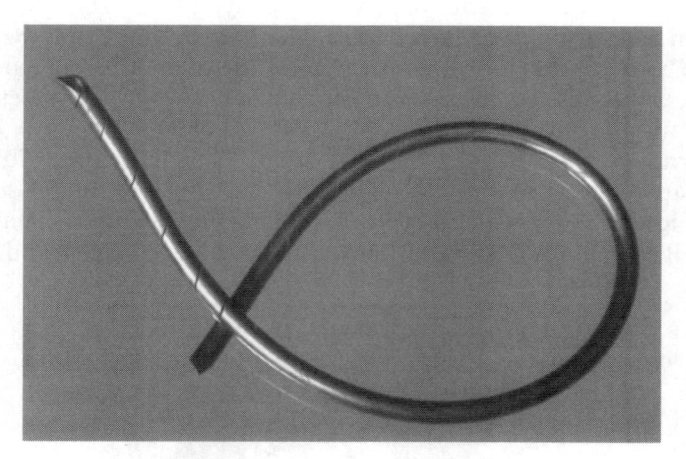

Figure 6.4 An O-ring that has spiralled in the groove. (Source: Freudenberg Simrit)

6.2.1.5 Hardening and compression set

An example is shown in Figure 6.5. Hardening is a common cause of ageing of elastomer materials. Heat and chemical attack cause a continuation of the formation of cross-links and the material retains the deformed shape. The seal may continue to operate satisfactorily provided there is no change of conditions or relative metal movement. Solutions to age hardening are either a more temperature resistant elastomer or some method of reducing the temperature of the seal.

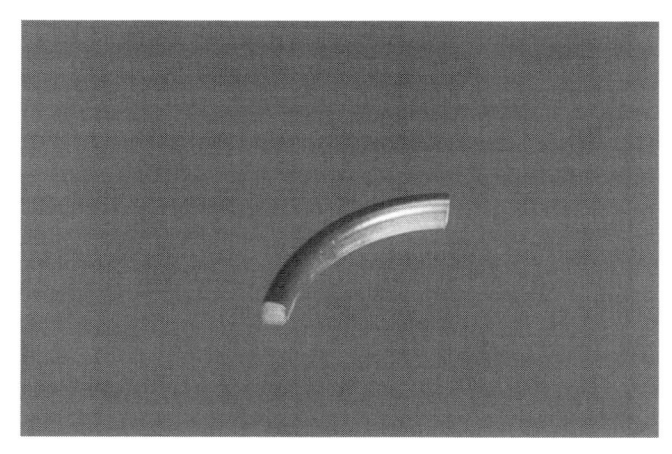

Figure 6.5 O-ring with high compression set and hardening. (Source: Freudenberg Simrit)

6.2.1.6 Softening and compression set

This is more common with fluid resistant elastomers such as fluoroelastomers. The fluid may react with the cross-links, directly break the polymer chain or the filler, leading to a loss of elastomeric properties. The material will become soft and probably be lacking in resilience. A change of material grade is required, potentially with a different cure system or fillers.

6.2.1.7 Compression set with no apparent change of properties

Many high temperature elastomers have poor recovery at room temperature. They may therefore have a high compression set if released at room temperature, Figure 6.6. Such seals may continue to seal satisfactorily at elevated

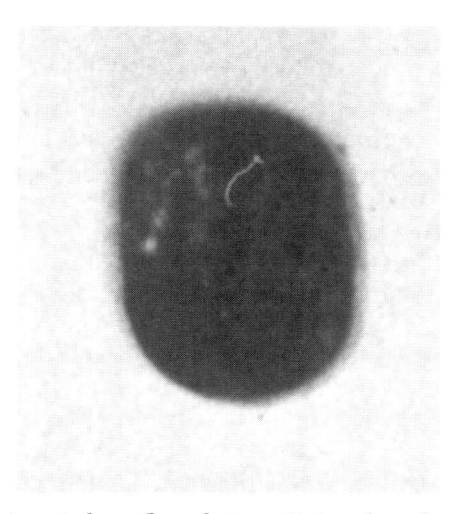

Figure 6.6 High compression set of a perfluoroelastomer O-ring when released at room temperature. It may continue to function at elevated temperature. (Source: BHR Group)

temperature and can be reactivated by warming the equipment before bringing into service. This is a variant of the problem described in section 6.2.1.4.

6.2.1.8 Fractures or blisters, gas service

Decompression of high pressure gas can cause severe damage to elastomers as the gas expands within the material. This is known as rapid gas decompression damage, also sometimes referred to as explosive decompression although no explosion takes place. Evidence may be either internal or external splits in the material or evidence of blisters on the seal surface, Figure 6.7. Specialist materials

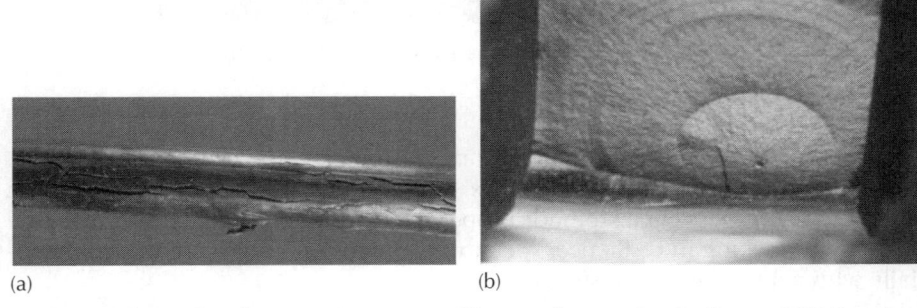

(a) (b)

Figure 6.7 Rapid gas decompression can cause blisters or fracture of seals. (Source: BHR Group)

are produced that are more resistant to decompression damage, they are typically very hard elastomers to provide the strength necessary to withstand the internal stress of expanding gas. Careful design of the seal groove can also benefit by creating a restraint to prevent overexpansion of the seal during decompression. The design criteria required to reduce this problem include:

- Select a proven decompression resistant elastomer from a reputable supplier.
- Use the smallest practicable seal cross-section to permit shorter degassing time and reduced internal stress.
- In conjunction with the seal supplier use the highest feasible groove fill to limit seal expansion during decompression, a figure of 85% nominal has been found to be effective in many cases. The groove fill must be analysed carefully as it is possible to overfill the groove causing damage on assembly, or create excessive seal strain on installation causing damage which can be easily confused with decompression fractures.

Detailed information on high pressure gas sealing with elastomers can be found in a guide from the UK Health and Safety Executive[1] as well as both manufacturers[2] and research organizations.[3]

6.2.1.9 Shrinkage

Shrinkage of elastomers may occur in conjunction with other failure modes described, such as sections 6.2.1.5, 6.2.1.6 or 6.2.1.7 or in isolation. Constituents may be dissolved out of a material by the sealed liquid. This is most common with low temperature formulations used in automotive, aerospace and refrigeration applications. Shrinkage can be up to 10 or 15% so can be sufficient to cause a loss of seal interference. The problem can be very specific both to elastomer grades and oil formulations. Sometimes oil formulations are adjusted to account for shrinkage by causing a complementary swell. A test for shrinkage requires a longer than standard soak test, so will not be obvious from a typical 72 hour or one week test, Figure 6.8.

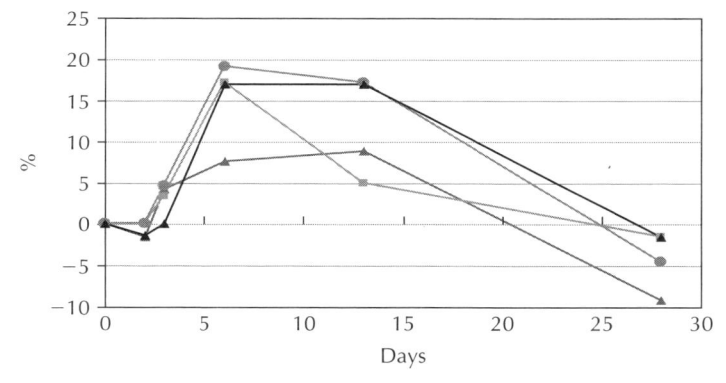

Figure 6.8 A graph of seal material that initially swells and then shrinks in a hydraulic oil. (Source: BHR Group)

6.2.2 Plastic seals

6.2.2.1 Leakage from new

This can be attributed to either design or fitting problems:

- If there is no damage to the seal then there is probably insufficient seal interference allowing leakage even with a new seal. A review of all dimensions and tolerances is required, including tolerance stackup with maximum potential eccentricity. If all dimensions are satisfactory check seals are of the correct specification.
- Damage to seals caused by faulty fitting or lack of correct provision of tapers, etc. Ensure that all potentially damaging edges have been considered including cross drillings, etc. Plastic seals are particularly susceptible to damage on assembly. Care must be taken with chamfers and if it is necessary to stretch the seals. Resizing may be necessary if seals have been stretched during fitting; consult the supplier for advice. It is also possible to dent the surface if insufficient care is taken handling the seals. These may recover with time but performance cannot be guaranteed.

6.2.2.2 Extrusion flash on atmospheric side of seal

Extrusion damage has been caused to the seal due to excess pressure for the extrusion gap, Figure 6.9. Solutions can include:

- Reduce the extrusion gap with closer tolerance metalwork. This is liable to involve higher cost machining. It may also be ineffective if metal movement or expansion is occurring due to pressure or temperature.
- Use a more extrusion resistant seal material, probably by use of either an increased proportion, or an alternative material, filler.
- Backup rings of a harder plastic may be used.
- Redesign of equipment to reduce potential for extrusion. Moving a seal position from for instance an axial sealing configuration to a flange seal where extrusion gap can be eliminated.
- Extrusion of plastics can be very temperature dependent. A change of material to one with a higher softening point may be effective. Alternatively even a small reduction in temperature at the seal can sometimes be very effective if close to a critical softening temperature.

Figure 6.9 Extrusion of a plastic seal. (Source: VDMA)

6.2.2.3 Extrusion flash on pressure side of the seal

This is very unlikely to be caused by swell with most plastic materials used for seals. It is almost certainly due to reverse pressure on the seal. Assess for problems such as pressure buildup between a pair of seals. This can occur even with static seals if care is not taken with the design and assembly of the seal arrangement. It is a particular problem if fluid is trapped between seals and temperature is then increased.

6.2.2.4 Wear of seal/fretting damage to counter-face

Small relative movements of the seal are occurring. This may be due to pressure fluctuations, vibration or similar causes. An example is secondary seals on mechanical seals, but other applications where vibration or relative component movement occurs will present a similar problem. The solution may be a combination of more wear resistant shaft component and a seal material that is less likely to cause fretting damage. Glass and other abrasive fillers should be avoided.

6.2.2.5 Static leakage or cold start leakage after a period of use

Flow of the plastic material has reduced the interference of the seal when cold. Options include:

- Change grade of plastic to one with less plastic flow at operating temperature. This may involve a less conformable plastic that does not offer equivalent sealing integrity at low temperature.
- Consider alternative designs that provide improved energization of the plastic. This could involve a different design of metal spring, or in the case of elastomer energized seals a change to metal or use of a higher grade of elastomer with less compression set.

6.2.3 Metal seals

6.2.3.1 Leakage from new

This can be attributed to either design or fitting problems:

- If there is no damage to the seal then there is probably insufficient seal interference allowing leakage even with a new seal. A review of all dimensions and tolerances is required, including tolerance stackup with maximum potential eccentricity. If all dimensions are satisfactory check seals are of the correct specification.
- Review all surfaces especially the counter-face for any scratches or marks that cross the sealing zone. Metal-to-metal seals require very good metal finishes to perform effectively.
- A high integrity gas seal can be difficult to achieve with many metal seal designs. A soft metal overlay can be used to flow into the metal surface texture and reduce leakage.

6.2.3.2 Leakage after a number of pressure or temperature cycles

Creep of the metal may have reduced the seal interference. Review the metal properties with respect to the application conditions.

6.2.3.3 Fretting damage to seal contact area

Vibration or rapid pressure fluctuations are causing small relative movements of the seal. It will be necessary to review the application and the housing design to eliminate potential movement.

6.2.3.4 Corrosion in the seal contact zone

The material specifications will require review, particularly taking into account any dissimilar metal combinations between seal and housings. Ensure that all fluid constituents have been included in the specification. In extreme situations even small particles of contaminant in a high grade alloy can lead to isolated crevice corrosion problems.[4]

6.2.3.5 Transient leakage in service

The seal design may be insufficiently flexible to provide sealing across all application conditions of temperature, pressure and component movement and vibration. A seal design with improved capability to seal under a variety of conditions and with relative movement of the sealed surfaces may be required; section 2.5 reviews a range of metal seal sections.

6.2.4 Gaskets

This section should be read in conjunction with the section on gaskets, 2.8, and also Appendix 1, sections 5.5 and 5.6 to ensure a thorough appreciation of the gasket failure. Independent work has shown that up to 80% of flange leakages are attributable to factors other than the gasket, Appendix 1. These other factors include improper installation, flange damage, loose bolts and flange misalignment. Complementary work to this has shown that careful attention to the specification of gaskets and correct application of assembly procedures can provide a very high reliability.[5]

It is necessary to assess both the type of damage to the gasket and also when it occurs. Early failures that exhibit localized damage to the gasket are very likely to be due to inadequate or uneven bolting, flange damage or misalignment. Damage or disintegration around the entire circumference are more indicative of the wrong type of gasket, such as excessively thick or inadequate material, or potentially overloaded and crushed gasket.

6.2.4.1 Leakage when initially pressurized
- Bolting incorrect, insufficient compression of gasket.
- Flanges misaligned giving uneven compression of gasket.
- Damaged flange providing leak path.
- Wrong type of gasket, insufficient pressure capability, review thickness and material.

6.2.4.2 Leakage during first temperature cycle
- Bolting incorrect, insufficient compression of gasket.
- Wrong type of gasket, insufficient pressure/temperature capability, review thickness and material.

These problems may be compounded by:

- Flanges misaligned giving uneven compression of gasket.
- Damaged flange providing leak path.

6.2.4.3 Premature failure after few hundred hours or limited number of thermal/pressure cycles

- Bolting incorrect, insufficient compression of gasket.
- Creep or ageing of gasket, review load compression and sealability data for gasket material.
- Creep of bolts, check material specification for creep at elevated temperature.
- Buckling of spiral wound gasket, see section 2.8.8.2. Fit an inner support ring or convert to a kammprofile.

These problems may be compounded by:

- Flanges misaligned giving uneven compression of gasket.
- Damaged flange providing leak path.

6.2.4.4 Leakage after an extended period, but inadequate service life

Failure after a long period of service is potentially due to either one or a combination of factors. Potential contributory factors include:

- Degradation of gasket material with time, see section 6.2.4.5 for further potential indicators.
- Long-term relaxation of gasket material, load compression data available from premium suppliers should assist with selection of an alternative with improved characteristics.
- Long-term creep of bolt material.
- Change of system conditions, fluid content, thermal/pressure cycles, vibration, external environment (e.g. removal of insulation, other causes of external heating/cooling).
- Individual event: Pressure surge, temperature cycle, external event causing temperature shock or vibration.
- Incorrect initial bolting may still be a factor due to less than ideal compression of the gasket.
- Misalignment can accelerate the effect of all the above factors and may be a contributory cause.

6.2.4.5 Indicators to type of gasket degradation

- Potential creep, gasket material has flowed outside original inside and/outside dimensions. Plastic flow, particularly with PTFE-based materials. Localized creep indicates uneven bolt loading. The pressure, temperature and flange loading need to be considered with a review of the gasket material.
- Entire gasket has hardened/disintegrated on removal. Degradation of binder and/or fibre has left aged material unable to provide an effective seal. Review both fibre type and binder material and their proportions. A high proportion of binder phase is unsuitable for all but low temperature applications.

- Localized damage such as blow out in one area. This is probably caused by local flange damage or uneven bolt loading.
- Degradation appears to progress from inside diameter of gasket. This is an indication of chemical attack of material by the sealed fluid.
- Degradation is most pronounced at the outside diameter of the gasket. This indicates that the external environment is the primary factor causing degradation. A prime example is the oxidation of graphite gaskets at high temperature, Figure 6.10. Similar but potentially less obvious problems could occur by for instance oxidation of binder phase if a fibrous gasket is used at elevated temperature.
- Localized degradation or corrosion on the atmospheric side of a gasket. A section of the exterior of the flange may be exposed to some environmental condition that creates a corrosive environment, such as salt spray or rain. Depending on operating conditions, gasket and flange materials this may cause additional degradation or potentially crevice corrosion around the gasket. Various protective shield devices are available to clad flange joints for protection from the environment.

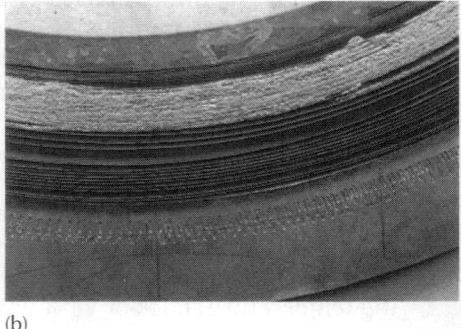

(a)   (b)

Figure 6.10 Oxidation of the atmospheric side of a graphite gasket. (Source: Flexitallic)

6.3 Rotary seal failure

6.3.1 Lip seals

Radial shaft seal leakage can be subdivided into two groups: the static leakage, which can occur at the static seal press fit on the outside diameter or at the sealing lip, and the dynamic leakage, which only occurs at the sealing lip.

A detailed analysis of prematurely malfunctioning radial shaft seals that had operated for less than 100 hours or a running performance of less than 10 000 km, by Freudenberg Simrit,[8] has shown that approximately:

- 30% of breakdowns are caused by incorrect shaft manufacture.
- 30% by incorrect assembly.
- 10% by a faulty seal.

- 15% by other causes such as lubricant incompatibility/excessively high temperatures/vibration/dirt, etc.
- 15% by apparent leakage/short-term leakage. (This may be short-term unexplained leakage, leakage from another source or excess application of lubricant on assembly.)

This emphasizes the importance of the correct procedures and training for the design, handling and assembly of seals. It is important to note that there is rarely only one cause of leakage. It is generally the interaction of several factors. In order to thoroughly analyse the problem it is necessary to have as a minimum both the seal and shaft available. The best results will be obtained by witnessing removal of the seals from the equipment.

A number of the following symptoms may be experienced by both elastomer and plastic lip seals although the actual effect may vary depending on the seal material. Plastic seals will not be expected to show symptoms in sections 6.2.1.4 and 6.2.1.5. Two useful documents to assist with the examination of seals and fault diagnosis are references 7 and 8.

6.3.1.1 No apparent damage to seal and no debris

- Insufficient seal interference or no spring.
- High viscosity oil causing excessive oil film.
- Machine lead on shaft is generating leakage flow, Figure 6.11.
- Excess application of grease on assembly may create the impression of leakage.

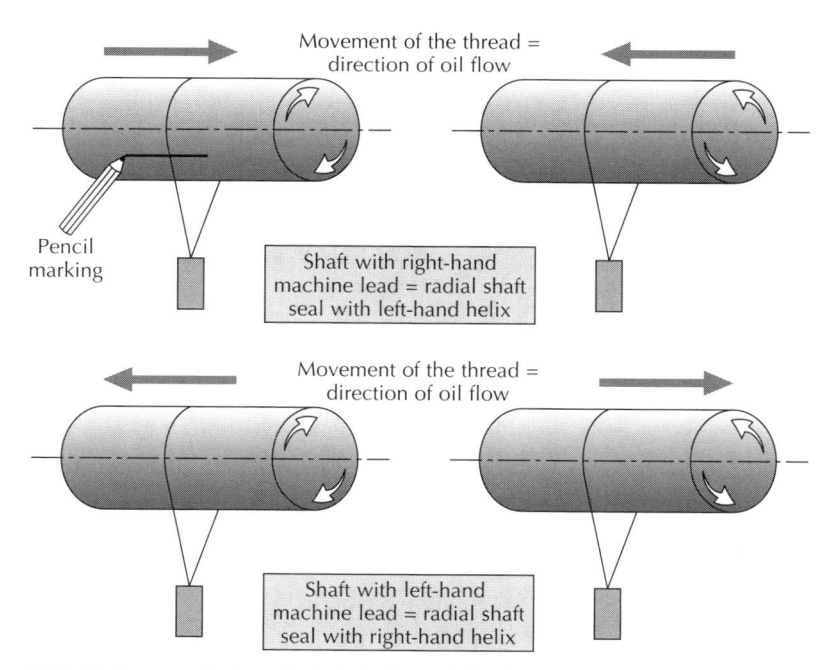

Figure 6.11 Shaft texture lead, method of checking, if detailed texture measurement methods are not available. (Source: Freudenberg Simrit)

6.3.1.2 Mechanical damage to seal lip

Seals are easily damaged on assembly. Common examples can include:

- Sharp edges on shafts or housing chamfers, Figure 6.12.
- Assembly via grooves, gear teeth or splines, Figures 6.13 and 6.14.
- Inadequate assembly tools.
- Damage to shafts by inadequate handling or protection, Figure 6.15.
- Plastic seals, such as PTFE, are more susceptible to damage on assembly than elastomers and if care is not taken the incidence of problems with these seals will increase.

Figure 6.12 Damage to sealing lip by use of incorrect assembly tools. (Source: Freudenberg Simrit)

Figure 6.13 Damage to the sealing edge caused by sharp-edged grooves. (Source: Freudenberg Simrit)

6.3.1.3 Carbon deposits on and around seal lip

This indicates that the oil has overheated in the seal contact and broken down to leave a carbon residue. A mild case may cause light deposits usually just to the atmospheric side of the lip, Figure 6.16. A more extreme case will cause heavy

Figure 6.14 Damage to the sealing edge caused by blind assembly using a spline shaft. (Source: Freudenberg Simrit)

Figure 6.15 Damage to the shaft surface caused by incorrect handling. (Source: Freudenberg Simrit)

Figure 6.16 Carbon deposits close to the atmospheric side of a seal lip, in this case in the helix pumping vanes. (Source: Freudenberg Simrit)

432 *Seals and Sealing Handbook*

Figure 6.17 Heavy deposits of carbon from the oil, around the sealing lip. (Source: Freudenberg Simrit)

deposits, Figure 6.17, that may directly interfere with the seal lip. The solution may involve improving heat dissipation of the shaft, reducing seal friction, or changing lubricant to a grade capable of higher temperatures.

6.3.1.4 Axial cracking of the seal lip

This may be a few large cracks, Figure 6.18, or a larger number of fine cracks, Figure 6.19. They may only be evident on very careful inspection of the seal lip. The seal lip has run at excessive temperature beyond the thermal limit of the material. This may be a combination of underlip temperature and ageing caused by the effect of the oil on the seal material. Cracks will often be evident in conjunction with carbon deposits, providing double evidence of excess temperature. Plastic seals may exhibit signs of plastic flow and extrusion if overheating has occurred.

Options available to remedy this include specifying a material with higher temperature capability, improving heat dissipation or reducing friction.

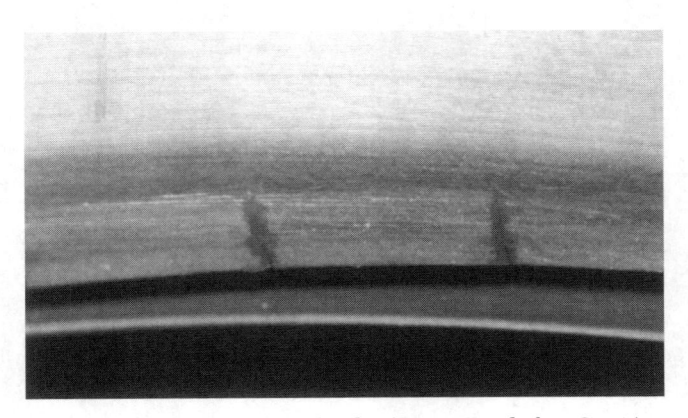

Figure 6.18 Cracks in the sealing lip. (Source: Freudenberg Simrit)

Figure 6.19 A series of fine cracks in the sealing lip in combination with carbon deposits on and around the sealing lip. (Source: Freudenberg Simrit)

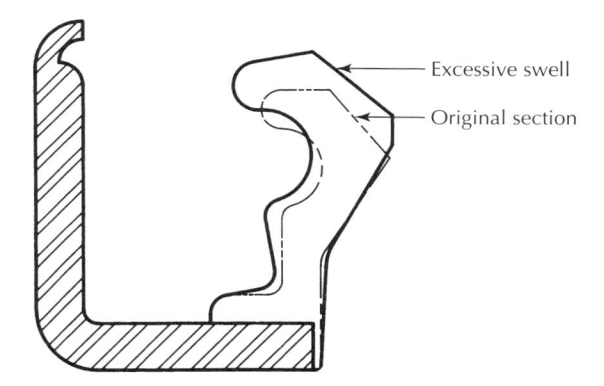

Figure 6.20 A seal can become distorted due to swell.

6.3.1.5 Seal/fluid compatibility problems

- Seal is swollen and distorted, Figure 6.20. Excessive swelling of the seal by the sealed fluid, an alternative seal material or a change of lubricant is required.
- Blisters on the seal surface, close to the seal lip, Figure 6.21, can be caused by chemical interaction of the sealed fluid and elastomer.

Many compatibility and also some wear problems can be caused by additives in the lubricant. These can become very chemically active at the increased temperature under the sealing lip. Some additives may also create solid deposits which can also cause wear. Test data carried out on a base or standard test oil will not be valid for a lubricant containing typical commercial lubricant additives.

Note that similar effects on the atmospheric side of the seal may occur if it is exposed to aggressive cleaning fluids or solvents, either during equipment manufacture, or as part of an industrial process.

434   *Seals and Sealing Handbook*

Figure 6.21   Blisters on the seal surface close to the lip caused by chemical attack to the elastomer.

6.3.1.6 Seal lip wear and debris in sealing area without signs of overheating

- Debris between seal lip and wiper or other exclusion lips indicates inadequate excluder design for the duty.
- Metallic debris in seal area, Figure 6.22. Potential causes are inadequate cleaning and assembly procedures or perhaps high metal wear due to misalignment.
- Rust and debris in seal area, particularly between seal and dust lip, Figure 6.23, are most probably caused by incorrect storage conditions allowing contamination of the seals.

6.3.1.7 Heavy wear and grooving

- Excessively wide wear track and mild grooving. Potentially either partial dry running or the seal is running under some pressure.

Figure 6.22   Metallic debris around the seal lip due to inadequate cleaning of components before assembly. (Source: Freudenberg Simrit)

Figure 6.23 Rust and dirt particles in the seal lip area caused by inappropriate storage facilities. (Source: Freudenberg Simrit)

- Heavy seal wear and wear to the shaft surface. This is probably caused by fine abrasive particles in the wear track, either from assembly or due to ingress of fine contaminant.
- Heavy wear and distorted seal. This can be caused by using the seal at excessive pressure, Figure 6.24. Plastic seals are generally more pressure resistant and in similar situations are more likely to show signs of plastic flow.
- Debris may become embedded in some types of plastic seals and cause excessive shaft wear.

Figure 6.24 Seal distorted by use at excessive pressure, in this example 2 bar. Original profile is on the right. (Source: BHR Group)

6.3.1.8 Circumferential splits

- Circumferential splits around the inside diameter of the metal casing at the end of the flex section, Figure 6.25. This is typically caused by fluctuating pressure in the oil system causing a fatigue fracture. Solutions are to modify the sealed system to prevent fluctuations or use a design of seal with increased pressure capability.

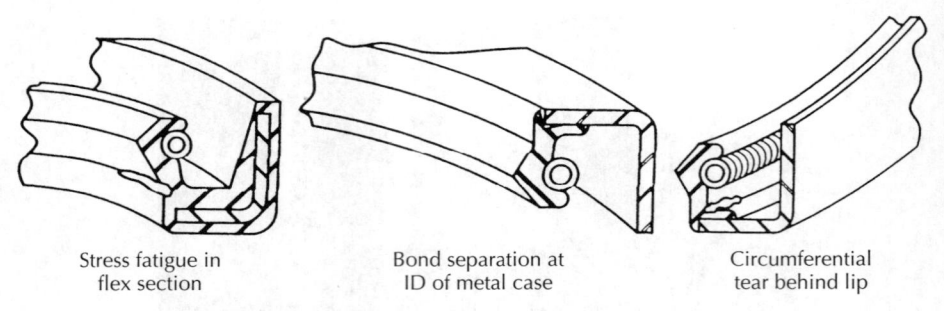

Stress fatigue in flex section   Bond separation at ID of metal case   Circumferential tear behind lip

Figure 6.25 Circumferential splits in a seal caused by fluctuating pressure.

- Plastic seals are generally more pressure resistant and in similar situations are more likely to show signs of plastic flow.

6.3.2 Other elastomer and plastic seal designs

- The alternative designs of elastomer and plastic seals described in section 3.3 will be susceptible to similar failure modes to those described for lip seals in section 6.3.1.

6.3.2.1 Mechanical damage to seal

- Seals will be equally liable to damage on assembly and the usual care will be required. Again this is particularly the case with PTFE and other plastic seals.

6.3.2.2 Signs of overheating

- As seals are often used at higher pressures the problems of overheating in the sealing zone are particularly relevant. Failure modes similar to items 6.2.1.3, 6.2.1.4 and 6.2.1.7 described for lip seals in section 6.3.1.4 are extremely relevant to seals at higher pressure. Plastic seals will show evidence of discolouration due to overheating and potentially plastic flow.

6.3.2.3 Circumferential splits at the heel

- Spring energized plastic U-rings are prone to failures similar to that described in section 6.3.1.8. Circumferential splits at the base of the U section will occur if a seal is used at a pressure for which it is unsuitable or if it overheats. Circumferential splits due to excessive stress on the plastic in the heel of the seal can cause failure.

6.3.2.4 Wear and grooving of the shaft

- With plastic seals this is a high probability unless a suitably hardened shaft is used. It is necessary to use a very hard shaft, or potentially change to a suitable filler material, see sections 3.3.4 and 5.2 including Table 5.8.

- Entrainment of any fine abrasive debris will cause heavy shaft wear with either elastomer or plastic seals. It can be a particular problem with plastic seals.

6.3.2.5 Squealing/stick slip

- Insufficient lubrication of elastomer seals at low speed and high pressure may cause friction, squealing and stick slip problems. A change of seal design or the use of a material with boundary lubrication additives may help.

6.3.3 Mechanical seals

Deciding the cause of failure of a mechanical seal, as opposed to viewing the obvious symptoms, can involve a complex investigation. It can be very misleading to take the direct obvious symptom without a thorough knowledge of the seal, machine and system plus the way that it has been operated. Mechanical seal failure analysis should therefore involve personnel with appropriate experience of all these relevant areas.

There are two important primary factors to consider before commencing an investigation:

- A large proportion of mechanical seal failures can be attributed to operational factors that are not directly a fault with the seal.[9,10] Investigations have shown that many failures can be attributed to maloperation of machinery such as allowing pumps to run dry, pumping against a closed valve and operation away from the design point which can cause cavitation and excessive vibration, especially on larger pumps.[11] Figure 6.26 is

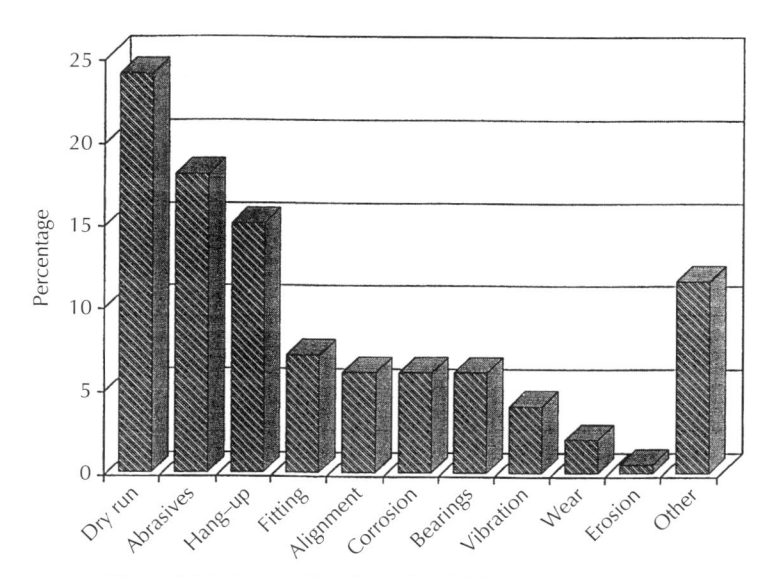

Figure 6.26 Causes of mechanical seal failure. (Source: Ref. 9)

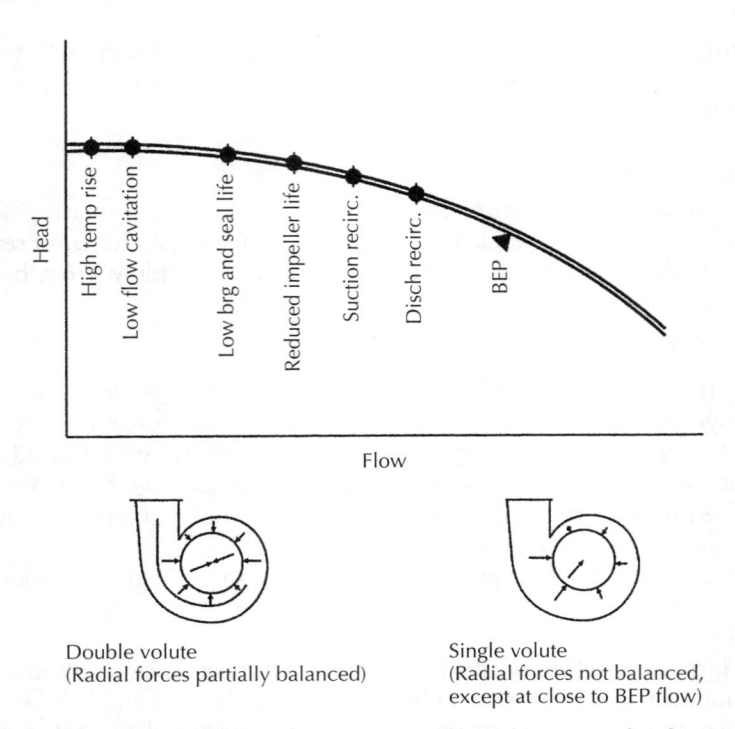

Figure 6.27 A typical pump head flow curve showing the problems that occur within the pump as flow is reduced below the best efficiency point. (Source: Ref. 11)

an example of the reasons for seal failure in one study[9] and Figure 6.27 is a graph showing the effects of a centrifugal pump being run progressively further away from the design point.[11] In addition there may also be auxiliary environmental control circuits, such as those described in section 3.4.6, and correct operation of these is also necessary to ensure that design conditions are maintained at the seal. Both Sheils[11] and Bachus and Custodio[12] describe many of the pump conditions that can contribute to problems with the seals.

- Mechanical seals are generally complex and contain both the dynamic seal faces and a number of subsidiary components and secondary seals. Each of these individual components of the seal may contribute to a failure. Figure 6.28 provides a summary of the different areas of a single spring process seal that may potentially contribute to seal leakage.

A combination of these two factors creates a particular problem with the operation of mechanical seals, the random nature of many of the seal failures that occur. This is a particular problem with the organization of maintenance. However, it has been shown that many of the random failures are attributable directly or indirectly to disturbances in machine operation. For this reason the first approach to a mechanical seal failure is to assess when the problem occurs.

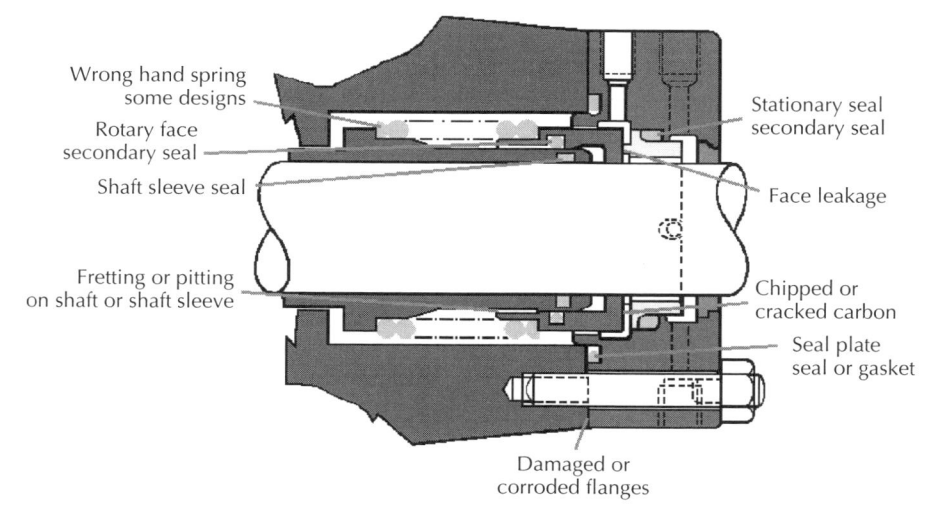

Figure 6.28 Examples of the types of problems that may occur with a mechanical seal. (Source: BHR Group)

6.3.3.1 Seal fails as soon as machine is pressurized or started

- Seal damaged during fitting, debris in faces, carbon broken, O-ring damaged.
- Components not fitted, e.g. O-ring.
- Incorrect setting, e.g. not sufficient spring compression, pump sleeve incorrectly located, pump shaft housing alignment faulty.
- Lubricant on faces causes high friction and carbon breakage.
- Most of these problems can be avoided by the use of cartridge seals.

6.3.3.2 Seal fails within a few hours or days of starting the machine

- Incorrect starting procedure, seal has been run dry, machine not vented before starting.
- Auxiliary circuit not restarted correctly.
- Accumulation of debris or solidified product in pipework when stopped has caused excessive abrasion on restart.
- Wrong seal design fitted.
- Wrong materials for duty fitted, or materials not to specification.
- Fitting errors, such as incorrect spring setting, fastenings working loose. The use of set screws on hard stainless steel shafts is a particular problem if correct precautions are not taken.
- Seal not suitable for full range of operating conditions as temperature, pressure, speed or pumped fluid change.

6.3.3.3 Sudden failure after period of satisfactory operation

- Thermal shock damage due to dry run followed by introduction of cool liquid.
- Machine fault such as shaft or bearing failure.
- Pressure spike overloading seal.

6.3.3.4 Sudden failure after restarting

- Product solidified between seal faces.
- Accumulation of solids around seal when stationary has damaged seal on restart.
- Auxiliary circuits not restarted.
- Pressure surge.
- Dry running of seal faces, especially hard/hard combinations.
- Transient conditions such as a sterilizing operation.

6.3.3.5 Random failures, thermal effects

Thermal damage to a seal may occur because of a single short-term event such as some of those listed above, or may affect the seal over a longer period of time such as product at a higher temperature than anticipated:

- Thermal cracking of seal faces. This may be evidenced as a series of small radial cracks in the seal face, or a few larger radial cracks. It is evidence of excess temperature in the seal face, usually transient causing thermal stress cracking. Faces with high conductivity are more resistant to this damage.
- Ageing and compression set of elastomers. Secondary seal O-rings may suffer excessive ageing, compression set, hardening or cracking. Note that operational conditions such as dry running can also cause ageing and set of elastomers as the heat generated can be conducted through the seal faces, a high conductivity seal face may survive a dry run, but cause ageing of the elastomer secondary seal.
- Vaporization damage, especially with aqueous liquids. If the sealed fluid vaporizes in the seal interface it can cause cavitation-like damage particularly to softer faces such as carbon. This will be evident as the damage will be limited to a band towards the atmospheric side of the seal.[13] Vaporization may also lead to thermal cracking, discussed above, of hard faces due to the rapid temperature fluctuations. It can also lead to thermal instability where the vaporization forces the faces apart which then permit the entry of cooler liquid. This can lead to puffs of leakage, severe seal damage and considerable noise.

6.3.3.6 Random failures, fluid effects

As with other seal types discussed it is necessary to consider all the fluids that will contact the seal. This can include cleaning and flushing cycles, minor constituents of a product, potential precipitates, products of degradation or evaporation, changes in sealed fluid with time due to process changes and both operational and static conditions. Operation at elevated temperature will accelerate most fluid related effects, such as corrosion of metals, and chemical attack of elastomers:

- Corrosion of metal parts. Highly stressed components such as springs are likely to be the most susceptible to corrosion.

- Chemical attack of elastomers, see section 5.1 on elastomers and section 6.2 on elastomer seal failures for assistance with determining the potential cause. The friction generated at the seal faces may create an effective temperature above the fluid temperature, especially if dry running or other upsets occur. Hence the secondary seals closest to the seal faces will usually be those most affected.
- Binder phases in ceramics and impregnants in carbon can be susceptible to chemical attack that does not affect the actual face material. It is necessary to select carbon grades with care to ensure that metal loading or resin impregnant is resistant to the pumped product. Similarly self-sintered silicon carbide may be preferred. The best corrosion resistance with tungsten carbide is achieved if a nickel binder is used.
- Dissolved solids can be deposited in areas where vaporization occurs. This is most problematic if it occurs in the seal face as the deposit will cause some face separation or damage. Dissolved solids can originate from obscure sources. They can be a very minor constituent of the pumped fluid that has been overlooked. System pipe work can be the source of minute quantities of dissolved metal or polymer material constituents that become deposited on the seal faces as evaporation occurs.
- Products that crystallize on heating or in the atmosphere can cause similar problems. Again it can sometimes be a very minor constituent of the product that has been overlooked during the specification phase.
- Leaked product can partially evaporate to cause viscous or solid deposits on the atmospheric side of the seal. This can impede the floating secondary seal, often known as 'hang-up'. This may be dispersed by using a 'quench' fluid on the atmospheric side of the seal. A widely used example is steam to soften and remove hydrocarbon deposits in refinery service. Other suitable fluids may be used such as water to remove sugar or similar water soluble deposits.

6.3.3.7 Abrasive solids

Abrasive solids may be present in a large proportion of sealed fluids. They may be a very low percentage and not taken into serious consideration at the specification stage, or a major proportion of the product in a slurry. Where the pumped product is slurry the seal design should have taken account of this at the design stage. Slurry seals are discussed in section 3.4.6. In many other situations a small proportion of solids may be present. They can cause a number of problems:

- Erosion of seal components, one radial location. Circulation flush contains abrasive solids, modifications to auxiliary flush circuit required.
- Erosion of seal, and potentially the housing circumferentially in limited axial locations. This indicates the presence of a standing vortex in the housing. Modifications to improve the housing design or otherwise improve flow in the seal housing, see section 3.4.5, will be necessary.
- Fretting damage to drive or anti-rotation lugs. The presence of fine abrasives combined with any inherent vibration can cause fretting damage to

the load transfer points of drive lugs, detents, etc. which then prevents the required relative movement of seal components as wear, temperature changes and other events occur.
- Abrasive wear to the seal faces. The entrainment of fine solids into the seal faces will cause heavy wear of anything other than the hardest of face materials. The incidence of solids close to the seal may be minimized by the design of the machine, seal housing or auxiliary circuits (seal plan) as discussed in section 3.4.6. The wear of the seal faces will normally be at the product side of the seal. Alternatively a face combination with two hard faces may be required. The potential options for this are discussed in section 5.8.
- Clogging of springs can occur, especially multi-spring designs where the small springs are particularly susceptible. Some designs move the springs outside the product, but this can introduce other problems; the springs are then usually much smaller creating more stringent setting requirements and are also prone to hang-up type problems as discussed in section 6.4.3.6. If the machine design and operational parameters permit then a large single spring design will be more resistant to clogging.

6.3.3.8 Wearout

Typical mechanical seals have a 'nose' on the carbon or other 'wearing' face of approximately 3 mm depth. In a low viscosity liquid such as water the typical wear rate of a standard design of conventional flat face seal will be of the order of 0.25 μm per hour or less. Specialist designed seals would exhibit an even lower wear rate. It is relatively unusual for seals to continue to operate until the wearout life is achieved. They more usually will have suffered one of the random event failures discussed in the sections above. Seal wearout may occur prematurely, this is most likely in clean, low viscosity fluids such as water, vehicle coolant or very light hydrocarbons in applications where the fluid is very close to vapour pressure. This can be prevalent in what may appear to be low pressure and fairly mundane applications such as vehicle cooling systems.[11] Wearout can be considered to have occurred if the seal nose has worn to a very shallow height, from an initial 3 mm, and there is no indication of overheating or abrasive wear grooves. To achieve a longer seal life will usually require attention to the seal face loading to prevent breakdown of the film. Many such applications will now use one of the enhanced lubrication options, section 3.4.4.10, to ensure adequate face lubrication.

6.3.3.9 Further sources of information

A detailed description of the individual symptoms of seal failures has been provided by Woodley.[14] However, this should be used within an overall assessment of the seal operating conditions and machine operation using appropriate information, such as that on pump operation.[11,12] The design of the pump to improve conditions at the seal can also be effective.[19]

Many of the seal failures that are attributable to poor face lubrication or disturbances to the lubrication regime causing overheating or dry running have been successfully addressed by the use of enhanced lubrication designs, section 3.4.4.10.[15, 16]

A number of other problems, particularly those associated with product deposition between the seal faces have been solved by the adoption of arrangements such as double face to face dry gas seals as described in section 3.4.6. This has been found to provide significant improvements, particularly in stop/start applications.[17, 18]

6.3.4 Dry gas seals

Dry gas seals are designed with many of the basic characteristics of liquid mechanical seals and can therefore be potentially subject to similar failure modes. However, as they are designed to operate on dry gas and usually at the high rotary speeds associated with turbo compressors they can be prone to rapid failure if conditions for seal operation are not correct.

A major survey into the reliability of dry gas seals in the UK[20] highlighted three key causes of dry gas seal failure, in descending order of frequency.

- Deposits.
- Mechanical damage.
- Secondary seals.

The presence of contaminants in the seal has proved to be a major contributor to failure of these seals and this is supported by evidence from both seal and compressor manufacturers.[21, 22] This can be the cause of 80% of seal failures predominantly due to hydrocarbon deposits.

The solution of dry gas seal problems is liable to involve a detailed knowledge and assessment of the seal, the buffer system, vent system, barrier backup seals and the compressor operation.

6.3.4.1 Deposits

In a dry gas seal deposits can be either solid or liquid. Major sources are caused by:

- Trace amounts of condensate in the gas stream that condenses in or around the gas seal.
- Oil from the compressor bearings.
- Trace amounts of solids in the gas stream.

This demonstrates the importance of the barrier seals and backup systems to the reliability of these seals. Reliability has been improved by the widespread adoption of the segmented ring barrier seals in place of labyrinths that were previously used. The barrier flush gas must also be as clean as practicable. A high incidence of seal failures is associated with stopping and restarting compressors. This can allow cooling which causes the condensation, and even freezing, of vapour. Failure to ensure that barrier systems are in operation correctly before startup will also allow contaminant into the seal area.

Deposits can also be a major factor in mechanical damage as liquid in the seal faces can cause rapid overheating that may create thermal stress of the components.

6.3.4.2 Mechanical damage

Mechanical damage to seal faces is likely to be caused by one or both of two factors:

- Rapid thermal fluctuations caused by liquid or other deposits disturbing the gas film and causing rapid overheating and an unstable film leading to thermal transients.
- Crevice corrosion or attack of the binder in ceramics leading to structural failure of the faces.

As dry gas seals typically operate at high rotational speeds failure can cause a large amount of damage. Most designs now incorporate a shroud to contain the debris if a seal face shatters.

Attention to the buffer gas system and backup seals plus analysis of all constituents of the gas stream is necessary to avoid these problems.

6.3.4.3 Secondary seals

Problems associated with secondary seals can include:

- Chemical attack of the elastomer.
- Extrusion at high pressure.
- Rapid gas decompression damage to elastomers.
- Wear of plastic secondary seals.

Elastomers are generally preferred for secondary seals as they can provide the best compromise of resilience and low leakage. However, application in gas seals can be a problem due to both resistance to chemical attack and more particularly to gas decompression damage. For this reason most secondary seals exposed to high pressure gas will be manufactured from filled PTFE or a similar polymer. The use of PTFE secondary seals can lead to increased problems with wear if there is any vibration or runout as these seals cannot accommodate the small movements and oscillations without sliding as may be the case with an elastomer.

The problems of chemical degradation and extrusion will be increased if there is any transient high temperature operation due to contaminant ingress. The secondary seals adjacent to the seal faces may experience temperatures considerably higher than the design specification making them more prone to extrusion and chemical attack.

As with corrosion of the seal faces it can be minor constituents of the gas stream, probably not provided on the process analysis, which can cause problems with chemical compatibility.

6.3.5 Compression packing

Packing may be used as a rotary, reciprocating or valve stem seal. This section can be used to ascertain the potential problems with packing from any of these applications.

6.3.5.1 Packing extruded at both ends of the box

This will be due to a combination of excess clearance or packing with insufficient extrusion resistance. If operation is satisfactory in other respects the solution is to provide backup rings at each end of the stuffing box.

Alternatively consider a packing with improved extrusion resistance but beware of other problems such as increased shaft wear.

6.3.5.2 Extrusion at gland follower end

If clearances at each end of the box are similar then extrusion is probably due to pressure. Fit backup ring next to gland follower.

6.3.5.3 Packing ring extruded into adjacent packing ring

This is caused by faulty fitting such as cutting the packing rings too short. Ensure that packing is fitted correctly as described in section 3.5.4 and Appendix 1.

6.3.5.4 Leakage around outside of gland follower

This indicates either that the packing has been fitted incorrectly or that the box bore is in poor condition, potentially with severe axial grooving.

6.3.5.5 Inside diameter of packing is black and hardened

Overheating of the packing has been caused either by lack of lubrication or overtightening. Investigate packing operation and if being used correctly a packing with improved lubricating properties may be required.

6.3.5.6 Worn shaft with excessive grooving

Abrasion of shaft by solids in sealed product or the packing. Some packing materials such as conventional Aramid are extremely abrasive. A very hard wear resistant shaft is required.

With any shaft wear problem it is necessary to investigate:

- Potential for providing a harder more wear resistant shaft. Note that some wear resistant coatings may not provide adequate conductivity and can lead to overheating.
- Alternative packing that will not cause as much abrasion on shaft.
- Other modifications to exclude abrasives from the stuffing box, such as clean flush water to a lantern ring, section 3.5.3.

6.3.5.7 Packing swollen, softened or decomposed

This is caused by chemical attack of the packing by the pumped product. An alternative material will be required. Check that all the constituents of the product have been considered in the material selection.

6.3.5.8 Packing stiffened and shrunk

Probably caused by the lubricants being washed out of the packing. An alternative grade with lubricant resistant to the process will be required.

6.3.5.9 Rapid onset of high leakage

Shaft eccentricity may be causing rapid wear and damage to the packing. A change to a more compliant packing grade such as one with an elastomer core should be considered.

6.3.6 Clearance seals

There are a wide variety of clearance seal designs, section 3.6. The primary cause of failure will be wear of the seal components leading to increased leakage. It is important to remember that particularly for liquid seals the leakage will be proportional to the cube of the clearance. Hence, even a relatively small increase in clearance gap can create a very significant increase in leakage.

The measures that can be taken to reduce the effect of wear include:

- More wear resistant material combination.
- Wear tolerant seal design, such as abradable seal to replace a conventional labyrinth.
- Assess clearances are realistic in relation to the bearing design, it may be necessary to provide improved shaft location.
- Temperature excursions may reduce clearance, assess whether better thermal compatibility of material characteristics is required or temperature control of the seal area.
- Change of fluid phase, such as condensate from a gas, can cause erosion damage. Assess temperature fluid phase characteristics of the seal.
- Contaminants can accelerate wear in small clearances, keep seal area clean.
- As clearance seals are often used at high speeds rotor stability is an important factor in avoiding wear due to contact at unstable running conditions. Assessment requires a thorough understanding of rotordynamics. Seal designs that improve rotor damping may be required. Section 3.6.2.4 describes some of the options available and includes references to authoritative work in this area.

Viscoseals present a particular problem of gas ingestion, as described in section 3.6.6. This will be a particular problem if the seal operates at varying speed as it will tend to ingest the gas at the higher speeds and then not seal adequately at lower speed. Two methods of reducing the incidence of this problem are providing a plain shaft section in the length of the seal and ensuring that the seal is mounted concentrically. Unfortunately the first option then requires a longer seal and the second is unlikely to be achieved with much consistency in most mechanical equipment.

6.4 Reciprocating seals

6.4.1 Polymer reciprocating seals

This section is intended to provide a guide to problem solving with elastomer and plastic-based reciprocating seals. The diagnosis of failures can be complex and very often a combination of factors will contribute to the failure. An investigation should therefore be approached with an open mind, no preconceived opinions and an investigation of all available information.

Interactions of the fluid and seal material may also be a contributing factor, and so potential problems discussed in the static polymer seals section, 6.2, should also be considered.

Wherever possible the investigation should include witnessing of the strip down of the equipment so that full information on the condition of the machinery involved, condition of the sealed fluid and mating components, etc., can be obtained. Further information on symptoms and probable causes of failure are available in literature provided by the industry such as references 23–26.

6.4.1.1 Leakage from new, no sign of seal damage

- Speed too high for seal design causing excessive fluid film.
- Viscosity of fluid too high causing excessive fluid film. This may be in combination with low temperature when the seal may have less interference and reduced elastomeric properties.
- Faulty seal, for example incorrect lip trimming.
- Housing dimensions incorrect, low interference.
- Full tolerance stack-up not taken into account, Figure 6.29.

6.4.1.2 Wear of seal lip

- Seal operating at low pressure with insufficient lubrication such as pneumatic seals with no or insufficient lubrication, or hydraulic seals with low viscosity fluid.
- Incorrect seal material.
- Counter-face surface texture incorrect.

6.4.1.3 Excessive wear at heel of seal

- Seal operating at high pressure with insufficient lubrication or low wear resistance. This will be evidenced as damage to fabric, wear of the seal heel or wear of a backup ring, Figure 6.30.
- Counter-face surface texture incorrect.

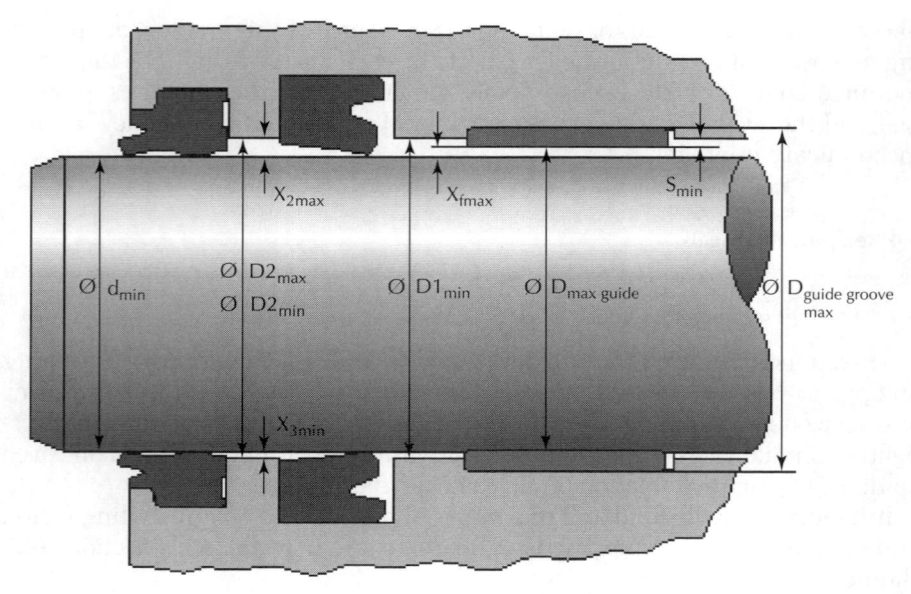

Figure 6.29 Example of tolerance stack-up assessment. (Source: Freudenberg Simrit)

Figure 6.30 Excessive wear at the heel of a reciprocating seal. This area has the least lubrication and the highest pressure differential. (Source: VDMA)

6.4.1.4 Extrusion

- Excess pressure. There may be pressure spikes in the system that cause it to be much higher than the rated pressure, Figure 6.31.
- Dynamic pressure buildup, see section 6.4.1.5, will also cause a higher pressure at the seal than that of the hydraulic circuit.
- Excessive temperature. Temperature above the design temperature will cause loss of mechanical properties of elastomers and softening of plastic or thermoplastic elastomer (polyurethane or polyester) seals.

6.4.1.5 Dynamic pressure generation

This is also sometimes known as 'drag pressure'.

- Axial motion of the rod through a small clearance such as the bearing will cause the generation of a hydraulic pressure along the length of the

(a)   (b)

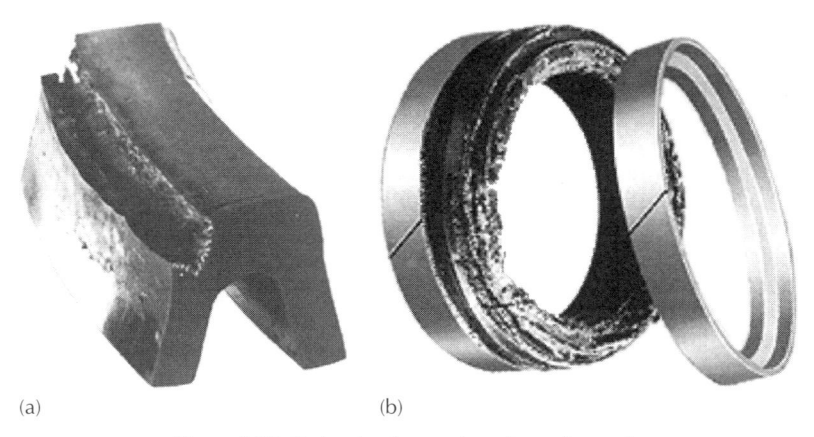

Figure 6.31 Extrusion damage to reciprocating seals.

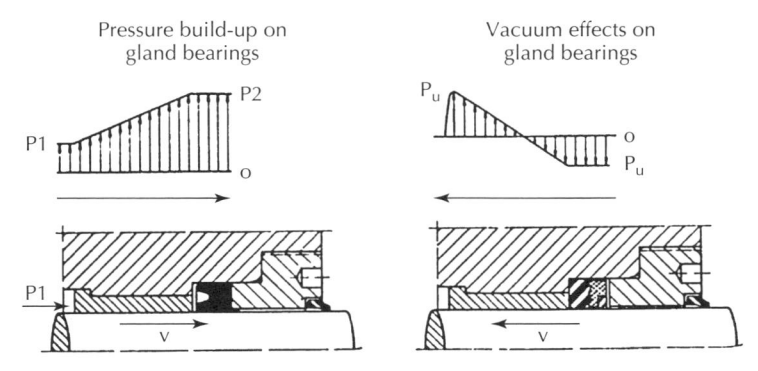

Figure 6.32 The generation of excess pressure due to hydrodynamic pressure buildup on a rod seal. (Source: BHR Group)

bearing. This will cause overpressure on the outwards stroke and a potential negative pressure on the return stroke, Figure 6.32. This can cause additional seal damage due to overpressure and also oscillation in the groove. On a piston seal it may cause both situations at once increasing the pressure differential.

Spiral grooves in bearings or the butt joint gap in plastic bearing rings will relieve this pressure.

6.4.1.6 Even series of longitudinal grooves
- Excessive dissolved air in the system. The air dragged in with the oil between seal and contact area expands the nearer they are to the non-pressurized side of the seal. This air bubble erosion results in longitudinal grooves in the surface of the seal, Figure 6.33.

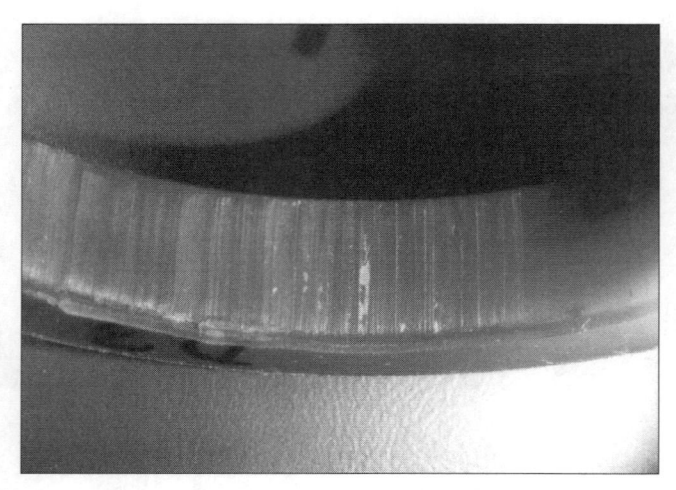

Figure 6.33 Effect of dissolved air causing damage to seal surface. (Source: VDMA)

6.4.1.7 Uneven scoring to the seal surface

- Debris in the fluid circuit or damage to the seal counter-face. This may be a single score on the surface or a series of uneven scores depending on the amount of debris or type of surface damage on the metal components. Examples are shown in Figure 6.34.

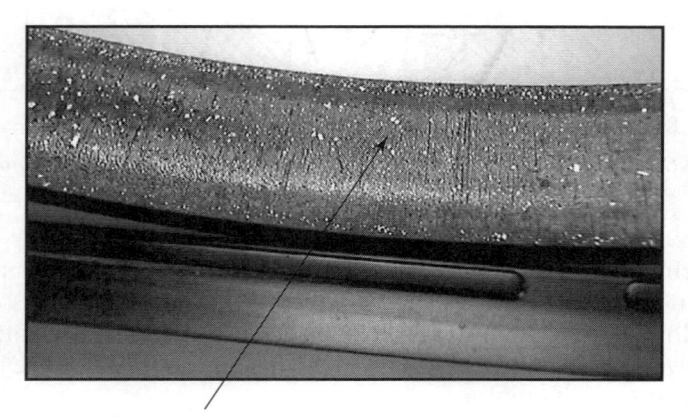

The metal particles embedded in the seal produce scores on the mating surface.

Figure 6.34 Scoring due to debris or metalwork damage. (Source: VDMA)

6.4.1.8 Extrusion damage to high pressure side of seal

- Excess pressure generation between a pair of seals. This can happen where a pair of seals is used in tandem and there is no provision for pressure relief. The film entrainment mechanism of the seal may create a pressure much higher than the system pressure between a pair of seals, Figure 6.35.

Failure guide 451

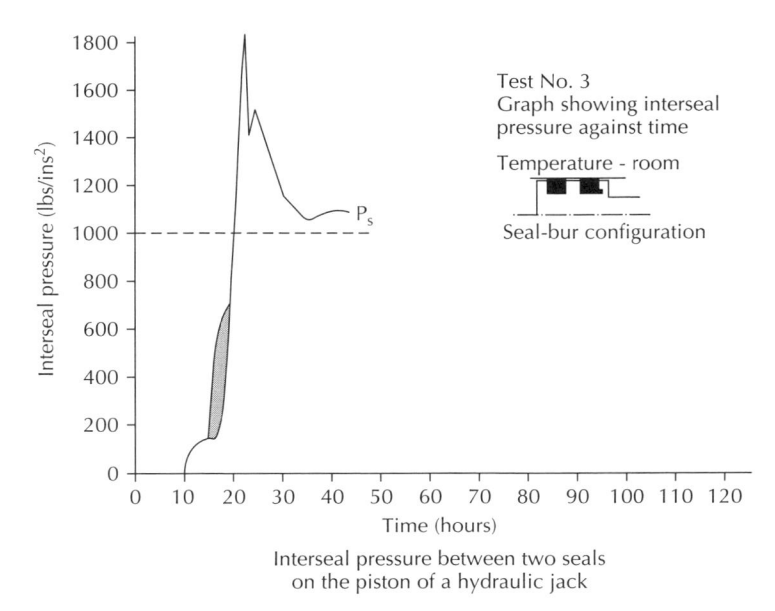

Figure 6.35 Buildup of interseal pressure between a pair of seals on a piston. (Source: BHR Group)

Select a seal design that will ensure pressure relief or incorporate facilities to ensure pressure cannot build up. This is discussed in detail in section 4.2.4.

6.4.1.9 Burnt areas of seal

- This may be evidenced as blackened or melted areas of plastic or charred and locally damaged elastomer that is very hard and carbonized, Figure 6.36. If there is a local area of a hydraulic circuit where air may accumulate at low pressure, and then if a sudden pressure increase occurs the mixture of air and local oil may detonate, effectively as in a diesel engine. Hence this

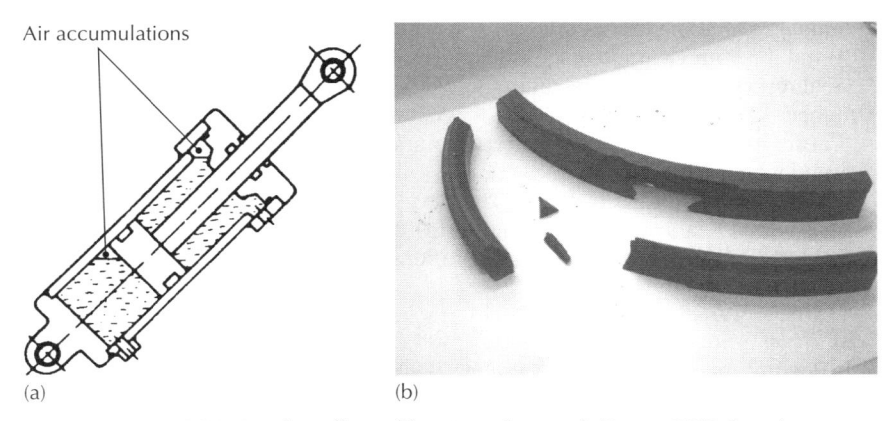

Figure 6.36 Dieseling effect and how it can be caused. (Source: BHR Group)

effect is known as dieseling. Effective venting of the hydraulic system is necessary. If a system has long stationary periods then a self-venting arrangement for critical points in the system, such as high points on large cylinders, may be necessary.

Hydraulic circuits with undersize reservoirs are particularly vulnerable to this problem.

Even if this effect does not directly burn the seal it may cause excessive carbonaceous debris in the system which can lead to seal wear.[25]

6.4.1.10 Spiral of an O-ring

- This can cause a sudden failure of an O-ring reciprocating seal. Excessive friction and/or uneven stress distribution can cause part or all of an O-ring to roll in the groove. The effect is to cause a spiral of the O-ring which can then easily leak. Generally caused by factors such as excessive friction due to poor lubrication, small O-ring section and uneven stress distribution around the O-ring due to eccentricity. Examples are shown in Figures 6.4 and 4.4. The most satisfactory method of resolving this problem is to select an alternative cross-section of seal that is less susceptible to rolling.

6.5 References

1. 'Elastomeric seals for rapid gas decompression applications in high pressure service: guidelines, Health and Safety Executive', BHR Group, 2006.
2. P. Embury, 'High-pressure gas testing of elastomer seals and a practical approach to designing for explosive decompression service', *Sealing Technology*, June 2004.
3. K. Edmond, 'Elastomer fatigue testing for explosive decompression cycling prediction', 17th Int. Conf. on Fluid Sealing, BHR Group, 2003.
4. 'High pressure, high temperature developments in the United Kingdom continental shelf', Report Number RR409, Health and Safety Executive, UK, 2005.
5. K.W. McQuillan, G. Milne and G. Smith, 'A leak-free start up: fantasy or reality', *Sealing Technology*, July 2003.
6. 'What if – your O-ring seal looks like this?', Freudenberg Simrit.
7. 'Sealing system leakage and analysis guide', Technical Bulletin, Oil Seal Subdivision, Rubber Manufacturers Association, USA.
8. 'Causes of failure, damage patterns, handling and assembly of Simmerrings', Freudenberg Simrit.
9. J.M. Plumridge and R.L. Page, 'The development of more tolerant mechanical seals', Shaft Sealing in Centrifugal Pumps, Seminar, I. Mech. E., February 1992.
10. S. Moore, 'Seals – problems and solutions', Seals and Sealing Today, Seminar, I. Mech. E., December 2005.
11. S. Shiels, 'Which is worse, specifying too much head or too much flow?' in *Stan Shiels on Centrifugal Pumps*, Elsevier, 2004.
12. L. Bachus and A. Custodio, *Know and Understand Centrifugal Pumps*, Elsevier, 2003.
13. D. Harrison and R. Watkins, 'Evaluation of Forties main oil line pump seals', 10th Int. Conf. on Fluid Sealing, BHR Group, 1984.

14. B.J. Woodley, 'Failure diagnosis', in Mechanical Seal Practice for Improved Performance, I. Mech. E., 1992.
15. L. Young, E. Roosch and R. Hill, 'Enhanced mechanical face seal performance using modified face surface topography', 17th Int. Conf. on Fluid Sealing, BHR Group, 2003.
16. J.L. Morton, J. Attard and J.G. Evans, 'Active lift seal technology impact on water injection services', 22nd Int. Pump Users Symposium, Texas A&M University, 2005.
17. R.K. Flitney, 'Excellence in environmental protection awards', *Sealing Technology*, January 2006.
18. 'Excellence in environmental protection', *Sealing Technology*, April 2004.
19. 'New pump design to improve seal life', *Sealing Technology*, July 2005.
20. 'Hydrocarbon release – dry gas seal integrity survey report', Offshore Technology Report 2000/070, Health and Safety Executive, UK, 2000.
21. G. Marsh, 'Dry gas seal system reliability, a seal manufacturer's viewpoint', Reliability of Rotating Machinery Sealing Systems Seminar, I. Mech. E., October 1999.
22. J.S. Stahley, *Dry Gas Seals Handbook*, PennWell Corp., 2005.
23. *Merkel Hydraulic Seals, Technical Principles*, Freudenberg Simrit, 2005.
24. N.A. Peppiatt, 'Reciprocating seals for hydraulic cylinders', Seals and Sealing Today, Seminar, I. Mech. E., December 2005.
25. *Sealing Handbook*, Parker Seals, Parker Hannifin GmbH, 1999.
26. 'Sealing systems for fluid power applications', Training Material, VDMA, 2005.

CHAPTER 7

General information

7.1 Glossary of sealing terms

This section contains an alphabetical listing of special features and technical terms which are of common usage in sealing technology terminology. It is divided into sub-sections dependent upon the specific sealing technology:

7.1.1 Elastomer and plastic seals, compression packing

Abrasion resistance   Ability of a material to resist abrasive wear.

Ageing   Change in characteristics of rubbers with time specifically influenced by environmental factors, e.g. light, heat, etc.

Anaerobic   A term used to describe materials that remain liquid while exposed to air, yet harden when confined between close fitting parts.

Anti-extrusion ring   A ring of material that is harder than the seal, installed on the low pressure side to prevent extrusion of the seal material into the clearance gap.

Anti-oxidant   An ingredient used when compounding rubber to suppress oxidation of the rubber.

Asperities   On a microscopic scale, 'high spots' of surface roughness produced by normal machining or finishing processes.

Backup ring   See Anti-extrusion ring.

Bedding-in   A period of running-in when sealing lips or surfaces develop intimate contact. Often a period of high seal wear rate.

Bloom   A cloudy or milky white discolouration appearing on the surface of elastomer components after storage.

Braid   Hollow or solid structure of round, square or polygonal section constructed from interlocking filaments or yarn strand laid obliquely to the axis of the braid.

Breakaway friction   See Static friction.

Breakout friction   See Static friction.

Cold flow   A continued deformation with time of a plastic material when subjected to a continuous load.

Composite seal   A bonded seal composed of two or more materials, often of a different hardness or construction, e.g. rubber and rubber/fabric. Sometimes incorporates an anti-extrusion ring.

Compression set   A permanent set or degree of permanent deformation of elastomer after a given time at a given temperature under a given strain.

Co-polymer   Polymer formed from the reaction of more than one species of monomer.

Cure   Vulcanization process applied to rubbers.

Dieseling   Self-ignition which causes the seal to burn or char, see section 6.4.7.9.

Double acting seal   Seal for reciprocating movements capable of sealing with both directions of movement.

Dry running   Rubbing contact without any liquid being present at the interface.

Durometer hardness   Arbitrary measurement of hardness related to the resistance to penetration of an indentor point on a durometer.

Elasticity   The ability of an elastomer to return to its original shape and size after deformation.

Elastomer   A synthetic material having rubber-like elastic properties.

Elastomer Compatibility Index   Result of a standard test to determine the swelling effect of a petroleum oil in relation to a standard elastomer. Procedure and details in ISO 6072.

Elongation   Percentage of stretch.

Excluder   Seal designed to exclude contaminant from the pressure seal and sealed system. See also Wiper ring and Scraper ring.

Extrusion   Distortion of part of a seal into the clearance gap on the low pressure side of the seal.

Ferromagnetic liquid   A liquid carrier composed of an ultrastable colloidal suspension of extremely small magnetic particles. See section 3.7.

Flash line   A raised ridge appearing on a moulding.

Flashing   A term used to describe the rapid change from liquid into a vapour.

Flinger   Also known as a slinger. A washer type disc fitted to a rotating shaft. Liquid leaking along the shaft is flung off by centrifugal action.

Flow mark Imperfection in a moulding due to incomplete flow of material in the mould.

Garter spring Helical wire spring of circular geometry fitted to a lip seal (specifically an oil seal) to enhance lip contact pressure.

Gland The cavity of a stuffing box.

Gland follower An axially moveable part of the stuffing box. Compresses the seals in the gland.

Hardness Rubber: Measured in International Rubber Hardness Degrees, 0 to 100° IRHD. (0° no measurable resistance to indentation, 100° no measurable indentation.) Similar to Shore A Durometer readings. For other materials Rockwell Hardness Numbers (RHN), Diamond Pyramid Hardness Numbers and Brinell Hardness Numbers (BHN) are used.

Heel Part of seal nearest to extrusion gap on low pressure side. (Reference to reciprocating seals.)

Housing Metal parts, other than shaft, bounding the annular recess in which a seal is installed.

Hydrodynamic seal Lip seal with pumping feature, also other seals with hydrodynamic pressure generation.

Hydrostatic seal Seal with external pressurization of interface.

Hysteresis Rubber: A phase difference between stress and strain representing a loss of energy put into the rubber during deformation. See also Resilience.

Interference Radial, diametral or axial. The geometric compression of the seal by the amount that the seal dimensions are greater than those of the groove to which it is fitted.

Interseal pressure Pressure buildup between a pair of seals in close proximity to each other, e.g. on a double acting piston with two seals.

Lantern ring A ring inserted between rings of packing to allow the introduction of lubricant or coolant.

Lead-in A chamfer on a shaft end or housing to effect an easy introduction of lubricant or coolant.

Lip opening pressure Air pressure required on the atmospheric side of a lip seal to lift the lip seal off its shaft and allow air leakage at the rate of 10 1/min.

Minor lip Auxiliary lip or wiper lip of a lip seal.

Ozone cracking Surface cracking of rubber due to the degrading effect of ozone.

Packing Used to denote a multitudinous variety of materials including leather, cotton, hemp, PTFE, rubber, rubber/fabric, etc., used to retain fluid under pressure in a housing or stuffing box. Best not used without qualification.

Permanent set   The amount of residual displacement in a rubber part after a distorting load has been removed. See also Compression set.

Permeability   A measure of the ease with which a liquid or gas can pass through a rubber film.

Plasticizer   Constituent of a rubber mix controlling the hardness and plasticity of the final product.

Polymer   Materials with long-chain molecules, e.g. natural rubber, elastomers and plastics.

Pressure-absolute   Pressure measure with respect to absolute zero pressure, i.e. gauge pressure plus atmospheric pressure.

Pressure-differential   Pressure difference between any two points of a system.

PV factor   Product of pressure and sliding speed. The units must be specified, e.g. $lbf/in^2 * ft/min$, $bar * m/s$.

Resilience   Ability to regain original size and shape after deformation.

Rider ring   Wear or load-carrying ring usually associated with reciprocating compressors.

Run-in   Period of initial operation and wear during which a seal becomes properly bedded down. See also Bedding-in.

Scorch   Premature curing of vulcanized rubber due to excessive heat.

Scraper ring   See Wiper ring and Excluder.

Screw seal   See Viscoseal.

Sealing interface   Refers to the area of rubbing contact between the sliding faces of a rotary or reciprocating seal.

Squeeze   Deformation of a seal product when assembled with an interference fit.

Static friction   Instantaneous or 'holding' friction of a seal under static conditions, also see Stiction.

Stick-slip   Phenomenon associated with rubber friction in which the rubber alternately moves with adjacent surface and then rapidly returns to its original position. Frequency can be high and can be heard as a squeal or seen as a juddering of the shaft. The effect is a function of speed and temperature. It is usually a non-lubricated or boundary-lubricated condition.

Stiction   Initial friction or breakout friction when motion is started.

Stress relaxation   The loss in stress when an elastomer is held at a constant strain over a period of time.

Stuffing box   Alternative name for a gland for containing packings or seal rings.

Surface roughness   The micro-geometry of any machined surface. See section 7.3.

Swell Increase in volume of a seal when immersed in fluid.

Thermoplastic The property of a material which plastically flows when the temperature is raised and reverts to its original state when cooled to the original starting temperature.

Thermosetting The property of a material which irreversibly solidifies at an elevated temperature.

Toroidal seal Alternative name for an O-ring.

Torque Resistance to shaft rotation expressed in newton-metres or pounds or feet.

- Breakaway or starting: Torque required to start a shaft rotating.
- Running: Torque required to keep a shaft rotating when conditions are stabilized.

Torr Unit of pressure used in vacuum measurement equivalent to 1/760 of a standard atmosphere or 1 mm mercury (Hg), $1 * 10^{-3}$ mm Hg $= 10^{-3}$ torr.

Viscoelastic A material that exhibits elasticity as well as a degree of plasticity is termed viscoelastic.

Viscoseal Also known as a 'screw seal' or 'windback seal', see section 3.6.6.

Vulcanisate The product of the vulcanization process in elastomer manufacture.

Vulcanization Introduction of cross-links usually by heating a mix containing vulcanizing (curing) ingredients, particularly elastomers and other polymers.

Windback seal See Viscoseal.

Wiper ring Elastomer or plastic ring which is an interference fit on the shaft. Its function is to wipe off dirt and prevent its ingress.

7.1.2 Expansion joints and flange gaskets terminology

Assembly pressure Pressure generated on a gasket during assembly.

Backup ring A ring (often metallic) around the outer periphery of a sealing material, usually to prevent extrusion.

Beater addition product Gasket material manufactured by a paper-making process.

Belleville washer Washer with a slightly conical shape, which acts as a spring when compressed axially.

Binder A substance (usually organic) used to bond the components of a gasket material into a matrix.

Blank flange Flange with no bore, used to provide a sealed closure to a flanged opening.

Blind flange   See Blank flange.

Bolt   Threaded fastener used to secure the members of a flange joint together and to apply compressive force to a flange.

Bolt load   Means of applying compressive pressure to make the gasket material flow into surface imperfections in the flange to create a seal.

Bolt tension   Tension (tensile stress) created in a bolt by assembly preloads and/or thermal expansion, service conditions, etc.

Calendered sheet   See Compressed fibre sheet.

Centring ring   An extension of a gasket for the purpose of locating it centrally on a flange.

Chemical compatibility   See Fluid resistance.

Class   An alpha-numeric designation related to a combination of mechanical and dimensional characteristics of a component of a pipework system. It comprises the word 'CLASS', followed by a dimensionless whole number and is used to identify ranges of related components in a number of different standards (for example, EN 1759).

Compressed fibre sheet   Gasket material, primarily containing fibres, rubber and fillers, manufactured on a calender under high load.

Compressibility   Percentage reduction of thickness under a compressive pressure, applied at a constant rate, at room temperature.

Compression set   Residual deformation of a gasket after it has been subjected to, and then released from a specified compressive pressure, over a defined time and at a given temperature.

Controlled swell   Property of gasket material to swell to a defined extent when in contact with the retained fluid, to provide additional sealing pressure.

Corrugated metallic gasket   Metal gasket, usually incorporating a filler material in the well of the corrugations, in which the seal is formed between the peaks of the corrugations and the mating flanges.

Creep deformation   Percentage loss of thickness over a specified time under constant load, applied at a specified rate, at a specified temperature.

Cure   Cross-linking reaction of elastomer with various chemicals, creating a matrix of greater stability.

DN   A designation of nominal size of components in a pipework system, defined in EN ISO 6708.

Double jacketed gasket   A gasket design in which the gasket material is enclosed within an outer metal cover.

Effective sealing width   That part of the actual width of a gasket considered to contribute to the performance of the gasket.

Elasticity Property of a body to recover its original size and shape immediately after removal of the external forces which cause it to deform.

Elastomer Generally long chain polymer molecules, which show elastic properties.

Envelope gasket A gasket design in which the gasket material is enclosed within an outer cover (typically PTFE) to minimize chemical degradation by the sealed fluid.

Eyelet Metallic cover around inner periphery of gasket material, to minimize chemical degradation by the sealed fluid. Depending on selection of geometry and metal, it may also improve sealability and blowout resistance.

Flange Basic component of a gasketed joint assembly, incorporating a substantially radially extending collar for the purpose of joining two or more items of process equipment.

Flanged joint See Gasketed joint.

Flange rotation Deformation of a flange caused by imposed forces.

Flat-face flange A flange where the entire mating faces are flat.

Fluid resistance Measure of the ability of the material to resist chemical attack.

Fugitive emission A chemical, or mixture of chemicals, in any physical form, which represents an unanticipated leak, from anywhere on an industrial site.

Full-face gasket A gasket which covers the entire flange surface extending beyond the bolt holes.

Garter ring A metal or hard elastomer material used to apply additional pressure to self-energizing seals.

Gasket Deformable material (or combination of materials) intended to be clamped between flanges to prevent leakage of contained fluid.

Gasketed joint The assembly of components (e.g. flanges, bolts, gaskets) required to join two or more items of process equipment and to prevent leakage.

Gasket load reaction Point at which the load on a gasket can be considered to react for moment calculation purposes.

Gasket pressure Effective compressive load per unit of gasket area.

Grip length Distance on a bolt between the inner face of a nut and the inner face of the bolt head.

Hard and soft gasket materials Differentiation between predominantly hard, metal-based gaskets (e.g. spiral wound) and softer or fibre-reinforced materials.

Hot creep during service Percentage reduction in thickness under constant compressive pressure at elevated temperature.

Hydrostatic end thrust　Relieving force caused by the pressure of the retained fluid, resulting in a reduction in gasket pressure and an increase in bolt load.

Initial preload　Tension created in a single bolt as torque load is applied. It is usually modified by subsequent assembly operations and service conditions.

Inside bolt circle (IBC) gasket　A gasket lying wholly within a ring of bolts.

Internal pressure　Fluid pressure applied to the joint.

Kammprofile gasket　Metal gasket with grooved faces, with or without a resilient sealing layer on surfaces.

Leakage rate　Quantity of fluid passing through the body and/or over the faces of a gasket per unit periphery of the gasket over a specified time.

Lip gasket　Gasket design which is self-tightening by virtue of a protruding lip, and which may alternatively be used for attachment.

Live loading　Application of a spring load to maintain seal surface pressure.

Load compression characteristic　Reduction of thickness under specified load and temperature conditions.

Maximum assembly pressure　Maximum allowable pressure during assembly to prevent unacceptable creep or failure of the gasket material under operating conditions.

Maximum gasket pressure under operating conditions　Maximum allowable pressure under operating conditions to prevent unacceptable creep relaxation or failure of the gasket material.

Minimum assembly pressure　Minimum pressure required on assembling the gasket in the flange to achieve the desired level of sealing under operating conditions.

Minimum gasket pressure under operating conditions　Minimum pressure required on gasket to remain within leakage class under operating conditions.

Nominal pipe size (NPS)　An alpha-numeric designation of size for components of a pipework system. For the purpose of class-designated flanges, it comprises of the letters NPS, followed by a number which is related to the physical size of the bore or outside dimensions of the pipe component (e.g. see EN 1759).

Operational gasket pressure　Pressure retained on the gasket under operating conditions (the situation after initial tightening when the flange has been pressurized, is at operational temperature, and creep and other relaxation mechanisms have occurred).

O-ring　A seal (often referred to as a packing or moulded ring in the USA), usually elastomeric or hollow metal, of circular cross-section, nipped in a groove.

Permeability　A measure of the ease with which a fluid can pass through a gasket material.

Pipe schedules Tables defining pipe thickness in relation to nominal bore and process pressure, according to ISO standard.

PN Alpha-numeric designation related to mechanical and dimensional characteristics of a component of a pipework system. It comprises of the letters 'PN', followed by a number. Used to identify ranges of related components in a number of standards (e.g. EN 1092) and is defined in EN 1333.

Porosity Percentage difference between the theoretical and actual density of a material (as a result of small voids or interstices within the material matrix).

Preload Clamping force which a bolt exerts on a joint when tightened.

Pressure Load per unit area on a body.

Proof load The maximum, safe, static, tensile load which can be placed on a fastener without causing it to yield. It is an absolute value, sometimes defined as force (N), or pressure (MPa).

p/T rating The rating of a flange manufactured from a specified material, indicating the allowable pressure (non-shock) at which it may operate at a specific temperature (e.g. see tables in EN 1092 and EN 1759).

PT value Numerical value resulting from the multiplication of the internal pressure by the temperature of the fluid being sealed. Provides only a rough guide for limiting gasket usage.

PVRC Pressure Vessel Research Committee (USA).

Raised-face flange A flange which makes contact with its mating joint member only in the region where the gasket is located. The faces of the flange do not make contact with each other at the bolt circle.

Recovery Increase of thickness over the compressed thickness, once the compressive load has been removed.

Reinforcement Material (such as fabric, cord and/or metal) within the gasket matrix, which imparts increased tensile strength or other desirable properties.

Residual stress Stress remaining in a gasket after service for a given time.

Ring-joint flange A flange system in which both flanges are grooved to accept a ring-joint gasket.

Ring-joint gasket A gasket machined from metal (usually oval or octagonal in cross-section) and used in conjunction with ring-joint flanges.

ROTT Room temperature operational tightness test, as defined by the PVRC.

Sealability Ability of a gasket material to prevent flow of fluid through the body and/or over the surfaces.

Soft gasket materials See Hard and soft gasket materials.

Spiral-wound gasket A gasket design which is formed by winding spring-like material, usually 'V'-shaped, plus a suitable filler material, into a spiral.

Spring constant Equivalent to the 'stiffness' of a bolt and defined as the initial preload divided by the elongation of the bolt after application of load.

Stiffness Ability of a body to resist deformation due to the action of external forces. Reciprocal of elasticity.

Strain Change in dimensions or shape of a body due to applied force or stress.

Stress Effect of load per unit area on a body.

Stress corrosion cracking A common form of stress cracking in which an electrolyte encourages the growth of a crack in a bolt under stress.

Stress relaxation Loss of stress at a constant gasket thickness as a function of time, after application of a specified compressive load at a specified rate, at constant temperature.

Stud Fastener which is threaded at both ends.

Surface roughness Fine irregularities of the flange surface finish.

Tensile strength Breaking tensile force divided by the original cross-sectional area.

Tightness class Maximum acceptable specific leakage rate for particular applications.

Tightness parameter Tp Mathematical relationship between the measured specific leakage rate and the internal fluid pressure causing it.

Tongue and groove flange A flange system in which one flange is provided with an anular tongue and the other with a complementary groove to accept it.

7.1.3 Mechanical seals terminology

Barrier fluid Externally supplied fluid at a pressure above the pump seal chamber pressure, introduced into a dual pressurized seal to isolate the process fluid completely from the environment.

Barrier liquid See Barrier fluid.

Buffer See Buffer fluid.

Buffer fluid Externally supplied fluid at a pressure lower than the pump seal chamber pressure, used as a lubricant and/or diluent in a dual unpressurized seal.

Buffer liquid See Buffer fluid.

Bushing Close-clearance restrictive bush around the shaft or sleeve. It may be fixed or flexible radially when used in the casing or gland plate.

CMA Chemical Manufacturers Association. A US-based industry group.

Containment chamber Component forming the cavity into which the containment seal fits.

Containment seal Mechanical seal design with one flexible element, seal ring and mating ring mounted in the containment chamber.

Data sheets A template used to list data, information and specifications applicable to a particular item of plant equipment.

Double seal See Dual pressurized seal.

Dual mechanical seal A dual pressurized seal or dual unpressurized seal of any kind.

Dual pressurized seal Seal configuration having two seals per assembly which utilize an externally supplied barrier fluid.

Dual seal See Dual mechanical seal.

Dual unpressurized seal Seal configuration having two seals per assembly with a containment chamber which is at a pressure lower than the seal chamber pressure.

Energized containment seal Lip seal mounted in the containment chamber and used in the manner of a containment seal.

EPA Method 21 US Federal Regulation 40 CFR 60, 1990, 'Determination of Volatile Organic Compound Leaks', Reference Method 21, Appendix A.

Flush Fluid which is introduced into the seal chamber on the process fluid side in close proximity to the seal faces and used typically for cooling and lubricating the seal faces.

Flush plan Configuration of pipe, instruments and controls designed to route the fluid concerned to the seals. Auxiliary piping plans vary with the application, seal type and arrangement.

General purpose mechanical seals Mechanical seals which have not had the benefit of recent technological advances in design, materials and tribology.

Live loading Method used to compress gland packing that is independent of any manual tightening of gland plate studs. Ordinarily, it comprises of a controlled spring force.

Mating ring Disc- or ring-shaped member, mounted either on the sleeve or in a housing such that it does not move axially relative to the sleeve or housing, which provides the mating seal face for the seal ring.

Mechanical containment seal See Containment seal.

Mechanical seal A device which prevents the leakage of fluids along rotating shafts. Sealing is accomplished by a seal ring, mounted flexibly on the shaft or the equipment casing, which bears against a radial face of a fixed mating ring. The seal faces are perpendicular to the shaft axis. Axial mechanical force and fluid pressure maintain the contact between seal faces.

Mechanical seal data sheets   See Data sheets.

Metal bellows   A series of metal convolutions or a stack of welded metal diaphragms used to provide secondary sealing and spring type loading in a mechanical seal design.

MTBR   Mean time between repairs. A statistical methodology used to measure reliability in equipment.

Non-contacting seal   Mechanical seal design in which the mating faces are designed intentionally to create aerodynamic or hydrodynamic separating forces in order to sustain a specific gap between the seal ring and the mating ring.

Primary seal   Mechanical seal which seals the process fluid in a dual unpressurized seal.

Rotodynamic pump   Pump which functions by adding energy to the pumped fluid through a rotating impeller. This may be an axial, mixed or radial flow pump.

Rotor   Assembly of all the rotating parts of a rotodynamic pump.

Seal chamber   Component, either integral with or separate from the pump case (housing), which forms the region between the shaft and casing into which the mechanical seal is installed.

Seal face   Side or end of a mating ring or seal ring which provides the sealing surface on the ring.

Seal ring   Seal face which contacts the mating ring; it is mounted flexibly using springs or bellows.

Secondary containment device   Component or seal used to restrict process leakage to the environment in the event of a malfunction of the primary seal.

Single mechanical seal   Seal configuration having only one mechanical seal per assembly.

Split seal   Mechanical seal which has the seal ring and mating ring, and in some designs the other parts of the seal assembly, supplied in two halves such that they can be assembled on or removed from the equipment without removal of adjacent parts of it.

Vapour pressure margin   The pressure difference between the seal chamber pressure and the pressure at which the process liquid changes to a vapour at the sealed temperature.

VOC   Volatile organic compound. A chemical compound of carbon, excluding carbon monoxide, carbon dioxide, carbonic acid, metallic carbides or carbonates, which vaporizes at or below 21°C.

7.2 Standards

This section contains a summary of many of the standards that may be of direct use for sealing applications. Where an established international standard is in regular use this has been listed. Equivalent national standards are also applied in many countries. For instance, ISO 6547 is available in the UK as BS ISO 6547.

It is essential to verify that the most recent edition of a standard is being used. This is signified by the inclusion of a year immediately after the standard number. For example, the current latest ISO 3601-5 is 2005. It is referenced as ISO 3601-5: 2005. The issue years have not been shown in the list below as this will cause confusion as new or revised versions of standards are issued.

7.2.1 General

There are many standards that impact on the design of equipment, the type of seal that may be fitted and the materials that may be used. A selection of these is listed here.

Pressure vessels and equipment
97/23/EC: EU Pressure Equipment Directive.
ASME BPVC Section VIII: Rules for Construction of Pressure Vessels. (Known as the ASME Pressure Vessel Code.)
BS PD 5500: Specification for unfired fusion welded pressure vessels.
API 6A: Specification for Wellhead and Christmas Tree Equipment.
Standards of the Tubular Exchanger Manufacturers Association (TEMA).

Fasteners
BS 4882: Specification for bolting for flanges and pressure containing purposes.
ISO 3506: Mechanical properties of corrosion-resistant stainless-steel fasteners.
ISO 898: Mechanical properties of fasteners made of carbon steel and alloy steel.
ASTM A 193/A 193M: Standard Specification for Alloy-Steel and Stainless Steel Bolting Materials for High Temperature or High Pressure Service and Other Special Purpose Applications.

Limits and fits
ISO 286: ISO system of limits and fits.

Rotary machines
ISO 9905: Technical specifications for centrifugal pumps – Class I.
ISO 5199: Technical specifications for centrifugal pumps – Class II.
ISO 9908: Technical specifications for centrifugal pumps – Class III.
ISO 13709 and API 610: Centrifugal pumps for petroleum, petrochemical and natural gas industries.
ISO 2858: End-suction centrifugal pumps (rating 16 bar) – Designation, nominal duty point and dimensions.

ASME B73.1: Specification for Horizontal End Suction Centrifugal Pumps for Chemical Process.
ASME B73.2: Specification for Vertical In-Line Centrifugal Pumps for Chemical Process.
ISO 10349: Petroleum, chemical and gas service industries – Centrifugal compressors.
API 617: Axial and Centrifugal Compressors and Expander-Compressors for Petroleum, Chemical and Gas Industry Services.

Hygiene
BS6920: Suitability of non-metallic products for use in contact with water intended for human consumption with regards to their effect on the quality of the water.
FDA approval: Food Contact Notifications.

Fluid power
BS 7714: Care and handling of seals for fluid power applications – Guide.
ISO 6020: Hydraulic fluid power. Mounting dimensions for single rod cylinders, 16 MPa (160 bar) series.
ISO 6022: Hydraulic fluid power. Mounting dimensions for single rod cylinders, 25 MPa (250 bar) series.
ISO 10762: Hydraulic fluid power. Cylinder mounting dimensions, 10 MPa (100 bar) series.

Emissions
ISO 15848: Industrial valves – fugitive emissions – measurement test and qualification procedures.
    Part 1: classification system and qualification procedures for type test of valve assemblies.
    Part 2: production acceptance test of valve assemblies, on–off valves.
    Part 3: production acceptance test of valve assemblies, control valves.
Shell MESC SPE 77/312: Industrial valves: fugitive emission (FE) measurement, classification system, qualification procedures and prototype and production tests of valves.
EPA method 21: Method for determination of Volatile Organic Compound leaks.
VDI Richtlinie 2440: Emissionsminderung Mineralölraffinerien.

Reliability
ISO 14224: Petroleum and natural gas industries – Collection and exchange of reliability and maintenance data for equipment.

7.2.2 O-rings

BS 1806: Specification for dimensions of toroidal sealing rings ('O'-rings) and their housings (inch series).

AS 568 B: Aerospace Size Standard for O-Rings.
AS 4716: Gland Design, O-Ring and Other Elastomeric Seals.
AS 5782: Aerospace Standard, Retainer, Backup Ring, Hydraulic, Pneumatic, Polytetrafluoroethylene, Resin, Uncut.
AS 5857: Gland Design, O-Ring and Other Elastomeric Seals, Static Applications.
AS 5860: Retainers (Back-Up Rings), Hydraulic and Pneumatic, Polytetrafluoroethylene Resin, Single Turn, Static Gland.
ISO 16032-1: Aerospace fluid systems – O-rings, inch series: Inside diameters and cross sections, tolerances and size-identification codes. Part 1: Close tolerances for hydraulic systems.
ISO 16032-2: Aerospace fluid systems – O-rings, inch series: Inside diameters and cross sections, tolerances and size-identification codes. Part 2: Standard tolerances for non-hydraulic systems.
All the above standards cover the same 'nominal' O-ring dimensions. The tolerances will vary between the individual documents. Note that BS 1806 is the only standard that includes both dimensions of the O-rings and also recommended groove design information within one document.
BS 4518: Specification for metric dimensions of toroidal sealing rings ('O'-rings) and their housings.
SMS 1586: Metric O-rings, dimensions. [Nominally similar to BS 4518 O-ring sizes.]
SMS 1587: Metric O-rings, materials.
SMS 1588: Metric O-rings, installation dimensions.
BS 5106: Back-up-rings: Specification for dimensions of spiral anti-extrusion back-up rings and their housings. [To suit BS 4518 seal dimensions.]
JIS B 2401: O-rings (Japanese standard O-rings).
ISO 3601-3: Fluid power systems – O-rings – Part 3: Quality acceptance criteria.
ISO 3601-5: Fluid power systems – O-rings – Part 5: Suitability of elastomeric materials for industrial applications.

7.2.3 Hygienic seals

EN 1672-2: 'Food processing machinery – Basic concepts – Part 2: Hygiene requirements.'
ISO 2853: Stainless steel threaded couplings for the food industry.
BS 4825 Part 4: Stainless steel tubes and fittings for the food industry and other hygienic applications. Specification for threaded (IDF type) coupling.
BS 4825 Part 5: Stainless steel tubes and fittings for the food industry and other hygienic applications. Specification for recessed ring joint type couplings.
DIN 11851: Fittings for food, chemical and pharmaceutical industry – Stainless steel screwed pipe connections.
DIN 11861: Drink and dairy fittings; sealing rings made of elastomeric materials, requirements testing.
ISO 14159: Safety of machinery. Hygiene requirements for the design of machinery.

7.2.4 Gaskets

Gasket dimensions

EN 1514-1: Flanges and their joints. Dimensions of gaskets for PN-designated flanges. Non-metallic flat gaskets with or without inserts.

EN 1514-2: Flanges and their joints. Dimensions of gaskets for PN-designated flanges. Spiral wound gaskets for use with steel flanges.

EN 1514-3: Flanges and their joints. Dimensions of gaskets for PN-designated flanges. Non-metallic PTFE envelope gaskets.

EN 1514-4: Flanges and their joints. Dimensions of gaskets for PN-designated flanges. Corrugated, flat or grooved metallic and filled metallic gaskets for use with steel flanges.

EN 1514-6: Flanges and their joints. Dimensions of gaskets for PN-designated flanges. Covered serrated metal gaskets for use with steel flanges.

EN 1514-7: Flanges and their joints. Dimensions of gaskets for PN-designated flanges. Covered metal jacketed gaskets for use with steel flanges.

EN 1514-8: Flanges and their joints. Dimensions of gaskets for PN-designated flanges. Polymeric O-ring gaskets for grooved flanges.

EN 12560-1: Flanges and their joints. Gaskets for class-designated flanges. Non-metallic flat gaskets with or without inserts.

EN 12560-2: Flanges and their joints. Gaskets for class-designated flanges. Spiral wound gaskets for use with steel flanges.

EN 12560-3: Flanges and their joints. Gaskets for class-designated flanges. Non-metallic PTFE envelope gaskets.

EN 12560-4: Flanges and their joints. Gaskets for class-designated flanges. Corrugated, flat or grooved metallic and filled metallic gaskets for use with steel flanges.

EN 12560-5: Flanges and their joints. Gaskets for class-designated flanges. Metallic ring joint gaskets for use with steel flanges.

EN 12560-6: Flanges and their joints. Gaskets for class-designated flanges. Covered serrated metal gaskets for use with steel flanges.

EN 12560-7: Flanges and their joints. Gaskets for class-designated flanges. Covered metal jacketed gaskets for use with steel flanges.

ASME B16.20: Metallic Gaskets for Pipe Flanges: Ring Joint Spiral Wound and Jacketed.

DIN 2696: Lenticular ring joint gaskets for flanged joints.

Design rules

EN 1591-1: Flanges and their joints. Design rules for gasketed circular flange connections. Calculation method.

EN 1591-2: Flanges and their joints. Design rules for gasketed circular flange connections. Part 2. Gasket parameters.

Further design rules are included in the pressure equipment standards.

Flange dimensions

EN 1759-1: Flanges and their joints. Circular flanges for pipes, valves, fittings and accessories, class-designated. Steel flanges, NPS 1/2 to 24.

EN 1759-3: Flanges and their joints. Circular flanges for pipes, valves, fittings and accessories, class designated. Copper alloy flanges.
EN 1759-4: Flanges and their joints. Circular flanges for pipes, valves, fittings and accessories, class designated. Aluminium alloy flanges.
ISO 15837: Ships and marine technology. Gasketed mechanical couplings for use in piping systems. Performance specification.
ISO 15838: Ships and marine technology. Fittings for use with gasketed mechanical couplings used in piping applications. Performance specification.
BS 3293: Specification for carbon steel pipe flanges (over 24 inches nominal size) for the petroleum industry.
ASME B16.5: Pipe flanges and flanged fittings NPS 1/2 through NPS 24 metric/inch standard.
ASME B16.47: Large Diameter Steel Flanges (26 to 60 inches).

Gasket material specifications and testing
EN 14772: Flanges and their joints. Quality assurance inspection and testing of gaskets in accordance with the series of standards EN 1514 and EN 12560.
EN 13555: Flanges and their joints. Gasket parameters and test procedures relevant to the design rules for gasketed circular flange connections.
ISO 4708: Composition cork. Gasket material. Test methods.
ISO 4709: Composition cork. Gasket material. Classification system, requirements, sampling, packing and marking.
BS 7531: Specification for compressed non-asbestos fibre jointing.
EN 12365-1: Building hardware. Gasket and weatherstripping for doors, windows, shutters and curtain walling. Performance requirements and classification.
EN 12365-2: Building hardware. Gasket and weatherstripping for doors, windows, shutters and curtain walling. Linear compression force test methods.
EN 12365-3: Building hardware. Gasket and weatherstripping for doors, windows, shutters and curtain walling. Deflection recovery test method.
EN 12365-4: Building hardware. Gasket and weatherstripping for doors, windows, shutters and curtain walling. Recovery after accelerated ageing test method.
EN 751-1: Sealing materials for metallic threaded joints in contact with 1st, 2nd and 3rd family gases and hot water. Part 1: Anaerobic jointing compounds.
EN 751-2: Sealing materials for metallic threaded joints in contact with 1st, 2nd and 3rd family gases and hot water. Part 2: Non-hardening jointing compounds.
EN 751-3: Sealing materials for metallic threaded joints in contact with 1st, 2nd and 3rd family gases and hot water. Part 3: Unsintered PTFE tapes.
BS 7786: Specification for unsintered PTFE tape. General requirements.

Further comprehensive lists of testing standards are published by ASTM and DIN.

Flanged joints and the matching gaskets are manufactured to a wide range of often obsolescent standards that have been in use on plant for many years. Some of these are included in the discussion on gaskets, section 2.8. Use of these gaskets will continue for many years but it is anticipated that new equipment will be designed around the latest standards listed in this section.

7.2.5 Rotary shaft lip seals

ISO 6194-1: Specification for rotary shaft lip type seals. Nominal dimensions and tolerances.
ISO 6194-2: Specification for rotary shaft lip type seals. Vocabulary.
ISO 6194-3: Rotary shaft lip type seals. Storage, handling and installation.
ISO 6194-4: Rotary shaft lip type seals. Performance test procedures.
ISO 6194-5: Specification for rotary shaft lip type seals. Identification of visual imperfections.
ISO 16589-1: Rotary shaft lip-type seals incorporating thermoplastic sealing elements. Nominal dimensions and tolerances.
ISO 16589-2: Rotary shaft lip-type seals incorporating thermoplastic sealing elements. Vocabulary.
ISO 16589-3: Rotary shaft lip-type seals incorporating thermoplastic sealing elements. Storage, handling and installation.
ISO 16589-4: Rotary shaft lip-type seals incorporating thermoplastic sealing elements. Performance test procedures.
ISO 16589-5: Rotary shaft lip-type seals incorporating thermoplastic sealing elements. Identification of visual imperfections.

7.2.6 Mechanical seals

ISO 21049 and API 682: Pumps – Shaft sealing systems for centrifugal and rotary pumps.
EN 12756: Mechanical seals. Principal dimensions, designation and material codes.
ISO 3069: End-suction centrifugal pumps – Dimensions of cavities for mechanical seals and for soft packing.
API 614: Lubrication Shaft-Sealing and Control-Oil Systems for Special-Purpose Applications.
ISO 10438: Petroleum, petrochemical and natural gas industries – Lubrication, shaft-sealing and control-oil systems and auxiliaries.

7.2.7 Compression packing

BS 4371: Specification for Fibrous Gland Packings.

7.2.8 Reciprocating seals

Fluid power seals

ISO 5597: Hydraulic fluid power – Cylinders – Housings for piston and rod seals in reciprocating applications – Dimensions and tolerances.
ISO 6547: Hydraulic fluid power – Cylinders – Piston seal housings incorporating bearing rings – Dimensions and tolerances.
ISO 7425: Hydraulic fluid power – Housings for elastomer-energised, plastic-faced seals – Dimensions and tolerances.
 Part 1: Piston seal housings.
 Part 2: Rod seal housings.
ISO 6195: Fluid power systems and components – Cylinder-rod wiper-ring housings in reciprocating applications – Dimensions and tolerances.

ISO 10766: Hydraulic fluid power – Cylinders – Housing dimensions for rectangular-section-cut bearing rings for pistons and rods.

ISO 7986: Hydraulic fluid power – Sealing devices – Standard test methods to assess the performance of seals used in oil hydraulic reciprocating applications.

Piston rings

ISO 6621-1: Internal combustion engines – Piston rings – Part 1: Vocabulary.

ISO 6621-2: Internal combustion engines – Piston rings – Part 2: Inspection measuring principles.

ISO 6621-3: Internal combustion engines – Piston rings – Part 3: Material specifications.

ISO 6621-4: Internal combustion engines – Piston rings – Part 4: General specifications.

ISO 6621-5: Internal combustion engines – Piston rings – Part 5: Quality requirements.

ISO 6622-1: Internal combustion engines – Piston rings – Part 1: Rectangular rings made of cast iron.

ISO 6622-2: Internal combustion engines – Piston rings – Part 2: Rectangular rings made of steel.

ISO 6623: Internal combustion engines – Piston rings – Scraper rings made of cast iron.

ISO 6624-1: Internal combustion engines – Piston rings – Part 1: Keystone rings made of cast iron.

ISO 6624-2: Internal combustion engines – Piston rings – Part 2: Half keystone rings made of cast iron.

ISO 6624-3: Internal combustion engines – Piston rings – Part 3: Keystone rings made of steel.

ISO 6624-4: Internal combustion engines – Piston rings – Part 4: Half keystone rings made of steel.

ISO 6625: Internal combustion engines – Piston rings – Oil control rings.

ISO 6626: Internal combustion engines – Piston rings – Coil-spring-loaded oil control rings.

ISO 6626-2: Internal combustion engines – Piston rings – Part 2: Coil-spring-loaded oil control rings of narrow width made of cast iron.

7.2.9 Material properties

7.2.9.1 Elastomer materials

ASTM D 1600: Abbreviations of Terms relating to Plastics.

ISO 1629: Rubber and lattices nomenclature.

Test methods

ISO 37: Rubber, vulcanized or thermoplastic – Determination of tensile stress-strain properties.

ISO 48: Rubber, vulcanized or thermoplastic – Determination of hardness (hardness between 10 IRHD and 100 IRHD).

ISO 815: Rubber, vulcanized or thermoplastic – Determination of compression set at ambient, elevated or low temperatures.
ISO 1432: Rubber, vulcanized or thermoplastic – Determination of low temperature stiffening (Gehman test).
ISO 3384: Rubber, vulcanized or thermoplastic – Determination of stress relaxation in compression at ambient and at elevated temperatures.
ISO 6072: Hydraulic fluid power. Compatibility between fluids and standard elastomeric materials.
ISO 6179: Rubber, vulcanized or thermoplastic. Rubber sheets and rubber-coated fabrics. Determination of transmission rate of volatile liquids (gravimetric technique).
ASTM D 395: Test Methods for Rubber Property – Compression Set.
ASTM D 471: Test Method for Rubber Property – Effect of Liquids.
NORSOK: M-710: Qualification of non-metallic sealing materials and manufacturers.
NACE: TM0187: Evaluating Elastomeric Materials in Sour Gas Environments.
NACE: TM0192: Evaluating Elastomeric Materials in Carbon Dioxide Decompression Environments.
NACE: TM0296: Evaluating Elastomeric Materials in Sour Liquid Environments.
NACE: TM0297: Effects of High-Temperature, High-Pressure Carbon Dioxide Decompression on Elastomeric Materials.

Material property specifications
EN 291: Rubber seals. Static seals in domestic appliances for combustible gas up to 200 m bar. Specification for material.
EN 549: Specification for rubber materials for seals and diaphragms for gas appliances and gas equipment.
EN 681-1: Elastomeric seals. Material requirements for pipe joint seals used in water and drainage applications. Vulcanized rubber.
EN 681-2: Elastomeric seals. Material requirements for pipe joint seals used in water and drainage applications. Thermoplastic elastomers.
EN 681-3: Elastomeric seals. Material requirements for pipe joint seals used in water and drainage applications. Vulcanized rubber.
EN 681-4: Elastomeric seals. Material requirements for pipe joint seals used in water and drainage applications. Cast polyurethane sealing elements.
ISO 3601-5: Fluid power systems – O-rings – Part 5: Suitability of elastomeric materials for industrial applications.
ISO 16010: Elastomeric seals – Material requirements for seals used in pipes and fittings carrying gaseous fuels and hydrocarbon fluids.

Aerospace materials
EN 2104: Acrylonitrile-butadiene rubber (NBR). Hardness 40 IRHD.
EN 2259: Silicone rubber (VMQ). Hardness 50 IRHD.
EN 2260: Silicone rubber (VMQ). Hardness 60 IRHD.
EN 2261: Silicone rubber (VMQ). Hardness 70 IRHD.
EN 2262: Silicone rubber (VMQ/PVMQ) with high tear strength. Hardness 50 IRHD.

EN 2428: Ethylene-propylene rubber (EPM/EPDM). Hardness 50 IRHD.
EN 2429: Ethylene-propylene rubber (EPM/EPDM). Hardness 60 IRHD.
EN 2430: Ethylene-propylene rubber (EPM/EPDM). Hardness 70 IRHD.
EN 2431: Ethylene-propylene rubber (EPM/EPDM). Hardness 80 IRHD.
EN 2432: Ethylene-propylene rubber (EPM/EPDM). Hardness 90 IRHD.
EN 3049: O-rings, in fluorocarbon rubber (FKM), low compression set. Hardness 80 IRHD.
EN 3050: O-rings, in fluorocarbon rubber (FKM), low compression set. Technical specification.
EN 3207: Rubber compounds. Technical specification.

7.2.9.2 Metals

ISO 15156 or NACE MR0175: Petroleum and natural gas industries – Materials for use in H_2S-containing environments in oil and gas production.
NACE MR0103: Materials Resistant to Sulfide Stress Cracking in Corrosive Petroleum Refining Environments.

There are many national and international standards for metals and alloys for either general materials or specific applications. Section 7.2.1 contains some on fastener materials.

7.2.10 Surface texture measurement

ISO 1302: Geometric Product Specification (GPS) – Drawing indication of surface texture.
ISO 3274: Geometric Product Specification (GPS) – Surface texture: profile method – Nominal characteristics of contact (stylus instruments).
ISO 4287: Geometric Product Specification (GPS) – Surface texture: profile method – Terms, definitions and surface texture parameters.
ISO 4288: Geometric Product Specification (GPS) – Surface texture: profile method – Rules and procedures for the assessment of surface texture.
ISO 13565-1: Geometric Product Specification (GPS) – Surface texture: profile method – Surfaces having stratified functional properties – Part 1. Filtering and general conditions.
ISO 13565-2: Geometric Product Specification (GPS) – Surface texture: profile method – Surfaces having stratified functional properties – Part 2. Height characterisation using the linear material ratio curve.
ISO 13565-3: Geometric Product Specification (GPS) – Surface texture: profile method – Surfaces having stratified functional properties – Part 3. Height characterisation using the material probability curve.

7.2.11 Standards organizations

International Standards Organization
ISO Central Secretariat, International Organization for Standardization (ISO), 1, rue de Varembé, Case postale 56, CH-1211 Geneva 20, Switzerland
Tel: +41 22 749 01 11, Fax: +41 22 733 34 30, Web: www.iso.org

European standards
CEN Management Centre, 36, rue de Stassart, B-1050 Brussels, Belgium
Tel: +32 2 550 08 11, Fax: +32 2 550 08 19, Web: www.cenorm.be

National standards organizations particularly involved in sealing standards
United Kingdom
BSI, 389 Chiswick High Road, London W4 4AL, UK
Tel: +44 20 8996 9000, Web: www.bsi-global.com

France
Association Française de Normalisation (AFNOR), 11, rue Francis de Pressensé, 93571 La Plaine Saint-Denis Cedex, France
Tel: +33 1 41 62 80 00, Fax: +33 1 49 17 90 00, Web: www.afnor.fr

Germany
Deutsches Institut für Normung e.V (DIN), Öffentlichkeitsarbeit, Burggrafenstrasse 6, 10787 Berlin, Germany
Tel: +49 30 2601 1113, Fax: +49 30 2601 1115, Web: www2.din.de

Japan
Japan Industrial Standards (JIS), 1-3-1 Kasumigaseki, Chiyoda-ku, Tokyo 100-8901, Japan
Web: www.jisc.go.jp

Norway
Standard Norge, Strandveien 18, Lysaker, Norway
Tel: +47 67 83 86 00, Web: www.standard.no

Sweden
Swedish Standards Institute (SIS), SE-118 80 Sankt Paulsgatan 6, Stockholm, Sweden
Tel: +46 8 555 520 00, Fax: +46 8 555 520 01, Web:www.sis.se

USA
American National Standards Institute (ANSI), 25 West 43rd Street, 4th floor, New York, NY 10036, USA.
Tel: +1 212 642 4900, Fax: +1 212 398 0023, Web: www.ansi.org

US organizations that also publish widely used industrial standards
American Society for Testing Materials (ASTM), 100 Barr Harbor Drive, West Conshohocken, Pennsylvania, USA
Tel: +1 610 832 9500, Fax: +1 610 832 9555, Web: www.astm.org
Material testing standards

American Petroleum Institute (API), 1220 L Street, NW, Washington, DC 20005-4070, USA
Tel: +1 202 682 8000, Web: www.api.org
Equipment standards for oil and gas exploration and refineries

American Society of Mechanical Engineers (ASME), Three Park Avenue, New York, NY 10016-5990, USA
Tel: +1 212 591 7159, Web: www.asme.org
Process plant pump and piping standards

Society of Automotive Engineers (SAE), SAE World Headquarters, 400 Commonwealth Drive, Warrendale, PA 15096-0001, USA
Tel: 1-724-776-4841, Fax: +1 724 776 0790, Web: www.sae.org
Produces many widely used automotive and aerospace standards. Automotive standards have the designation J and aerospace standards AS or AMS

Aerospace Industries Association, 1000 Wilson Boulevard, Suite 1700, Arlington, VA 22209-3928, USA
Tel: +1 703 358 1000, Web: www.aia-aerospace.org
Produces NAS standards which include some specific O-ring size and material specifications

NACE International (formerly the National Association of Corrosion Engineers), 1440 South Creek Drive, Houston, Texas 77084-4906, USA
Tel: +1 281-228-6200, Web: www.nace.org
Some widely used material testing or application standards particularly for corrosive or other aggressive service

Food and Drug Administration, 5630 Fishers Lane, rm. 1061, Rockville, MD, 20852, USA
Tel: +1-888-463-6332, Web: www.fda.gov
Approves materials that are to be used in contact with food or pharmaceuticals

US EPA, Office of Air Quality Planning and Standards (OAQPS), Research Triangle Park, NC 27711, USA
Web: www.epa.gov
Emission testing standards

7.3 Surface texture measurement

7.3.1 Introduction

The surface texture of the sealed surfaces is fundamental to reliable operation of seals. This is true for both static and dynamic seals. An understanding of the various parameters used to define surface texture, and how they are derived, is therefore important to achieving the optimum sealing solution and for liaison with component suppliers and production engineers.

The situation often becomes more confused than necessary because different parameters have traditionally been used in various parts of the world. An understanding of the relative values and merits of the different parameters is extremely valuable in ensuring effective communication. (The author has witnessed what very nearly became a genuine fight across the table at a meeting over what was considered to be a significant difference in the surface texture required for seal grooves. It was extremely difficult to persuade the protagonists that one was using Ra and the other Rt, and in fact they were effectively talking about a very similar surface.)

This section describes the main surface texture parameters that are liable to be encountered when specifying surfaces for sealing counter-faces. It is important to understand a little of the method by which the surface is sampled as this is fundamental to the detail results.

The machine should be set to sample correctly for the type of surface being measured. This will set the stroke length of the machine. Two terms are used within the calculation of the parameters:

- Sampling length (also familiarly known as 'cutoff'). This is the basic length of surface that is used to calculate the parameter values. For most sealing surfaces, with Ra in the range 0.1 to 1.6 μm this will be 0.8 mm. The prescribed values recommended for individual surface texture ranges are specified in ISO 4288.
- Evaluation length (also known as 'assessment length'). This is a multiple of the sampling length which is used to average the sampling length values to provide the resulting parameter value. It is conventionally five times the sampling length.

With most instruments the sampling length (cutoff) is selected. This will then set the machine stroke to just over five times this value to provide the evaluation length plus a short run-in and -out. The various parameters and the procedures that should be used are described in ISO 4287 and ISO 4288, or the national equivalents. The parameter names vary quite significantly between successive versions of these standards and so it is important to ensure reference to a current edition. The current editions during 2006 are listed in section 7.2.10.

When viewing a graphical trace of surface texture it is also important to remember that it is not a direct representation of the surface exposed to the seal. The vertical scale is magnified by 5000 or 10 000 but the horizontal scale by only 50 or 100. The apparently sharp peaks and valleys are therefore 100 times lower than they appear on the trace.

7.3.2 Ra arithmetic mean

Definition
Mathematically, Ra is the arithmetic average value of the profile departure from the mean line, within a sampling length.

A method of visualizing how Ra is derived can be described using the graphs in Figure 7.1.

Graph A: A mean line X–X is fitted to the measurement data.
Graph B: The portions of the profile within the sampling length l and below the mean line are then inverted and placed above the line.
Graph C: Ra is the mean height of the profile above the original mean line.

This parameter has also been known in the past as centre line average (CLA), or in the USA, arithmetic average (AA).

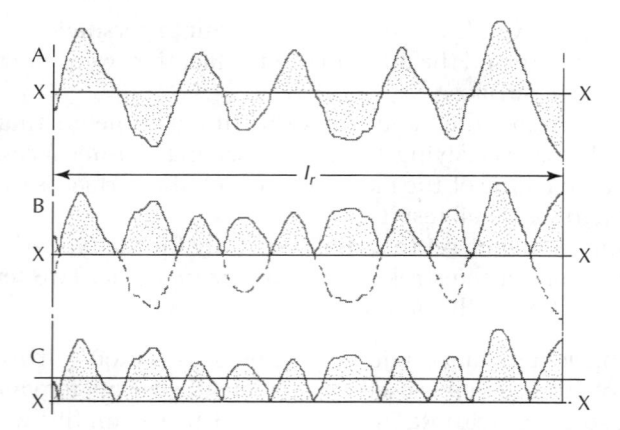

Figure 7.1 The calculation of arithmetic mean, Ra, of a surface profile. (Source: Taylor Hobson)

Applications

Ra is a controlling parameter, if the Ra value changes then the process producing the surface has changed, e.g. cutting tip, speeds, feeds and cutting fluid (lubricant). Ra is the most commonly used parameter in industry and is available in the simplest and lowest priced instruments from all manufacturers. The averaging nature of Ra makes it a stable parameter which is not influenced by odd or spurious spikes or scratches. When measuring extremely fine surfaces Ra is not sensitive enough to pick out the odd or infrequent defects that are important and Rq, described below, is more effective. The primary version Pa is often used on very short surfaces such as O-ring grooves where filtering would remove relevant detail affecting the performance of the seal.

The waviness version Wa can be used on larger-scale sealing faces such as those for cylinder head gaskets. There is not enough compliance in the components structure to flatten out large waviness features and thus the seal may fail.

Limitations

The Ra parameter is often misused due to lack of understanding of its limitations, although the parameter itself is a stable parameter due to its averaging of the surface. The use of Ra in isolation does not tell us anything about the surface characteristics of the component, quite different surfaces can have exactly the same Ra value and consequently perform in a different manner. Figure 7.2 illustrates four different surface textures that will all have the same Ra value.

It is therefore important to use additional parameters to define the overall texture.

7.3.3 Root mean square deviation of the profile from the mean line

Mathematically, Rq is the square root of the mean of all the 'Z' values after they have been squared.

Rq is defined over sampling length l as shown in Figure 7.3. It is also often known, particularly in the USA, as rms.

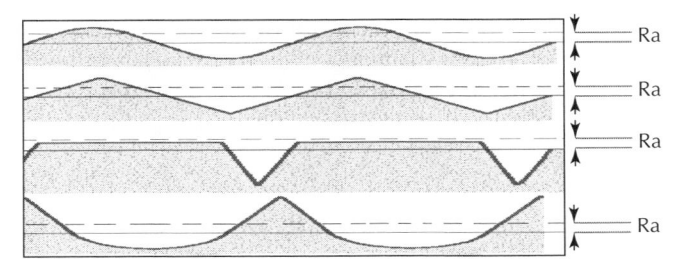

Figure 7.2 Ra is independent of the surface texture profile and machining operation. In this example three different surface profiles all have the same value of Ra. (Source: Taylor Hobson)

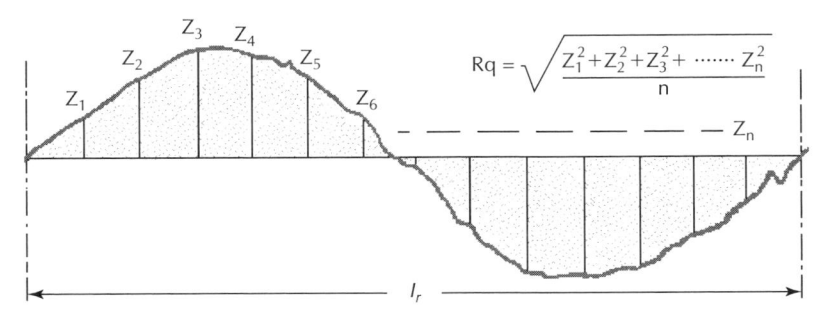

$$Rq = \sqrt{\frac{Z_1^2 + Z_2^2 + Z_3^2 + \cdots Z_n^2}{n}}$$

Figure 7.3 The calculation of the root mean square of the surface profile, parameter Rq. (Source: Taylor Hobson)

Applications
Rq has the effect of magnifying single or odd spikes and valleys, thus making it a parameter which differentiates between very smooth surfaces with similar surface with non-typical marks or defects. For statistical work, Rq values are more meaningful than arithmetic average ones. This parameter is not used very much in general engineering, but is used more in the optical and electronic substrates industry as it is able to detect spurious peaks and valleys. High Rq values on mirrors or lens surfaces will signal potential image quality reduction and/or local distortions. The primary version Pq is used to assess the quality of sealing surfaces in O-ring grooves and the interaction between the rubber and metal surfaces. Such short length surfaces do not require the normal roughness filters as they would remove relevant detail.

Limitations
Compared with the Ra parameter, Rq (rms) has the effect of giving extra weight to the higher values. This can be illustrated with three groups of values:

3, 4, 5   2, 4, 6   1, 4, 7

The arithmetic average is 4 in each case, the successive increase of one in the higher value being exactly balanced by the decrease in one in the lowest value. The respective rms values are 4.07, 4.3 and 4.7, showing that the increase in the highest figure outweighs the decrease in the lowest. Although this ability is useful for very fine 'optical' class surfaces it gives rise to large variations on general machined components when compared to Ra. These variations are not usually relevant to the performance of the parts and would thus be misleading if used for process control.

7.3.4 Rt – total height of the profile

Rt is the maximum peak to valley height of the profile in the assessment (evaluation) length (ln) as shown in Figure 7.4. It is often confused with the old Ry and Rmax parameters.

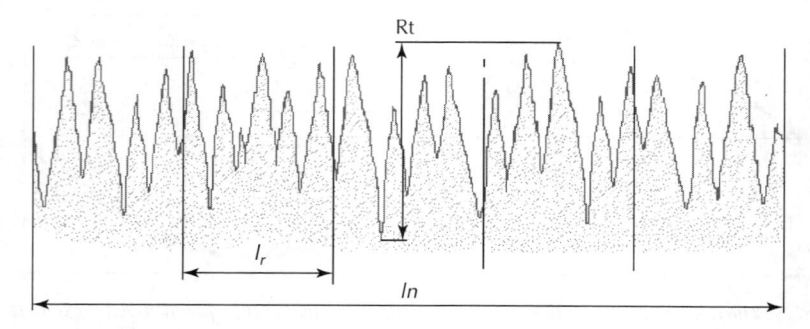

Figure 7.4 The derivation of the Rt parameter. (Source: Taylor Hobson)

Applications
Rt is the maximum peak to valley height of the profile in the evaluation length (ln); however, because this is a peak parameter it is subject to large variations and can be unstable. It shows the extreme limits of the profile, but they may not be coincident. Rt is used as a controlling parameter, particularly useful where components are subjected to high stresses, any large peak to valley could suffer from crack propagation. Singular large peaks can also penetrate oil lubrication films increasing wear, debris and damage to sliding surfaces. Electrical contact effectiveness and the risk of arcing or sparking from singular non-typical peaks can also be identified.

The primary version Pt is used to quantify the overall form error of a component or surface.

Pt is also used in O-ring grooves where the surface is very short and filtering becomes meaningless, to assess the adequacy of the surface to seal properly.

Limitations
The Rt parameter is extremely sensitive and will be greatly influenced by one speck of dirt or a very deep scratch. Surfaces should be very carefully cleaned to remove all dirt prior to measurement. It is best to confirm the result by graphical display.

Rt, as with all extreme parameters, is divergent, the more measurements that are taken on the surface the greater the possibility of finding a higher value of Rt.

7.3.5 Rz (ISO) – maximum height of profile

Mathematically, the highest peak to valley within a sampling length, usually analysed as a mean over a minimum of five sampling lengths, Figure 7.5. This parameter has also been known in the past as Rz(din), Rz, Ry and Rtm.

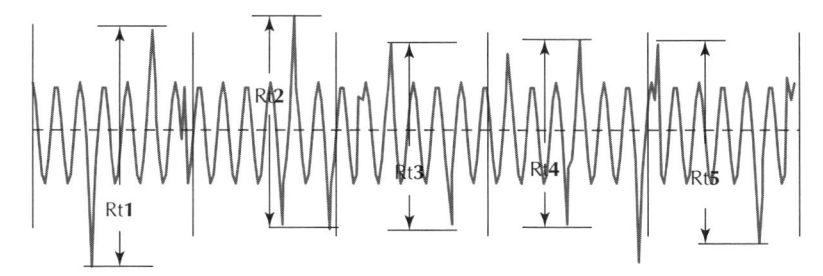

Figure 7.5 The current Rz (ISO) parameter is a mean of the five highest and five lowest valleys in the sampling length. (Source: Taylor Hobson)

Applications
This parameter has similar uses to the Rt parameter but is a little more stable as it has averaging involved when assessed over a number of sampling lengths. Rz is an alternative to Rt as a controlling parameter. It has applications for assessing oil film penetration probability in sliding contact bearings and electrical contact effectiveness and likelihood of arcing or burning.

Limitations
This is the most commonly used height parameter and gives a fairly stable reading but can be influenced by dirt in the positive direction and deep scratches in the negative direction.

7.3.6 History of peak parameters

The peak parameters were developed and promoted by Germany and the USSR as they lent themselves to optical techniques that were preferred at that time. Averaging parameters, however, such as Ra and Rq were developed and promoted in the measurement and analysis of surface roughness by the UK, USA and other European countries. This was due to the availability and development of stylus-based instruments.

The ISO 4287 (-1997) description of Rz is the highest peak to valley within a sampling length and is usually analysed as a mean over a minimum of five sampling lengths.

Rz (ISO) – ISO 4287-1984 Average value of the heights of the five highest peaks and the depths of the five deepest valleys within a sampling length – This is the equivalent to Rz (JIS) (JIS B 0601-1994) which is the preferred name. This parameter was also known as the 'ten point height' parameter.

The industry non-standard definition of Rz, before 1997, was the average value of the heights of the five highest peaks and the depths of the five deepest valleys within an evaluation length of an unfiltered profile (primary profile). There is no equivalent modern parameter name. This was never an ISO parameter. Originally this was used as a method of assessing very short surfaces from an unfiltered graph. There were very many problems with this parameter as it requires five peaks and five valleys within evaluation length. It is better to recommend the use of Pt as the method of assessing very short surfaces from an unfiltered graph.

Rtm – The average of the highest peak to valley readings in each sampling length. This is usually assessed over five sampling lengths. It is equivalent to Rz (ISO) (ISO 4287-1997 terminology) which is the preferred name.

7.3.7 Rsk skewness

Mathematically, Rsk is a measure of the symmetry of the profile about the mean line. This parameter indicates whether the spikes on the surface are predominately negative or positive or if the profile has an even distribution of peaks and valleys. This is demonstrated by the three examples in Figure 7.6. These show a surface with an even distribution of peaks and valleys, Figure 7.6(a), such as may be produced by grinding, a surface with dominant peaks, Figure 7.6(b), and one with predominant valleys, Figure 7.6(c). This latter surface is typical of a surface that undergoes a secondary finishing operation, such as honing.

Applications

Rsk indicates the nature of the surface as produced by the manufacturing process, whether it has a majority of peaks, valleys or a balance of the two. A negatively skewed surface will be produced by the plateau honing process for instance. Used in engine cylinder bores this predominance of valleys is good for the retention of oil and thus has good lubrication properties. Analysing skew during the phases of engine oil and engine wear testing records the changing character of the surfaces.

Conversely a negative skew in the corner radius of a crankshaft main bearing would point to stress raising areas, crack propagation and potential early failure. The positive skew surface produced by turning, however, does not retain liquid under sliding contact conditions and is good for high friction applications such as the discs and brake drums of vehicles. A negative skew means the surface will have good 'wetability' which is needed for oil and lubrication or surface coating processes. Conversely, negative skew surfaces are hard to clean and give rise to contamination in pharmaceutical, food and chemical transport systems. When measuring skew non-phase-corrected filters should also be avoided due to possible distortion. Skidded instruments should be avoided if Rsk is positive.

Limitations

Skew has no units and is best viewed in conjunction with one or more other parameters to give a complete understanding of the surface and its magnitude. If the Rsk value exceeds ± 1 averaging parameters such as Ra and Rq are unlikely to be good discriminators as they will have large variations on the same surface sample.

General information 483

(a)

(b)

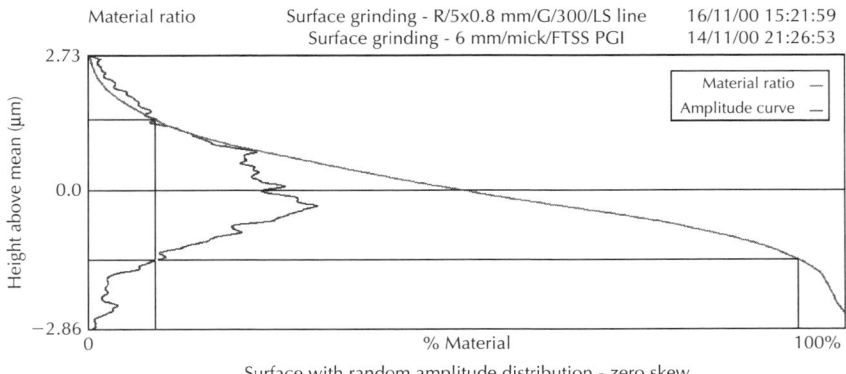

(c)

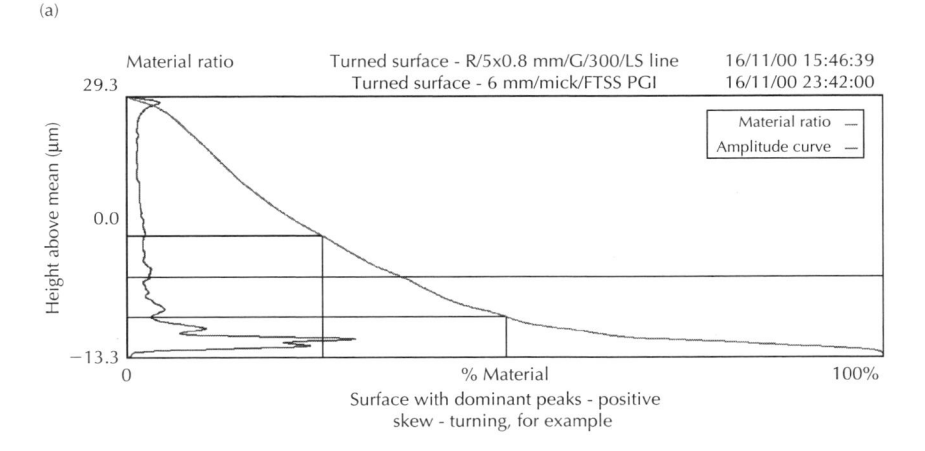

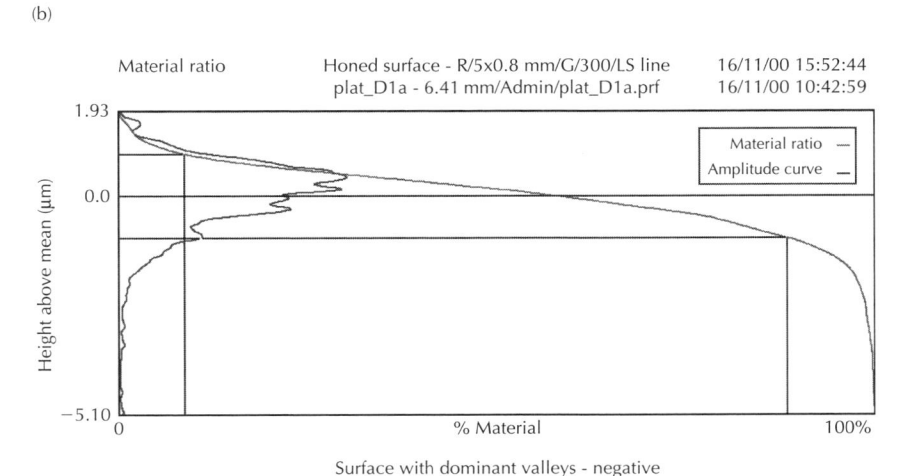

Figure 7.6 The skew parameter defines whether the surface has regular peaks and valleys of either has dominant peaks or is polished. (Source: Taylor Hobson)

7.3.8 Material ratio curve

Material ratio curve is a plot of depth into the surface 'V', the ratio, of the length of the bearing surface at that depth into the profile, with respect to the profile length and is quoted as a percentage.

The analysis takes two horizontal slices through the surface at defined depths, the reference level and the determination level, and measures the proportion of material in the slice. A good bearing surface will have a high proportion of material in the slice, i.e. it has few peaks but relatively deeper valleys.

The relative material ratio, from the MR curve, is determined by the evaluation length, Figure 7.7.

Material ratio value is determined at a depth d below a reference level c. The reference level may be 'below the peak', 'above the valley' or 'above or below the mean line'.

The material ratio curve replaces tp% – bearing ratio as defined in ISO 4287-1984.

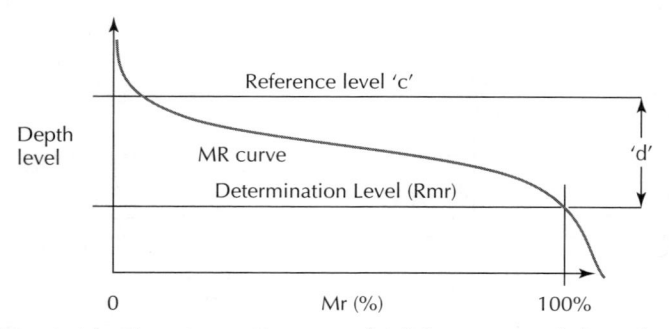

Figure 7.7 The material ratio curve provides a more detailed assessment of the surface profile which permits improved definition of the machining process. (Source: Taylor Hobson)

History of the material ratio curve

This parameter was originally devised in 1933 and described in a paper by E.J. Abbott and F.A. Firestone.[1] It is also known by a number of alternative names including Abbott Firestone curve, Abbott curve, bearing area curve, bearing ratio curve, material ratio curve and Traganteil.

Applications

This parameter has been used for the specification of critical lubricated surfaces such as cylinder bores for a number of years. The ability to measure the bearing area and hence infer the amount of secondary machining or surface treatment provides a method of specifying and monitoring the overall type of surface texture. It therefore provides a method of defining the type of characteristics required of the surface in a reasonable format. However, when comparing measurements it is important to ensure that they are directly equivalent, in that the depth into the profile is the same in each case.

7.3.9 Conclusion

To accurately define and monitor a surface for potential seal applications it is necessary to use at least one height parameter and a further parameter that is dependent on the finishing operation. This conclusion was also reached in a major study of surface texture for seal application.[2]

For height parameters Ra will provide the most stable and consistent readings, but alternatives are quite appropriate provided that the overall equivalence is appreciated between the values obtained.

If Rz (ISO) is monitored in parallel this will indicate whether there is a tendency for spurious peaks that could be damaging to a seal installation.

As discussed in section 7.3.7 skew will indicate whether a surface has predominantly peaks or valleys, and is thus an indicator of the final machining or finishing operation. However, as it has no units it is difficult to quantify.

It is therefore preferable to use the material ratio curve to prescribe the overall characteristics of the texture. This will help to ensure that a texture appropriate to the seal duty is obtained and also demonstrate if the manufacturing method has changed.

This method of specifying sealing installations is now used in some ISO and equivalent national standards. An example is that given in the reciprocating seal standard ISO 5597.

- Sliding surfaces: The surface texture of components that are in sliding contact with the seal shall not exceed 0.4 μm Ra, 1.6 μm Rz and have a bearing ratio Rmr of between 50 and 90% at a depth of 25% of Rz, from a reference line of 5% bearing ratio, when measured in accordance with ISO 4287 and ISO 4288.
- Static surfaces: The surface texture of the piston head groove in the case of a piston seal or the seal housing bore in the case of a rod seal shall not exceed 1.6 μm Ra, 6.3 μm Rz and have a bearing ratio Rmr of between 50 and 90% at a depth of 25% of Rz, from a reference line of 5% bearing ratio, when measured in accordance with ISO 4287 and ISO 4288.

This specification permits the use of two height parameters to align with common practice in different countries and also sets out a material ratio curve limit that will ensure appropriate machining practice to achieve suitable surfaces for the static and dynamic seals.

For critical dynamic surfaces it is possible to carry out a more detailed analysis of the profile to determine the overall characteristics. This is used for components such as engine cylinder bores and diesel injector pumps and components. This additional analysis is described in ISO 13565 Parts 1–3.

References
1. E.J. Abbott and F.A. Firestone, 'Specifying surface quality', *Mechanical Engineering*, September 1933, Vol. 55, 569–572.
2. R.K. Flitney, I. Hansford and B.S. Nau, 'The effect of surface texture on reciprocating seal performance', BHR Group Report CR3069, 1989.

Acknowledgement

This section has been prepared using information and diagrams provided by Taylor Hobson Ltd and has provided a brief introduction to this subject. For further information it will be necessary to consult the standards discussed or contact the technical or training departments of surface texture equipment manufacturers, such as Taylor Hobson.

7.4 Organizations with a direct interest in sealing technology

See also organizations involved with standards, section 7.2.

7.4.1 Trade and industry organizations

Aerospace and Defence Industries Group – Europe (ASD)
Gulledelle 94-b.5, B-1200 Brussels, Belgium. Tel: +32 2 775 81 10, Fax: +32 2 775 81 11, Web: www.asd-europe.org
ASD is a merger of the former European Association of Aerospace Industries (AECMA), European Defence Industries Group (EDIG) and the Association of the European Space Industry (EUROSPACE).

European Sealing Association
Tegfryn, Tregarth, Gwynedd LL57 4PL, UK. Tel: +44 1248 600 250, Fax: +44 1248 600 250.
Web: www.europeansealing.com

The Fluid Sealing Association
994 Old Eagle School Road, Suite 1019, Wayne, Pennsylvania 19087, USA. Tel: +1 610 971 4850, Fax: +1 610 971 4859.
Web: www.fluidsealing.com

Gasket Cutters' Association
105 St Peter's Street, St Albans, Hertfordshire AL1 3EJ, UK. Tel: +44 1727 896084, Fax: +44 1727 896026.
Web: www.gcassociation.co.uk

Gasket Fabricators Association
994 Old Eagle School Road, Suite 1019, Wayne, Pennsylvania 19087, USA. Tel: +1 610 971 4850, Fax: +1 610 971 4859.
Web: www.gasketfab.com

Hydraulic Institute
9 Sylvan Way, Parsippany, NJ 07054, USA. Tel: +1 973 267 9700.
Web: www.pumps.org

Independent Sealing Distributors (ISD)
105 Eastern Avenue – Suite 104, Annapolis, Maryland 21403, USA. Tel: +1 410 263 1014, Fax: +1 410 263 1659.
Web: www.isd.org

International Tribology Council
The Institution of Mechanical Engineers, 1 Birdcage Walk, Westminster, London
    SW1H 9JJ, UK.
Tel: +44 20 7222 7899, Fax: +44 20 7222 4557.
Web: www.itctribology.org

The Japan Society of Industrial Machinery Manufacturers (JSIM)
Kikai Shinko Bldg, 3-5-8 Shiba-koen, Minato-ku, Tokyo 105, Japan. Tel: +81 3
    3434 6825, Fax: +81 3 3434 4767.
Web: www.jsim.or.jp

National Fluid Power Association (NFPA)
3333 North Mayfair Road, Suite 101, Milwaukee, Wisconsin 53222-3219, USA.
    Tel: +1 414 778 3344, Fax: +1 414 778 3361.
Web: www.nfpa.com

Society of Tribologists and Lubrication Engineers (STLE)
840 Busse Highway, Park Ridge, Illinois 60068-2376, USA. Tel: +1 847 825
    5536, Fax: +1 847 825 1456.
Web: www.stle.org

British Fluid Power Association (BFPA)
Cheriton House, Cromwell Park, Chipping Norton, Oxfordshire OX7 5SR, UK.
    Tel: +1 44 1608 647900.
Web: www.bfpa.co.uk

VDMA
Lyoner Straße 18, 60528 Frankfurt am Main, Germany. Tel: +49 69 66 03
    1318, Fax: +49 69 66 03 2318.
Web: www.vdma.org

European Hygienic Equipment Design Group
Avenue Grand Champ 148, 1150 Brussels, Belgium. Tel: +32 2 761 7408,
    Fax: +32 2 763 0013.
Web: www.edehg.org

China Hydraulics, Pneumatics and Seals Association (CHPSA)
No. 46 Sanlihe Rd, Beijing 100823, China. Tel: +86 10 68595199, Fax:
    +86 10 68595199.
Web: www.chpsa.org.cn

Rubber Manufacturers Association
1400 K Street, NW, Suite 900, Washington, DC 20005, USA. Tel: +1 202 682
    4800.
Web: www.rma.org

7.4.2 Research organizations

BHR Group
The Fluid Engineering Centre, Cranfield, Bedford MK43 0AJ, UK.
Tel: +44 1234 750422, Fax: +44 1234 750074.
Web: www.bhrgroup.com
Long history of R&D on hydraulic cylinder seals, mechanical seals and static seals. Series of 18 International Conferences.

CETIM
74 Rue De La Rainiere BP 957, 44076 Nantes Cédex 3, France.
Tel: +31 40 37 36 35.
Web: www.cetim.fr
Testing of packings and gaskets.

Georgia Institute of Technology
School of Mechanical Engineering, 801 Ferst Drive, NW, Atlanta, GA 30332-0405, USA.
Tel: +1 404 894 3200, Fax: +1 404 894 8336.
Web: www.me.gatech.edu
R&D on rotary seals, mechanical seals and polymer lip seals.

IMA
University of Stuttgart, Pfaffenwaldring 9, 70569 Stuttgart, Germany.
Tel: +49 711 685 6170, Fax: +49 711 685 6319.
Web: www.ima.uni-stuttgart.de
R&D on rotary and reciprocating seals primarily for automotive and fluid power.

MERL
Wilbury Way, Hitchin, Hertfordshire SG4 0TW, UK.
Tel: +44 1462 427850, Fax: +44 1462 427851.
Web: www.merl-ltd.co.uk
Elastomer and composite materials, seal testing for oil and gas applications.

Mechanical Seal Technology
1006 Tramway Lane NE, Albuquerque, NM 87122, USA.
Tel: +1 505 821 7264, Fax: +1 505 856 1704.
Mechanical seals R&D.

MPA
University of Stuttgart, Pfaffenwaldring 32, 70569 Stuttgart, Germany.
Tel: +49 711 685 62604, Fax: +49 711 685 63144.
Web: www.mpa.uni-stuttgart.de
Testing of gaskets and valve packing.

RAPRA Technology
Shawbury, Shrewsbury, Shropshire SY4 4NR, UK.
Tel: +44 1939 250383, Fax: +44 1939 251118.
Web: www.rapra.net
Elastomer and plastic materials testing.

Turbomachinery Laboratory
Texas A&M University System, College Station, TX 77843-3254, USA.
Tel: +1 979 845 7417, Fax: +1 979 845 7417.
Web: http://turbolab.tamu.edu
Turbomachinery sealing R&D.

WRc-NSF
Frankland Road, Blagrove, Swindon, Wiltshire SN5 8YF, UK.
Tel: +44 1495 248 454, Fax: +44 1495 249 234.
Web: www.wrcnsf.com
Testing of equipment and materials for drinking water approval.

APPENDIX 1

Sealing Technology – BAT guidance notes

European Sealing Association e.V.

Sealing Technology – BAT guidance notes

Guidance notes to the best available techniques for sealing technology used in equipment on industrial installations covered by the EU IPPC Directive

ESA Publication No. 014/05   2005 June

ESA Publication No. 014/05

Sealing Technology – BAT guidance notes

Guidance notes to the best available techniques for sealing technology used in equipment on industrial installations covered by the EU IPPC Directive

This document is the copyright © 2005 of the European Sealing Association (ESA).

All rights reserved.

Members of the ESA may copy this document as required.

No part of this publication may be reproduced in any form by non-members without prior written permission of the ESA.

European Sealing Association
Tegfryn
Tregarth
Gwynedd LL57 4PL
United Kingdom
☎: +44 1248 600 250
Fax: +44 1248 600 250
www.europeansealing.com

The **European Sealing Association** is a pan-European organisation, established in 1992 and representing over 80% of the fluid sealing market in Europe. Member Companies are involved in the manufacture, supply and use of sealing materials, crucial components in the safe containment of fluids during processing and use.

The ESA is a non-profit-making trade association, registered in Neu-Ulm (D) as VR 713.

Acknowledgements

The ESA is pleased to recognise the co-operation of Member Companies and others in the preparation of this document. Without their support, this document would not have been possible. Individuals who have made a particularly significant contribution to this publication include:

David Edwin-Scott	James Walker & Co. Ltd.
Brian S Ellis	European Sealing Association
Chris J Fone	John Crane EAA
Ian Forsyth	Chesterton International GmbH
Marco Hanzon	Chesterton International GmbH
Heinz-Dieter Hehle	Kempchen Dichtungstechnik GmbH
John R Hoyes	Flexitallic Ltd.
Adrian Jefferies	James Walker & Co. Ltd.
Jörg Latte	Klinger AG
John Southall	Beldam Crossley Ltd.
Ralf Vogel	Burgmann Packings Ltd.

The ESA is also pleased to acknowledge the assistance of the **Fluid Sealing Association** (Wayne, PA, USA) in the preparation of this document. The Fluid Sealing Association (FSA) is an international trade association, founded in 1933. Members are involved in the production and marketing of virtually every kind of fluid sealing device available today. FSA membership includes a number of companies in Europe and Central and South America, but is most heavily concentrated in North America. FSA Members account for almost 90% of the manufacturing capacity for fluid sealing devices in the NASFTA market.

Fluid Sealing Association
994 Old Eagle School Road
Suite 1019
Wayne, PA 19087 – 1802
United States of America
☏: +1 610 971 4850
Fax: +1 610 971 4859
www.fluidsealing.com

In particular, we thank Rick Page (John Crane Inc.) for the development of the seal life-cycle cost estimator tool.

This publication is intended to provide information for guidance only. The European Sealing Association has made diligent efforts to ensure that recommendations are technically sound, but does not warrant, either expressly or by implication, the accuracy or completeness of the information, nor does the Association assume any liability resulting from the reliance upon any detail contained herein. Readers must ensure products and procedures are suitable for their specific application by reference to the manufacturer. Also, the document does not attempt to address compliance requirements of regulations specific to a particular industrial facility. Readers should consult appropriate local, regional, state, national or federal authorities for precise compliance issues.

Contents

	Page
1. Executive summary	**497**
2. Preface	**502**
Relevant legal obligations of the IPPC Directive	502
Definition of BAT	502
Objective of this document	503
Information sources	503
How to understand and use this document	503
3. General introduction	**505**
Fugitive emissions	505
Other European approaches to the control of fugitive emissions	506
Sources of fugitive emissions	506
Volatile organic compounds (VOC's)	508
4. Generic BAT for sealing technologies	**510**
Good operating practices	510
Management systems	510
Training	511
Process design	511
Maintenance	512
Monitoring to determine leaking losses	512
Key sources of leaks	515
Spill and leak prevention	517
Volatile Organic Compounds	518
Relative costs of generic BAT for sealing technologies	520
5. BAT for bolted flange connections	**521**
Gaskets	521
Current emission levels	524
Gasket selection	525
Storage and handling of gaskets and gasket materials	527
Assembly procedures	528
Safety aspects and joint failure	535
BAT for bolted flange connections	538
Relative costs of BAT for bolted flange connections	539
Emerging techniques	540
6. BAT for rotodynamic equipment	**542**
Pumps	542
Emission management in pumps	542
Pump reliability	543
Mechanical seals	544
BAT for pumps	549
Life-cycle cost (LCC)	550
Relative life-cycle cost guide	550
Compressors	553
Emission management in compressors	553
BAT for compressors	554

Appendix 1: Sealing Technology – BAT guidance notes   495

Other rotodynamic equipment	*555*
Emission management in other rotodynamic equipment	*555*
BAT in other rotodynamic equipment	*556*
Emerging techniques	*557*
7. BAT for reciprocating shafts	**558**
BAT for reciprocating compressors	*558*
8. BAT for valves	**560**
Valve leakage	*560*
Compression packings	*560*
Installation of compression packings	*563*
Current emission levels	*565*
Application guide for BAT in valves	*565*
Valve live-loading	*566*
BAT for valves	*567*
Relative costs of BAT for valves	*568*
Emerging technologies	*570*
9. Conversion factors	**574**
SI units	*574*
Multiples of SI units	*574*
Units of common usage in sealing terminology	*574*
Conversion factors (SI units)	*576*
10. Further reading	**579**
11. References	**581**

1 Executive summary

This Sealing Technology – BAT (Best Available Techniques) Guidance Notes document reflects an information exchange carried out within the sealing industry under Article 16(2) of Council Directive 96/61/EC. This Executive Summary is intended to be read in conjunction with BAT chapters which are relevant for any particular application.

For the purposes of BAT information exchange, this document has been divided into sectors of equipment found typically at industrial sites which are likely to be covered by Council Directive 96/61/EC. It has not been possible to carry out a detailed information exchange on sealing technology used in every industrial process because the scope would be so large. Consequently, this BAT document contains a mixture of generic and detailed information on typical industrial processes. The remainder of the document is arranged as follows:

Preface (Chapter 2)

This chapter provides basic details and obligations about Council Directive 96/61/EC, including the definition of BAT and how to use this document. For guidance throughout this document, bullet points are used as follows:

- *warning – this is where the challenges may arise*
- *recommendation – use these approaches to achieve BAT*

General introduction (Chapter 3)

This chapter describes the challenge of industrial emissions in general, with a focus on, and definition of, fugitive emissions. This chapter also has particular reference to fugitive emissions of Volatile Organic Compounds (VOC's).

Generic BAT for sealing technologies (Chapter 4)

This chapter provides an overview of generally available techniques and their application to generic processes. It includes details of good operating practices and focuses especially on **key sources of leaks**. Generic BAT for sealing technologies includes:

- *comprehensive training of all appropriate personnel on correct installation plays a vital role*
- *reverse the pressure gradient by operating the plant at below ambient pressure (this is probably most feasible at the design stage)*
- *obviate the need for vessel opening through design modifications (e.g. cleaning sprays) or change the mode of operation (e.g. spray anti-caking reagents directly into vessels)*
- *convey leaks from compressor seals, vent and purge lines to flares or to flameless oxidisers*

- ☑ *enclose effluent drainage systems and tanks used for effluent storage/treatment*
- ☑ *identify all hazardous substances used or produced in a process*
- ☑ *identify all the potential sources/scenarios of spillage and leakage*
- ☑ *assess the risks posed by spills and leaks*
- ☑ *review historical incidents and remedies*
- ☑ *implement hardware (e.g. containment, high level alarms) and software (e.g. inspection and maintenance regimes) to ameliorate the risks*
- ☑ *establish incident response procedures*
- ☑ *provide appropriate clean-up equipment (e.g. adsorbents for mopping up spills after small leaks or maintenance works)*
- ☑ *establish incident reporting procedures (both internally and externally)*
- ☑ *establish systems for promptly investigating all incidents (and near-miss events) to identify the causes and recommend remedial actions*
- ☑ *ensure that agreed remedial actions are implemented promptly*
- ☑ *disseminate incident learning, as appropriate, within the process, site, company or industry to promote future prevention*
- ☑ *install low-emission valve stem packing on critical valves (e.g. rising-stem gate-type control valves in continuous operation)*
- ☑ *use alternative low-release valves where gate valves are not essential (e.g. quarter-turn and sleeved plug valves both have two independent seals)*
- ☑ *fit high performance sealing systems (especially on dynamic equipment and for critical applications)*
- ☑ *fit blind flanges to infrequently used fittings to prevent accidental opening during plant operation*
- ☑ *minimise the number of flanged connections on pipelines (e.g. by using welded pipes)*
- ☑ *fit double isolation at any points with high risk of leakage*
- ☑ *use balanced bellows-type relief valves to minimise the valve leakage outside of design lift range*
- ☑ *use advanced technology mechanical seals in pumps*
- ☑ *use zero leakage pumps on critical applications (e.g. double seals on conventional pumps, canned pumps or magnetically driven pumps)*
- ☑ *use containment seals in centrifugal compressors to channel leakage through vent and purge lines to flares or to flameless oxidisers*
- ☑ *use end caps or plugs on open-ended lines and closed loop flush on liquid sampling points*
- ☑ *losses from sampling systems and analysers can be reduced by optimising the sampling volume/frequency, minimising the length of sampling lines, fitting enclosures and venting to flare systems*
- ☑ *quantify VOC emission sources with the idea of identifying the main emitters in each specific case*
- ☑ *focus on the high emitters first of all*
- ☑ *execute an ongoing LDAR programme*

- ☑ route relief valves to flare and add rupture disks
- ☑ route off-gases to flare system (also as part of an odour abatement programme)
- ☑ install flare gas recovery
- ☑ install a maintenance drain-out system to eliminate open discharges from drains

BAT for bolted flange connections (Chapter 5)

This chapter focuses on the sealing of pipe connections, providing advice on the variety of technologies available and the key details about how to ensure good sealing performance. BAT includes:

- ☑ minimise the number of flanged connections
- ☑ use welded joints rather than flanged joints where possible
- ☑ fit blind flanges to infrequently used fittings to prevent accidental opening
- ☑ use end caps or plugs on open-ended lines
- ☑ ensure gaskets are selected appropriate to the process application
- ☑ ensure the gasket is installed correctly
- ☑ ensure the flange joint is assembled and loaded correctly
- ☑ instigate regular monitoring, combined with a repair or replacement programme
- ☑ focus on those processes most likely to cause emissions (such as gas/light liquid, high pressure and/or temperature duties)
- ☑ for critical applications, fit high-integrity gaskets (such as spiral wound, kammprofile or ring joints)

BAT for rotodynamic equipment (Chapter 6)

This chapter focuses on the sealing of rotating shafts (such as in pumps, compressors and agitators). BAT includes:

- ☑ proper fixing of the pump unit to its base-plate or frame
- ☑ connecting pipe forces to be within those recommended for the pump
- ☑ proper design of suction pipe work to minimise hydraulic imbalance
- ☑ alignment of shaft and casing within recommended limits
- ☑ alignment of driver/pump coupling to be within recommended limits when fitted
- ☑ correct level of balance of rotating parts
- ☑ effective priming of pumps prior to start-up
- ☑ operation of the pump within its recommended performance range. The optimum performance is achieved at its best efficiency point
- ☑ the level of net positive suction head available (NPSHA) should always be in excess of the pump design's net positive suction head required

(NPSHR). This can vary dependent upon the operating position on the pump performance curve
- ☑ regular monitoring and maintenance of both rotating equipment and seal systems, combined with a repair or replacement programme
- ☑ exchange gland packings in VOC services for mechanical seals where feasible
- ☑ selection of appropriate mechanical sealing technology based on required maximum leakage control levels and with consideration of process fluid characteristics
- ☑ use mechanical seals designed to accommodate large radial and angular misalignments ('mixer seals')
- ☑ use mechanical seals with bearing(s) integrated into their assembly, to constrain equipment run-out
- ☑ use advanced compression packing designs from reputable manufacturers only
- ☑ re-engineer the gland arrangement where necessary to accommodate shaft misalignment, run-out and equipment wear
- ☑ use 'live loading'

close collaboration between the user and seal manufacturer can provide the most economical sealing solution

BAT for reciprocating shafts (Chapter 7)

This chapter focuses on the sealing of equipment with reciprocating shafts. BAT includes:

- ☑ select packing case, packing ring and piston ring design appropriate for operating conditions
- ☑ please consult the manufacturer

BAT for valves (Chapter 8)

This chapter focuses on the sealing of valves, usually considered to be the greatest challenge for fugitive emissions. BAT includes:

- ☑ correct selection of the packings material and construction for the process application
- ☑ correct installation of the packings material into the stuffing box
- ☑ regular monitoring, combined with a repair or replacement programme
- ☑ focus on those processes most likely to cause emissions (such as gas/light liquid, high pressure and/or temperature duties)
- ☑ focus on those valves most at risk (such as rising stem control valves in continual operation)
- ☑ for critical valves fit high-integrity packings. Many of these are available in special constructions, using advanced technology materials, often specifically formulated for environmental performance

- ☑ *use live-loading, in combination with low emission, fire safe packings in VOC or hazardous services*
- ☑ *where toxic, carcinogenic or other hazardous fluids are involved, fit diaphragm, ball or bellows valves*

Conversion factors (Chapter 9)

This chapter covers the International System of Units (Le Système International d'Unités, or SI units) and gives equivalent conversions into SI units (and other units where appropriate) for **non-SI units** which are used regularly in connection with sealing terminology.

Further reading (Chapter 10)

This chapter provides a listing of appropriate ESA technical documents which form the basis for this particular publication on BAT.

References (Chapter 11)

This chapter is a listing of references cited throughout this publication.

2 Preface

This document is a publication of the European Sealing Association. It is not an official publication of the European Communities and does not necessarily reflect the position of the European Commission. Unless otherwise stated, the terms '*the Directive*' and '*the IPPC Directive*' in this document refer to the Council Directive 96/61/EC on integrated pollution prevention and control.

2.1 Relevant legal obligations of the IPPC Directive

Some of the most relevant provisions of the IPPC Directive, including the definition of the term 'best available techniques', are described in this preface. This description is inevitably incomplete and is given for information only. It has no legal value and does not in any way alter or prejudice the actual provisions of the Directive.

The purpose of the Directive is to achieve integrated prevention and control of pollution arising from the activities listed in its Annex I, leading to a high level of protection of the environment as a whole. The legal basis of the Directive relates to environmental protection. Its implementation should also take account of other Community objectives such as the competitiveness of the Community's industry thereby contributing to sustainable development.

More specifically, it provides for a permitting system for certain categories of industrial installations requiring both operators and regulators to take an integrated, overall look at the polluting and consuming potential of the installation. The overall aim of such an integrated approach must be to improve the management and control of industrial processes so as to ensure a high level of protection for the environment as a whole. Central to this approach is the general principle given in Article 3 that operators should take all appropriate preventative measures against pollution, in particular through the application of best available techniques enabling them to improve their environmental performance.

2.2 Definition of BAT

The term '*best available techniques*' is defined in Article 2(11) of the Directive as 'the most effective and advanced stage in the development of activities and their methods of operation which indicate the practical suitability of particular techniques for providing in principle the basis for emission limit values designed to prevent and, where that is not practicable, generally to reduce emissions and the impact on the environment as a whole.' Article 2(11) goes on to clarify further this definition as follows:

> '*techniques*' includes both the technology used and the way in which the installation is designed, built, maintained, operated and decommissioned;

> '*available*' techniques are those developed on a scale which allows implementation in the relevant industrial sector, under economically and technically viable

conditions, taking into consideration the costs and advantages, whether or not the techniques are used or produced inside the Member State in question, as long as they are reasonably accessible to the operator;

'*best*' means most effective in achieving a high general level of protection of the environment as a whole.

Furthermore, Annex IV of the Directive contains a list of 'considerations to be taken into account generally or in specific cases when determining best available techniques... bearing in mind the likely costs and benefits of a measure and the principles of precaution and prevention'. These considerations include the information published by the Commission pursuant to Article 16(2).

Competent authorities responsible for issuing permits are required to take account of the general principles set out in Article 3 when determining the conditions of the permit. These conditions must include emission limit values, supplemented or replaced where appropriate by equivalent parameters or technical measures. According to Article 9(4) of the Directive, these emission limit values, equivalent parameters and technical measures must, without prejudice to compliance with environmental quality standards, be based on the best available techniques, without prescribing the use of any technique or specific technology, but taking into account the technical characteristics of the installation concerned, its geographical location and the local environmental conditions. In all circumstances, the conditions of the permit must include provisions on the minimisation of long-distance or trans-boundary pollution and must ensure a high level of protection for the environment as a whole. Member States have the obligation, according to Article 11 of the Directive, to ensure that competent authorities follow or are informed of developments in best available techniques.

2.3 Objective of this document

The aim of this document is to provide reference information for industrial sectors, to enable them to comply with the requirements of the IPPC Directive. The document also aims to provide reference information for appropriate permitting authorities to take into account when determining permit conditions. By providing relevant information concerning best available techniques, this document should act as a valuable tool to drive environmental performance.

2.4 Information sources

This document represents a summary of information collected from a number of sources, including in particular the expertise of the sealing industry in Europe. All contributions are gratefully acknowledged.

2.5 How to understand and use this document

The information provided in this document is intended to be used as an input to the determination of BAT in specific cases. When determining BAT and setting

BAT-based permit conditions, account should always be taken of the overall goal to achieve a high level of protection for the environment as a whole.

The determination of appropriate permit conditions will involve taking account of local, site-specific factors such as the technical characteristics of the installation concerned, its geographical location and the local environmental conditions. In the case of existing installations, the economic and technical viability of upgrading them also needs to be taken into account. Even the single objective of ensuring a high level of protection for the environment as a whole will often involve making trade-off judgements between different types of environmental impact, and these judgements will often be influenced by local considerations. Consequently, the techniques and levels presented in each section will not necessarily be appropriate for all installations.

On the other hand, the obligation to ensure a high level of environmental protection including the minimisation of long-distance or trans-boundary pollution implies that permit conditions cannot be set on the basis of purely local considerations. Therefore, it is of the utmost importance that the information contained herein is taken into account fully by permitting authorities.

For guidance throughout this document, bullet points are used as follows:

- *warning – this is where the challenges may arise*
- *recommendation – use these approaches to achieve BAT*

Other bullet points are used as appropriate.

Since the best available techniques inevitably will vary over time, this document will be reviewed and updated as appropriate. All comments and suggestions should be made to the European Sealing Association.

3 General introduction

It is recognised that industry must reduce its impact on the environment if we are to continue global development for future generations (the so-called 'sustainable development' option). A major contributory factor will be through the lowering of industrial emissions, which has been catalysed by a combination of public pressure, environmental legislation and the internal requirement to minimise the loss of valuable feed-stocks. Large proportions of the emissions to atmosphere are represented by the by-products of combustion (notably the oxides of carbon, nitrogen and sulphur), along with known losses of volatile hydrocarbons and steam. In general, these are all emissions *anticipated* from the industrial process, under the *control* of the plant operator, and will not be considered in detail here.

However, a proportion of industrial emissions occurs through unanticipated or spurious leaks in process systems. From this, it is apparent that sealing systems play a vital role in the environmental performance of industrial installations, and yet the sealing technology itself is usually given scant consideration! It must be emphasised that sealing technology can perform at its peak only after careful selection (appropriate for the specific application), correct installation, operation according to the performance envelope, regular inspection and maintenance. **These areas are the key focus for this document.**

The best available techniques for sealing technology are described, together with the best practices for their selection, installation and use, in order to enable the plant operator to achieve the requirements of the IPPC Directive.

3.1 Fugitive emissions

The term 'fugitive emissions' covers all losses of (usually volatile) materials from a process plant, through evaporation, flaring, spills and unanticipated or spurious leaks. It is often defined as: *any chemical, or mixture of chemicals, in any physical form, which represents an unanticipated or spurious leak, from anywhere on an industrial site*.

To put the scale of the challenge into perspective, fugitive emissions in the USA have been estimated to be in excess of 300 000 tonnes per year, accounting for about **one third of the total organic emissions from chemical plants**, and inevitably mirrored in Europe. Irrespective of any environmental impact which it may cause, this is a tremendous financial burden on industry because it represents a huge loss of potentially valuable materials, and cause of plant inefficiency. Yet in most instances, the true costs are not appreciated, since many of the costs associated with fugitive emissions are invisible.

The values of fugitive emissions will depend upon:

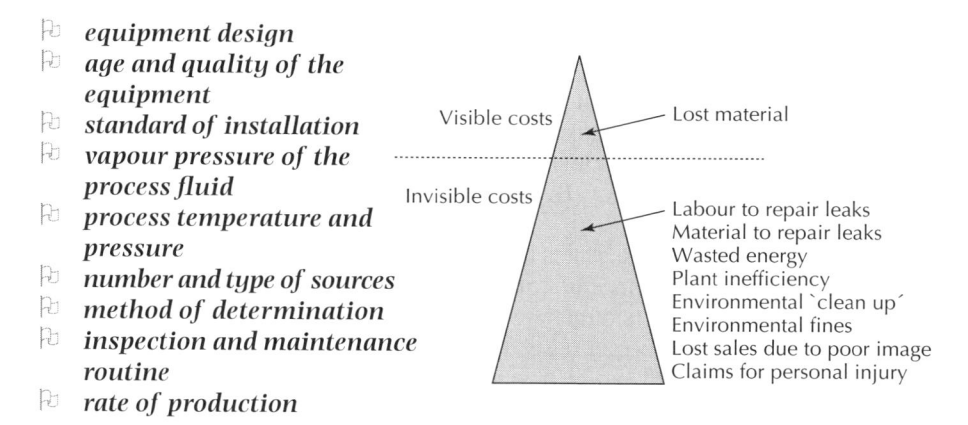

- *equipment design*
- *age and quality of the equipment*
- *standard of installation*
- *vapour pressure of the process fluid*
- *process temperature and pressure*
- *number and type of sources*
- *method of determination*
- *inspection and maintenance routine*
- *rate of production*

Visible costs — Lost material

Invisible costs — Labour to repair leaks
Material to repair leaks
Wasted energy
Plant inefficiency
Environmental `clean up´
Environmental fines
Lost sales due to poor image
Claims for personal injury

Many process streams in petrochemical refineries are 'light' (containing at least 20% of substances with a vapour pressure greater than 0.3 kPa at 20°C) and at high pressure (1500–3000 kPa), conditions which encourage fugitive losses. On the other hand, in some aromatics operations with lower operating temperatures and pressures and where the fluid vapour pressures are lower, fugitive emissions are considerably less.

3.2 Other European approaches to the control of fugitive emissions

Some EU Member States have introduced other legislation to control fugitive emissions, much of which is complementary to the IPPC Directive and in some cases this may be necessary in the transposition of the Directive into national legislation. On the other hand, some legislation or guidelines have been introduced which go further than the IPPC Directive. These include the latest refinements of the TA-Luft (D), and VDI 2440 (D) on emission reduction in mineral oil refineries.[27] The reader is advised to refer to these items as appropriate.

3.3 Sources of fugitive emissions

A significant proportion of fugitive emissions can be losses from unsealed sources, including storage tanks, open-ended (non-blanked) lines, pressure-relief valves, vents, flares, blow-down systems, spills and evaporation from water treatment facilities. These are part of the industrial process, anticipated (usually) by the process operator, and will not be considered further here.

In other cases, these losses may be caused by **leaks in the sealing elements of particular items of equipment**, such as:

- *agitators/mixers*
- *compressors*
- *flanges*

- *pumps*
- *tank lids*
- *valves*

As unsealed point sources have become well controlled in recent years, equipment leaks are often the greatest source of fugitive emissions. **These equipment leaks are where the sealing industry is playing a crucial role, through the development and application of innovative sealing technology appropriate to low or zero emission requirements.**

The primary purpose of a seal is to contain a fluid and so protect the immediate environment from contamination (and vice versa), which may vary in significance from innocuous fluid loss (such as steam, water, etc.) up to nauseous, toxic or hazardous fluid loss. In the former case, the loss of such innocuous fluid will lead primarily to lack of plant efficiency for the operator, although such leakages may still present hazards (such as leakages of high pressure water or steam). Clearly, in the latter case it is not only financially inefficient but also environmentally dangerous; for employees, members of the public and for nature at large! Consequently, the correct selection and use of the appropriate sealing technology for the application is just part of the environmental responsibility of the plant operator.

Although losses per piece of equipment are usually very small, there are so many items of equipment on a typical Large Volume Organic Chemical (LVOC) plant or petrochemical refinery that the total loss via fugitive routes may be very significant. For example, fugitive emissions from European refineries range from 600 to 10 000 tonnes of VOC's per year. In some plants in the Netherlands, 72% of VOC emissions were attributed to leakage losses from equipment, 18% from flaring, 5% from combustion, 1% from storage and 4% from other process emissions. In these plants, leakage is the greatest challenge and therefore it is crucial that programmes are established to identify leak sources and to instigate actions to minimise them.

It has been estimated that for every pump on an average plant, there will be 32 valves, 135 flanges, 1 safety valve and 1.5 open-ended lines. Hence, with so many potential sources, leaking losses are often hard to determine. They are also very dependent on the age of the equipment and how well the installation is maintained. Some important causes of leaking losses are:

- *ill-fitting internal or external sealing elements*
- *installation or construction faults*
- *wear and tear*
- *equipment failure*
- *pollution of the sealing element*
- *incorrect process conditions*

Leaking losses are generally higher from dynamic equipment (compared with static equipment) and from older equipment.

Valves are considered to account for approximately 50–60% of fugitive emissions. Furthermore the major proportion of fugitive emissions comes from only a small fraction of the sources (e.g. less than 1% of valves in gas/vapour service can account for more than 70% of the fugitive emissions in a refinery).

Some valves are more likely to leak than others such as:

- *Valves with rising stems (gate valves, globe valves) are likely to leak more frequently than quarter turn type valves such as ball and plug valves.*
- *Valves which are operated frequently, such as control valves, may wear quickly and allow emission paths to develop. However, newer, low leak control valves provide good fugitive emissions control performance.*

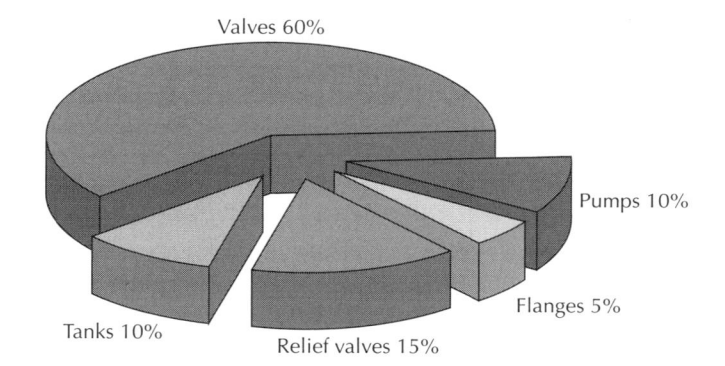

Valves 60%
Pumps 10%
Flanges 5%
Relief valves 15%
Tanks 10%

Although out of the main scope of this document, it must be remembered that energy is consumed whenever heating or pumping action is applied to a process. This creates additional emissions at the power generation plant, emphasising the need for utilising efficient pumping and heating technologies, as part of pollution prevention and good operating practice.

3.4 Volatile Organic Compounds (VOC's)

VOC's emissions are of significant environmental concern because some have the potential for Photochemical Ozone Creation Potential (POCP), Ozone Depletion Potential (ODP), Global Warming Potential (GWP), toxicity, carcinogenicity and local nuisance from odour. These properties mean that VOC's are a major contributor to the formation of 'summer smog'. The prevention of VOC emissions is therefore one of the most important issues facing the operation of many industrial processes.

VOC is the generic term applied to those organic carbon compounds which evaporate at ambient temperature, and is defined usually as '*a substance having a vapour pressure of greater than 0.3 kPa at 20°C*' (this is close to the US definition for the application limits of systematic LDAR). The term covers a diverse group of substances and includes all organic compounds released to air in the gas phase, whether hydrocarbons or substituted hydrocarbons. Their properties, and hence need for control, vary greatly and so systems have been developed to categorise VOC's according to their harmfulness.

For example, a system developed in the UK Environment Agency identifies three classes of VOC and requires a commensurate level of prevention and control for each class. The three classes are:

- ***extremely hazardous to health*** *(such as benzene, vinyl chloride and 1,2 dichloroethane)*
- ***class A compounds****, which may cause significant harm to the environment (for example, acetaldehyde, aniline and benzyl chloride)*
- ***class B compounds****, which have lower environmental impact.*

Some VOC's may also be highly odorous, for example aldehydes, amines, mercaptans and other sulphur-containing compounds. This may necessitate additional stringency in the prevention measures (e.g. high integrity equipment to reduce fugitives) and the abatement of losses.

VOC emissions typically arise from: process vents; the storage and transfer of liquids and gases; fugitive sources and intermittent vents. Losses are greatest where the feedstock or process stream is a gas; in these cases VOC losses can exceed 2% of total production.

On many installations, there has been a focus on the control of point sources of VOC's over recent years and losses of fugitives as leaks (from pumps, valves, tanks etc.) have become the major source of VOC emissions from many plants. This emphasises the importance of **best available techniques for sealing technology** and reinforces the need for this document.

4 Generic BAT for sealing technologies

In a formal BREF note (also known as BAT Reference note) approved by the European Commission, usually this chapter provides a catalogue of techniques which can be used to prevent and control emissions from the specific process in question. However, this is not possible for the many sealing technologies employed across the variety of industrial processes covered by IPPC. Instead, this section provides an overview of generally available techniques and their application to generic processes. In reading this chapter, reference should also be made to relevant horizontal BREF's, especially the BREF on emission monitoring, describing the techniques which may be used across all industrial sectors.

In most cases, processes will maximise environmental protection with sealing technology by using a combination of management commitment, training, plant design, selection of the optimum sealing technology for the application, monitoring, inspection, maintenance and repair. Some of these concepts are described in detail throughout the industry-specific BREF notes, and so will be mentioned only briefly here. The chapter therefore considers techniques involving: **good operating practices, management systems, training, process design, maintenance, monitoring, determining leaking losses, key sources of leaks, spill and leak prevention, and volatile organic compounds**.

4.1 Good operating practices

The IPPC Directive definition of best available techniques strongly emphasises the presumption for preventative techniques over other methods. The prevention of pollution is just part of good operating practices, which are techniques involving management, organisation or personnel. Good operating practices can often be implemented quickly, at little cost, and bring efficiency savings with a high return on investment.

Prevention offers a *precautionary*, rather than curative, approach to environmental protection. As such, it is compatible with the principle of 'sustainable development'. Many companies have already shown that the creative use of pollution prevention techniques not only minimises environmental impact, but also improves efficiency and increases profits.

The points which follow in this section on generic BAT for sealing technologies are **all** part of good operating practices.

4.2 Management systems

In order to minimise the environmental impact of any process, it is necessary to appreciate the central role of effective management systems. **The purchase of state-of-the-art hardware does not automatically guarantee the best environmental performance since it must also be installed and operated correctly.** Likewise, the limitations of older equipment can often be mitigated by diligent operation. The best environmental performance is usually achieved by the installation of the best technology and its operation in the most effective and efficient manner. This is recognised by the IPPC Directive definition of 'techniques' as *'both the technology used and the way in which the installation is designed, built, maintained, operated, and decommissioned'*.

An Environmental Management System (EMS) is that part of the overall management system which includes the organisational structure, responsibilities, practices, procedures, processes and resources for developing, implementing, achieving, reviewing and monitoring the environmental policy. Environmental Management Systems are most effective and efficient where they form an inherent part of the management and operation of a process. Effective environmental management involves a commitment to continuous environmental improvement through a cyclical system of: gathering and analysing data, establishing objectives, measuring progress and revising the objectives according to results.

Management action is crucial in the motivation of personnel and this appears to be an important factor in the overall emission abatement of leaking losses. Management practices should also include a clear specification to employees of what *good practice* actually entails.

4.3 Training

Appropriate training should be given to all staff involved in process operation to ensure that they are competent for their duties. The training should include the environmental implications of their work and the procedures for dealing with incidents. This is crucial for the correct selection, installation and performance of sealing systems. It is equally important that sub-contractors, who may be responsible for maintenance work, or installation of sealing systems during plant shutdown, should be fully trained in the handling and installation of sealing materials. A number of relevant guidebooks on the selection, handling and installation of sealing materials are available from the European Sealing Association (see *10, Further reading*).

Ensure environmental awareness is included in training programmes. Records should be kept of the training given to staff and these should be reviewed periodically to ensure that they reflect the needs of the job and the latest technologies.

Importantly, the in-service performance and life of any sealing technology is very reliant upon correct installation. As an example, the major cause of flange sealing failure is due to assembly errors. Consequently:

- ☑ *comprehensive training of all appropriate personnel on correct installation plays a vital role*

4.4 Process design

Operators should work to written standards and procedures when modifying existing installations or designing new plant. As a minimum this should follow the requirements of any national and international technical codes for materials, equipment design and fabrication. All design decisions or modifications should be recorded in order to provide an audit trail. Environmental protection should be an inherent feature of the design standards since techniques incorporated at the design stage are both more effective and more economical. Initial process design should consider how fundamental principles may be applied to process materials, process variables and equipment in order to prevent releases. For example, consideration should be given to identify opportunities for using new, low emission sealing developments or other appropriate sealing technology options.

Other possible operational and maintenance techniques include:

- ☑ *reverse the pressure gradient by operating the plant at below ambient pressure (this is probably most feasible at the design stage)*
- ☑ *obviate the need for vessel opening through design modifications (e.g. cleaning sprays) or change the mode of operation (e.g. spray anti-caking reagents directly into vessels)*
- ☑ *convey leaks from compressor seals, vent and purge lines to flares or to flameless oxidisers*
- ☑ *enclose effluent drainage systems and tanks used for effluent storage/ treatment*

4.5 Maintenance

The maintenance of process plant and equipment is an essential part of good operation and will involve both pro-active (preventative) and reactive approaches.

Preventative maintenance plays a very significant role in optimising environmental performance and it is often the preferred approach. A structured programme of preventative maintenance should be established after detailed consideration of equipment failure frequencies and consequences. The maintenance programme should be supported by appropriate record keeping systems and diagnostic testing. There should be clear responsibility for the planning and execution of maintenance.

The need for reactive maintenance can be minimised by employee vigilance in relation to imminent problems (e.g. process upsets and leaks). Leak Detection and Repair programmes can also play an important role.

Equipment modifications during maintenance are a frequent occurrence on many plants and should be covered by procedures which give authorisation only after a suitable level of risk assessment. Subsequent process start-up should be dependent upon suitable post-modification checks.

4.6 Monitoring to determine leaking losses

Leaking losses are often hard to determine since there are many potential sources and they are very dependent on how well the installation is operated, maintained and inspected. Some important causes of leaking losses are:

- *ill-fitting sealing elements*
- *installation faults*
- *construction faults*

- *wear and tear*
- *ageing*
- *equipment failure*
- *contamination of the sealing element*
- *excursions out of normal process conditions*
- *poor maintenance procedures*

Leaking losses are generally higher from dynamic equipment (compared with static equipment) and from older equipment.

A structural reduction of leaking losses is only possible when insight on the leaking losses is gained. There are various methods to determine the leaking losses. The simplest way to estimate the leaking losses is by multiplying the number of each type of equipment by an *emission factor* for that type of equipment. This method can be applied to obtain a general estimation of the emissions **without measurements**. Emission factors are not intended as an accurate measure of a single piece of equipment, and do not reflect the site-specific conditions of process units.

Many companies determine their leaking losses by calculations or estimations based on measurements, but it is hard to measure all possible sources in a large plant (possibly tens of thousands) and not all sources are accessible. In most cases, a representative sampling of sources will suffice to estimate or calculate the leaking losses of the plant. The number of samples depends on the kind of process fluids in the plant and the kind of equipment (the sources). However, to provide the best estimate of emissions, every potential 'source' on a site must be **monitored** (usually using a 'sniffing' process such as EPA Method 21).

Monitoring has been identified as a common activity across IPPC processes and is the subject of a horizontal BREF note, entitled, '***Monitoring of Emissions***'. The document provides generic information on sampling and analysis, and should be read in conjunction with other industry-specific BREF notes.

Monitoring is often expensive and time consuming, so the objectives should be clear when a programme is established. Process operators and regulators may use monitoring to provide information on a wide range of topics. For this BAT guidance note on sealing technologies, the key objectives of monitoring are:

- Process control and optimisation; monitoring is the way used to control a process by means of following-up significant physical and chemical parameters. By control of the process, it is meant the application of conditions in which the process operates safely and economically.
- Emission monitoring; emissions to air and water are characterised and quantified to provide a check on compliance with permit requirements (or other performance measures). This also provides a check of whether all significant emissions are covered by the permit and can indicate the effectiveness of abatement techniques and sealing technologies employed. For the latter, emission monitoring can give an assessment of leaking losses and will indicate equipment where attention is required. Wherever possible, data should be collected on flow rates to enable the calculation of mass discharges.

- Occupational health and safety; tests to identify the short and long term risks to personnel from work place exposure.
- Troubleshooting; intensive, short duration programmes may be used to study specific topics.

A monitoring programme to address any of these topics will need to stipulate the frequency, location and method of both sampling and analysis. Monitoring usually involves precise quantitative analysis, but simple operator observations (either visually or by smell) can also play an important role in the detection of abnormal releases. The results of monitoring programmes should be actively utilised; records of results should be kept for trend analysis and diagnostic use.

Leaking losses from equipment and fittings can be significantly reduced by the use of monitoring and maintenance programmes such as LDAR (Leak Detection and Repair). Leaks are detected by monitoring equipment and repairs must be carried out if the leakage rate exceeds certain levels. A leak detection and repair programme consists of using a portable VOC detecting instrument to detect leaks by 'sniffing' (usually, according to EPA Method 21) during regularly scheduled inspections of valves, flanges, and pump seals. Leaks are then repaired immediately or are scheduled for repair as quickly as possible. An LDAR programme could reduce fugitive emissions by 40 to 60%, depending on the frequency of inspections, the process conditions and the fluid emitted.

LDAR can be structured to meet local requirements using appropriate techniques, frequencies and priorities, but in all cases the largest losses should be tackled first. Such programmes have shown that gland leaks on valves and pumps are often responsible for the majority of leaking releases.

Some estimates have been made of the costs of monitoring schemes. For example, a simple LDAR scheme, involving the annual inspection of gas and volatile liquid service components, is estimated to have a net annualised cost of over €15 K per year (for a typical plant handling 20 000 tpa of gaseous hydrocarbon streams and 30 000 tpa of volatile liquids).

A strategy to reduce VOC emissions may include a complete inventory and quantification by a DIAL LIDAR technique[26] (differential absorption light detection and ranging). In some cases, emissions estimates using 'sniffing' methods give lower emissions than estimates based on the DIAL monitoring. In some cases, the discrepancies are very large. For example, by using the method for estimating fugitive emissions proposed by EPA 'Workbook for estimating fugitive emissions from petroleum production operations 1992', the emissions from the process area at an average European refinery have been estimated to be 125 tonnes per year. Extrapolations of the DIAL measurements to a yearly emission give emissions of 500–600 tonnes per year.

Note that most reported fugitive emissions are calculated rather than monitored (measured), but unfortunately, correlations are often dubious! Equally, not all calculation formats are comparable. For example, monitoring at well-maintained plants in the Netherlands shows that the average emissions factors are generally higher than measured (monitored) values.

4.7 Key sources of leaks

The main potential sources of leaks, and possible reduction techniques, are considered below, in alphabetical order:

4.7.1 Compressors

Lower speed, **positive displacement compressors** are typically sealed by a barrier oil lubricated, mechanical seal and emissive leakage is low; containment sealing arrangements are used in many services. The larger, higher speed **centrifugal compressors** may have high leaking losses. The change to gas lubricated single seals with an outer containment seal has enabled improved reliability and the management of primary seal emissive leakage to a flare or recovery system. Various externally supplied gas purges are used with both types of machinery. Regular control and maintenance is indispensable.

4.7.2 Flanges

Individual flanges generally do not have very large leaking losses but, since plants utilise so many flanges, they can make a major contribution to the overall leaking losses. Preventative measures, among which correct selection of the gasket and regular maintenance (e.g. controlled tightening of the flange), are very important. The regular control and replacement of the gaskets is also necessary, especially for those gaskets exposed to temperature fluctuation or vibration (where gasket load may be lost). When a removable connection is not necessary, flanges can be replaced by welded piping. When welding is not possible, a conventional gasket may be replaced by one providing a higher level of sealing integrity.

4.7.3 Open-ended lines

Emissions from open-ended lines can be controlled by installing a cap, plug or second valve at the open end. If a second valve is installed, the upstream valve should always be closed first after use of the valves to prevent the trapping of fluids between the valves.

4.7.4 Pipe work

Leaking losses in pipe work can be reduced at the design stage by arranging equipment to minimise the pipe run length, eliminating underground piping (or design with cathodic protection) and using lined pipe. Once in place, pipe work leaks can be minimised by painting to prevent external corrosion and regular monitoring for corrosion and erosion. Releases when cleaning lines should be minimised by using 'pigs' for cleaning, sloping pipe work to low point drain, using heat tracing and insulation to prevent freezing and flushing to product storage tank or treatment facility.

4.7.5 Pumps

The relatively low process leakage levels emitted from pumps and their relatively low numbers in a plant result in the overall leakage contribution from pumps being relatively small. As there are few pumps it is relatively simple to find and repair leaking pumps. Pump leaking losses occur mainly where the rotating shaft penetrates the casing. The sealing technologies employed are:

Gland Packing
Gland Packing with a barrier flush
Single Mechanical Seals
Single Mechanical Seals with a mechanical containment seal and leakage collection (dual unpressurised seals)
Double Seals with a separate barrier fluid (dual pressurised seals)
Sealless drive systems

Gland packing leaks more than the mechanical seals in rotodynamic pumps, and for this reason and reliability issues, is not used generally in services where emissions must be avoided. It is sometimes used in some slow speed equipment. Single mechanical seals can provide adequate sealing of emissive VOC process liquids but improvements can be achieved with the addition of a mechanical containment seal and even better results are obtained with a dual mechanical seal (a barrier liquid between the two mechanical seals which almost completely prevents leaking losses). Leakage losses may also be reduced by replacing conventional pumps with sealless pumps. These pumps have a completely closed construction which prevents leaking losses almost completely but they may be restricted in application by the process properties. Their lower energy efficiency compared to conventional pumps and the upstream influence on power generation may restrict the wider usage of this technology.

4.7.6 Safety valves

Safety valves can be responsible for 10% of a plant's leaking losses. Losses are higher where safety valves are exposed to pressure fluctuations, and when a safety valve has activated. Therefore, safety valves should be checked after an emergency situation. Leaking losses via safety valves may be reduced by the installation of rupture discs prior to the safety valve to damp small pressure fluctuations. However, these fluctuations may pollute the valve, making complete closure impossible. An additional measure is to connect safety valves to a central flare system or another type of dedicated collection system (e.g. vapour recovery/destruction unit).

4.7.7 Sampling points

Emissions from sampling connections can be reduced by using a closed-loop sampling system or by collecting the purged process fluid and transferring it to a control device or back to the process.

4.7.8 Tank lids

There are numerous static sealing applications, such as tank lid seals, which are used extensively in ships and lorries (primarily for the transportation of media) and they are also used in housing seals in chemical plants and refineries. In general, these static seals do not have high leakage amounts, but are likely to be a larger contributor than flanges and, as such, they do contribute to the overall emission level. As they are opened and closed from time to time, these seals should be checked for integrity of fit and wear, and replaced on a regular basis.

4.7.9 Valves

Valves, and especially control valves, are an important source of leaking losses, and may account for 75% of the leaking losses in a plant. The chance of leakage depends on the kind of valve. Diaphragm valves, ball valves and, above all, bellows-sealed valves, offer better sealing characteristics than other types.

The stuffing box packing has a dominant influence on valve leaking losses, especially in disc valves. For processes containing hazardous fluids, conventional packings can be replaced with *low emission packings* made from advanced technology materials and constructions.

4.8 Spill and leak prevention

Precautionary modifications should be made to ensure that spills and leaks do not occur, and that they are dealt with promptly when they do arise. The following techniques may be applicable:

- ☑ *identify all hazardous substances used or produced in a process*
- ☑ *identify all the potential sources/scenarios of spillage and leakage*
- ☑ *assess the risks posed by spills and leaks*
- ☑ *review historical incidents and remedies*
- ☑ *implement hardware (e.g. containment, high level alarms) and software (e.g. inspection and maintenance regimes) to ameliorate the risks*
- ☑ *establish incident response procedures*
- ☑ *provide appropriate clean-up equipment (e.g. adsorbents for mopping up spills after small leaks or maintenance works)*
- ☑ *establish incident reporting procedures (both internally and externally)*
- ☑ *establish systems for promptly investigating all incidents (and near-miss events) to identify the causes and recommend remedial actions*
- ☑ *ensure that agreed remedial actions are implemented promptly*
- ☑ *disseminate incident learning, as appropriate, within the process, site, company or industry to promote future prevention*

Along with improved process design, operation, monitoring and maintenance (all identified above), one of the key techniques which may be used to minimise leaking losses is to install **high integrity equipment**. For example:

- ☑ *install low-emission valve stem packing on critical valves (e.g. rising-stem gate-type control valves in continuous operation)*

- ☑ use alternative low-release valves where gate valves are not essential (e.g. quarter-turn and sleeved plug valves both have two independent seals)
- ☑ fit high performance sealing systems (especially on dynamic equipment and for critical applications)
- ☑ fit blind flanges to infrequently used fittings to prevent accidental opening during plant operation
- ☑ minimise the number of flanged connections on pipelines (e.g. by using welded pipes)
- ☑ fit double isolation at any points with high risk of leakage
- ☑ use balanced bellows-type relief valves to minimise the valve leakage outside of design lift range
- ☑ use advanced technology mechanical seals in pumps
- ☑ use zero leakage pumps on critical applications (e.g. double seals on conventional pumps, canned pumps or magnetically driven pumps)
- ☑ use containment seals in centrifugal compressors to channel leakage through vent and purge lines to flares or to flameless oxidisers
- ☑ use end caps or plugs on open-ended lines and closed loop flush on liquid sampling points
- ☑ losses from sampling systems and analysers can be reduced by optimising the sampling volume/frequency, minimising the length of sampling lines, fitting enclosures and venting to flare systems

4.9 Volatile Organic Compounds

As mentioned in the **General Introduction** section, volatile organic compounds (VOC's) are high priority species for emission reduction, as a consequence of their specific chemical and physical properties. They present a major challenge to sealing technology and, for this reason, a separate section has been included here on general minimisation techniques.

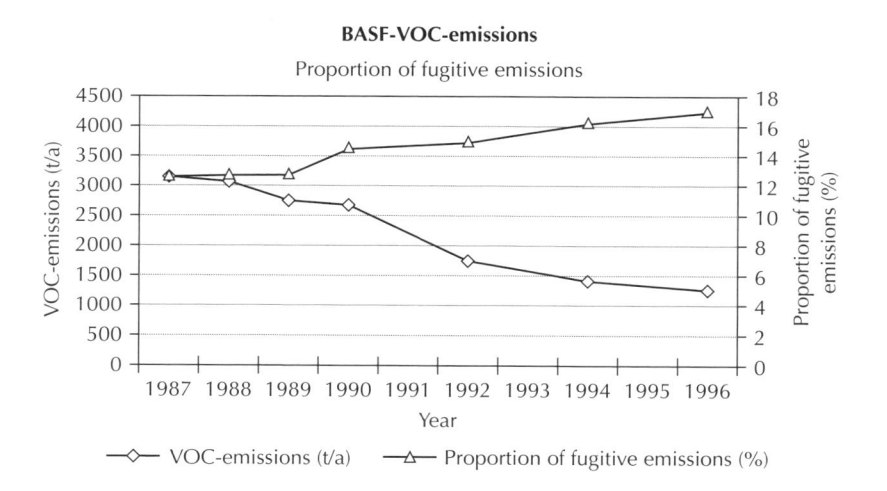

BASF-VOC-emissions

Proportion of fugitive emissions

―◇― VOC-emissions (t/a) ―△― Proportion of fugitive emissions (%)

During the ESA's 2nd European Fugitive Emissions Conference in Düsseldorf in 1998, BASF reported[34] that in 1996, VOC fugitive emissions amounted to 17% of their total VOC emissions, so emphasising the priority for control.

The effectiveness and costs of VOC prevention and control will depend on the VOC species, the VOC concentration, the flow rate, and the source. Inevitably, resources are targeted initially at high flow, high concentration, process vents, but recognition should be given also to the cumulative impact of low concentration diffuse emissions.

In general, toxic VOC's should be replaced by less harmful substances as soon as possible, where this is technically and economically feasible. Where possible, it is also good practice to substitute volatile compounds with compounds that have a lower vapour pressure. Where this is not possible, the initial efforts should be to minimise losses and then to recover the calorific value of unavoidable emissions.

In many cases the VOC's in question are used as solvents in the process and the Solvents Directive provides useful guidance on the prevention and reduction of air pollution from solvent emissions.

In general, VOC's as such do not offer challenging sealing problems but every leak has to be considered in the context of the environment of the process equipment or pipeline which is being considered. The interaction between that environment and the seal may make the creation of a reliable seal far more complex than the medium alone would require.

In addition to the points noted above (*sections 4.7 and 4.8*), some general techniques to reduce VOC emissions are:

- *quantify VOC emission sources with the idea of identifying the main emitters in each specific case*
- *focus on the high emitters first of all*
- *execute an ongoing LDAR programme*
- *route relief valves to flare and add rupture disks*
- *route off-gases to flare system (also as part of an odour abatement programme)*
- *install flare gas recovery*
- *install a maintenance drain-out system to eliminate open discharges from drains*

The investment cost for those techniques is negligible and the operating cost is around 0.1 M€/yr (accounting for 190 € per tonne of VOC recovered).

About 93% of the sources of fugitive emission are accessible. The achievable emission reduction depends on the current conditions of the components, with typical reduction rates of at least 50 to 75% related to average emission factors. The estimated efficiency for quarterly inspection and maintenance is 80 to 90%. Higher efficiencies may be reached, when more intensive inspection and maintenance programmes are implemented.

A good inspection and maintenance programme for valves and flanges is a very cost-effective way to reduce VOC emissions in a refinery. Savings may reach 0.19 €/kg of hydrocarbon reduced.

4.10 Relative costs of generic BAT for sealing technologies

In most cases, the cost of the actual sealing technology is infinitesimally small when compared with the investment made in the plant as a whole. Indeed, for many sealing technologies, the cost per unit may be in the region of a few cents, completely insignificant when the total plant costs are considered.

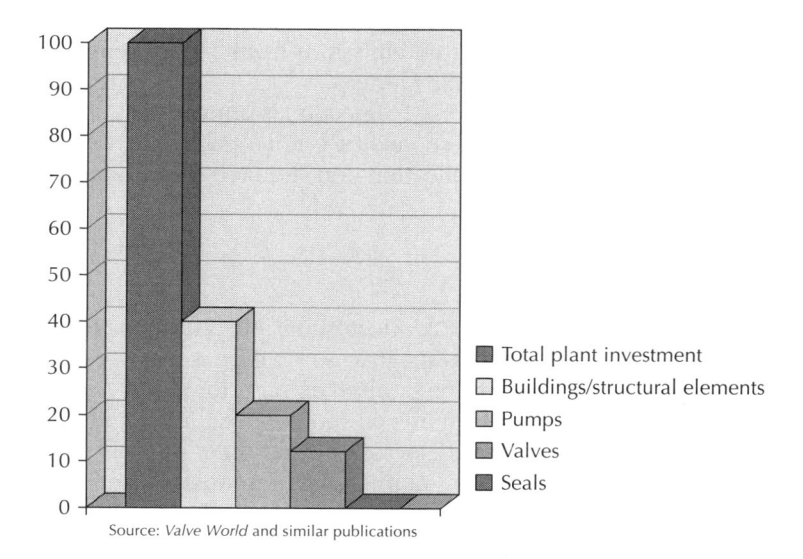

Source: *Valve World* and similar publications

Legend:
- Total plant investment
- Buildings/structural elements
- Pumps
- Valves
- Seals

Importantly, the unit cost of the sealing technology is overwhelmed completely by the labour costs required to fit the seal, let alone the downtime of the plant. Consequently, the actual cost of the sealing device is immaterial in terms of economic considerations for BAT. However, indications of relative costs are provided as examples in specific sections in this document.

5 BAT for bolted flange connections

Where pipe work and process equipment on an industrial installation need to be inspected, maintained and/or repaired on a regular basis, connections are usually in the form of flanges, so enabling easy removal and replacement. Individual flanges generally do not have very large leaking losses, but since plants utilise so many flanges, they can make a major contribution to the overall leaking losses. For example, BASF has reported[28] that emissions from flanges represent 28% of the total fugitive emissions from BASF plants.

As a general rule, where a removable connection is unnecessary, flanges should be replaced with welded piping. Where welding is not feasible, the *flange joint system* must be appropriate for the application and should be maintained by trained personnel only.

5.1 Gaskets

Historically, *c*ompressed *a*sbestos *f*ibre sheet material (CAF) has been the material of choice for 'soft' gasket materials. It was regarded as easy to use and very tolerant of abuse, for which it was recognised as very 'forgiving'. Consequently, the material was used to seal almost all common applications, and usually gave performance which was deemed satisfactory at that time. A broad experience of the material was established over many years amongst manufacturers and users alike.

More recently, with the ban on the use of asbestos fibres, a new generation of asbestos-free substitutes has been developed by the sealing industry. These provide improved levels of sealing performance, although they are usually more application specific than the earlier asbestos materials. Equally, handling of these new materials requires more care in general. Overall, these new materials can outperform their asbestos predecessors, but are usually less forgiving; users must exercise more care in selecting the right material for the job and assembling the seal.

Over time, alternative gasket styles have been developed, especially for more severe services, and these include the 'hard' gasket types, primarily of metallic or semi-metallic construction. These offer users even more choice in higher performance sealing technology for the application.

Useful guidance notes on gasket selection and installation are available in a recent publication from the ESA, *Guidelines for safe seal usage – Flanges and Gaskets* (ESA publication no. 009/98), available in several language versions.

A gasket is used to create and retain a static seal between two stationary flanges, which may connect a series of mechanical assemblies in an operating plant, and which may contain any one of a wide range of fluids. These static seals aim to provide a complete physical barrier against the fluid contained within, and so block any potential leakage path. To achieve this, the gasket must be able to flow into (and fill) any irregularities in the mating surfaces being sealed, while at the same time be sufficiently resilient to resist extrusion and creep under operating conditions. The seal is effected by the action of force upon the gasket surface, which compresses the gasket, causing it to flow into any flange imperfections.

The combination of contact pressure between the gasket and flanges, and densification of the gasket material, prevents the escape of the contained fluid from the assembly. As such, gaskets are vital to the satisfactory operation of a broad range of industrial equipment and must be regarded as an integral design element of the whole plant.

On seating, a gasket must be capable of overcoming minor alignment and flange imperfections, such as:

- *non-parallel flanges*
- *distortion troughs/grooves*
- *surface waviness*
- *surface scorings*
- *other surface imperfections*

When assembled, a flange gasket seal or 'joint' is subject to compressive pressure between the faces of the flanges, usually achieved by bolts under tension. In order to ensure the maintenance of the seal throughout the lifetime of the assembly, sufficiently high pressure must remain on the gasket surface to prevent leakage. Under operating conditions, this pressure will be relieved by *hydrostatic end thrust*, the force produced by internal pressure which acts to separate the flanges. The gasket itself is also subject to a side load due to the internal fluid pressure tending to extrude it through the flange clearance space. To maintain seal integrity, the effective compressive pressure on the gasket (that is, the assembly load minus the hydrostatic end thrust) must be greater than the internal pressure by some multiple, dependent upon the gasket type, manufacturing process involved and level of tightness required. For soft gaskets, there must also be adequate friction between the gasket and flange faces to help prevent extrusion (blow-out) of the gasket from the joint. To allow for any relaxation of gasket compressive pressure which is normally inevitable, a factor of at least two is usually recommended between the compressive pressure on the assembly and that required to maintain a seal. A number of publications[3, 4, 5, 6] provide more detail of the flange/gasket interaction.

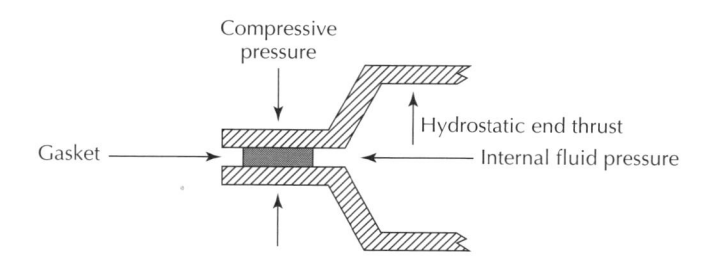

So, the primary function of a gasket is to create and maintain a seal between flanges, under conditions which may vary markedly from one joint to another, dependent upon the nature and type of application. To meet these varying conditions, a number of flange/fastener/gasket systems have been developed, and many factors must be considered when selecting the most appropriate assembly, including:

Application	Flange arrangement	Gasket
Pressure of media	Configuration/type	Blow out resistance
Temperature of media	Surface finish	Creep resistance
Chemical reactivity of media	Material	Stress relaxation
Corrosive nature	Available bolt load	Ability to recover/elasticity
Searching ability of media	Likelihood of corrosion/erosion	Expected service life
Viscosity	Flange strength/stiffness	Comparative cost
pH of media (acidity)	Alignment tolerance	Chemical and physical compatibility
Concentration		Ease of handling/installation/removal
		Fire resistance
		Sealability
		Combined pressure temperature resistance

Importantly, for all of these systems, the performance of the seal depends upon the interaction of the **various elements of the flange joint system:**

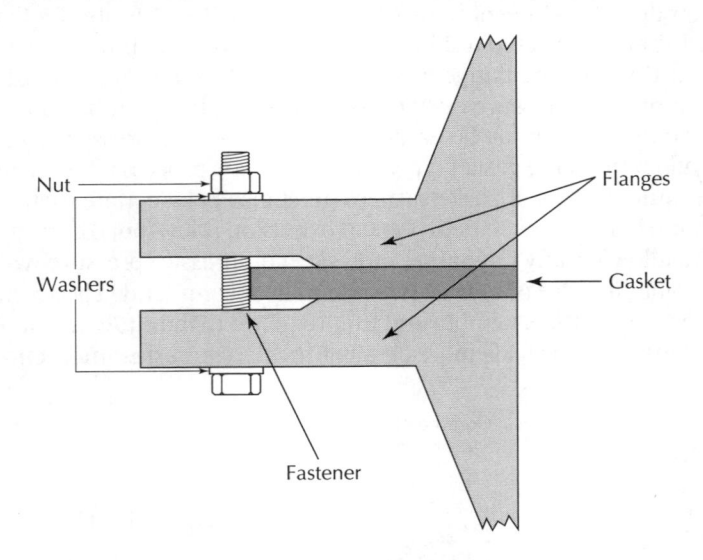

Only when all the components of the system are working together in harmony can the seal be expected to provide a good performance over a reasonable lifetime. The integrity of a safe seal depends upon:

- ☑ *selection of correct components appropriate for the application*
- ☑ *careful preparation, cleaning, installation and assembly*
- ☑ *correct bolt tightening and loading*
- ☑ *regular inspection*

The behaviour of a flanged joint in service depends on whether or not the tension created in the fasteners will clamp the joint components together with a force great enough to resist failure of the seal, but small enough to avoid damage to the joint components (fasteners, gasket etc.). The clamping load on the joint is created on assembly, as the nuts on the fasteners are tightened. This creates tension in the fastener (often referred to as *preload*). Although there may be some plastic deformation in the threads when a fastener is tightened normally, especially on the first tightening, most of the joint components respond elastically as the nuts are tightened. Effectively, the entire system operates as a spring, with the fasteners being stretched and the other joint components being compressed.

5.2 Current emission levels

Typical emission levels experienced today were reported[1] in the ESA's 2nd European Fugitive Emissions Conference, which was held in Düsseldorf in September 1998. The project was conducted by a team from the University of Dortmund.

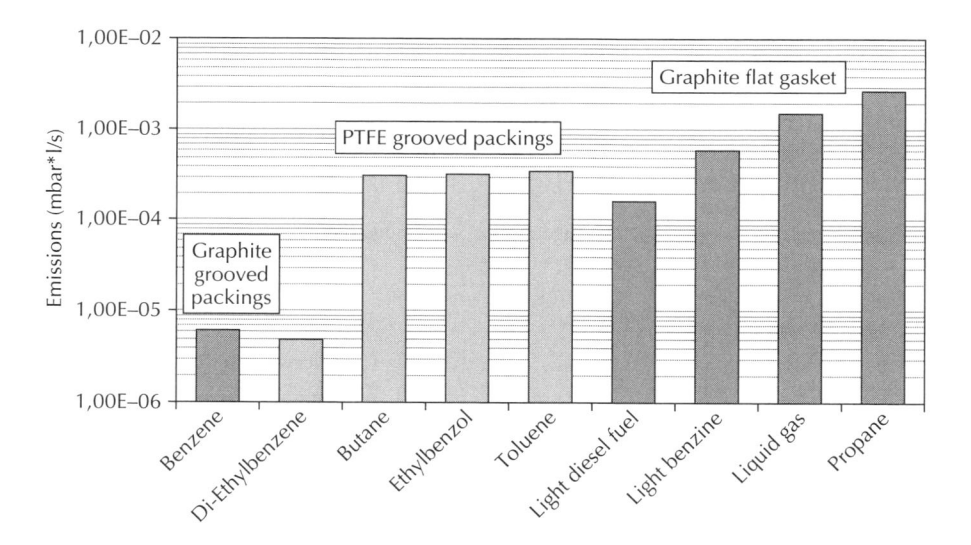

Readings were taken on flanges of various process installations at a chemical plant.

These results were compared with laboratory checks on the same type of flange gasket systems:

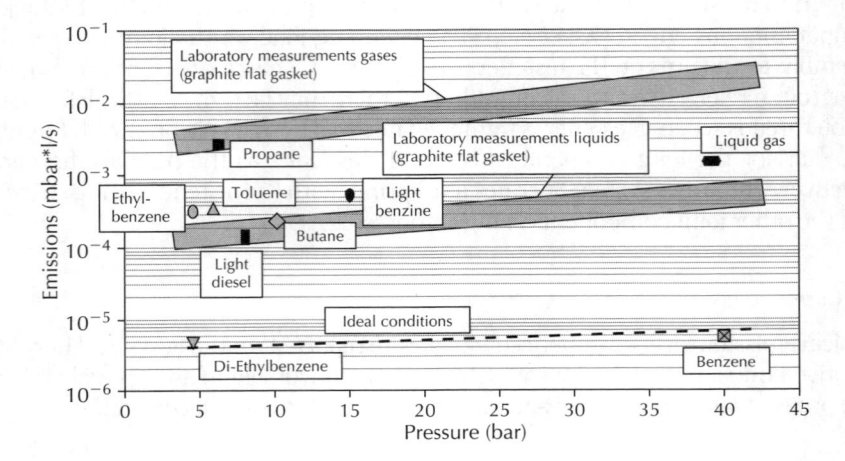

[Figure: Emissions (mbar*l/s) vs Pressure (bar) chart showing laboratory measurements of gases and liquids with graphite flat gaskets, with data points for Ethyl-benzene, Propane, Toluene, Butane, Light benzine, Light diesel, Di-Ethylbenzene, Benzene, Liquid gas, and Ideal conditions]

A similar study was conducted by BASF and the results were also reported[28] during the ESA's 2nd European Fugitive Emissions Conference. This study had run in parallel to the Brite Euram Project, which focussed on gathering data on asbestos-free gasket materials (note, in this chart IT-C and IT-400 refer to particular grades within the former DIN 3754):

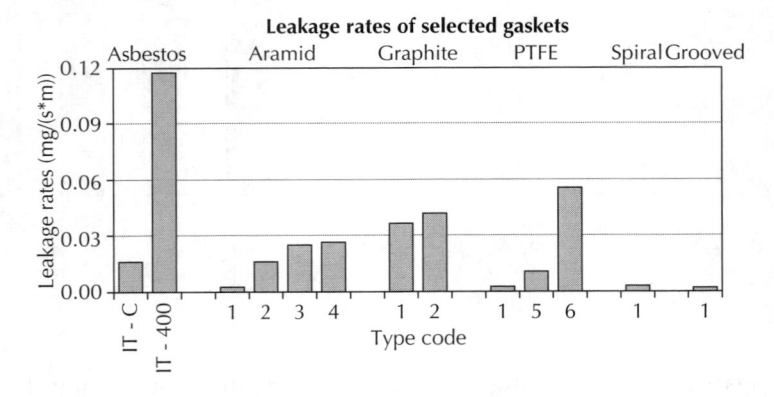

[Figure: Leakage rates of selected gaskets — bar chart showing leakage rates (mg/(s*m)) for Asbestos (IT-C, IT-400), Aramid (1,2,3,4), Graphite (1,2), PTFE (1,5,6), and Spiral Grooved (1,1) type codes]

For more information about leakage rates of gasket types, please refer to EN 1591, part 2 (data generated according to EN 13555). For specific performance details and recommendations for particular applications, **please consult the manufacturer**.

5.3. Gasket selection

Primarily, selection must be based upon:

- *sealing capability, appropriate to the application*
- *compatibility with the operating medium (process fluid)*

- operating temperature and pressure
- variations of operating conditions (for example, during cycling)
- the type of joint involved

A word of caution; despite the similarity of many materials, the properties of the seal, and performance achieved, will vary from one manufacturer to another. **Always consult the manufacturer for detailed guidance on specific products and their applicability.**

> ☑ *importantly, always use a good quality gasket from a reputable supplier, because the cost of a gasket is insignificant when compared to the cost of downtime or safety considerations.*

Gaskets can be defined into 3 main categories:

- soft (non-metallic)
- semi-metallic
- metallic

The mechanical characteristics and sealing performance capabilities of these categories will vary extensively, depending on the type of gasket selected and the materials from which it is manufactured. Obviously, mechanical and sealing properties are important factors when considering gasket design, but the selection of a gasket is usually influenced primarily by:

- temperature and pressure of the medium to be contained
- chemical nature of the medium
- mechanical loading affecting the gasket
- sealing characteristics of the gasket

Soft gaskets (non-metallic)	Often composite sheet materials, suitable for a wide range of general and corrosive chemical applications. Generally limited to low to medium pressure applications. Types include: fibre reinforced sheet, exfoliated graphite, sheet PTFE in various forms and high temperature sheet materials based upon forms of mica.
Semi-metallic gaskets	Composite gaskets consisting of both metallic and non-metallic materials, the metal generally providing the strength and resilience of the gasket. Suitable for both low and high temperature and pressure applications. Types include: covered serrated metal cored, covered metal jacketed, covered corrugated metal, metal eyelet, metal jacketed, metal reinforced soft gaskets (including tanged graphite and wire reinforced fibre materials), corrugated metallic and spiral wound gaskets.

Metallic gaskets — Can be fabricated from a single metal or a combination of metallic materials, in a variety of shapes and sizes. Suitable for high temperature and pressure applications.
Types include: lens rings, ring type joints and weld rings.

The gasket must be resistant to deterioration from the fluids being sealed, and it must be compatible chemically and physically. For gaskets which are electrically conductive, consideration must be given to electrochemical (or 'galvanic') corrosion, which can be minimised by selecting gasket and flange metals which are close together on the electrochemical series. This type of corrosion is an electrochemical process occurring in the presence of an ion-conducting medium, which may be an aqueous solution made conductive by dissolved ions. The base element is dissolved in a redox process, in which electrons emitted by the base element (anode) are taken into solution and deposited on the noble element (cathode).

For gaskets cut from sheets, always use the thinnest material which the flange arrangement will allow, but thick enough to compensate for unevenness of the flange surfaces, their parallelism, surface finish and rigidity etc. The thinner the gasket, the higher the bolt load which the gasket can withstand, the less the loss of bolt stress due to relaxation, and hence the longer the service life of the gasket. Also, the lower the gasket area which will be exposed to attack from the internal pressure and aggressive media.

5.4 Storage and handling of gaskets and gasket materials

Although many gasket materials can be used safely after storage for many years, ageing will have a distinct effect on the performance of certain types of gasket materials, resulting from chemical degradation which occurs over time. Primarily, this is a concern with materials which are bonded with elastomers, which, in general, should not be used after about 4 years from the date of manufacture. Those materials with elastomeric binders will inevitably deteriorate over time, and even more quickly at higher ambient temperatures. Degradation is also catalysed by intense sunlight. Although this is of little concern with metallic gaskets, it may have an effect on semi-metallic gaskets (specifically, those which are combined with elastomer-bound materials). Since graphite and PTFE materials contain no binders, sheets and gaskets of these materials have a virtually indefinite shelf life.

The condition of the gasket plays an important part in its performance. Some gasket materials are relatively robust (such as metallic gaskets), others are reasonably forgiving (such as PTFE), but others may be very brittle or prone to cracking. Consequently, all gaskets and gasket materials are best handled with the same care and attention. Bent, nicked, gouged, scratched or hammered gaskets will rarely seal effectively! When working in the field, carry cut gaskets carefully, ideally within some form of protective cover. Although carrying small gaskets in a pocket is a common practice, it is best avoided! If you bend the gasket it will be damaged. If it picks up debris from the inside of your pocket or elsewhere, it may scratch the surface and so create a leak path.

Never re-use a gasket, since it may have been modified dramatically under operating conditions and hence cannot guarantee the required level of sealing performance. Even if the gasket appears to be okay, it is not worth taking a risk! The cost of a new gasket is minuscule when compared with the cost of down time caused by a leak or blow-out and the considerations of safety and environmental protection.

Similarly, bolts or studs may have been damaged due to corrosion, or may have lost ductility by being tightened past yield; if you cannot be sure – do not take the chance! Consequently, never re-use gaskets or fasteners.

5.5 Assembly procedures

For the seal to perform as designed, proper assembly of the joint is crucial. This process is subject to a large number of variables, including the condition of all the components, the smoothness, the hardness, the lubricity of surfaces, the calibration of the tools, the accessibility of the fasteners, the environment in which the engineers must operate . . . .

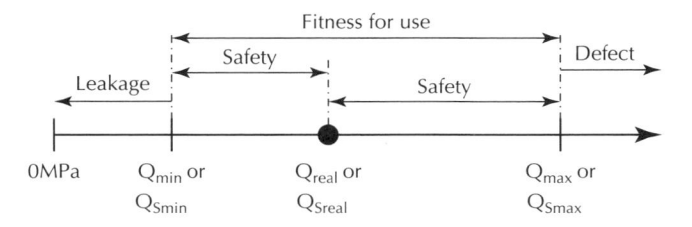

```
                        Fitness for use
                    Safety                    Defect
        Leakage                    Safety

        ├──────┼──────┼──────┼──────┼──────→
        0MPa   Q_min or    Q_real or      Q_max or
               Q_Smin      Q_Sreal        Q_Smax
```

Most importantly, it is a good idea to be consistent. If your present practices follow recommended procedures (according to the **Guidelines for safe seal usage**) and have proved robust, then don't change them! You should aim to keep the number of variables to a minimum. If possible, use the same tools in the same manner.

5.5.1 Tools required

You will require tools to both clean the flange and tension the fasteners. The tensioners will require regular calibration and may include torque wrench, hydraulic or other tensioners. Instruments to measure tension may include a micrometer, or ultrasonics.

Generally, this can be a pretty messy job! Therefore you will need appropriate clothing (protective clothing where necessary), safety helmet, safety goggles, gloves, and a security pass to the area, as appropriate.

5.5.2 Cleaning

To ensure good seal performance, all load-bearing surfaces must be clean:

- **Fasteners/nuts/washers** – clean with a wire (ideally brass) brush to remove dirt on the threads
- **Flange assembly** – clean gasket seating surfaces with suitable implements (see below)

On opening the flange and removing the previous gasket, the flange faces will often be contaminated with fragments of the old gasket material, which must be removed before a new gasket can be safely installed. Suitable implements for cleaning the faces of a flange may include a wire brush (use stainless steel bristles on alloy components). However, always brush in the direction of the grooves (rather than perpendicular to them), in order to minimise undue wear. Inevitably, use of a wire brush will result in wear across the faces over time. Consequently, other tools have been developed, such as the *brass drift*. This concept is based upon the use of a softer material (brass) than the flange surface (usually steel) to avoid damage. A suitable drift can be made from a sheet of brass, ~5 mm (0.2 in) thick × 50 mm (2 in) wide, which is filed and shaped to a 45° chisel across the width. Using a hammer, lightly tap the drift into the flange grooves to remove debris.

5.5.3 Visual inspection

All load-bearing surfaces must be free from any serious defects. Even a perfect gasket will be unable to seal a badly damaged or warped flange:

▣	**Fasteners/nuts/ washers**	– examine after cleaning to assure freedom from defects, such as burrs or cracks
▣	**Flange assembly**	– inspect the flange surfaces for defects, such as radial scores and warping
		– ensure the flange surfaces are sufficiently flat and parallel
▣	**Gasket**	– check that the correct gasket is available (suitable for the service, size, thickness)
		– examine the gasket prior to installation to ensure it is free from defects

If any defects are observed, don't take chances!

☑ *replace defective components with a good alternative. If in doubt, seek advice*

Note that for *spiral wound gaskets* in particular, the flatness and parallelism of the flanges are important factors for good sealing performance:

- **flange surface flatness** should vary by less than 0.2 mm over the gasket seating width
- **flange surface parallelism** should be less than 0.4 mm total out of parallel across the whole flange

5.5.4 Lubrication

It is estimated that, in the absence of a suitable lubricant, up to 50% of the torque effort may be used to merely overcome friction. Effectively, this would mean that the same torque applied to non-lubricated fasteners on a joint might provide markedly different loads on each one! Therefore, lubrication is essential when torque is used as the control for setting tension in the joint.

When selecting a lubricant, the following factors should be considered:

- *lubricity* — the better the lubricant, the lower will be the effect of friction
- *compatibility* — the lubricant must be compatible with the fastener materials (including nuts and washers), and ideally also with the process fluid. For example, copper-based lubricants may contaminate the process fluid, while chlorides, fluorides and sulphides may contribute to corrosion of the fastener materials (including nuts and washers)
- *temperature* — ensure the recommended service temperatures of the lubricant are within the process service temperature limits

The following procedures are recommended:

- ☑ *lubricate fastener threads and all bearing surfaces (underside of bolt heads, nuts, washers)*
- ☑ *use only specified or approved lubricants*
- ☑ *apply the lubricant in a consistent manner as a thin, uniform coating (avoid 'lumps' of lubricant as this may reduce the efficiency)*
- ☑ *ensure lubricant does not contaminate either flange or gasket faces*
- ☑ *avoid contamination of the lubricant (store in a closed container). After use, store in a 'clean' area*
- ☑ *avoid contaminating the gasket with the lubricant*

5.5.5 Gasket installation and centralisation

Prior to installation, ensure that the flange components are correctly assembled and the flange mating surfaces are parallel.

- ☑ *carefully insert the new gasket between the flanges to prevent damage to the gasket surfaces*
- ☑ *for large diameter spiral wound gaskets, seat the gasket in its mounting on the flange, remove securing straps, then slide the gasket from its mounting onto the flange using an appropriate number of persons to avoid damage to the gasket*
- ☑ *ensure the gasket is central in the flange*
- ☑ *do not use tape to secure the gasket to the flange. If it is necessary to secure the gasket to the flange, use a light dusting of spray adhesive (e.g. 3 M type 77)*
- ☑ *do not use jointing compounds or release agents, as these would reduce the friction between flange faces and gaskets, leading to extrusion of the gasket*
- ☑ *line up the joint components (including the flanges and the gasket) and examine them to ensure that an acceptable fit has been obtained*
- ☑ *take care when bringing the flanges together, to ensure that the gasket is not pinched or otherwise damaged*

5.5.6 Calculation of torque

Despite the number of developments to improve the reproducibility of fastening flanged joints (such as tension control fasteners, hydraulic tensioning devices, ultrasonic fastener analysis and simultaneous torque/turn methods), torque remains the most popular method to control joint tightening. When using torque tightening methods, there are 3 main factors to take into account in order to ensure that the required forces are produced:

Factor 1 +	**Factor 2** +	**Factor 3**
(torque applied to load the fastener)	(torque applied to overcome thread friction)	(torque applied to overcome friction at the nut)

These factors include the pre-load on the fastener spot face. Factors 1 and 2 include the dimension of the thread and Factor 3 includes the dimension of the nut. Factors 2 and 3 also include the coefficient of friction between these surfaces, which is dependent upon the type of lubricant used.

It must be emphasised that friction makes a significant contribution to the torque which must be applied, and hence the use of specified lubricants is crucial for good torque control. Values for the coefficient of friction provided by the lubricant must be known, in order to establish the fastener load accurately. Torque may be represented (in either metric or imperial units):

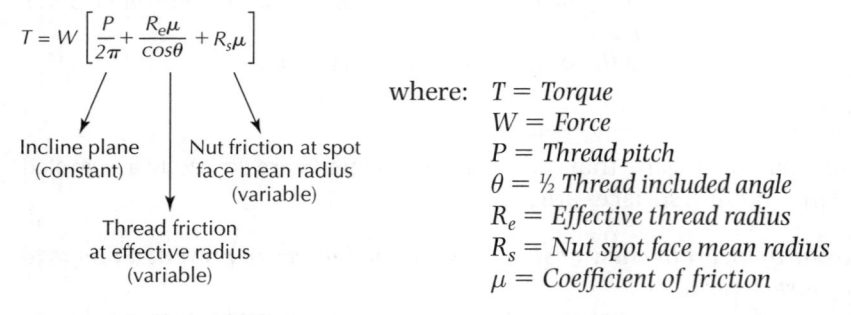

$$T = W \left[\frac{P}{2\pi} + \frac{R_e \mu}{\cos\theta} + R_s \mu \right]$$

- $\frac{P}{2\pi}$: Incline plane (constant)
- $\frac{R_e \mu}{\cos\theta}$: Thread friction at effective radius (variable)
- $R_s \mu$: Nut friction at spot face mean radius (variable)

where: T = Torque
W = Force
P = Thread pitch
θ = ½ Thread included angle
R_e = Effective thread radius
R_s = Nut spot face mean radius
μ = Coefficient of friction

In simplified form, for lubricated fasteners, washers, nuts, etc., the relationship between torque and fastener load may be represented as:

$$T = L \times 0.2 \times db$$

where: T = torque per fastener in N.m (in-lbf)
L = load per fastener in kN (lbf)
db = nominal diameter of fastener in mm
0.2 = factor of loss due to friction

Note also that the factor of 0.2 may vary considerably. It may be increased to 0.3 for non-lubricated systems, or reduced to 0.15 for lubricants with a low coefficient of friction.

The performance of the seal is largely dependent upon the correct level of tension in the fastener. Remember that for maximum effectiveness, the load on the fastener should be kept within its elastic region.

Other points to consider:

- *the crushing strength of the gasket material*
- *hydrostatic end thrust will increase the fastener tension under the operating internal pressure*
- *using a fastener stress which represents less than 50% of yield may cause problems*
- *most flanges are tightened by ordinary wrenching methods, and it is advantageous to have design stresses which can be achieved with this! (often impossible for larger diameter fasteners)*

5.5.7 Bolt/stud tightening pattern

One of the most difficult jobs facing the specifying engineer is to produce the correct assembly pressure on the gasket, low enough to avoid damaging the gasket, but high enough to prevent a leak in the seal. The gasket supplier will always be happy to assist in this task. Asbestos sheet materials are sufficiently robust usually to resist damage from overloading, but the same is not always true for asbestos-free alternatives! Consequently, when tightening up fasteners on a flange with any gasket type not incorporating a metal stop (such as a sheet gasket), *never* use an impact tool or scaffolding pole! It is vitally important to control accurately the amount of force applied to any particular flange arrangement, and hence:

- ☑ *always use a torque wrench or other controlled-tensioning device (recently calibrated)*

The sequence in which bolts or studs are tightened has a substantial bearing upon the distribution of the assembly pressure on the gasket. Improper bolting may move the flange out of parallel. A gasket will usually be able to compensate for a small amount of distortion of this type, but serious difficulties may be encountered if the flanges are substantially out of parallel. Consequently:

- ☑ *always torque nuts in a cross bolt tightening pattern*

Always run the nuts or bolts down by hand initially. This gives an indication that the threads are satisfactory (if the nuts will not run down by hand, then there is probably some thread defect – check again and, if necessary, replace defective parts).

Now torque the joint using a minimum of 5 torquing passes, using a cross-bolting sequence for each pass, as shown. The following procedure is recommended:

- ☑ *Pass 1 – Tighten nuts loosely by hand in the first instance, according to the cross bolt tightening pattern, then hand-tighten evenly*
- ☑ *Pass 2 – Using a torque wrench, torque to a maximum of 30% of the full torque first time around, according to the cross bolt tightening pattern. Check that the flange is bearing uniformly on the gasket*

- ☑ **Pass 3** – *Torque to a maximum of 60% of the full torque, according to the cross bolt tightening pattern*
- ☑ **Pass 4** – *Torque to the full torque, according to the cross bolt tightening pattern*
- ☑ **Pass 5** – *Final pass at full torque, in a clockwise direction on adjacent fasteners*

After the five basic torquing passes are completed, it may be beneficial to repeat pass 5 until no further rotation of the nut is observed. The final tightening must be uniform, with each bolt pulling the same load.

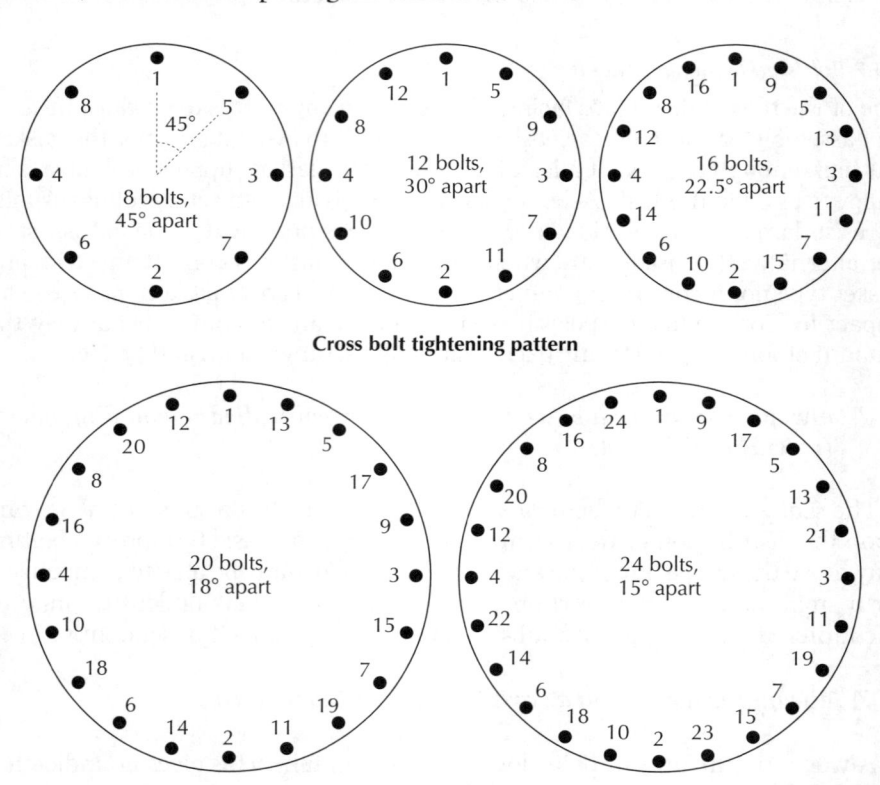

Cross bolt tightening pattern

Hydraulic tensioners are often used to preload fasteners. Although a number of engineers believe that these tensioners provide very good control (because the hydraulic ram exerts an accurate force on the fastener during the assembly operation), some load on the fastener is lost when the tensioner is removed as a result of elastic recovery. When the tensioner load is applied, the nut is run down against the joint (finger tight). The hydraulic pressure is then released and the tensioner removed. The nut and fastener now carry the full load, although there may be some embedment of material at the thread surfaces and at the nut bearing surfaces, which will reduce the load. Although hydraulic tensioners provide a consistent result, they require particular care, because the fasteners must be given a higher load to compensate for relaxation when the tensioner is removed. This may create hydraulic overload, which can cause fastener yielding (despite the apparent safety margin below 0.2% proof stress), especially a risk with certain fastener alloys (such as duplex stainless steels and other cupro nickel alloys), where the true elastic limit can be over 30% below the 0.2% proof stress value.

Alternatively, there are tension control fasteners available, which are pre-set to the required load.

Another way to tighten large bolts is to insert a heating rod in a hole drilled down through the centre of the bolt. As it heats up, the bolt expands length wise, and the nut can be run down against the joint (finger tight). The heater is now removed, and as the bolt cools, it shrinks, so developing tension. The method is relatively slow, but inexpensive (heaters are cheaper than high torque tools, for example). However, by itself, heating is not an accurate way to control a specified tension, and it should be combined with a measure of the residual stretch of the fasteners (such as with ultrasonics), which will then provide much more accuracy. There is some danger that heaters may alter the surface characteristics of the fastener, leaving them more susceptible to fatigue and stress corrosion cracking. If you plan to use heaters, use several at once at cross points around the joint, go for the final stretch in a single pass, measure the residual stretch after the bolts have cooled, re-heat and re-tighten those which are not correct.

5.5.8 Tagging

During a major overhaul, many thousands of flanges are required to be disassembled and then reassembled, usually involving teams of fitters who work their way through the plant. After the event, if a poorly assembled joint is discovered, it can be difficult (if not impossible!) to identify the root cause of failure. Consequently, tagging has been employed successfully on a number of industrial installations to encourage fitters to assemble the flange joints with care. It relies upon the following principles:

- each fitter is given a unique identification (letters and/or numbers)
- each fitter is provided with metal or plastic tags ('dog tags') stamped with his/her unique identification number
- one tag is connected (by wire or cord) to each flange joint as the fitter completes the assembly
- tag may vary from one metal to another, one colour to another, one shape to another, at every overhaul

Hence, as the installation comes back on stream, the plant operator will be able to identify which fitter has assembled which joint. Obviously, a fitter who is working sloppily will know that he/she can be readily identified after the event. However, on a positive note, the procedure will encourage some competition between fitters to be the best! It will identify those fitters who may need more training in one (or more) aspect of the job, and it will identify those fitters who are particularly good at their job – can they help to train others?

Equally, it may also highlight inferior quality gaskets.

With tags which vary from one overhaul to another, the operator can easily identify the date of assembly.

5.5.9 Re-tightening

For the majority of materials in the flange system (including gaskets, fasteners, nuts, washers), relaxation sets in after a fairly short time. For soft gasket materials, one of the major factors is usually the creep relaxation of the gasket. These effects are accentuated at elevated temperatures, with the net result that the compressive load on the gasket is reduced, thus increasing the possibility of a leak. Consequently, some engineers recommend that fasteners should be re-tightened (to the rated torque) 24 hours after the initial assembly, and again after 48 and 72 hours (care: always re-tighten at ambient temperature). However, this is an area of conflicting views (!): depending on temperature and duration, soft gasket materials cure until they have become too brittle to withstand any additional force.

Certainly, care must be exercised with repeated re-torquing in order to avoid damage to the gasket. This is especially important in the case of gaskets with relatively narrow sealing areas, as the stress on the gasket is liable to be high and therefore closer to the limit which the gasket can withstand.

Elastomer-based 'it' gasket materials (the German abbreviation for CAF) continue to cure in service, especially on start up as the operating temperature is reached. Once fully cured, gasket materials may become embrittled and liable to cracking under excessive load, and this is especially the case with elastomer-based asbestos-free materials. It is impossible to predict the time for embrittlement to start, as it will depend on the application temperature and the gasket composition. **Always consult the manufacturer for advice about re-tightening**, but as a general rule:

- ☑ *do not re-torque an elastomer-based asbestos-free gasket after it has been exposed to elevated temperatures*

5.6 Safety aspects and joint failure

Joints fail, not just gaskets! Low bolting torques, over-tight bolt loads, weak bolt materials, inadequate bolt/washer/nut lubrication, poor flange design or materials, poor gasket cutting or storage, improper installation practices, may each and all contribute to seal failure, even though the gasket material itself may be correctly specified! This publication will attempt to provide solutions to all of the above challenges.

Seal failure can occur when any component of the flange/fastener/gasket system is not performing correctly. The normal result is leakage from the joint, which may be virtually undetectable at first and build up over time, or may be a sudden, drastic failure. It is mainly observed when the fasteners fail to perform their clamping function, usually when they provide too little force, but occasionally when they exert too much!

A study commissioned by the Pressure Vessel Research Council (PVRC) of the USA, indicated that most flange joint failures resulting in leaks are due to:

- *improper installation*   *(26%)*
- *flange damage*   *(25%)*
- *gasket*   *(22%)*
- *loose bolts*   *(15%)*
- *flange misalignment*   *(12%)*

Although this list is by no means exhaustive (further details are available in a number of publications[5]), some common failure modes follow.

5.6.1 Failure due to the fastener

Fasteners which are insufficiently tight provide the most common cause of joint failure, which may result from:

- *incorrect assembly*
- *fastener failure*
- *self-loosening*
- *fatigue/relaxation over time*

On the other hand, when a **fastener is too tight** (usually as a result of an over-enthusiastic fitter or mechanic during assembly!), the joint may fail because the excessive load has:

- *crushed the gasket*
- *encouraged stress corrosion cracking*
- *increased fatigue*

Fastener failure occurs when the applied load exceeds the ultimate strength of the fastener or threads, and for a variety of reasons, typically:

- *fasteners do not meet design specifications (ruptured during assembly or at elevated temperature)*
- *over-tightened during assembly*
- *corrosion*
- *stress corrosion cracking*
- *fatigue*

5.6.2 Failure due to the gasket

This may result from a number of causes, such as:

- *selection of incorrect gasket for the application conditions*
- *selection of incorrect gasket thickness, particularly for soft gaskets*
- *excursions outside normal operating envelope or bending moments on pipe work*

- gasket damaged in storage, handling or on installation
- gasket crushed by excessive load during assembly
- deterioration over time
- gasket reused
- re-tightening after exposure to service (elevated) temperature

5.6.3 Failure due to the flange

Fairly unusual, but may result from:

- *flange surfaces damaged*
- *flanges warped*
- *flanges not parallel*
- *corrosion*
- *flanges not clean on assembly*

5.6.4 Minimising the chances of joint failure

From the above list of key causes of joint failure, it is obvious that the selection of the correct materials for the application is fundamental. Make sure that all components of the joint are compatible with each other and with the conditions which they will face during service. Allow an additional safety margin, just in case the application conditions move outside of the expected operating envelope (known as 'excursions!').

Follow the gasket storage and handling (and cutting recommendations, where appropriate) key recommendations throughout this publication.

Follow the cleaning and visual inspection key recommendations, to ensure that the joint components are free from defects and fit for subsequent use.

The above list also emphasises the requirement for good assembly practices. Unless the joint is put together with sufficient care, it cannot be expected to provide a safe seal. Ensure that the engineers involved are thoroughly trained in assembly procedures and briefed about the challenges they will face on site. Follow the key recommendations on installation, assembly and bolt tightening.

Corrosion is one of the most common challenges in the field! It can affect the integrity of the clamping force and will reduce the life of the joint components. It requires all four of the following conditions:

- *an anode*
- *a cathode*
- *an electrolyte*
- *an electrical connection between anode and cathode*

If any one of the conditions can be eliminated, corrosion will not occur. A solution is to keep the area dry by providing drainage holes (not always feasible) or, more commonly, by selecting fasteners manufactured from corrosion-resistant material. Most popular of all, by providing some form of protective coating on the fasteners and/or the flange.

Stress corrosion cracking (SCC) is the result of a combination of stress and electrochemical attack. Just humid air, or a dirty fingerprint, may be sufficient to initiate SCC! It is a specific form of corrosion and requires:

- *a susceptible material*
- *an electrolyte*
- *an initial flaw*
- *stress levels above a threshold*

All metallic fasteners are susceptible to SCC under certain conditions, but most of the problem can be minimised with suitable heat treatment. As with corrosion, provision of a suitable coating (aluminium, ceramics, or graphite) on the fasteners can minimise contact with the electrolyte. However, stress control is the most common way to reduce SCC, by keeping the stress level in the fasteners below a given limit (specific for the material).

Fatigue is time dependent and requires:

- *a susceptible material*
- *elevated stress levels above an endurance limit*
- *cyclic tensile stress*
- *an initial flaw*

In general, the higher the loads, the faster fatigue will set in. The item which usually has the greatest impact on reducing fatigue of the joint is the reduction of load excursions. Therefore, identify and achieve the correct preload in the fasteners. Note the differences in maximum preload between fasteners with rolled versus machined threads. Also, periodically replace the fasteners before they fail (it is advisable to keep records of how long they have lasted between failures, and then reduce the time frame somewhat to provide a reasonable safety margin). Ideally of course, always replace the fasteners when reassembling the joint!

Self-loosening is usually experienced in the presence of vibration, requiring:

- *relative motion between fastener, nut and joint components*
- *cyclic loads perpendicular to the fastener axis*

This is often countered by preventing slip between the fastener, nut and/or joint components by mechanical lock nuts or washers, or by the use of adhesives.

5.7 BAT for bolted flange connections

Individual flanges generally do not have very large leaking losses but, since plants utilise so many flanges, they can make a major contribution to the overall leaking losses. Preventative measures, among which regular maintenance (e.g.

controlled tightening of the flange), are very important. Regular inspection and replacement of the gaskets is also necessary, especially for those gaskets exposed to temperature fluctuations or vibrations as they age rapidly. When a removable connection is not necessary, flanges can be replaced by welded piping.

Best available techniques for bolted flange connections include:

- ☑ *minimise the number of flanged connections*
- ☑ *use welded joints rather than flanged joints where possible*
- ☑ *fit blind flanges to infrequently used fittings to prevent accidental opening*
- ☑ *use end caps or plugs on open-ended lines*
- ☑ *ensure gaskets are selected appropriate to the process application*
- ☑ *ensure the gasket is installed correctly*
- ☑ *ensure the flange joint is assembled and loaded correctly*
- ☑ *instigate regular monitoring, combined with a repair or replacement programme*
- ☑ *focus on those processes most likely to cause emissions (such as gas/light liquid, high pressure and/or temperature duties)*
- ☑ *for critical applications, fit high-integrity gaskets (such as spiral wound, kammprofile or ring joints)*

Alongside the introduction of asbestos-free gasket materials, there have been many developments for alternative gasket technologies. As a result, there is a vast range of gasket surface pressure and temperature limits in which the various gasket types may be used safely and readers should be aware that gasket performance may be improved or reduced under specific application conditions. For specific performance details and recommendations for particular applications, **please consult the manufacturer.** Members of the **ESA Flange Gaskets Division** may be contacted via the ESA web site at: www.europeansealing.com/divisions/flange_gaskets.htm

For the maximum and minimum sealing surface pressures for specific types of gaskets, please refer to EN 1592, part 2 (data generated according to EN 13555).

5.8 Relative costs of BAT for bolted flange connections

As mentioned in the section covering generic BAT, the cost of the actual sealing technology is infinitesimally small when compared with the investment made in the plant as a whole. Indeed, for many sealing technologies, the cost per unit may be in the region of a few cents, completely insignificant when the total plant costs are considered. Importantly, the unit cost of the sealing technology is overwhelmed completely by the labour costs required to fit the seal, let alone the downtime of the plant. **Consequently, the actual cost of the sealing device is immaterial in terms of economic considerations for BAT.** However, for the sake of completeness, the following table provides an overview of the relative cost of the gasket and its installation.

Gasket type	Gasket cost index	Installation cost index[#]
Compressed fibre sheet	1–10	300–1000
Elastomeric	1–5	300–1000
Micaceous (fibre free) sheet	4–60	300–1000
Reinforced graphite sheet	2–25	300–1000
Spiral wound	5–65	300–1000
PTFE	4–80	300–1000
Covered corrugated metal	4–60	300–1000
Kammprofile	8–90	300–1000

Note[#] installation cost index includes process shut down, removal of the gasket to be replaced and installation of the new gasket. This index is very dependent upon the local labour rate involved.

Besides these common types of gaskets, there may be special designs for specific sealing solutions, but these tend to be rare and usually well above the cost ranges mentioned here.

For some applications, only certain types of gasket options may be appropriate. The specific parameters of the application may preclude certain gasket types in order to attain the required safety margin, and this is likely to be reflected in the overall cost range.

Relative cost of gasket technology is dependent upon:

- plant type (for example, chemical, power generation, pulp and paper, etc.)
- plant size (for example, power station 200 MW, 500 MW, etc.)
- process type (for example, continual, batch, etc.)

The relative cost of the actual gasket itself is dependent upon:

- raw material
- energy costs to manufacture the gasket material
- energy cost to manufacture the gasket
- resource costs (for example, running costs of buildings and equipment, labour costs, etc.)

5.9 Emerging techniques

Sealing materials were previously based upon asbestos fibre and the manufacturing processes which evolved for those materials were therefore dependent upon aspects of that fibre for their successful operation. Thus, when the manufacturers of sealing materials started the development of asbestos-free sealing materials, it was natural that the use of fibres other than asbestos in conjunction with those processes should be the starting point.

However, as we now know, that approach was not entirely successful as the available fibres with the characteristics required were limited in their capabilities compared with asbestos. Consequently attention turned to trying to create sealing materials by alternative processes and using alternative materials.

This approach has turned out to be very successful. Exfoliated graphite and processed PTFE sealing materials have been the outcome and these have been proved to be very successfully applied to a whole range of gasket styles providing the user with sealing performance way beyond that of the original asbestos-based gaskets.

It is to be expected that such novel development will continue as the sealing potential of further new materials are exploited.

6 BAT for rotodynamic equipment

Modern process equipment with rotating shafts (such as pumps and compressors) is equipped with gland packings, mechanical seals or 'sealless' systems to eliminate (or at least minimise) emission of the process fluid into the atmosphere.

Gland packings, mechanical seals and sealless drive systems all require fluid for lubrication; in the majority of arrangements the process fluid is used for this lubrication and a very small level of leakage is inherent in pumps and compressors. Emissions from centrifugal pumps can be reduced by an order of magnitude by replacing packed glands with mechanical seals or sealless drive systems.

A glossary of sealing terms used in this section is contained in section 9.2.

6.1 Pumps

6.1.1 Emission management in pumps

The relatively low process leakage levels emitted from pumps and their relatively low numbers in a plant result in the overall leakage contribution from pumps being relatively small. As there are few pumps it is relatively simple to find and repair leaking pumps. Pump leaking losses occur mainly where the rotating shaft penetrates the casing. The technologies employed are:

Gland Packing
Gland Packing with a barrier flush
Single Mechanical Seals
Single Mechanical Seals with a mechanical containment seal and leakage collection (dual unpressurised seals)
Double Seals with a separate barrier fluid (dual pressurised seals)
Sealless drive systems

Gland packing leaks more than the mechanical seals in rotodynamic pumps, and in general, for this reason and reliability issues, is not used in VOC services that are emissive. It is used in some slow speed applications, which are discussed in section 6.3.

Historically, field surveys have investigated both liquid and gas phase leakage from mechanical seals[6, 7, 8] and, in general, these results have been excessively high by today's standards, indicating for example, that 25% of pumps equipped with mechanical seals were leaking more than 10 g/h. Continual developments by seal suppliers, pump manufacturers and end users however have resulted in significant improvements in sealing performance, such that the results of these earlier studies do not reflect the impact of new and current seal design technology or improved design, maintenance and operation of rotating machinery.

Recent field studies[2] in the USA, using EPA Method 21 to measure the VOC fugitive emissions from a variety of manufacturing facilities, show that 83% of single mechanical seals currently in use meet the Phase III Level of 1000 ppm in the US Regulations. From this survey, pumps with excessive leakage represent only **11% of the pump population, but 93% of the total emissions**, and for this reason it is clear that priority action must be focussed on these relatively few, but excessive leakers!

Detailed sampling[3] in the USA has indicated that low emission single mechanical seals can operate in the field for over three years and after this time they remain in compliance with the 1000 ppm regulatory limit. Indeed, most are still emitting less than 100 ppm at the exit point of the shaft. The same study showed that **dual unpressurised** mechanical seals can operate reliably in the field for over three years and after this time **most are still emitting less than 10 ppm**.

The CMA/STLE joint survey[9] of leak rates from pumps in a variety of services in chemical and petrochemical plants found typical leak rates of around 1 g/h for single mechanical seals with good face materials. Of all the pumps surveyed, 91.7% were emitting less than 1000 ppm (using EPA Method 21) and were thus within the Phase III standards of the US Regulations[10]. The survey reports that a substantial proportion of the 8.3% outside of the standard may be brought into compliance through the implementation of improved maintenance programmes, upgrades of seal face materials, secondary seal materials and selection of more appropriate seal design. The study concludes with the statement that *'single mechanical seals can perform to meet the requirements set forth by the United States Environmental Protection Agency's current and proposed future standards'*.

A 1991 survey[11] in the USA of 1112 pumps using *single* mechanical seals found that 94% were producing emissions of less than 1000 ppm, 92% were below 500 ppm and **84% were below 100 ppm (~0.7 g/h)**.

Double Seals with a separating barrier fluid (dual pressurised seals) eliminate process leakage to the atmosphere, as do sealless pumps. Sealless rotodynamic pump designs are available in two formats, 'canned' and magnetic drive, both of which enclose the rotor in the casing and provide the drive energy magnetically through a thin-walled region of the casing. The technology, in general, uses the process fluid for lubricating the rotor bearings, resulting in poorer reliability in some services.

6.1.2 Pump reliability

Single mechanical seals provide low levels of leakage but these levels increase significantly when they fail. The design, installation and operation of the pump influence heavily the life potential and reliability of the seal. The following are some of the main factors which constitute best practice:

- proper fixing of the pump unit to its base-plate or frame.
- connecting pipe forces to be within those recommended for the pump.
- proper design of suction pipe work to minimise hydraulic imbalance.
- alignment of shaft and casing within recommended limits.
- alignment of driver/pump coupling to be within recommended limits when fitted.
- correct level of balance of rotating parts.
- effective priming of pumps prior to start-up.

- operation of the pump within its recommended performance range. The optimum performance is achieved at its best efficiency point.
- the level of net positive suction head available (NPSHA) should always be in excess of the pump design's net positive suction head required (NPSHR). This can vary dependent upon the operating position on the pump performance curve.
- regular monitoring and maintenance of both rotating equipment and seal systems, combined with a repair or replacement programme.

Pump Standards provide the current best practice for all these subjects. Examples are ISO 9908, ISO 5199, ISO 9905, ANSI B73.1 & 2, API 610, ISO 13709.

6.1.3 Mechanical seals

Sealing technology has been continually improved, in anticipation of increasing environmental regulations, and over the past few years a new generation of mechanical seal products has been developed to provide cost-effective solutions for the control of emissions. Cost-effectiveness is an important consideration here, because the vast majority of low emission applications are covered successfully by mechanical seals.

The oil production and Refining Industries have been driven by the need for greater reliability and a lower level of VOC emissions from pumps. A group of major USA users came together within the structure of the American Petroleum Institute and produced the first standard on Mechanical seals, API 682[20] in the mid-90's. The scope of the standard requires that the sealing systems supplied, *'have a high probability of meeting the objective of at least three years of uninterrupted service while complying with emission regulations'*. It revolutionised the Industry by specifying rationalised seal designs and materials of which users had good field experience, and required seal suppliers to carry out rigorous qualification tests on a variety of fluids before being able to market their products to the standard. It is the default seal selection in the renowned pump standard API 610 and has become the base standard for most global oil refiners and producers.

A second Edition of API 682 has been issued which extends the original sealing philosophy to include the Chemical Industry and has added newer technologies for improved seal emission management and elimination. This new document has been developed with the International Standards Organisation and published as ISO 21049[27]. A third Edition of API 682 will mirror this standard.

Correct seal selection is dependent on the clarity and detail of the information provided to the seal vendor. Pump and mechanical seal data sheets as described in the above international standards are effective structures for ensuring the information needed is supplied.

The chart below (Adapted from STLE SP-30, revised in April 1994[14], with the kind permission of the Society of Tribologists and Lubrication Engineers, Illinois) can be used as a guideline for selecting the recommended sealing solution based on EPA Method 21 emission levels or mass emission rate. A detailed explanation of the recommended sealing solutions referenced in the three segregated areas is given in sections 6.1.3.1 to 6.1.3.4.

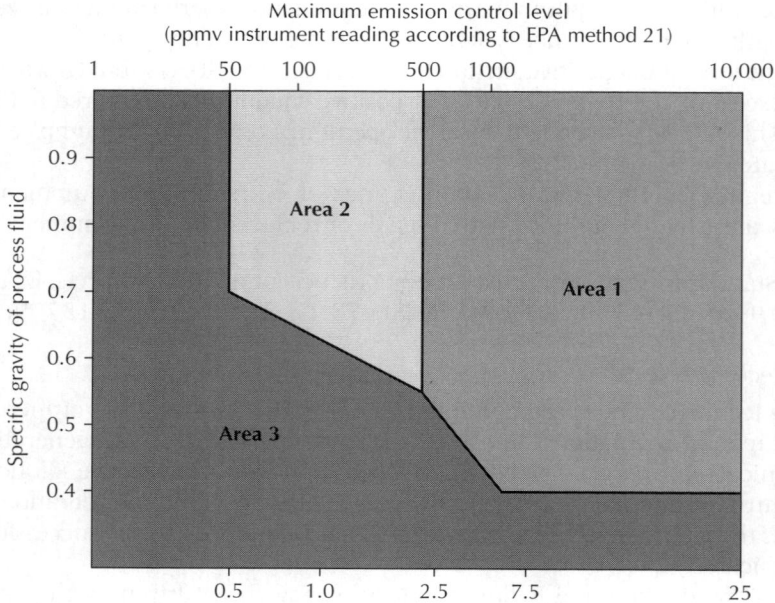

Maximum emission control level
(ppmv instrument reading according to EPA method 21)

[Chart showing Specific gravity of process fluid (y-axis: 0.4 to 0.9) vs Maximum emission control level (top x-axis: 1, 50, 100, 500, 1000, 10,000) and Maximum mass emission control level g/h (bottom x-axis: 0.5, 1.0, 2.5, 7.5, 25), with Area 1, Area 2, and Area 3 regions marked.]

Maximum mass emission control level (g/h) - from CMA STLE correlation data
Note: guideline only-emission level will vary with vapour pressure at fluid temperature

Area	Leak rate maximum (g/h)	Specific gravity of process fluid	Acceptable sealing solution
1	2.5–24.0	>0.4	General purpose single seals, advanced technology single seals, dual unpressurised (tandem) or dual pressurised (double) seals
2	0.5–2.5	>0.5–0.7	Advanced technology single seals, dual unpressurised (tandem) or dual pressurised (double) seals
3	<0.5	>0.4	Advanced technology single seals vented to a closed vent system, dual unpressurised (tandem) seals vented to a closed vent system, dual pressurised (double) seals, or sealless systems
		<0.4	Dual pressurised (double) seals, or sealless systems

This general guideline describes typical performance of mechanical seals of less than 150 mm (6 inch) shaft size, at pressures of less than 4100 kPa, with speeds of less than 28 m/sec and temperatures between −40°C and +260°C. Readers should be aware that local emission and/or hazardous fluid legislation may dictate a particular sealing solution, and operators should consult the appropriate regulatory authorities for precise compliance details.

When initially introduced by STLE, the above '*Application Guide*' was derived from user experience in the field, incorporating the performance of mechanical seals from a wide variety of suppliers. The guide represents a reliable performance profile for mechanical seals from **all** manufacturers, each of whom has access to differing levels of technological sophistication. Clearly, **those manufacturers with**

access to more advanced technology are able to provide products with generally higher performance levels.

Many of today's single mechanical seals, using modern materials and advanced technology, are reliably performing within Area 2 of the Application Chart, with emissions typically below 1 g/h under normal operating conditions in the field[15]. For this reason, **where low emission rates are essential, operators should approach only respected mechanical seal manufacturers in order to be assured of the use of modern materials and advanced technology.** These 'qualified' suppliers will ensure that today's mechanical seals incorporate 'design-in' capabilities to offset the combined effects of pressure distortion, thermal distortion and heat generation.

6.1.3.1 Advanced technology single seals

The original mechanical seal, still employed widely today, is a single mechanical seal installation aimed at general purpose applications, and consists of a fixed ring in the casing held in tight contact with a rotating ring on the shaft to form a seal. More recently, the application of **advanced sealing technology** has enabled the development of reliable, **low emission single mechanical seals**, which can give leak rates close to those of some dual seal installations. The technologies employed include highly sophisticated finite element and other modelling techniques in the optimisation of component shapes, computational fluid dynamics, specialised material developments, improved tribological properties rubbing face surface profile adjustments and pre-set packaged assemblies to eliminate fitting errors. A further essential factor, in support of the enhanced performance and reliability of new seal technologies, is the performance testing capability of the reputable seal manufacturers.

Additionally, for applications where hazard containment is required from the single seal arrangement, it is usual to include some form of external containment device to allow collection of any abnormal levels of vapour leakage and, where required, warn operators through a pressure induced alarm system. There are many kinds of secondary containment devices, including fixed or floating bushing and lip seals (spring energised or pressure energised). The space between the mechanical seal and some types of secondary containment device can be filled with a fluid to provide an environment where degradation or crystallisation of leakage is prevented.

A single mechanical seal provides the most economical form of seal, with emission values typically below 1 g/h under normal operating conditions in the field.[15] Single mechanical seals provide cost effective, reliable sealing for most VOC services[4], in line with API Standard 682 specifications[20] and ISO 21049[27], provided the following conditions are satisfied[3]:

process fluid specific gravity >0.4

vapour pressure margin in the seal chamber is sufficient for seal face lubrication

process or flush fluid provides adequate lubrication and cooling of the seal faces

For advanced technology single mechanical seals, users report[21] leak rates of between 0.42 and 1.25 g/h on one petrochemical plant in the Netherlands and between 0.63 and 1.67 g/h on a chemical plant in Germany.

This experience and data has been consolidated into the German guideline VDI 2440[27] which recommends that operators use 1 g/h as the mean leakage rate from single mechanical seals on process pumps.

6.1.3.2 Single seals with a mechanical containment seal (dual unpressurised seals)
The simple sophistication of a single seal (which contains the process fluid) is attractive to operators but where the process fluid is a VOC and the low emissive leakage to the atmosphere requires minimising it is common to include a second mechanical seal outboard of this primary seal. This provides a far more effective containment device than the bushings described in 6.1.3.1. The VOC leakage entering the containment chamber between the two seals can then be effectively channelled to a plant flare or vapour recovery system. Dual unpressurised seal arrangements will provide very low levels of process emissions to the atmosphere (see section 6.1.3).

Dual unpressurised seals use two different technologies for the lubrication of the outer mechanical containment seal. The numerically highest proportion fills the containment chamber with a separate buffer liquid that is piped to and from an adjacent reservoir. Flow is induced around the circuit, both lubricating the containment seal and assisting the channelling of VOC leakage into the buffer-fluid reservoir, where it is able to separate from the carrier buffer liquid. Ordinarily there is a connection from the top of the reservoir to a plant flare or vapour recovery system together with an orifice and an alarm to warn of deterioration in the sealing performance of the primary seal. This is referred to as flush Plan 52 in ISO 21049[27].

Dual seal arrangements with unpressurised buffer liquid provide emission values typically below 0.01 g/h, achieving emission levels less than 10 ppm[17] (<1 g/day).

Engineers on a hydrocarbon plant in the USA report emissions of less than 10 ppm (<1 g/day) from most dual unpressurised seals with buffer liquids on site after 12 months operation from start-up[22].

The alternative and more recent technology has been created by advances in high speed gas lubrication of mechanical seals (see section 6.2); no liquid buffer is required and the VOC gas, now at atmospheric conditions in the containment chamber, itself provides the lubrication of the containment seal. The containment chamber is directly connected to a plant flare or vapour recovery system with an orifice and a pressure alarm to warn of deterioration in the sealing performance of the primary seal. The benefit to the operator is a lower investment and operating cost. This is referred to as flush Plan 76 in ISO 21049[27].

A recent assembly of data from European and USA plants[26] studying single seals with gas lubricated mechanical containment seals concluded that 93.8% had Method 21 emission levels less than 1000 ppm and over 70% less than 50 ppm.

To achieve near complete elimination of emission to the atmosphere some plant operators connect a flow of Nitrogen buffer gas to purge the gas lubricated, mechanical containment seal of process VOC's and help channel them to the recovery/disposal system. This is referred to as flush Plan 72 in ISO 21049[27].

6.1.3.3 Double seals with a separate barrier system (dual pressurised seals)
This solution consists of two seals with a barrier fluid (liquid or gas) between them operated at a pressure greater than the process stream. Any leakage (outboard to atmosphere or inboard to the process stream) is of the barrier fluid, and therefore, selection of a safe barrier fluid compatible with the process stream is essential. This type of seal arrangement is useful for sealing process fluids with poor lubricating properties, on services where single seals are unreliable, or where the process fluids may change frequently (such as in pipeline services) and is selected[4] when the process fluid is particularly hazardous.

Dual pressurised systems virtually eliminate leakage of the process fluid into the environment and typically have emission values approaching zero, usually described as '*not measurable with existing instrument technology*'. Liquid lubricated mechanical seals typically use water or a light lubricating oil as the barrier fluid supplied from a self-contained support system and gas lubricated designs utilise a convenient plant gas source such as nitrogen managed by a control system. This former is referred to as flush Plan 53 or 54 and the latter flush Plan 74 in ISO 21049[27]. The simplicity and very low energy consumption of dual pressurised gas seals has been a strong driver in the growth of this technology in recent years.

Dual non-contacting seals with a pressurised nitrogen barrier fluid are showing near zero emissions in field applications[3, 25]. Double seals on a hydrocarbon plant in the USA are emitting between 0–5 ppm (<0.5 g/day) after 12 months operation from start-up.[22]

The potential of a failure of the Barrier system to maintain a pressure greater than the process stream, although unlikely, is a scenario that must be considered by the operator. The system can be configured to warn the operator of the problem. In addition, modern dual pressurised mechanical seals can be provided with componentry that will withstand a failure of the barrier system and continue to effectively contain the process for a period of time; most International Pump Standards now require features that provide this capability. The features are also required if the seals are supplied to API 682[20] or ISO 21049[27].

In all installations of mechanical seals, users should refer to the appropriate machinery and mechanical seal manuals for specific tolerances and guidance. In addition, a number of independent publications[12, 13] offer good advice on best practice.

6.1.3.4 Sealless pump drive systems
This technology also provides a zero emission capability but may be restricted in application by the process properties (see section 6.1.1). Users state that, to date, there is no universal sealing solution capable of handling satisfactorily all of the varying conditions of every application on a refinery, petrochemical or chemical installation. Consequently, although it is recognised that mechanical seals are the cost-effective solution in achieving emissions control, **on hazardous processes in**

general a combination of mechanical seal and 'sealless' systems is employed, with individual seal selection dependent upon the particular operating parameters.

It should be recognised that sealless pumps typically have significantly lower levels of efficiency compared to conventional pumps, requiring more energy for the same service. Energy consumption is the largest element in the 'Total Life Cost' of the pump and must be considered in the context of potential stack (CO_2, SO_2, NO_2) emissions from power generation equipment. In the context of the integrated pollution strategy advocated by the IPPC Directive this 'apparent' transfer of emissions from the pump to a power plant should be considered if sealless drive systems are being considered.

Some of the key distinguishing features of these sealing options are:

Mechanical seals	'Sealless' systems
Low capital investment[16, 17] – cost effective (especially single seals) – particularly advantageous as power ratings increase[22]	
Low repair and maintenance costs[16]	Potentially lower repair frequency[16] (although very dependent upon the service)
Long working lifetime[17]	
	Less frequent monitoring required[10]
Operator confidence in well-known technology[16]	
Wide process applicability: – allows flexibility of equipment use – widely used on existing plant – preferred for batch process plant – preferred for majority of applications (see Section 4 for specific suitability guidance)	Preferred for process streams where health hazard risk is very high (for example, carcinogens etc.). However, the technology may not be practicable on certain services.
	Lower risk of process stream contamination
Operator experience with technology parameters: – safely enabling performance optimisation[16] – minimal training required	
Simplicity (especially single seals)	
Energy efficient: – low operating costs[16] – no transfer of emission burden off-site[18]	
Special seals can be run dry: – preferred for 'dirty' processes[19] – preferred for process fluids with high viscosity or where fluid may undergo a high rate of change of vapour pressure[19] – preferred for batch processes	

6.1.4 BAT for pumps

- ☑ *proper fixing of the pump unit to its base-plate or frame*
- ☑ *connecting pipe forces to be within those recommended for the pump*
- ☑ *proper design of suction pipe work to minimise hydraulic imbalance*

Appendix 1: Sealing Technology – BAT guidance notes 549

- ☑ alignment of shaft and casing within recommended limits
- ☑ alignment of driver/pump coupling to be within recommended limits when fitted
- ☑ correct level of balance of rotating parts
- ☑ effective priming of pumps prior to start-up
- ☑ operation of the pump within its recommended performance range. The optimum performance is achieved at its best efficiency point
- ☑ the level of net positive suction head available (NPSHA) should always be in excess of the pump design's net positive suction head required (NPSHR). This can vary dependent upon the operating position on the pump performance curve
- ☑ regular monitoring and maintenance of both rotating equipment and seal systems, combined with a repair or replacement programme
- ☑ exchange gland packings in VOC services for mechanical seals where feasible
- ☑ selection of appropriate mechanical sealing technology based on required maximum leakage control levels and with consideration of process fluid characteristics (section 6.2.2)

6.1.5 Life-cycle cost (LCC)

The energy demand in certain process systems can be a huge component of the total costs of operating the plant. For example, some pumping systems may account for over 25% of the total energy usage on certain industrial installations. Yet, initial procurement and installation costs are used widely as the primary criteria for equipment or system selection. In these cases, the initial purchase price of a piece of equipment may be insignificant when compared with the total lifetime operating cost. Under these circumstances, procurement and installation costs in isolation may be simple to use, but will lead to poor long term decisions! This is where **life-cycle cost (LCC) analysis**[35] can be a valuable management tool to help maximise capital investment and plant efficiency.

The life cycle cost of a piece of equipment is the total 'lifetime' cost to purchase, install, operate, maintain and dispose of the item. LCC analysis is a useful comparative tool between a series of technology or operating alternatives and can indicate the most cost-effective solution, in order that the least long term cost of ownership can be achieved.

Components of such an LCC analysis include usually initial purchase costs, installation and commissioning costs, operating costs (including energy costs), environmental costs, maintenance costs, decommissioning and disposal costs. LCC uses net present value (NPV) concepts, which consider discount factors, cash flow and time. Consideration must be given to whenever costs occur during the life cycle of the equipment or project, while statistical equipment failure rates add further economic reality.

6.1.6 Relative life-cycle cost guide

The following matrix is intended as a guide for the consideration of lifetime costs, including initial investment, operating and maintenance costs for different mechanical sealing configurations/technologies and sealless solutions and their impact on the environment.

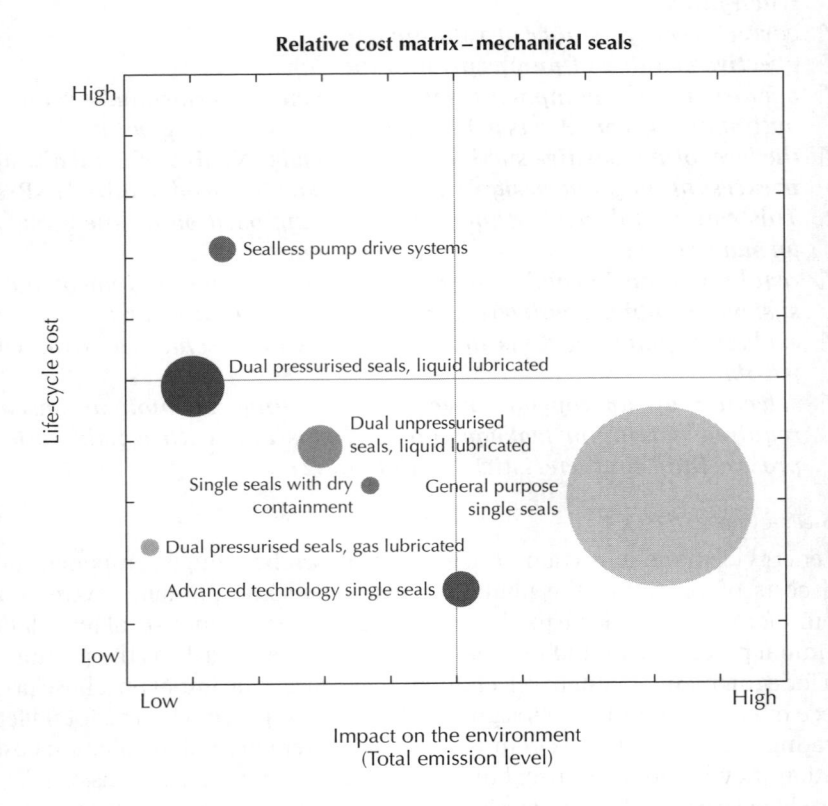

Relative cost matrix – mechanical seals

Y-axis: Life-cycle cost (Low to High)
X-axis: Impact on the environment (Total emission level) (Low to High)

Bubbles plotted:
- Sealless pump drive systems
- Dual pressurised seals, liquid lubricated
- Dual unpressurised seals, liquid lubricated
- Single seals with dry containment
- General purpose single seals
- Dual pressurised seals, gas lubricated
- Advanced technology single seals

The matrix is intended to be used with typical rotodynamic pumps but cannot be assumed to be universally applicable. The bubble size reflects population of seal type (hence, general purpose single seals are by far the majority of the installed pump population, whereas gas lubricated dual pressurised seals, being a more recent technology development, are the smallest).

Although often overlooked, the energy efficiency of the particular sealing technology can make the most significant contribution to the life cycle cost (LCC) overall. In addition reduced energy efficiency contributes to more emissions at the power generation plant, hence more impact on the environment in considering **total** emission level.

The general rationale behind the matrix guide is as follows:

1. General Purpose Single Seals:

Rotary seals, predominantly unbalanced. Initial investment is smallest of all sealing technologies considered, but higher energy usage and shorter MTBF increase the LLC. Emission range typically 500–1000 ppm (applying EPA Method 21).

2. Advanced Technology Single Seals:

Balanced stationary seals with advanced face designs. Initial investment higher than for (1) but with a potential for extended MTBF and therefore lower LCC. Emissions ranges 100–500 ppm.

3. Dual Seals, unpressurised:

Different containment technologies may be used (6.1.3.2) but MTBF values should be equivalent to Advanced Technology single seals. Investment costs are somewhat lower than pressurised dual seals and therefore have a lower LLC. Emission range 50–500 ppm depending on the containment technology applied.

4. Dual Seals, pressurised, liquid barrier:

Normally higher capital investment and running costs than dual unpressurised seals. Emissions approaching zero ppm.

5. Dual Seals, pressurised, gas barrier:

Lower capital and operational costs than liquid barrier technologies. Emissions levels comparable to sealless pump technologies at lower investment and maintenance costs.

6. Sealless pump technologies:

The chart position is based primarily on magnetic drive pumps[29]. This outlined MTBF's for seals and sealless pumps as being on average comparable. Repair costs of sealless pumps are typically higher than dual mechanical seals.

A detailed and separate examination of the initial investment and operating cost, comparing the differing mechanical sealing technologies, is shown in the table below:

Relative costs for different mechanical sealing technologies

	General purpose single seal	Advanced technology single seals	Single seal with dry containment	Dual unpressurized seals with liquid buffer[3]	Dual pressurized seals with liquid barrier[3]	Dual pressurized seals with gas barrier[4]
Emission level[1]	1000	500	50	10	0	0
Initial investment	100%	125%	250%	500%	500%	435%
Operating cost	100%	65%	100%	120%	190%	100%
Total life cycle cost[2]	100%	80%	135%	205%	260%	170%

Notes:
1 - ppm level measured using EPA method 21
2 - Based on a discounted 10 year operating life
3 - Includes reservoir
4 - Includes gas seal panel

552 *Seals and Sealing Handbook*

The information in the table above has been generated using the **Seal Life-Cycle Cost Estimator**[36] tool, which has been developed by the ESA and FSA Mechanical Seals Divisions. This tool allows you to estimate life-cycle costs for different sealing solutions on a comparative basis to assist in decision-making when specifying capital projects or upgrading existing rotating equipment technology.

The Seal Life-Cycle Cost Estimator allows comparison of a variety of sealing arrangements including single seals, dual seals, single seals with liquid lubricated and gas lubricated secondary containment, non-contacting gas seals, compression packing, and sealless pumps.

Life Cycle Costs are influenced strongly by the **reliability** of the selected sealing solution. Users are advised to think carefully about the individual MTBR (Mean Time Between Repair) values which are most appropriate for the different sealing solutions considered and, if necessary, contact your mechanical seal, packing, or pump supplier for guidance.

6.2 Compressors

6.2.1 Emission management in compressors

The sealing of two types of rotodynamic compressor will be discussed in this section. The first grouping is lower velocity, positive displacement designs, operating typically at 50/60 cycle synchronous speeds. They are used with many different types of gases but are commonly used in smaller refrigeration cycle services. The same technology is applied on some process gases. The shaft bearing assemblies are at either end of the shaft and mounted in-board of the seal assembly. Equipment leakage losses occur mainly where the rotating shaft penetrates the casing. The technologies employed are similar to pumps;

Single mechanical seals
Single mechanical seals with an energised containment seal
Single mechanical seals with a mechanical containment seal and leakage collection (dual unpressurised seals)
Double seals with a separate barrier fluid (dual pressurised seals)

Centrifugal process compressors are commonly applied on VOC process gases but typically operate at much higher velocities to achieve their performance efficiencies. The shaft bearing assemblies are at either end of the shaft and mounted out-board of the seal assembly. Equipment leakage losses occur mainly where the rotating shaft penetrates the casing at its drive and non-drive ends. The technologies employed are:

Labyrinth seals
Single mechanical seals
Single mechanical seals with a mechanical containment seal and leakage collection (dual unpressurised seals)
Tandem mechanical seals with a mechanical containment seal and leakage collection (triple seals)
Double seals with a separate barrier fluid (dual pressurised seals)

6.2.2 BAT for compressors

Equipment reliability is equally important in minimising emissions from compressors and the techniques which are recommended for pumps in section 6.1.2 are similarly applicable. Lower velocity rotodynamic positive displacement compressors are typically sealed by single mechanical seals lubricated by oil which jointly flows through the inboard bearing assembly. The oil is separated and recycled. To minimise process gas leakage when the equipment is static and the barrier oil has drained back to the supply reservoir, an energised lip seal or inboard mechanical seal and by-pass gallery can be employed as a damming assembly, retaining the barrier oil locally around the mechanical seal face.

To avoid the potential oil/process chemical degradation of dynamic elastomeric components used in some types of mechanical seal there is a preference for designs which utilise a metal bellows to manage its axial and alignment flexibility requirements.

It is common practice to use an energised lip seal outboard of the primary seal to contain any oil leakage. This helps channel process contaminated oil into a suitable collection chamber. This concept is improved by the addition of a gas lubricated mechanical containment seal in the same configuration as that described in section 6.1.3.2. A nitrogen buffer gas is occasionally used with this arrangement to purge the outer containment seal and assist the collection and separation of lubricating oil and process gas.

A double seal with a separate barrier fluid, as described in 6.1.3.3, is required where no emissive process leakage is permitted.

Some centrifugal compressors with an integrated gearbox are successfully sealed using similar technologies to the positive displacement type machines. These machines have relatively low velocities at the seal faces because the sealing point is at the input shaft to the gearbox.

Centrifugal compressors are traditionally sealed by labyrinth seals (fixed or floating carbon bushings) or oil lubricated mechanical seals as described above for positive displacement compressors. High leakage levels from labyrinth seals however negate their use in VOC emissive services and they should be exchanged for mechanical seal assemblies.

The high capital investment and relatively poor reliability of oil lubricated mechanical seals at the high velocities employed in centrifugal compressors encouraged the development of gas lubricated mechanical seals in the 1970's. The technology utilises macro-topographical alteration of the rubbing surface profile to encourage hydrodynamic and hydrostatic gas film forces; this enables the maintenance of a dynamic, non-contact gap a few microns thick. In most circumstances a single mechanical seal is employed to seal the process pressure and an outer containment seal, using the same technology, minimises the emissions to the atmosphere and channels the primary seal process leakage to a flare or recovery system. It is important for the reliability of the outer containment seal to exclude from it lubricating oil from the outer shaft bearing and, where practical, a gas purge (air or nitrogen) between two labyrinths or floating bushings is used to separate the mechanical seal assembly from the bearing. This purge assists with channelling process leakage through the containment seal into a flare or recovery system and minimising escape to the atmosphere.

In circumstances where the process gas is contaminated by a toxic impurity (e.g. H_2S in a sour hydrocarbon gas), an inert buffer gas such as nitrogen, if practical, can be used to purge the process side of the containment seal as described in section 6.1.3.2. Where this is not practical, inert gas flush can be added.

In very high process pressure services the dual configuration is also employed utilising metal spring energized polymer rings as secondary sealing elements. Such seals have been operating at over 30 000 kPa since 2001.

6.3 Other rotodynamic equipment

6.3.1 Emission management in other rotodynamic equipment

Other rotodynamic equipment includes agitators, reactors, de-waxing filters, dryers, mixers, rotary kiln furnaces and rotary pumps for drinking water distribution, where heavy wear and shaft misalignment are notorious. Whilst leaking valves are rightly considered to be the largest **total** source of fugitive emissions, large rotary vessels will generally suffer greater mass losses as a single point source. The emissions from one rotary vessel can often exceed the total emissions from a large number of valves, costing the operator a significant amount in lost product and wasted energy.

As expected, emission performance in these applications relies upon the sealing of the rotating shaft or cylinder against the stationary sections of the equipment. Yet, in many cases, accurate alignment of the rotating shaft may be difficult to achieve, particularly as differential thermal expansion and contraction takes place during the process cycle. In addition to thermal effects, both shaft and bearings can become worn, even though speed of rotation is often low, and the net result is that the shaft can move radially by amounts in excess of 2 mm, particularly on start-up. Operating equipment under these conditions may cause excessive emissions during operation and result in excessive leakages caused by premature failure and/or rapid degradation of the sealing system. It is evident that such operation should be avoided if possible by adequate equipment (re-) design and operation within the design specification.

Mechanical seals are used widely as sealing technology for agitators and mixing equipment, providing maximum control of fugitive losses in critical applications. Both single and double mechanical seals are available which are designed specifically to handle larger radial and angular misalignments ('mixer seals') or mechanical seals with integrated bearings which effectively constrain radial run-out directly at the point of sealing.

In other instances however, especially on existing equipment, the investment required for more sophisticated sealing technologies cannot be justified. In these cases, compression packings may provide an economical alternative. Such a sealing system often comprises a sealing flange fixed to the rotating shaft, over which sealing carriers are clamped. According to requirement, these are fitted with two or four packing rings. The appropriate contact pressure for the packing rings is achieved by means of adjustable springs or hydraulic guidance. Rotation of the sealing carriers is prevented by suitable torque brackets. As size increases, the entire sealing system (except for the sealing flange) may be freely suspended, so that the system can accommodate small misalignments while retaining its sealing performance. In all of these cases, the compression packings must be selected carefully to accommodate shaft misalignment (transient and permanent), vibration and shock loadings. Readers should consult reputable sealing material manufacturers for advice on the best choice for their specific application.

6.3.2 BAT in other rotodynamic equipment

Losses from large rotating machines may be reduced dramatically by a combination of approaches:

- ☑ *use mechanical seals designed to accommodate large radial and angular misalignments ('mixer seals')*
- ☑ *use mechanical seals with bearing(s) integrated into their assembly, to constrain equipment run-out*
- ☑ *use advanced compression packing designs from reputable manufacturers only*
- ☑ *re-engineer the gland arrangement where necessary to accommodate shaft misalignment, run-out and equipment wear*
- ☑ *use 'live loading' (see below)*
- ☑ *close collaboration between the user and seal manufacturer can provide the most economical sealing solution*

In many agitators and mixing equipment, the sealing arrangement operates in either a dry environment or in media which provide little or no lubrication to the seal faces. Therefore, the applicability of single mechanical seals is limited to very slow rotating equipment where the dry run limit of the mating faces is not exceeded, and applications where an external lubricating system can be employed (such as a liquid flush, greasing system or a quench arrangement). In cases where such alternatives are neither feasible nor allowed, double pressurised seals should be considered. In critical equipment where the lowest possible emission levels need to be warranted, only double, pressurised mechanical seals provide emissions levels approaching zero.

Whilst 'live-loading' is generally considered to reduce the total requirement for maintenance and adjustment, it also serves to provide cushioning for the effects of growth and contraction during thermal cycling. This is where live-loading has far greater benefit than simply for taking up wear of the packing. It also ensures that the correct initial gland load was applied by compressing the spring stacks evenly by a known amount. Inevitably, as the packing wears and the springs open-up, then their applied force decreases and eventually some re-tightening will be required.

It is evident that the variety of equipment, operating conditions and allowable emission levels require a close collaboration between the user and seal manufacturer to determine the best sealing arrangement a given application.

6.4 Emerging techniques

Split seal technology is used increasingly on large rotodynamic equipment as an alternative sealing solution for compression packings. Although split seal seals should not be installed on equipment where emissions control is critical, advances in split seal technology provide emission containment levels exceeding those of conventional packings. The higher investment required for such technology is often offset by more favourable sealing efficiency, for instance when equipment maintains a vacuum. In other instances, the application of split seal technology may offer longer MTBR and lower overall operating costs.

The sealing of high vapour pressure liquids, including cryogenic liquid gases, can be sealed reliably and with lower levels of leakage using shaped, recessed regions between the seal running faces. This new technology has enabled much improved sealing Mean Time Between Failures (MTBF) of liquefied gases with corresponding lower overall emission volumes.

7 BAT for reciprocating shafts

Process equipment with reciprocating shafts is usually equipped with gland packings to minimise emissions into the atmosphere.

7.1 BAT for reciprocating compressors

Reciprocating compressors are mainly applied in the process industry for gas transportation, and pressure increase of various gases.

7.1.1 Packing cases
Packing cases create the seal between piston rod and cylinder. The cross-head guided piston rod is sealed towards the cylinder with a packing case design. The packing cases cover a broad range of operating conditions and are used widely in lubricated and oil-free compressors. They consist of a series of angular plates with cup-like recesses which house the sealing elements. These angular cups and a mounting flange are combined by tie rods. The number of sealing elements in a packing case is determined by the operating conditions of the compressor.

In order to obtain enhanced running time, the use of cooled packing cases is often applied for oil-free as well as lubricated compressors, depending on operating temperatures. Cooled packing cases have internal passages for circulating coolant for heat dissipation. The coolant channel design enables cleaning during packing case reconditioning. The closed coolant passage meets API standards.

The packing case designs are tailored to cope with the full spectrum of pressure, temperature, venting, purging and lubrication requirements. Customised solutions for fugitive emissions reduction are offered from reputable manufacturers.

7.1.2 Packing rings
The sealing rings are mostly segmental rings or multiple ring assemblies. Materials are either compounds or metal for lubricated service or combinations of both.

7.1.2.1 Segmental packing rings
The full floating segmental ring seal fully compensates for normal wear on the rings throughout their lifetime.

Radial or tangential cut ring segments balance the wear at the contact face.

7.1.2.2 Multiple packing rings
These are free floating packing rings, which follow the radial piston rod movement to assure a positive seal in the packing cups. As this ring configuration has a smaller cross section, they are suitable for restricted packing case dimensions.

7.1.3 Piston rings

The clearance between the piston and the cylinder liner is sealed with piston rings. The material is mostly a non-metallic composition, which provides a unique combination of sealing and bearing properties. They satisfy the operational requirements of many reciprocating compressors in oil-free service and in applications permitting various levels of lubrication, including min-lube.

Metallic piston rings consist of one piece, multiple piece or segmental rings to meet the needs of lubricated reciprocating compressors.

The design features must be selected in accordance with the operating conditions and should be discussed with the manufacturers.

- ☑ *select packing case, packing ring and piston ring design appropriate for operating conditions*
- ☑ *please consult the manufacturer*

8 BAT for valves

Valves are used widely on installations for controlling (or preventing) the flow of fluids. The choice and design of valves is very specific to the application, although in general terms the most common valve types are gate, globe, plug and control. Valve internal parts are usually actuated externally and this necessitates an operating stem. The loss of process fluid from valves is prevented usually by the use of a packed gland seal, in a similar manner to pumps.

The integrity of the seal is crucial and may be affected by poor selection, poor installation, heat, pressure, vibration, corrosion and wear, any of which may significantly reduce performance. Valves which fail to perform as designed may have severe environmental implications, either for fugitive emissions or catastrophic failure. The risk of mechanical failure can be minimised by an appropriate regime of inspection and maintenance. However, valve failure is more frequently due to incorrect operation, which underlines the need for effective operating procedures.

8.1 Valve leakage

Valves, and especially control valves, are an important source of leaking losses, and may account for more than 60% of the fugitive emissions in a plant. Furthermore, the major proportion of fugitive emissions comes from only a small fraction of the sources (e.g. less than 1% of valves in gas/vapour service can account for more than 70% of the fugitive emissions in a refinery).

Some valves are more likely to leak than others:

- valves which are operated frequently, such as control valves, may wear quickly and allow emission paths to develop. However, newer, low leak control valves provide good fugitive emissions control performance. Valves with rising stems (gate valves, globe valves) are likely to leak more frequently than quarter turn type valves such as ball and plug valves.

The valve packing performs the role of the shaft seal and hence is a major influence on valve leaking losses.

8.2 Compression packings

Control of fluid loss is essential to the successful operation of mechanical equipment used in fluid handling. Various methods are utilised to control leakage at shafts, rods, or valve stems and other functional parts of equipment requiring containment of liquids or gases.

The original and still most common of these sealing devices is the compression packing, so called because of the manner in which it performs the sealing function. Made from relatively soft, pliant materials, compression packings consist of a number of rings which are inserted into the annular space (stuffing box) between the rotating or reciprocating member and the body of the pump or valve. By tightening a follower against the top or outboard ring, pressure is transmitted to the packing set, expanding the rings radially against the side of the stuffing box and the reciprocating or rotating member, effecting a seal.

Compression packings are used in all process industry sectors to seal all types of media. They are used in rotary, centrifugal and reciprocating pumps, mixers, agitators, dryers, valves, expansion joints, soot blowers, and many other types of mechanical equipment. In this document, the focus of attention will be on their use as valve packings.

Compression packings are relatively easy to install and maintain. With proper attention, a high degree of successful operation can be anticipated.

Successful sealing with compression packings is a function of several important and related factors:

- careful selection of packing materials to meet the specific application requirements
- complete consideration of surface speeds, pressures, temperatures, and medium being sealed
- proper attention to good installation and break-in procedures
- high standards of equipment maintenance

These factors are discussed in other segments of this publication and are covered in detail in most of the product bulletins of the major packing manufacturers.

Compression packings are made from various materials in a variety of shapes, sizes, and constructions:

8.2.1 Diagonal-interlock braided yarn packings

This braid is designed for general service application or as braided end rings on the top and bottom of graphite die-formed tape rings for critical or control valve applications. When used alone (straight-sets), required compression rates generally fall within a range from 25 to 30. When used as braided end rings on the top and bottom of a set of die-formed graphite tape rings, the required compression rates are reduced to approximately 20. These rings act as anti-extrusion rings, wiper rings, and they compensate for any surface irregularities in the bottom of the stuffing box as well as add resiliency to the set. For packings of 5 mm and under, square or plait braid is normally used.

8.2.2 Flexible graphite products

Graphite packings are available in 3 different forms, as described below:

a. Flexible Graphite Die-Formed Rings

Valve stem packing rings manufactured from flexible graphite are die-formed from flexible graphite ribbon. A predetermined length of flexible graphite tape is compressed in a properly dimensional mould to the desired density resulting in a solid ring.

Traditionally, the rings are of square cross section and have a density of about $1.6 \, gm/cm^{-3}$ but square section rings can be manufactured in density ranges from $1.2–1.8 \text{ g.cm}^{-3}$. For such square section rings, especially those of high density, the clearances of the ring with respect to the dimensions of the valve stem and stuffing box are crucial for good sealing.

The recent trend in die-formed graphite rings has been to move away from such rings as described above. One alternative consists of sets which are made up of rings of different densities, with the density used depending upon the position of the ring in the set. Another alternative is to use rings which are **not** of square section. In both cases the purpose of the design modification is to allow more of the effort from the gland follower nuts to be concentrated in radial force on to the stem and box wall to ensure a better seal, rather than being lost as friction down the set.

Whatever form is used, for ease of installation when the valve bonnet cannot be removed, split rings should be specified. These are available in a range of forms of cut construction. All the forms of cut rings are designed to assist the user in obtaining a good quality seal.

Die-formed flexible graphite rings are available in commercial grade (95–98% purity) and nuclear grade (99+% carbon purity, while low sulphur content forms are also available). The commercial grade material makes up the vast majority of usage for industrial applications. Nuclear grades are almost exclusively specified by the power generation industry for service in PWR installations. The nuclear grade rings are made under stringent control procedures during manufacturing.

Active or passive inhibitors may be applied during or after fabrication of each ring to insure against corrosion and pitting of the valve stem.

Rings of a compatible braided material are recommended as anti-extrusion and wiper rings. These latter rings should be selected from a material that will not relax or creep excessively during service and thus allow the load from the follower nuts to be lost.

A prime consideration of the installation of such sets is that the load imposed upon the sets of rings should not be so high that the valve stem cannot be operated!

Similarly, and perhaps of critical importance in an emergency, the hysteresis created by the packing set of a control valve should not be significant in terms of the positioning of the stem. To reduce this problem, various forms of modified design of die formed graphite rings are available. Low density braided materials allowing the formation of rectangular rather than square section anti-extrusion rings are also available for the same reason.

b. Expanded Graphite Tape

When flexible graphite die-formed tape rings are not available, one solution is to actually die-form flexible graphite tape into packing rings in the stuffing box itself. Using this method, a length of tape is wrapped around the valve stem or pump shaft until the build-up of the material is sufficient to completely fill the packing space. The wrapped tape ring is then eased down into the box and individually compressed to approximately 50% of the original tape width until the stuffing box is completely filled. A top and bottom end ring of a compatible braided material should be used to eliminate tape extrusion during the compression operation and in subsequent service.

c. Other Graphite Packing Forms

In addition to die-formed graphite ring packings and expanded graphite tape packings, flexible graphite packings are available in several other forms, including: braided packings, laminated rings, and injection moulded rings.

8.3 Installation of compression packings

The importance of packing the valve correctly cannot be over-emphasised. Many packing failures are due to incorrect installation of the packing. The first step in getting the most out of a valve packing is correct installation. The following steps have been devised to ensure effective installation of packings on valves:

8.3.1 Remove the old packing from the stuffing box

Make sure the stuffing box is cleaned out and all remnants of the old packings are removed. Ensure both shaft and sleeve are not damaged.

- ☑ *clean stuffing box and shaft thoroughly*
- ☑ *examine shaft or sleeve for wear and scoring*
- ☑ *replace shaft or sleeve if wear is excessive*

8.3.2 Use the correct cross section

To determine the correct packing size, measure the diameter of the shaft (inside the stuffing box area if possible) and then measure the diameter of the stuffing box (to give the OD of the ring). Subtract the ID measurement from the OD measurement and divide by two. The result is the required size.

- ☑ *use correct cross section for stuffing box and shaft size*

8.3.3 When using coil or spiral packing, always cut the packing into separate rings

Never wind a coil of packing into a stuffing box. Rings can be cut with butt (square), skive (or diagonal) joints, depending on the method used for cutting. The best way to cut packing rings is to cut them on a mandrel with the same diameter as the shaft in the stuffing box area. If there is no shaft wear, rings can be cut on the shaft outside the stuffing box.

Hold the packing tightly on the mandrel, but do not stretch. Cut the ring and insert it into the stuffing box, making certain it fits the packing space properly. Ensure the first ring is cut carefully and tested on the stem. Each additional ring can be cut in the same manner, or the first ring can be used as a master from which the balance of the rings is cut.

If the butt cut rings are cut on a flat surface, be certain that the side of the master rings, and not the OD or ID surface, is laid on the rings to be cut. This is necessary so that the end of the rings can be reproduced.

When cutting diagonal joints, use a mitre board so that each successive ring can be cut at the correct angle.

It is necessary that the rings be cut to the correct size. Otherwise, service life is reduced. This is where die-cut rings are of great advantage, as they give you the exact size ring for the ID of the shaft and the OD of the stuffing box. There is no waste due to incorrectly cut rings.

- ☑ *cut coil or spiral into separate rings*
- ☑ *cut the packing rings on a mandrel*
- ☑ *fit the first ring carefully and test on the stem*
- ☑ *use a mitre board to cut diagonal joints*
- ☑ *alternatively, use die-cut rings of the correct size*

8.3.4 Install one ring at a time

Make sure it is clean and has not picked up any dirt in handling. Seat rings firmly (except PTFE filament and graphite yarn packings, which should be snugged up very gently, then tightened gradually after the pump is operating). Joints of successive rings should be staggered and kept at least 90° apart. Each individual ring should be firmly seated with a tamping tool, or suitable split bushing fitted to the stuffing box bore. When enough rings have been individually seated so that the nose of the gland will reach them, individual tamping should be supplemented by the gland.

- ☑ *install one ring at a time*
- ☑ *seat ring correctly*

8.3.5 Last ring

After the last ring is installed, take up gland bolts finger tight or very slightly snugged up. Do not jam the packing into place by excessive gland loading.

- ☑ *after the last ring is installed, take up carefully*
- ☑ *do not load excessively*

8.3.6 Slide gland forward until it makes contact with the packing

Make sure gland bolts are tightened up evenly. Tighten to the point when heavy resistance is felt. During this time, turn the valve stem back and forth to determine ease of turning. Do not torque down to the point where the stem won't turn.

- ☑ *tighten the gland bolts evenly*
- ☑ *take care not to tighten too much; ensure the stem can still turn!*

8.3.7 Inspect the valve after it has been on line

If leakage is observed, adjust the gland in accordance with safe maintenance procedures and manufacturer's recommendations.

- ☑ *inspect and adjust gland bolts if necessary*

8.3.8 Live loading of the valve stem gland

In its simplest form, live loading is the application of a spring load to the gland follower of a packed valve. Live loading may enable a seal to be maintained for a longer period. A belleville spring between the gland follower and its fastening studs and nuts provides an effective way to establish and maintain a controlled amount of stress in the packing set. The amount of the packing stress in a live loaded system can be controlled by the size of the belleville spring used and how far it is compressed or deflected.

In a live loaded packing system, the follower will continue to push against the packing even when packing volume is lost (by friction, extrusion, consolidation, etc.). The spring load will be slightly reduced as the springs expand, but this reduction in load will be much less than the load which would be lost if the packing set were not live loaded. This remaining load allows the packing stress to remain at a level above the minimum sealing stress and may enable the packing to remain leak free.

8.4 Current emission levels

Each valve type performs differently in terms of emissions, as indicated in this study by Cetim:

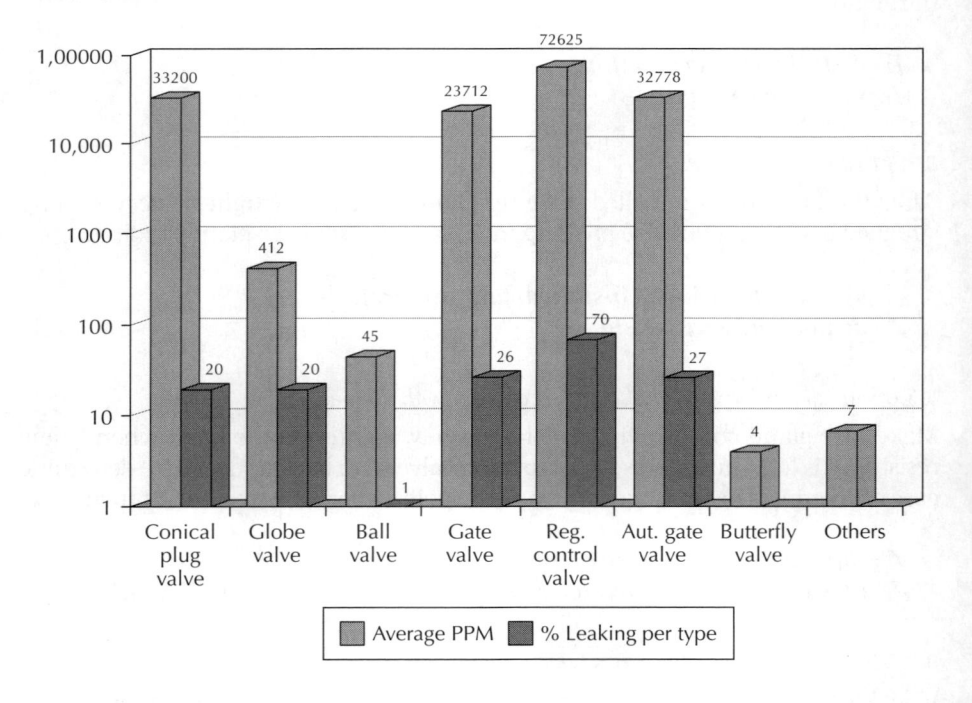

Valve type	Average PPM	% Leaking per type
Conical plug valve	33200	20
Globe valve	412	20
Ball valve	45	1
Gate valve	23712	26
Reg. control valve	72625	70
Aut. gate valve	32778	27
Butterfly valve	4	—
Others	7	—

8.5 Application guide for BAT in valves

Although stuffing box packings are one of the oldest forms of sealing technology, new developments have been introduced continually. This has been the case particularly since the advent of low emission valves, where high integrity packings materials and constructions have been provided to meet the tighter controls. As a simple guide:

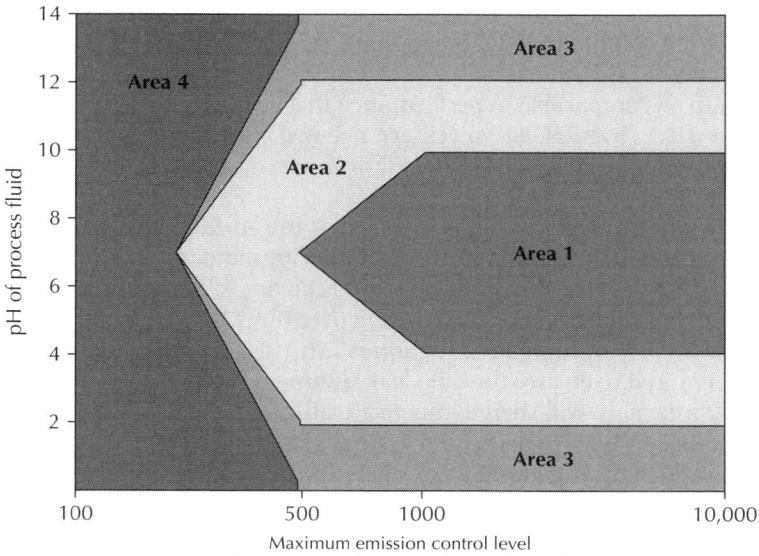

Maximum emission control level
(ppmv instrument reading according to EPA method 21)

Area	pH of process fluid	Maximum emission (ppm)	Sealing solution options
1	4–10	1000	Simple packings materials, often of traditional materials and constructions plus any of the options below....
2	2–12	500	Advanced packings constructions, including acrylic, aramid, glass, melamine, novoloid and polyphenylene materials plus any of the options below....
3	0–14	500	Braided flexible graphite (but not with oxidising media), PTFE packings materials (various constructions)
4	0–14	<500	Low emission/high integrity packings, generally of graphite (but not with oxidising media) or PTFE materials
Seek specialist solution from the manufacturer

Note that this is a general guide only. Many of the sealing options will give improved performance under certain conditions. For specific performance details and recommendations for particular applications, **please consult the manufacturer.**

8.6 Valve live-loading

Live-loading involves the use of disc spring assemblies mounted above the gland on the gland bolts or between the packing gland and gland nose. The live-load assemblies transfer the bolt force to the packing set. The major benefit of valve live-loading is the amount of elastic energy that is stored in the spring assemblies, which is typically 10–30 times that of the bolts themselves. This ensures that packing relaxation over time is fully compensated for, so that the reliability of a packed application is significantly increased. A properly designed live-loading arrangement in combination with a low emission packing is deemed equivalent to a 'bellows valve'. Live-loaded valves with low emission packing can be and often are proven to be comparable in performance to a bellows valve (TA-Luft). In addition, in the USA, live-loaded valves are deemed MACT (Maximum Achievable Control Technology) 'equivalent to a bellows valve' in terms of leak tightness and reliability.

Normally when valve packing is installed in the stuffing box, a high preload is required to ensure deformation of the packing rings and minimize in service consolidation of the packing. Preloads up to 40 MPa are sometimes recommended to minimize further relaxation of the packing over time. Yet these high preloads will result subsequently in high stem friction. With valve live-loading, a much lower preload is required to ensure that the leak tightness requirement is achieved. The live-loading assembly will then compensate automatically for the packing relaxation over time, without excessive valve stem friction.

Whilst live-loading is generally considered to reduce the total requirement for maintenance and adjustment, it also serves to provide cushioning for the effects of growth and contraction during thermal cycling. This is where live-loading has far greater benefit than simply for taking up wear and relaxation of the packing. It also ensures that the correct initial gland load was applied by compressing the spring stacks evenly by a known amount. Inevitably, as the packing wears and the springs open up, then their applied force decreases and eventually some re-tightening will be required.

Live-load spring assemblies should be designed for each application, based on the packing materials and operating parameters involved. Some packings require higher preloads than others and therefore one spring assembly design will not suit all packing materials and combinations.

Valve live-loading, in combination with a low emission, fire safe packing is the best available technique for fugitive emissions control in VOC or hazardous services.

8.7 BAT for valves

As these can provide such an impact on plant emissions, valves should be a high priority for attention. Best available techniques for valves include:

- ☑ *correct selection of the packings material and construction for the process application*
- ☑ *correct installation of the packings material into the stuffing box*
- ☑ *regular monitoring, combined with a repair or replacement programme*
- ☑ *focus on those processes most likely to cause emissions (such as gas/light liquid, high pressure and/or temperature duties)*

- ☑ *focus on those valves most at risk (such as rising stem control valves in continual operation)*
- ☑ *for critical valves fit high-integrity packings. Many of these are available in special constructions, using advanced technology materials, often specifically formulated for environmental performance*
- ☑ *use live-loading, in combination with low emission, fire safe packings in VOC or hazardous services*
- ☑ *where toxic, carcinogenic or other hazardous fluids are involved, fit diaphragm, ball or bellows valves*

Note that safety valves can be responsible for 10% of a plant's leaking losses. Losses are higher where safety valves are exposed to pressure fluctuations, and when a safety valve has activated. Therefore, safety valves should be checked after an emergency situation. Leaking losses via safety valves may be reduced by the installation of rupture discs prior to the safety valve to damp small pressure fluctuations. However, these fluctuations may pollute the valve, making complete closure impossible. An additional measure is to connect safety valves to a central flare system or another type of dedicated collection system (e.g. vapour recovery/destruction unit).

8.8 Relative costs of BAT for valves

As mentioned in the section covering generic BAT, the cost of the actual sealing technology is infinitesimally small when compared with the investment made in the plant as a whole. Indeed, for many sealing technologies, the cost per unit may be in the region of a few cents, completely insignificant when the total plant costs are considered. Importantly, the unit cost of the sealing technology is overwhelmed completely by the labour costs required to fit the seal, let alone the downtime of the plant. Consequently, the actual cost of the sealing device is immaterial in terms of economic considerations for BAT. However, for the sake of completeness, the following diagram provides an overview of the relative cost of the best available sealing technologies for valves and the environmental impact of the sealing systems.

The challenges associated with valves are dependent upon the valve type, the application and the sealing technology employed. The first matrix below represents the relative life-cycle costs versus impact on the environment for **on-off valves**:

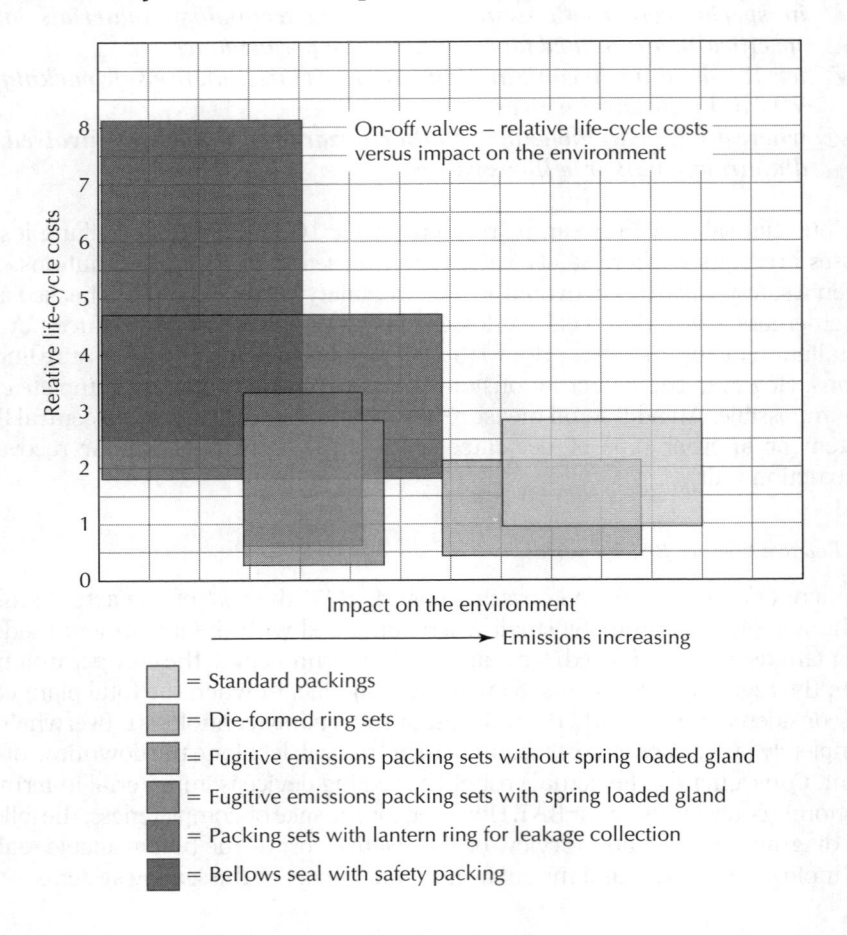

On-off valves – relative life-cycle costs versus impact on the environment

Relative life-cycle costs (y-axis, 0–8)

Impact on the environment → Emissions increasing

- = Standard packings
- = Die-formed ring sets
- = Fugitive emissions packing sets without spring loaded gland
- = Fugitive emissions packing sets with spring loaded gland
- = Packing sets with lantern ring for leakage collection
- = Bellows seal with safety packing

The matrix is intended to be used with **typical valves** but cannot be assumed to be applicable universally.

The second matrix below represents the relative life-cycle costs versus impact on the environment for **control valves**:

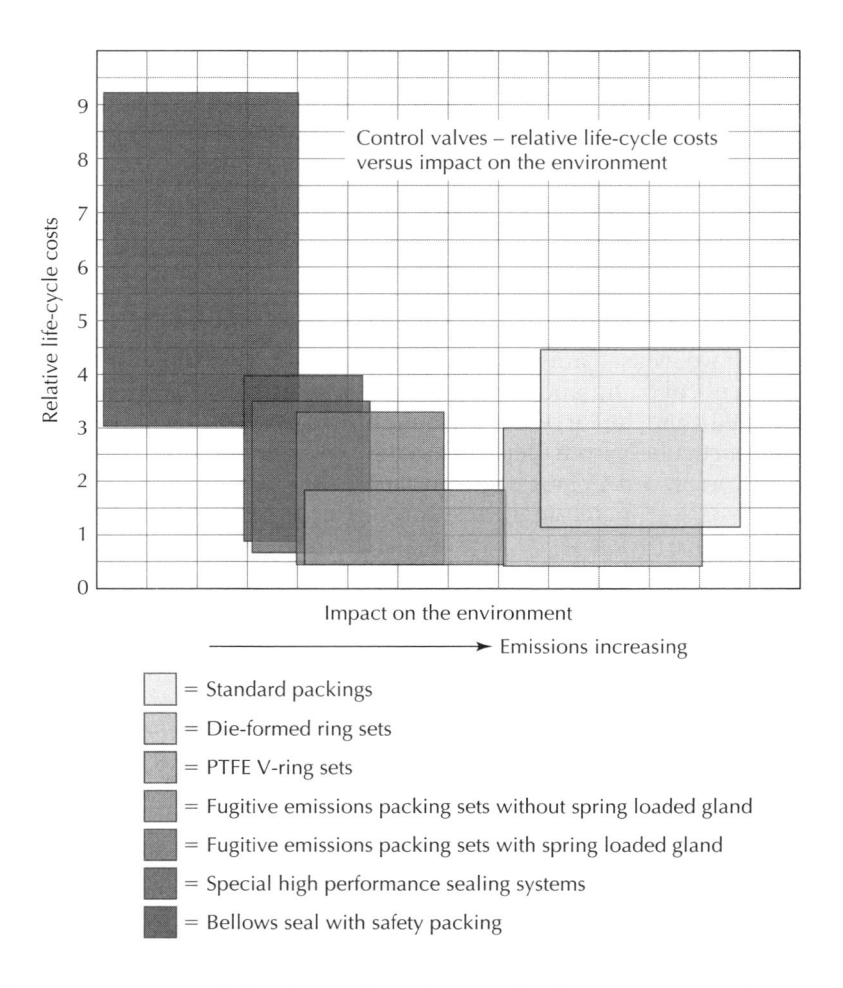

Control valves – relative life-cycle costs versus impact on the environment

Y-axis: Relative life-cycle costs (0–9)
X-axis: Impact on the environment → Emissions increasing

- ☐ = Standard packings
- ☐ = Die-formed ring sets
- ☐ = PTFE V-ring sets
- ☐ = Fugitive emissions packing sets without spring loaded gland
- ☐ = Fugitive emissions packing sets with spring loaded gland
- ■ = Special high performance sealing systems
- ■ = Bellows seal with safety packing

The matrix is intended to be used with **typical valves** but cannot be assumed to be applicable universally.

8.9 Emerging technologies

The ***valve stem leak tightness testing methodologies*** project of the ESA (financed under the EU's Standards Measurement and Testing Protocol), may provide some interesting spin-offs. This SMT Project (SMT4-CT97-2158) was initiated to provide guidance on the various methodologies for measuring *fugitive emissions*, and especially those fugitive emissions from *valves*. Importantly, these methodologies were aimed at ***valve qualification and QA tests***, rather than the choice of test methods in the field.

This collaborative project started in October 1997, to investigate valve stem seal leak-tightness testing methodologies.

Part funded by the European Commission under the Standards, Measurement and Testing (SMT) programme, this 30 month, k€uro 500 project brought together independent R&D organisations (BHR Group and Cetim), a petrochemical end user (Elf Antar), a valve manufacturer (Neles Automation), a sealing material manufacturer (Chesterton) and two manufacturers of leakage detection equipment (Inficon and Alcatel). Widespread support for the work was ensured by the involvement of European sealing industry (ESA) and valve industry (AFIR) umbrella organisations.

8.9.1 Background and scope

With increased concerns over environmental issues, plus legislation in the USA and Europe have focused attention on fugitive emissions of VOC's from chemical and petrochemical plant, and over leakage from particular sources. Site surveys have indicated that the largest contribution to plant VOC leakage comes from valves, and particularly from their stem seals[30]. In an attempt both to reduce product loss, to comply with plant emission targets and to rationalise procurement, plant operators are increasingly insisting on procuring valves demonstrating acceptable leakage performance. Two important messages coming from the chemical and petrochemical end-users[31] are:

It is not enough to know that the valve, as purchased, gives acceptable leakage. The real test is: how will it perform after a representative number of mechanical and, especially, thermal cycles?

This highlights the necessity for a *leak-tightness qualification test*.

Assuming that the valves purchased are of a design which has been qualified against such a test, the end-user who buys large quantities wants to know what proportion of these are likely to leak in service.

This highlights the necessity for a systematic *quality assurance strategy*.

A valve stem leak-tightness test comprises three essential elements:[32]

- the test protocol (specification of test temperature, pressure, duration, mechanical and thermal cycles, etc.)
- the leak-tightness criterion (specification of what level of leakage is deemed acceptable)
- the leakage measurement method

This project focussed on *measurement methods for valve qualification and QA tests*, and the interpretation of readings thus obtained. It must be emphasised that the project was not concerned with the choice of test methods *in the field*.

The emergence of standards on leak-tightness of industrial valves has given rise to a set of problems associated with the way in which leak-tightness is both measured and expressed. Issues surround:

- ***The specified gas;*** End users prefer a leakage target in gases representative of the duty. Valve manufacturers and test houses prefer the use of helium as

a safe, detectable, inert test gas, widely used for leak testing in many other contexts.
- *The measurement method;* Many end users prefer a leak-tightness result expressed in terms of ppm measured by sniffing, since this is the format of Environmental Protection Agency (EPA) targets in the USA and is widely used across the world for on-site leakage detection and repair (LDAR) programmes.

Unfortunately, ppm is a not a measure of leakage rate, but rather provides an indication of leakage severity. Thus, the requirement is for means of 'translating' both between ppm and leak rate and between leakage of one gas and leakage of another under similar conditions.

'Translation' between ppm and leakage is usually accomplished by means of empirical power law correlations. However, due to the wide range of data sources on which they are based, scatter in the data is very broad, rendering this approach, as it stands, unsuitable for valve qualification and QA test standards.

'Translation' between leakage rates of different gases can either adopt a similar empirical approach or, alternatively be based on predictive relationships derived from an understanding of which leakage mechanism predominates. Until now, little directly relevant work has been undertaken in this area.

The project work programmes provide relevant technical data and analysis to help resolve these issues by:

- *investigating measurement methods and recommending means by which scatter in the relationships between ppm and leak rate can be dramatically reduced*
- *evaluating measurements of leakage of different gases under similar conditions to identify means of predicting leakage of various gases from measurements of helium leakage*

Experimental work was concentrated on valves representative of typical volatile organic compound (VOC) duties over a range of sizes and types (but particularly 4" Class 300 gate valves) with several different packing types (but mostly graphite) over a range of leakage rates representative of this duty.

8.9.2 Key conclusions and recommendations

- ☑ *The use of a single detector type and elimination of significant local air currents can dramatically reduce scatter in the accuracy of power law relationships between ppm and leakage rate (from 2 orders of magnitude to a factor of <3).*
- ☑ *Sniffer probe flow rate has a major first-order effect on this relationship.*
- ☑ *Under well-controlled circumstances, the typical near-proportionality between leakage rate and ppm suggests the basis for a predictive relationship based on an empirical 'sniffing factor' (percentage of leaking gas taken up by the detector) and known probe flow rate. The first is not a constant, but can be averaged over a range of measurement circumstances,*

with some increase in the resulting scatter. The second is a function of the detector employed. By specifying these quantities, use of this approach removes significant sources of scatter. Further work is required for both helium and VOC detectors across a wide range of temperature valve types and sizes, packing types, etc.

- ☑ Flushing represents a suitable, reasonably accurate alternative to sniffing over the range of leakage rates for which it was evaluated and is a suitable technique for VOC leak rate measurement. Lantern ring ports may facilitate the use of this technique.
- ☑ The vacuum method is suitable over a wide range of helium leakage rates. It may involve minor modifications to a valve, however, to ensure a good vacuum seal. Again, a lantern ring port may circumvent that problem.
- ☑ Molecular flow appears to predominate at room temperature in many (not all) cases studied, but some data are contradictory. An assumption of laminar leakage flow could not be supported. Conclusions are not generally applicable to other valve types, sizes, packing types and test gases. Further work is required to confirm results, generalise and determine error bands.
- ☑ A significant data set was obtained for high temperature helium across a range of pressures and under well-controlled conditions. This represents a useful database for future studies.
- ☑ VOC ppm measurements at elevated temperature confirmed the scatter involved in this approach and contributes to a significant margin of error in any correlation from helium leak rate to VOC – typically a decade or more.
- ☑ Leakage measurements of methane, ethane and propylene by flushing demonstrated the feasibility of this approach for VOC's at elevated temperature. Results enabled selective checks on the relationship between leak rates of helium and VOC's.
- ☑ Whilst molecular flow appeared to be the predominant leakage mechanism in the high temperature helium tests reported here, the role of temperature on the leak rate versus pressure relationship was not as expected: more work is required.
- ☑ The leakage mechanism in VOC's at elevated temperatures remains unknown: more work is required.

The full conclusions and recommendations from the project are documented in the Publication Report.[33]

9 Conversion factors

The International System of Units (Le Système International d'Unités, or SI units) was first adopted by the 11th General Conference of Weights and Measures in 1960. This list is not exhaustive, and more details of the SI system can be found in publications such as ISO 31, ISO 1000, DIN 1301, BS 5555, BS 5775.

9.1 SI units

Quantity	Name of unit	Symbol	Expressed in terms of other SI units
Energy (work)	joule	J	$J = N.m = kg.m^2.s^{-2}$
Force	newton	N	$N = kg.m.s^{-2}$
Length	metre	m	
Mass	kilogram	kg	
Pressure	pascal	Pa	$Pa = N.m^{-2} = MN.mm^{-2}$
Power	watt	W	$W = kg.m^2.s^{-3}$
Temperature (thermodynamic)	kelvin	K	$K = °C + 273.15$
Time	second	s	

9.2 Multiples of SI units

The multiples are expressed by orders of magnitude, which are given as a prefix to the SI unit:

Prefix name	Prefix symbol	Factor by which the primary unit is multiplied	
exa	E	10^{18}	1 000 000 000 000 000 000
peta	P	10^{15}	1 000 000 000 000 000
tera	T	10^{12}	1 000 000 000 000
giga	G	10^{9}	1 000 000 000
mega	M	10^{6}	1 000 000
kilo	k	10^{3}	1 000
hecto	h	10^{2}	100
deca	da	10^{1}	10
deci	d	10^{-1}	0.1
centi	c	10^{-2}	0.01
milli	m	10^{-3}	0.001
micro	μ	10^{-6}	0.000 001
nano	n	10^{-9}	0.000 000 001
pico	p	10^{-12}	0.000 000 000 001
femto	f	10^{-15}	0.000 000 000 000 001
atto	a	10^{-18}	0.000 000 000 000 000 001

As an example, the multiple unit MPa (megaPascal = 10^6 Pa) is often used when referring to pressure in fluid systems, such as those in the process industries.

9.3 Units of common usage in sealing terminology

The following list covers **non-SI units** which are used regularly in connection with sealing terminology, and gives equivalent conversions into SI units (and other units where appropriate). The list is in alphabetical order (for conversion factors for SI units, please refer to **section 9.4**).

Unit	SI equivalent	Other non-SI unit equivalents			Various other units or conversions
		bar	$kp.cm^{-2}$	$N.mm^{-2}$	psi	
1 at	0.1013 MPa	1.013 bar	1.033 $kp.cm^{-2}$	0.1013 $N.mm^{-2}$	14.695 psi	0.987 atmospheres
1 bar	0.1 MPa			0.10 $N.mm^{-2}$	14.504 psi	
1°C	−273.15 K					
1°F						(°C × 1.8) + 32
1 ft (foot)	0.305 m					
1 in (inch)	0.025 m					
1 in²	645.2 mm²					
1 kgf	9.81 N					2.2046 lbf
1 kg/cm²	0.098 MPa	0.981 bar	1 $kp.cm^{-2}$	0.098 $N.mm^{-2}$	14.223 psi	
1 N/mm²	1 MPa	10.0 bar	10.197 $kp.cm^{-2}$	1 $N.mm^{-2}$	145.038 psi	
1 lb (pound)	4.45 N					0.4536 kp
1 lbf.ft	1.355 N.m					
1 lbf.in	0.113 N.m					
1 mm Hg	0.133322 kPa					
1 ppm	$35.92^{-0.733}$ g.h⁻¹					#
1 psi	6.895 kPa	0.0689 bar	0.0703 $kp.cm^{-2}$	0.00689 $N.mm^{-2}$		

This follows from the standard US field measurement technique, known as EPA Reference Method 21, which was introduced by the US Environmental Protection Agency (US EPA) for the monitoring of fugitive emissions in parts per million (ppm). This approach was established to provide a 'go'/'no go' method (i.e. there is either a **leak** or **no leak**). While this is useful as a **qualitative** measure of emissions, ppm cannot be converted directly into **quantitative** units. Accordingly, the US EPA has developed a series of correlations for the prediction of mass flow rate. These resemble closely a later joint study in the USA by the Chemical Manufacturers Association (CMA) and the Society of Tribologists and Lubrication Engineers (STLE), in which bagging data were analysed to determine the following relationship:

Leakage rate (lb.h⁻¹) = $6.138 \times 10^{-5} \times (SV)^{0.733}$, where SV is the screening value in ppm

When converted into metric units (453.6 g = 1 lb):

Leakage rate (g.h⁻¹) = $0.02784 \times (SV)^{0.733}$

9.4 Conversion factors (SI units)

Quantity	SI unit	Non-SI unit	Conversions
Acceleration	$m.s^{-2}$	$ft.s^{-2}$	$1\ m.s^{-2} = 3.281\ ft.s^{-2}$
			$1\ ft.s^{-2} = 0.305\ m.s^{-2}$
	$9.806\ m.s^{-2}$	$32.174\ ft.s^{-2}$	= Standard acceleration of gravity
Area	ha (hectare)	acre	$1\ ha = 10\,000\ m^2 = 2.471\ acres = 3.86 \times 10^{-3}\ mile^2$
			$1\ acre = 0.405\ ha = 4046.86\ m^2$
	m^2	ft^2	$1\ m^2 = 10.764\ ft^2$
			$1\ ft^2 = 9.290 \times 10^{-2}\ m^2$
	m^2	in^2	$1\ m^2 = 1.550 \times 10^3\ in^2$
			$1\ mm^2 = 1.550 \times 10^{-3}\ in^2$
			$1\ in^2 = 6.452 \times 10^{-4}\ m^2 = 645.2\ mm^2$
	m^2	$mile^2$	$1\ m^2 = 3.861 \times 10^{-7}\ mile^2$
			$1\ mile^2 = 2.589 \times 10^6\ m^2 = 259\ ha$
	m^2	yd^2	$1\ m^2 = 1.196\ yd^2$
			$1\ yd^2 = 0.836\ m^2$
Density	$kg.m^{-3}$	$lb.ft^{-3}$	$1\ kg.m^{-3} = 6.243 \times 10^{-2}\ lb.ft^{-3}$
			$1\ lb.ft^{-3} = 16.018\ kg.m^{-3}$
	$kg.m^{-3}$	$lb.gal^{-1}$	$1\ lb.gal^{-1} = 0.099\ kg.dm^{-3}$
	$kg.m^{-3}$	$lb.in^{-3}$	$1\ lb.in^{-3} = 27.679\ g.cm^{-3}$
Energy (work)	J	Btu	$1\ J = 9.478 \times 10^{-4}\ Btu$
			$1\ Btu = 1.055 \times 10^3\ J$
	J	ft.lbf	$1\ J = 0.738\ ft.lbf$
			$1\ ft.lbf = 1.356\ J$
	J	kcal	$1\ J = 0.238 \times 10^{-3}\ kcal$
			$1\ kcal = 4.19 \times 10^3\ J$
	J	kgf.m	$1\ J = 0.102\ kgf.m$
			$1\ kgf.m = 9.810\ J$
	J	kWh	$1\ J = 0.278 \times 10^{-6}\ kWh$
			$1\ kWh = 3.6 \times 10^6\ J$
Force	N	kgf	$1\ N = 0.102\ kgf$
			$1\ kgf = 9.81\ N = 2.205\ lbf$
	N	lbf	$1\ N = 0.225\ lbf$
			$1\ lbf = 4.448\ N$
	N	tonf	$1\ N = 1.003 \times 10^{-4}\ tonf$
			$1\ tonf = 9964\ N$
Length	m	ft	$1\ m = 3.281\ ft$
			$1\ ft = 0.305\ m$
	m	in (1″)	$1\ m = 39.37\ in$
			$1\ in = 0.025\ m$
	m	mile	$1\ m = 6.214 \times 10^{-4}\ mile$
			$1\ mile = 1.609 \times 10^3\ m$
	m	milli-inch ('thou')	$1\ 'thou' = 25.4\ \mu m$
	m	yd	$1\ m = 1.094\ yd$
			$1\ yd = 0.914\ m$
Mass	kg	cwt	$1\ kg = 1.968 \times 10^{-2}\ cwt$
			$1\ cwt = 50.802\ kg$
	kg	oz	$1\ kg = 35.274\ oz$
			$1\ oz = 28.349\ g$
	kg	pound (lb)	$1\ kg = 2.203\ lb$
			$1\ lb = 0.454\ kg$

Quantity	SI unit	Non-SI unit	Conversions
	kg	ton	$1\,kg = 9.842 \times 10^{-4}\,ton$ $1\,ton = 1.016 \times 10^3\,kg = 1.016\,tonne$ $1\,tonne\,(= 1\,metric\,tonne) = 1000\,kg$
Moment of force (torque)	N.m	kgf.m	$1\,N.m = 0.102\,kgf.m$ $1\,kgf.m = 9.807\,N.m$
	N.m	ozf.in	$1\,ozf.in = 7061.55\,\mu N.m$
	N.m	lbf.ft	$1\,N.m = 0.738\,lbf.ft$ $1\,lbf.ft = 1.356\,N.m$
	N.m	lbf.in	$1\,N.m = 8.85\,lbf.in$ $1\,lbf.in = 0.113\,N.m$
	N.m	tonf.ft	$1\,kN.m = 0.329\,tonf.ft$ $1\,tonf.ft = 3.037\,kN.m$
Moment of inertia	$kg.m^2$	$oz.in^2$	$1\,kg.m^2 = 5.464 \times 10^3\,oz.in^2$ $1\,oz.in^2 = 0.183 \times 10^{-4}\,kg.m^2$
	$kg.m^2$	$lb.ft^2$	$1\,kg.m^2 = 23.730\,lb.ft^2$ $1\,lb.ft^2 = 0.042\,kg.m^2$
	$kg.m^2$	$lb.in^2$	$1\,kg.m^2 = 3.417 \times 10^3\,lb.in^2$ $1\,lb.in^2 = 2.926 \times 10^{-4}\,kg.m^2$
Power	W	$ft.lbf.s^{-1}$	$1\,W = 0.738\,ft.lbf.s^{-1}$ $1\,ft.lbf.s^{-1} = 1.356\,W$
	W	hp	$1\,W = 1.341 \times 10^{-3}\,hp$ $1\,hp = 7.457 \times 10^2\,W$
	W	$kgf.m.s^{-1}$	$1\,W = 0.102\,kgf.m.s^{-1}$ $1\,kgf.m.s^{-1} = 9.81\,W$
Pressure	Pa	bar	$10^6\,Pa = 1\,MPa = 10\,bar = 1\,N.mm^{-2}$ $1\,bar = 0.10\,MPa = 14.504\,psi$
	Pa	ft H_2O (feet of water)	$1\,kPa = 0.335\,ft\,H_2O$ $1\,ft\,H_2O = 2.989\,kPa$
	Pa	in Hg (inch of mercury)	$1\,kPa = 0.295\,in\,Hg$ $1\,in\,Hg = 3.386\,kPa$
	Pa	$kgf.m^{-2}$	$1\,Pa = 0.102\,kgf.m^{-2}$ $1\,kgf.m^{-2} = 9.81\,Pa$
	Pa	$kp.cm^{-2}$	$1\,MPa = 10.194\,kp.cm^{-2}$ $1\,kp.cm^{-2} = 0.0981\,MPa = 0.981\,bar = 14.223\,psi$
	Pa	$N.mm^{-2}$	$1\,MPa = 1\,N.mm^{-2} = 1\,MN.m^{-2} = 10.197\,kp.cm^{-2}$
	Pa	$lbf.ft^{-2}$	$1\,kPa = 20.885\,lbf.ft^{-2}$ $1\,lbf.ft^{-2} = 47.880\,Pa$
	Pa	psi $(lbf.in^{-2})$	$1\,Pa = 1.450 \times 10^{-4}\,lbf.in^{-2}$ $1\,lbf.in^{-2} = 6.895\,kPa = 0.0703\,kp.cm^{-2} = 0.689\,bar$
	Pa	$ton.in^{-2}$	$1\,MPa = 6.477 \times 10^{-2}\,ton.in^{-2}$ $1\,ton.in^{-2} = 15.44\,MPa = 15.44\,N.mm^{-2}$
	$1.013 \times 10^5\,Pa$	$14.696\,lbf.in^{-2}$	Standard atmosphere $= 1.013\,bar = 1.033\,kp.cm^{-2}$
Rate of flow (volumetric)	$m^3.s^{-1}$	$ft^3.s^{-1}$ (cusec)	$1\,m^3.s^{-1} = 35.314\,ft^3.s^{-1}$ $1\,ft^3.s^{-1} = 0.028\,m^3.s^{-1} = 28.317\,dm^3.s^{-1}$
	$m^3.s^{-1}$	imperial $gal.h^{-1}$	$1\,m^3.s^{-1} = 7.919 \times 10^5\,imp\,gal.h^{-1}$ $1\,imp\,gal.h^{-1} = 1.263 \times 10^{-6}\,m^3.s^{-1} = 4.546\,dm^3.h^{-1}$
	$m^3.s^{-1}$	$in^3.min^{-1}$	$1\,m^3.s^{-1} = 0.366\,in^3.min^{-1}$ $1\,in^3.min^{-1} = 2.731 \times 10^{-7}\,m^3.s^{-1}$

Quantity	SI unit	Non-SI unit	Conversions
	$m^3.s^{-1}$	US gal. min^{-1}	$1\ m^3.s^{-1} = 1.585 \times 10^4$ US gal. min^{-1}
			1 US gal. $min^{-1} = 6.309 \times 10^{-5}\ m^3.s^{-1}$
Temperature	K	°C	K = °C + 273.15
			°C = K − 273.15
		°F	°C = (°F − 32) × 0.556
			°F = (°C × 1.8) + 32
Velocity	$m.s^{-1}$	$ft.s^{-1}$	$1\ m.s^{-1} = 3.281\ ft.s^{-1}$
			$1\ ft.s^{-1} = 0.305\ m.s^{-1}$
	$m.s^{-1}$	$km.h^{-1}$	$1\ m.s^{-1} = 3.6\ km.h^{-1}$
			$1\ km.h^{-1} = 0.278\ m.s^{-1}$
	$m.s^{-1}$	$mile.h^{-1}$	$1\ m.s^{-1} = 2.237\ mile.h^{-1}$
			$1\ mile.h^{-1} = 0.447\ m.s^{-1} = 1.467\ ft.s^{-1}$
Viscosity (dynamic)	Pa.s	P (poise)	1 Pa.s = 10 P
			1 P = 0.1 Pa.s
	Pa.s	$lbf.s.ft^{-2}$	1 Pa.s = $2.089 \times 10^{-2}\ lbf.s.ft^{-2}$
			$1\ lbf.s.ft^{-2} = 47.880$ Pa.s
Viscosity (kinematic)	$m^2.s^{-1}$	$ft^2.s^{-1}$	$1\ m^2.s^{-1} = 10.764\ ft^2.s^{-1}$
			$1\ ft^2.s^{-1} = 9.290 \times 10^{-2}\ m^2.s^{-1}$
	$m^2.s^{-1}$	$in^2.s^{-1}$	$1\ in^2.s^{-1} = 6.452\ cm^2.s^{-1} = 645.16$ cSt
	$m^2.s^{-1}$	St (stokes)	$1\ m^2.s^{-1} = 10^4$ St
			$1\ St = 10^{-4}\ m^2.s^{-1}$
Volume (capacity)	m^3	ft^3	$1\ m^3 = 35.315\ ft^3$
			$1\ ft^3 = 0.028\ m^3$
	m^3	imperial fl oz	1 fl oz = $28.413\ cm^3$
	m^3	imperial gal	$1\ m^3 = 2.199 \times 10^2$ imp gal
			1 imp gal = $4.546 \times 10^{-3}\ m^3$
	m^3	imperial pt (pint)	1 pt = $0.568\ dm^3$
	m^3	in^3	$1\ m^3 = 6.102 \times 10^4\ in^3$
			$1\ in^3 = 1.639 \times 10^{-5}\ m^3$
	m^3	litre (L)	$1\ L = 10^{-3}\ m^3 = 0.220$ imp gal = 0.264 US gal
	m^3	US gal	$1\ m^3 = 2.642 \times 10^2$ US gal
			1 US gal = $3.785 \times 10^{-3}\ m^3$

10 Further reading

The **European Sealing Association** has produced a wide variety of technical publications, often in collaboration with colleagues throughout the world, and focused primarily on helping users to achieve and maintain good sealing performance. These documents form the basis for this particular publication on BAT and are available from the ESA, either as hard copy or electronically. As part of good operating practice, the following ESA documents should be consulted where appropriate:

- ☑ *Expansion Joints – Engineering Guide – fabric expansion joints for ducting systems*, (ESA publication no. 011/01), published January 2001.
  Available in the following language version: English (other language versions in preparation).

- ☑ *Glossary of Sealing Terms, part 1, Flanges and Gaskets*, (ESA publication no. 008/97), published November 1997.
  Available in the following language versions: English, Italiano.

- ☑ *Guidelines for safe seal usage – Flanges and Gaskets*, (ESA + FSA publication no. 009/98), a joint publication of the European Sealing Association (ESA) and Fluid Sealing Association (FSA), published September 1998:
  Available in the following language versions: Deutsch, English, Español, Français, Italiano.

- ☑ *Guidelines for the use of Compression Packings – revised edition*, a joint publication of the FSA and ESA, published 1997.
  Available in the following language versions: Deutsch, English, Español, Français, Italiano.

- ☑ *Meeting emission legislation requirements with today's advanced technology mechanical seal systems*, (ESA publication no. 005/95), published November 1995.
  Available in the following language versions: Deutsch, English, Français.

- ☑ *Seal Forum – case studies in pump performance*, a joint publication of the ESA and FSA, published 2003.
  Available in the following language versions: Deutsch, English.

In addition, the following **leaflets/pamphlets** provide installation guidance for engineers in the field:

- ☑ *Fabric Expansion Joints – Installation Guide*, (ESA publication no. 015/04), a joint publication of the European Sealing Association (ESA), Gütegemeinschaft Weichstoff Kompensatoren e.V. (RAL) and Fluid Sealing Association (FSA), published January 2004.

Available in the following language version: English (other language versions in preparation).

☑ *Gasket Installation Procedures*, a joint publication of the FSA and ESA, published 2000.
Available in the following language versions: Deutsch, English, Español, Français, Italiano, Nederland, Portuguese, Turkish.

☑ *Pump and Valve Packing Installation Procedures*, a joint publication of the FSA and ESA, published 2003.
Available in the following language versions: Deutsch, English, Español, Français, Italiano.

11 References

1. *Determination of emissions of flange joints in a chemical plant*, K. Kanschik and H. Schmidt-Traub, proceedings of the 2nd European Fugitive Emissions Conference, Düsseldorf (8–9 September 1998), published by VDI-Verlag as VDI Berichte 1441(ISBN 3-18-091441-6), Düsseldorf 1998.
2. *Control of Fugitive Emissions of Mechanical Seals*, E Vanhie (Durametallic Europe NV), presented at the Shaft Sealing Seminar of the I Mech E, London, February 1992.
3. *Seal Systems for Emissions Control*, W Key (BW/IP International Inc.), presented at the 11th International Pump Users' Symposium, Houston, March 1994.
4. *Seal Technology, a Control for Industrial Pollution*, James P Netzel (John Crane Inc.), presented at the 45th STLE Annual Meeting, Denver, May 1990.
5. *Controlling Emissions to Atmosphere through the use of a Dry-sliding Secondary Containment Seal*, P M Flach, J E Sandgren and D P Casucci (EG & G Sealol), proceedings of the 10th International Pump Users' Symposium.
6. *Emissions of Hydrocarbons to the Atmosphere from Seals on Pumps and Compressors*, B J Steigerwald, Joint District, Federal and State Project for the Evaluation of Refinery Emissions, Report No. 6, NTIS No- PB 216582, 1958.
7. BHRA Seals Survey: Part 1, *Rotary Mechanical Face Seals*, R K Flitney, B S Nau, BHRA Report CR 1396, February 1977.
8. *Assessment of Atmospheric Emissions from Petroleum Refining*, **Radian Corporation, Vol. 1–4, Research Triangle Park, North Carolina, USA, Pub. No. EPA – 600/2-80-075a-d, 1980.**
9. *CMA/STLE Pump Seal Mass Emissions Study*, T Kittleman, M Pope, W Adams, proceedings of the 11th International Pump Users' Symposium, Houston, March 1994.
10. *National Emission Standards for Organic Hazardous Air Pollutants for Equipment Leaks*, Federal Register, Vol. 59, No. 78, Subpart H, pages 19568–19587, April 1994.
11. *Effectively Managing your Pump Seal Emissions*, C J Fone (John Crane EAA), proceedings of the Pump Performance and Reliability Conference, Aberdeen, October 1993.
12. *Mechanical Seal Practice for Improved Performance*, I Mech E, Mechanical Engineering Publications, Second edition, 1992. ISBN 0 85298 806 0.
13. *Guidelines for Meeting Emission Regulations for Rotating Machinery with Mechanical Seals*, Society of Tribologists and Lubrication Engineers, Special Publication SP-30, September 1990.
14. *Guidelines for Meeting Emission Regulations for Rotating Machinery with Mechanical Seals*, Society of Tribologists and Lubrication Engineers, Special Publication SP-30, revised April 1994.
15. *Measurement of Vaporised Leakage*, Dr A Voigt (John Crane GmbH), presented at the STLE/I Mech E Environmental Forum, Pittsburgh, May 1994.
16. *Canned Motor and Magnetically-coupled Pump Applications, Operations, and Maintenance in a Chemical Plant*, H Vollmtiller, W Seifert (Hoechst AG, Frankfurt) and K B Fischer (Hoechst-Celanese, Houston), proceedings of the 10th International Pump Users' Association.
17. *The increasing pressures on Process Plant Operators and how Mechanical Seal Technology is keeping pace*, C J Fone (John Crane EAA), proceedings of the Process Pump Seminar, I Mech E, London, June 1991.
18. *Zero Emission Solutions for Mechanical Seals on Light Hydrocarbons*, N M Wallace and JAM ten Houte de Lange (Flexibox International), proceedings of the 9th International Pump Users' Symposium, College Station, March 1992.

19. *A User's Engineering Review of Seailless Pump Design Limitations and Features*, T Hernandez (Exxon Chemical Company), proceedings of the 8th International Pump Users' Symposium.
20. *Shaft Sealing Systems for Centrifugal and Rotary Pumps, API Standard 682*, American Petroleum Institute, September 1992 (First and Second Edition).
21. *Die Bestimmung der Leakage von Gleitringdichtungen*, G Knoll, H Peeken, R Schroder (Rheinisch-Westfalische Technische Hochschule Aachen), VDMA-Projekt Literaturrecherche, February 1993.
22. *Meeting New Safety Emission Standards in a Hydrocarbon Plant*, D Brandt (Quantum Chemical Corporation) and J Netzel (John Crane Inc.), proceedings of the 66th Annual Fall Conference, Pacific Energy Association, Irvine, California, October 1991.
23. *Dry Gas Compressor Seals*, P Shah, proceedings of the 17th Turbomachinery Symposium, College Station, 1988.
24. *Application of Dry Gas Seals on a High Pressure Hydrogen Recycle Compressor*, D R Carter, proceedings of the 17th Turbomachinery Symposium, College Station, 1988.
25. *Design and Development of Gas Lubricated Seals for Pumps*, J R Wasser (John Crane Inc.), R Sailer and G Warner (GE Plastics), proceedings of the 11th International Pump Users' Symposium, Houston, March 1994.
26. *Containment Seals for API 682 Second Edition*, Peter E Bowden and Chris J Fone (John Crane EAA), proceedings of the 19th International Pump Users' Symposium, Houston, February 2002.
27. *ISO 21049, Pumps-Shaft Sealing Systems for Centrifugal and Rotary Pumps*, Published by Association Francaise de Normalisation (AFNOR) for and on behalf of ISO, February 2004.
28. *VDI Richtlinie 2440 Emissionsminderung Mineralölraffinerien*, Gründruck, VDI-Verlag, Düsseldorf, 1999.
29. *Ohne Pumpen läuft es nicht*, Dr. Ing. Friedrich-Wilhelm Hennecke (Leiter der Fachwerkstatt für Pumpen, Getriebe und Motoren, BASF AG, Ludwigshafen), Seal Forum, Issue 3, published by the European Sealing Association and Fluid Sealing Association, Summer 2002.
30. *Guidelines for controlling fugitive emissions with valve stem sealing systems*, STLE Special Publication SP-33, Society of Tribologists and Lubrication Engineers, October 1992.
31. *Valve World Conference*, Den Haag, November 1998, Workshop on Standardisation. Convenor: Barrie Kirkman (BP Amoco Chemicals).
32. *SMT project on valve emission measurement methods*, by Simon E. Leefe, 2nd European Fugitive Emissions Conference – Controlling Fugitive Emissions from Valves, Pumps and Flanges, VDI Verlag GmbH (on behalf of the ESA), 1998, pp. 237–249.
33. *Valve stem leak-tightness test methodologies, Publication Report*, published September 2000 by BHR Group Ltd, Report CR7 132.
34. *Control of fugitive emissions – activities and attitudes of the chemical and petrochemical industry*, K. Herrmann, and H.-J. Siegle, proceedings of the 2nd European Fugitive Emissions Conference, Düsseldorf (8–9 September 1998), published by VDI-Verlag as VDI Berichte 1441(ISBN 3-18-091441-6), Düsseldorf, 1998.
35. *Life-cycle cost and reliability for process equipment*, H. Paul Barringer, Proceedings of the 8th Annual Energy Week Conference and Exhibition (28–30 January 1997), organised by the American Petroleum Institute, published by the American Society of Engineers and Penn Well Publishing.
36. *Life-cycle cost estimator*, software tool developed jointly by the Mechanical Seals Divisions of the Fluid Sealing Association and European Sealing Association. The tool can be accessed directly at http://65.215.75.3/fsa/seallife.asp

APPENDIX 2

The application of European ATEX legislation to mechanical seals

The European Directive ATEX 94/9/EC (ATEX 95) requires that suppliers of equipment into potentially explosive atmospheres satisfy essential safety criteria and confirm this by support documentation that may vary depending on the level of risk. This appendix summarizes how the legislation is applied to rotating equipment and specifically mechanical seals. The Essential Health and Safety Requirements (EHSR) require an Ignition Hazard Assessment and guidance is provided on the possible content of such a review plus issues related to the maximum surface temperature. Marking of ATEX components is also discussed and proposals made. The position of mechanical seals in the context of ATEX 1999/92/EC (ATEX 137) is also summarized and advice on relevant issues provided.

This appendix was prepared in August 2006 and is a current assessment of the legislation and how it applies to mechanical seals at that date.

A2.1 Introduction

The term ATEX is derived from the French words, **At**mosphères **Ex**plosibles. The primary legislation, Directive 1999/92/EC, sometimes referred to as ATEX 137, is re-enacted into UK law by what is referred to as the DSEAR Regulations 2002. These are primarily applicable to plant operators and require them to make judgements about the occurrence and level of potential risk of an explosive atmosphere and how they should manage the risk. Comparable national implementation will be undertaken by other European national legislative bodies.

A2.2 ATEX zones and the products used therein

An important element of the DSEAR Regulations is the allocation of areas or zones based on the level of risk of a potentially explosive atmosphere occurring, see Figure A2.1.

New equipment installed into these zones, whether electrical or mechanical, is required to conform to a sister EU Directive, 94/9/EC, sometimes referred to as ATEX 95. This has been enacted in UK law through the Equipment and Protective

Appendix 2: The application of European ATEX legislation to mechanical seals 583

Gas/Air zone	Dust/Air zone	Likelihood of an explosive mixture occurring	ATEX 95 category
0	20	Highly likely to be explosive	1
1	21	Occasionally explosive	2
2	22	Unlikely to be explosive	3

Figure A2.1 Relationship between ATEX zones and equipment categories.

Systems Intended for Use in Potentially Explosive Atmospheres Regulations 1996 (EPS). It is this legislation which directly affects the rotating equipment into which mechanical seals are installed. The intended physical applicability is described in Article 1 and explanatory definitions are given. Four kinds of products are described, 'Equipment', 'Protective Systems', 'Components' and 'Safety, controlling or regulating devices' and it is the reference to 'Equipment' that covers most of the machinery using mechanical seals.

A2.3 ATEX Groups and Categories

The legislation splits the applicable industry and degree of explosive risk into Groups and Category sub-sections as presented in Figure A2.2. The level and complication of accreditation will depend on the required Group and Category but higher level approvals exclude the need for lower category accreditation. Unclear from the legislation but captured in supporting EN Standards (see section A2.7) is the need to understand and qualify whether the explosive risk is from a gas/air mix, dust/air mix or could be either. Group I covers the mining

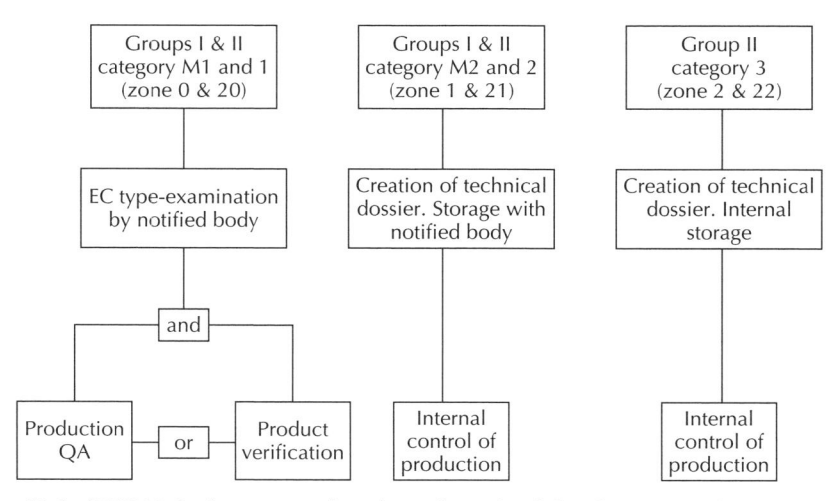

Figure A2.2 ATEX 95 Conformity procedures for mechanical seals based on group and category.

industry, whereas in the case of mechanical seals the major areas of application are in other sectors and hence where explosive atmospheres are a risk they are mainly used with equipment classed as Group II. There is a direct correlation between the zone defined in ATEX 137 and the category of the required equipment in ATEX 95 defined in Figure A2.1. In the context of mechanical seals they are primarily used with equipment located within the potentially explosive zones where the risk is from a gas/air mixture and classed as zone 1 and zone 2 depending on the level of risk. Where dust particles constitute an explosive risk the equivalent levels of risk are referred to as zone 21 and 22. In both these cases ATEX Category 2 and 3 accreditation respectively will be required if ATEX is considered applicable to the sealing application as described in section A2.4.

In some exceptional cases mechanical seals are used when the casing of the machine contains a fluid mixture that is classified as an explosive zone 0 or 20 and thus requires an ATEX 95 Category 1 product level. These are unusual circumstances and ordinarily have to be treated as specific ATEX applications.

A2.4 Applicability of ATEX 95 to mechanical seals

The ATEX Standing Committee is a committee of experts legislatively tasked with arbitrating in areas of interpretation of the legislation and is the author of the only formal EC Guide to ATEX, 'ATEX Guidelines', now published as a second edition.[2]

The original introduction of the ATEX regulations caused some confusion amongst equipment vendors and users with respect to how equipment and components should be specified. This has been resolved by the ATEX Standing Committee in a paper referred to as 'Considerations by the ATEX Standing Committee',[5] this made a clarification of the term 'ATEX component' and categorized mechanical seals as *machine elements*, not to be considered under ATEX and states:

Machinery Element:
These are parts of machinery not defined within 94/9/EC.
  Most mechanical seals are machinery elements. Typically these seals are:

- Catalogue mechanical seals and their parts, selected by the equipment manufacturer alone or with assistance from the mechanical seal manufacturer.
- Mechanical seals stocked by the equipment manufacturer or end user for general applications.
- Mechanical seals used for applications where the service conditions are not closely specified.
- Non cartridge-seals and parts.
- Standard cartridge-seals.

Mechanical seals will also be machinery elements if a risk assessment by the mechanical seal or equipment manufacturer shows that the seal is not expected to be an ignition source even in the event of fault conditions.

However, it also states that at an exceptional level of assembly and use a mechanical seal should be considered as requiring ATEX 'component' accreditation.

The 'Consideration' has been reinforced in the EC published 'ATEX Guidelines (Second Edition)', July 2005,[2] wherein section 3.9 of the document the clarified definition of an ATEX component has been confirmed. Mechanical seals are used in the same section as an example of a product used for 'general engineering purposes' when assembled in another product and the ATEX conformity assessment must be made for the complete integral product.

A2.4.1 Interpretation of the Considerations by the ATEX Standing Committee

In the text of the 'Considerations by the ATEX Standing Committee' there is reference to the possibility that some 'Engineered mechanical seals may be classified and sold as ATEX components' and two examples are given where this might occur. The use of the term 'Engineered mechanical seal' is often very loosely applied in the seal industry and has no clear definition. In order to help clarify the term and embracing the intent of the Standing Committee with regard to risk assessment the seal industry have agreed that, 'Mechanical seals that use standard parts or modifications thereof are considered machinery elements. Only in exceptional circumstances should a mechanical seal be classified as an ATEX component. An engineered mechanical seal, in the context of ATEX, should be when a specifically designed mechanical seal (which meets the criteria of an ATEX component) has its design features based on the ignition potential of a particular service.'

The Consideration paper makes it clear that zone 0 or 20 environments require mechanical seals to be classified as ATEX components, Category 1. These circumstances, however, in rotating equipment are very rare and almost exclusively relate to the atmosphere inside the equipment. There is a temptation to be conservative in judging the probability of a potentially explosive atmosphere and commercial considerations have been known to reinforce this bias. Unreasonable judgements, however, do not benefit industry in the longer term. A guideline for classifying zones is contained in EN 60079-10 and for zone 0, which are typically described as 'highly likely to be potentially explosive' is defined as an area in which 'the explosive atmosphere typically exists for more than 1000 hours per year'. There is a very similar definition for dust/air mixtures. When an item of rotating equipment is classified as requiring a Category 1 ATEX component it should be judged in the context of this guideline.

A2.5 Auxiliary seal systems

Mechanical seals are sometimes supplied with auxiliary fluid systems, as discussed in section 3.4.6, which help lubricate the contacting surfaces and assist in improving levels of process fluid containment.

Auxiliary systems are ordinarily considered as assemblies of parts supplied by different manufacturers, some of which could be individually classified as ATEX equipment or components, but which through their assembly into a auxiliary seal system *do not add any additional ignition risk*. These auxiliary system assemblies can thus be marketed as conforming to the Directive by supplying all the appropriate attestations of Conformity issued by the specific original suppliers of the equipment or components that individually constitute a potential ignition risk. Assemblies fitting this description need not have a Technical File and an overall conformity assessment. The interpretation of ATEX in regard to 'assemblies' is supported and confirmed in the EC ATEX Guideline, section 3.7.5. It is thus important that the purchaser, in his order documentation, warns the system supplier of the local ATEX requirements.

A2.6 ATEX 137 and DSEAR; 2002

Process industry employers, in the UK, are required by ATEX 137 and the DSEAR Regulations to:

- Carry out a risk assessment of any work activities involving dangerous substances;
- Provide technical and organizational measures to eliminate or reduce to as far as is reasonably practicable the identified risks;
- Provide equipment and procedures to deal with accidents and emergencies;
- Provide information and training to employees;
- Classify places where explosive atmospheres may occur into zones and mark these zones where necessary.

A2.6.1 Risk assessment

A major consequence thus of this legislation is that operators of machinery have to carry out a risk assessment of any existing installations involving dangerous substances. As a result they must provide technical and organizational measures to eliminate or reduce, to as far as is reasonably practicable, the identified risks from potential sources of ignition in existing rotating machinery.

The work carried out by the EC and the seal industry to clarify the position of mechanical seals in regard to their definition as an ATEX component helps operators understand the consideration of mechanical seals in regard to their ignition source risk analysis.

- There may, in exceptional cases, be mechanical seals assemblies that require a review of their specification as a result of the risk assessment. These will primarily be seals in contact with a zone 0 or 20 environment.
- Mechanical seals must be considered as a potential risk in the context of the complete machine and how that machine is operated and maintained.
- The Consideration makes it clear that the mechanical seal supplier has a responsibility for clarifying the mitigating provisions needed to prevent

the seal from becoming an ignition risk. The Machinery Directive, 98/37/EC, and ATEX, 94/9/EC, require the equipment's instructions to include 'useful instructions, in particular with regard to safety'. Although mechanical seals are not typically covered by the scope of these two Directives there is a responsibility by the industry to ensure that, as a supplier of equipment that is covered, instruction manuals or supporting documentation contain advice about potential risk and safety issues. The operator carrying out the risk assessment should contact his seal supplier for the latest version of those instructions.
- The information summarizing the EHSR in section A2.7 reinforces the more practical instructions referenced above.

A2.6.2 Spare and alternative seals

A spare mechanical seal assembly is normally not required to comply with ATEX 95 unless the spare seal represents a component as defined by the Directive when the original item of rotating equipment was supplied. Spare seals supplied for equipment delivered to the operator before 1 July 2003 are outside the scope of ATEX 95 but would still have to be considered in his risk assessment described above in section A2.6.1.

If the manufacturer of the seal offers an alternative seal, as long as there has been no substantial modification to the original design, no further review with regard to ATEX is required. ATEX Guidelines[2] make it clear that a substantially modified product is 'any modification affecting one or more of the health and safety characteristics covered by the EHSRs'. This could be:

- Changes in the nature of an existing hazard or the introduction of a new hazard.
- An increase in the risk from an existing hazard.

In the case of mechanical seals these would primarily occur if the seal configuration were changed, e.g. from a dual pressurized seal to a single seal, or there was a change in the rating of the seal. If this is being considered on an item of rotating equipment conforming to ATEX 95 then the operator would have to carry out an ATEX 137 (DSEAR) risk assessment.

A2.6.3 Repairs to equipment

When rotating equipment is repaired it is normal to replace or repair a worn or defective part to restore the machine's functionality. As this has been done after the original machine has been placed on the market and it is not being sold as a 'new product' ATEX does not apply. If in the course of the repair a 'substantial modification' is needed the precautions described in section A2.6.1 would then be applied.

A2.7 Essential Health and Safety Requirements (ESHR) considerations by equipment manufacturers

To meet the requirements of ATEX 95 it is necessary to carry out a risk assessment of the ignition hazard of the equipment and after due analysis of the possible technical and operating faults ensure the product is designed and manufactured as far as possible to preclude any dangerous situation. Two ATEX supporting standards, EN 1127-1 and EN 13463-1, provide an understanding of the requirements of the Ignition Hazard Assessment and the latter document offers a format that might be applied. Different ATEX Category levels require attention to ignition potential during what is described as 'normal operation' (Category 3) and at different frequencies of equipment or component malfunction, differentiated in EN 13463-1 by the terms 'expected malfunction' (Category 2) and 'rare malfunction' (Category 1). The legislation and the ATEX Guidelines use different wording to describe the same criteria and use the equivalent terms 'frequently occurring disturbances' and 'rare incidents'.

All suppliers of machine parts or elements have a responsibility to advise ATEX equipment manufacturers using their products of their potential ignition hazards. This should be formally managed through instruction manual documentation. Advice, however, in a format that directly supports an Ignition Risk Assessment, is difficult to manage at a detail level; the summaries in Appendices 2.1 and 2.2 are intended to provide some guidance in this respect.

A2.8 Marking of ATEX components

As published there is no requirement to mark parts defined as 'components'. The organizations offering guidance, notably the EC ATEX Standing Committee, appear to feel this is an omission and have recommended through their official document 'ATEX Guidelines' that marking should be applied to 'components'.

Marking requirements for equipment exists in the legislation but the supporting standard EN 13463-1 clarifies and modifies the recommendation. In addition the 'ATEX Guidelines' acknowledge its deviation from the ATEX Directive 94/9/EC and further modifies this in an attempt to bring more consistency with other industrial practice and Directives.

An aspect of marking that has been consistently agreed within the legislation, related standards and guidelines is that ATEX 'components' should not include the CE mark. This, however, continues to be a commonly requested feature even though the mechanical seal supplier cannot comply.

With many mechanical seals there are practical difficulties in the availability of suitable space for marking. They are produced in high numbers but in a multitude of shapes, sizes and materials such that very small batch quantities are typical; this is not a product easily and ordinarily serialized or allocated a batch reference. The method of assembly usually applied to this type of product in the supply process means that marking would significantly add to the total cost of production. A further contributing issue is that reliability is one of the key purchasing drivers for mechanical seals and a part supplied to a purchaser with an ATEX specification would likely be identical to an equivalent product in a less hazardous area of the plant. Most suppliers, prior to the mechanical seal 'Consideration by the ATEX Standing Committee' (section A2.4), have thus chosen a strategy of not marking. This has raised some conflict with the views of Notified Bodies.

In other products supplied as ATEX 'components' there is a clearer need to have a visible differentiation between these and other similar parts. This has presumably driven the enforcers of the legislation and the Notified Bodies to steer in the direction of marking ATEX 'components' even though this is not overtly driven by the legislation. In the same context, when considering the mechanical seal specification, for some equipment users there might be concern about unchecked changes to the specification during the life of the equipment in which it is installed.

The improved clarity provided by the EC concerning the specification of when the definition of an ATEX 'component' applies to mechanical seals as discussed in section A2.4 has initiated a review of the marking strategy of mechanical seals. Marking will be adopted on most mechanical seals supplied as an ATEX component. There will be some compromise on the content which will be to minimize, where practical, the cost impact to the purchaser without compromising the intent of ATEX. It is likely the following marking will be applied;

- Name of the manufacturer.
- The manufacturer's type or code description of the product.
- The symbol of the equipment group and the category. The source of the explosive atmosphere (gas or dust).
- The symbol for the type of protection employed.
- The use of the symbol 'TX' to indicate that the maximum surface temperature or Temperature Class is advised in supporting documentation.

The example below indicates a typical, proposed mechanical seal marking.

```
John Crane UK Ltd
   Type 8620
  ⟨Ex⟩ II 2GD c TX
```

Where ⟨Ex⟩ indicates a symbol for explosion protection as laid down in Directive 84/47/EEC

    II indicates Group II as defined by Directive 94/9/EC
    2 indicates Category 2 as defined by Directive 94/9/EC
    GD indicates suitability for both gas/air and dust/air mixtures
    c indicates an ignition protection method by 'constructional safety' – see EN 13463-5
    TX indicates the maximum surface temperature information contained in the supporting documentation

A2.9 Conclusions

- The Directive 94/9/EC (ATEX 95), EPS and supporting EN standards endeavour to bring non-electrical equipment into the same industrial control methodology as has been established for electrical equipment in potentially explosive atmospheres. In the process there have been errors and misunderstandings introduced which this appendix has endeavoured to deal with in the context of mechanical seals.
- Mechanical seals are ordinarily classed as machinery elements that are not defined within the scope of 94/9/EC. The seal manufacturer still has a responsibility to provide complete documentation for the safe use of his product and when requested, specific safety aspects in the context of a potentially explosive atmosphere.
- It has been established that mechanical seals are only classed as ATEX 'components' when intended for use in a zone 0 or 20 atmosphere or where a close cooperation between seal manufacturer and equipment manufacturer is required to include specific seal design features to eliminate an ignition risk. This will often result in a specifically designed mechanical seal.
- Guidance is offered in this appendix for operators implementing ATEX 1999/92/EC (ATEX 137) and DSEAR legislation on mechanical seals.
- The responsibility for assessing the ignition risk from mechanical seals has been formally moved to the equipment industry using these products and there is summarized an Ignition Risk Assessment and explanation of the issues related to maximum surface temperature that may be considered. This is included in Appendix 2.1 and Appendix 2.2.
- The marking of ATEX 'components' is not adequately managed by the legislation. The mechanical seal industry will support the EC 'ATEX Guidelines' and when it is appropriate for the seal a pragmatic but shortened version of the full marking specification will be applied.

A2.10 References

1. EN ISO 21049: 2004, 'Pumps – Shaft Sealing Systems for Centrifugal and Rotary Pumps', International Organization for Standardization, Geneva, Switzerland.
2. ATEX Guidelines (Second Edition), July 2005, Directorate General – Enterprise and Industry of the European Commission.
3. C. Fone, 'ATEX 95 – a view from a mechanical seal supplier', Seminar Proceedings, ATEX and DSEAR Regulations – The Implications for Fluid Machinery, London, UK, 2005.
4. J. Lewis, 'DSEAR: assessment of existing mechanical equipment in classified areas', Seminar Proceedings, ATEX and DSEAR Regulations – The Implications for Fluid Machinery, London, UK, 2005.
5. 'When a Mechanical Seal is a Machinery Element and when an ATEX-Component', http://europa.eu.int/comm/enterprise/atex/rotating.htm.

A2.11 Acknowledgement

This appendix is an updated version of a paper presented by Chris Fone BSc, CEng, MIMechE and John Crane EAA at 'Seals and sealing today' a seminar organized by the Institution of Mechanical Engineers, 6 December 2005. It is republished by permission of the IMechE, London, UK.

APPENDIX 2.1

Ignition Hazard Assessment

Potential ignition source	Normal operation (NO) or potential disturbance (PD)	Frequency of disturbance	Control measure applied to prevent the source becoming active	Type of protection
High surface temperature	NO		Seal supplier to advise (see Appendix 2.2). Instruction manual	EN 13463-1
Leakage if process > auto-ignition	NO		Continuous atmospheric steam or inert quench. Dual seal option. International standards	EN 1127-1
Leakage influencing zone specification	NO		Guidelines exist for consideration of single and dual seals. Dual seal option reduces risk. Instruction manual. International standards	EN 1127-1
Barrier/buffer liquid leakage contacting adjacent hot surfaces	NO		Barrier fluid properties are part of the seal system selection. Instruction manual. International standards	EN 1127-1
Barrier gas may contribute to an explosive atmosphere	NO		Barrier fluid properties are part of the seal system selection. Instruction manual	EN 1127-1
Static electricity with Teflon bellows seal on non-conductive process	NO		Seal performance limitation or specific earth connection. Equipment casing is earthed. Instruction manual. International standards	EN 13463-1
Friction and impact overheating	NO		Exclusion of materials >7.5% Mg and Ti or Zi unless protected. Contact containment bushings require inert quench	EN 1127-1
Exothermic reaction with process	NO		Exclusion of copper in acetylene and antimony carbon graphite in hydrogen peroxide	EN 1127-1
Wear or damage to seal ring	PD	Rare	Performance monitoring. Auxiliary system monitoring alarms on dual seals. Instruction manual	EN 13463-1

Insufficient face lubrication	PD	Rare	Operational control – Service conditions within the operational range of the seal. Effective venting of equipment and auxiliary systems. Appropriate auxiliary system monitoring alarms. Proper control of process flow system and operational range of equipment. International standards. Instruction manual	EN 13463-1
Spray seal leakage	PD	Rare	Design – safe collection and disposal of leakage. Instruction manual	EN 809
Operation of equipment outside allowable range	PD	Rare	Operational control. Instruction manual. International standards	EN 1127-1
Incorrect operation of seal auxiliary system	PD	Rare	Design. Auxiliary system monitoring alarms. Instruction manual	EN 13463-1
Change or exceeding planned process conditions	PD	Rare	Operational control. Instruction manual	EN 13463-1
Provision of services outside of allowable range	PD	Rare	Operational control. Instruction manual	EN 13463-1
Damage from transportation	PD	Rare	Design and packaging	EN 13463-5
Inadequate assembly on equipment	PD	Rare	Design and instruction manual. International standards	EN 13463-1
Damage through corrosion	PD	Rare	Conservative material selection. International standards	EN 13463-1
Properties of process influencing reliability	PD	Rare	Enquiry documentation, choice of materials, seal configuration. International standards	EN 13463-1
Incorrect maintenance and repair	PD	Rare	Instruction manual	EN 13463-1
Process escape during maintenance	PD	Very rare	Correct isolation. Design and proper draining. Sometimes purging of the equipment is necessary. Instruction manual	EN 13463-1

(*Continued*)

Potential ignition source	Normal operation (NO) or potential disturbance (PD)	Frequency of disturbance	Control measure applied to prevent the source becoming active	Type of protection
Transient process shock pressures	PD	Very rare	Operational control. Instruction manual	EN 13463-1
Steel tools in Explosion Group IIc atmospheres	PD	Very rare	Instruction manual	EN 13463-1
Rubbing contact resulting from misalignment/eccentricity	PD	Very rare	Design – e.g. cartridge seals, spigotting. Minimum radial clearance 0.5 mm. International equipment standards. Dynamic testing of equipment	EN 13463-1
Seal overcompression through thermal movements in equipment	PD	Very rare	Operational control. Design. International standards. Instruction manual	EN 13463-1
Seal overcompression through shaft movements	PD	Very rare	Design. International standards	EN 13463-1
Shaft bearing failure	PD	Rare	Instruction manual. Monitoring. Design. Choice of materials	EN 13463-1

APPENDIX 2.2

Maximum Surface Temperature

The ATEX legislation requires a declaration of the maximum surface temperature as advice in the instruction manual but the supporting standard EN 13463-1 advises the vendor state of this as part of the descriptive code used in the Attestation of Conformity and marking process (see section A2.8). The latter has been common practice with electrical equipment prior to the ATEX legislation.

The purpose in declaring the temperature is to permit a judgement on the potential for ignition of an explosive atmosphere when in contact with a hot surface. This requires input from the user of the equipment advising the vendor what the source of the explosive risk is (gas/air or dust/air) and a temperature at which there is a specific margin on the minimum auto-ignition for this combination. This subject is dealt with in EN 1127-1 and EN 13463-1 with the latter advising the use of a Temperature Class classification that has been commonplace in electrical equipment for many years.

In the case of mechanical seals this is an important issue because in many circumstances the rubbing surface of the seal dictates the maximum surface temperature of the whole equipment in which it is being used. The seal supplier's advice is thus important whatever the circumstances with respect to formal ATEX compliance. The subject is complicated by a number of issues.

Seals are supplied in broadly four different types of rotating shaft machinery and in pumping equipment this can be in more than 25 different seal arrangements. Each combination can differently affect the temperature in contact with an external ATEX zone area (see Figure A2.3).

Location of maximum surface temperature

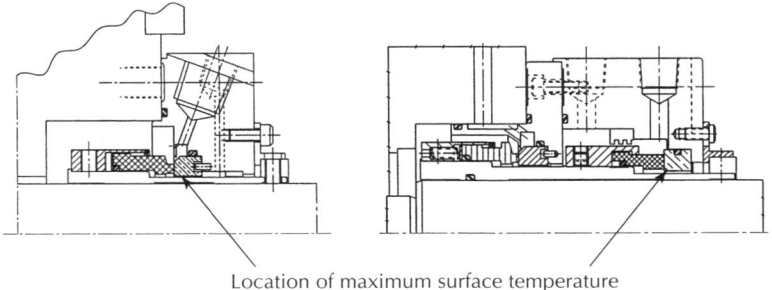

Figure A2.3 Maximum surface temperatures on single and dual unpressurized mechanical seals.

In addition to the above options the factors affecting the maximum surface temperature for a mechanical seal are influenced strongly by the prevailing service conditions such as the process fluid physical properties, temperature, pressure, shaft diameter and speed, making virtually every pump application unique. To further complicate the issue in many circumstances these are not known at the time of supply.

In the case of dual seals using an auxiliary system the design of the system will directly affect the maximum surface temperature and where more than one vendor is involved the system vendor should be responsible for defining this temperature.

In some large or high performance items of rotating equipment the mechanical seal surface temperature is dictated by the internal shape, materials and fluid dynamics of the equipment and the seal vendor is unable to predict the maximum surface temperature.

The maximum surface temperature of the mechanical seal, whilst consistent in normal operation, is affected by disturbances in the operation of the equipment itself. There are many methods applied to prevent these disturbances but there is a major need for effective communication and more attention to applying these methods by operators.

Research in other circumstances has indicated a relationship between explosive auto-ignition through hot surface temperatures and the surface area involved. Small areas require higher temperatures. The area of the mechanical seal that might be considered as the source of the maximum surface temperature and exposed to an external explosive zone would itself be considered small. No detail research exists to quantify this important issue.

Inevitably seal suppliers have adopted a conservative approach and to manage the above issues they have formulated complicated guidelines for use by equipment manufacturers for predicting the maximum surface temperature. Historically it has been commonplace for the industry to supply ATEX mechanical seals for orders without clarification of the service conditions. This has been resolved by supplying them with a 'TX' marked Attestation (see section A2.8) but with the availability of information sufficient for the equipment manufacturer to estimate the maximum surface temperature himself. This general guidance on maximum surface temperature should continue to be available with mechanical seals defined as 'machine elements'.

INDEX

Abbreviations: ISO 1629, 382
Abradable labyrinth seals, 246–8, 251, 252
Abradable materials, 249
Abradable seals, 253
Abrasives:
  mechanical seals, 441–2
  rotary seals, 274
Accessibility: static seals, 102
Accumulators, 355–7
Acetal (POM) properties, 396, 397
ACM see Polyacrylic
Acrylonitrile polymer, 362, 365
Actuators: piezo-electric, 42
Addresses: standards organizations, 475–6
Advanced technology single seals, 545–6, 551
Advantages:
  bellows seals, 172
  centrifugal seals, 266
  magnetic fluid seals, 268–70
AEM see Ethylene acrylic
Aerospace materials, 473
Aggressive fluid seals, 321
  see also Chemical attack
Agitator emissions, 554–5
Agricultural equipment, 131
Air purges, 243–4
Airzet piston seals, 318–19
AISI 3xx stainless steels, 411–12
Alloys: nickel alloys, 412–14
Alternative designs/sections:
  elastomers, 25–43
  metal seals, 45–8
  plastic, 43–4
Alternatives:
  chemical seals, 174–5
  elastomer bellows, 187–8
  elastomer seals, 148–59
  plastic seals, 148–59
  rotary seals, 148–59
Alternative seals: ATEX, 587
Alumina ceramic, 409
Anaerobic sealants, 52–6
API Arrangement 3 Dual Seals, 203–4
API plans:
  auxiliaries, 192–6
  Plan 11, 193
  Plan 12, 192–3
  Plan 13, 192
  Plan 14, 192–3
  Plan 21, 193
  Plan 23, 193–4
  Plan 31, 194–5
  Plan 32, 194–5
  Plan 41, 195–6
  Plan 52, 199–200
  Plan 62, 196–7
  Plan 72, 201, 204
  Plan 74, 219
  Plan 75, 202
  Plan 76, 202
Applicability: European ATEX legislation, 582–98
Applications:
  arithmetic mean, 478–9
  clearance seals, 345
  diaphragms, 352–3
  elastomers, 368, 381–93
  ferrofluid magnetic seals, 271–2
  gaskets, 78–92, 522
  hydraulic, 282–314
  material ratio curves, 485
  mechanical seals, 186–219
  metal O-rings, 21–3
  piston rings, 339
  plastics, 397–403

598 Index

Applications: (*continued*)
  pneumatic, 318–19
  positive lubrication rotary seals, 155–8
  PTFE, 399–401
  reciprocating seals, 282–314
  rod seals, 304, 305–6
  root mean square, 480
  valves BAT, 564–5
Application techniques, 55–6, 59–60
Aramid fibres, 231–2
  brushes, 256
  gaskets, 80
Arithmetic mean (Ra), 477–8
Aromatic polymer fibres, 231–2, 343–4
ASME flanges, 66
ASME VIII calculation method, 67–8
Asperities, 110–11
Assembly procedures:
  cleaning, 527–8
  flanges, 527–34
  lubrication, 528–9
  tools, 527
  visual inspection, 528
Assessment: ignition hazards, 592–4
ASTM *see* Standards
ATEX legislation, 582–98
  'ATEX 137', 586–7
  categories/groups, 583–4
  marking components, 588–90
  mechanical seal applicability, 582–98
  Standing Committee, 585
  zones, 582–3
AU *see* Polyurethane (polyester)
Automobiles, 125–6
Automotives:
  cartridge assemblies, 189
  cylinder heads, 25
  metal O-ring grooves, 25
  piston rings, 323
  water pumps, 165, 166, 189
Auxiliaries: API plans, 192–6
Auxiliary fluid systems, 585–6
Axial cracking, 432–3
Axial excluders, 129, 130–1
Axial scoring, 298
Axial space limitation, 170
Axially constrained seals, 157

Back-to-back seals:
  double, 199, 204
  dry gas seals, 218–19
  pistons, 302, 303
  U-rings, 302, 303
Backup flange seals, 30–1
Backup rings:
  elastomer O-rings, 18
  T-seals, 29
Backup single seals, 196–8
Balance piston labyrinth, 246, 248
Ball bearing seals, 139–40
Barrel faced piston rings, 325–6
BAT guidance notes, 490–581
  bolted flange connections, 520–40
  compressors, 552–4
  definition, 501–2
  document objectives, 502
  emissions, 504–5
  executive summary, 496–500
  flanges, 520–40
  further reading, 578–9
  gaskets, 520–3
  generic BAT, 509–19
  information sources, 502
  introduction, 504–8
  preface, 501–3
  pumps, 541–52
  reciprocating shafts, 557–8
  rotodynamic equipment, 541–56
  valves, 559–72
  volatile organic compounds, 507–8, 517–18
BAT Reference note, 509
Beads, 317–18
Bearings:
  excluders, 239–41
  self-aligning, 241, 242
Bearing seals, 137–40, 175–9
  ball type, 139–40
  deep groove ball type, 139–40
  earthmoving equipment, 176–9
  ferrofluid magnetic seals, 270
  general purpose, 238
  labyrinth type, 138, 238
  mechanical, 139–40, 175–9
  process plant, 239
  pumps, 175–6, 239–41
  rolling element type, 137
  sealed, 138–9
  shielded, 137–8, 238
  steel rolling mills, 132–3

Bellows:
  accumulator type, 356–7
  convolution width, 354
  elastomers, 356
  fabric reinforced, 356
  Hipres, 356
  hydroform type, 354–5
  metal bellows, 354–7
  polymer, 353
  reciprocating seals, 347–57
  sewn fabric, 356
Bellows seals, 160–1, 165–6
  advantages/disadvantages, 172
  corrosive applications, 174
  formed metal, 172–3
  hygienic seals, 172–3
  industrial, 167–8
  PTFE, 174
  welded metal, 170–2
  see also Diaphragms
Best available techniques see BAT guidance notes
Best practice: gasket installation, 72
Bidirectional pumping aids, 115–16
Blind assembly: lip seals, 430, 431
Blisters, 422, 433–4
Bolted flange connections see Flanges
Bolted joints/gaskets, 61–96, 520–40
  assemblies, 61–2
  bolting, 73–8
  compact flanges, 92–6
  compressed fibre gaskets, 64–5
  fibre gasket performance, 64–5
  forces, 62
  gasket applications, 78–92
  gasket behaviour, 62–5
  installations, 72–3
  materials, 63, 73
  surface finish, 63–4
Bolts/bolting, 73–8
  built-in strain measurement, 77
  cartridge assemblies, 190
  formulae, 74–5
  hydraulic tensioning, 75
  load discs, 75–6
  materials, 73, 74
  patterns, 531–3
  strain measurement, 77–8
  stresses, 71
  tightening, 72–6, 531–3
  torque vs tension, 73–4
  ultrasound extension measurement, 75
  see also Fasteners
Bonded seals, 31–4
  recesses, 32
  self centring, 32–3
Bonded washers, 31–2
Braided packing:
  carbon packing, 234–5, 236
  fibre, 228–9
  valves, 560
BREF note, 509
Brush seals, 252–6
  assemblies, 253, 254
  benefits, 253, 255
  labyrinths, 255
  manufacture, 253, 254
  MTU Aero Engines, 256
  PEEK housing, 256
  retrofitted, 255
  shoed, 254–5
BS see Standards
Buckled spiral wound gaskets, 85
Buffers: pneumatic cylinder pistons, 319–20
Built-in strain measurement: bolts, 77
Burnt seal areas, 451–2
Butadiene, 365
Butt joints: piston rings, 324–5
Butyl (*IIR*), 383
BX joints, 91

Carbon, 403–5
  lip seals deposits, 430–3
  manufacture, 403–4
  properties, 405, 406
  tribology, 404–5
  see also Graphite ...
Carbon fibre gaskets, 81
Carboxylated nitrile (XNBR), 384
Carpenter 36 (nickel alloy), 414
Cars see Automobiles
Cartridges see Cassettes
Cartridge seals:
  assemblies, 188–90
  automotives, 189
  bolts, 190
  gland plates, 190
  internal, 174–5
Cassettes, 133–4

Cellular seals, 253
CEN calculation method, 69–71
Centralisation: gaskets, 529
Centrifugal compressors, 246–7
Centrifugal excluders, 131
Centrifugal ring seals, 264–7
  advantages/disadvantages, 266
  with lip seals, 267, 268
  pressure, 265, 267
  pumps, 265
Ceramics, 409
CFS system *see* Compact flanges
Chamfers: piston rings, 329
Chemical attack:
  aggressive fluid seals, 321
  alternative seals, 174–5
  elastomers, 433–4
  lip seals, 433–4
  mechanical seals, 440–1
  rotary, 274
Chicago Rawhide Waveseal, 114
Chloroprene (CR), 383
CIP *see* Cured in place seals
Circulation:
  housing design, 190–1
  impellers, 190–1
  plus auxiliaries to API plans, 192–6
  single seal cartridge assemblies, 190–2
  vortex flow, 185
Circumferential seals, 262–4
  gas compressors, 264, 265
  segmented, 262–4
  wastewater pumps, 265
  see also Segmented ring seals
Circumferential splits, 435–6
Clamp connectors, 94, 95
Clamps:
  diaphragms, 350
  hygienic fittings, 36–8
  V-ring seals, 136
Cleaning: assembly procedures, 527–8
Clearance ratios: viscoseals, 260–1
Clearance seals, 237–67, 345–7
  applications, 345
  brush seals, 252–6
  centrifugal ring seals, 264–7
  circumferential seals, 262–4
  failure guide, 446–7
  hole slot seals, 249–52
  honeycomb seals, 249–52

labyrinth seals, 238–49
leaf seals, 256–8
leakage, 345–6
liquid ring seals, 264–7
use, 238
viscoseals, 258–62
Clip over piston seals, 320
Coatings:
  hydraulic applications, 299, 301
  piston rings, 324–5, 331–2
  soft metal overlay, 415
Cold start leakage: plastic seals, 425
Compact flanges, 92–6
Comparison testing: flange seals, 58
Composite materials, 213–14
Composite seals, 157–8
Compounds: elastomers, 365–6
Compressed fibre gaskets, 80
  graphite fibre rings, 235–6
  leak rate vs operating stress, 64
  performance, 64–5
Compression: elastomer O-rings, 11
Compression packing, 219–37
  Aramid fibres, 231–2
  aromatic polymer fibres, 231–2
  braided carbon, 234–5, 236
  braided fibre, 228–9
  control valves, 237
  cotton, 230
  emissions, 237
  exfoliated graphite material, 229, 234–5
  failure guide, 445–6
  filaments, 232–3
  fitting, 220, 225–8
  flax, 230
  gland arrangements, 221–5
  glass packing, 232
  glossary, 454–8
  graphite material, 229, 232–6, 560–2
  Hornet packing, 232
  installation, 562–4
  live loading, 226–8
  metallic foil, 233
  operation method, 221–2
  PTFE, 230–1
  pumps, 223
  ramie, 230
  reciprocating seals, 341–5
  ribbon, 236
  rotary shafts, 229–33

standards, 471
tightening, 226
types, 228–37
using, 225–8
valves, 233–7, 559–64
yarn braided packing, 560
Compression piston rings, 325–8
Compression set:
 elastomers, 369–72, 420–2
 failures, 420–2
 measurement, 371
Compressors, 246, 247, 248
 abradable seals, 249, 253
 BAT guidance notes, 552–4
 emission management, 552
 hole slot seals, 251
 hydrogen, 340
 inert gases, 340
 leakage sources, 514
 lubricated pistons, 332
 material temperatures, 249
 oxidizing gases, 340
 piston rings, 332–40
 reciprocating, 557–8
 reducing gases, 340
 see also Dry compressors
Cone face geometry, 213
Connectors, 92–6
 clamps, 94
 see also Flanges
Construction industry equipment, 131
Contact band distortion, 110–11
Contact forces: flanges, 93
Contaminants: double acting wipers, 311, 312
Control valves, 237, 569
Conversion factors, 573–7
 SI units, 573, 575–7
Convex metal gaskets, 92
Convolution width: bellows, 354
Cork gaskets, 79
Corrosion:
 flanges, 536–7
 metal seals, 425–6
Corrosive applications, 174–5
Corrugated metal gaskets, 87–8
Costs:
 bolted flange connections BAT, 538–9
 gasket BAT, 539
 generic BAT, 519

installations BAT, 539
mechanical seals, 549–52
PTFE lined elastomer seals, 126
pumps, 549–52
valves BAT, 567–9
see also Relative costs
Cotton packing, 230, 342
CR *see* Chloroprene
Cracking:
 lip seals, 432–3
 see also Stress corrosion cracking
C-rings, 45–6
 see also Metal O-rings
Cured in place (CIP) seals, 48–51
Current emission levels, 564
Cylinders:
 hydraulic standards, 309
 pneumatic seals, 315–21
 powder materials, 299, 300

Dead end single seals, 186–8
DEAR Regulations, 582, 586–7
Debris:
 lip seals, 434
 mixer seals, 209, 211
 wells, 209, 211
Deep groove ball bearing seals, 139–40
Degradation indicators: gaskets, 427–8
Deposits: dry gas seals, 443–4
Design:
 codes, 99–100
 mechanical seals, 160–1
 piston rings, 325–31
 PTFE lined elastomer seals, 124–7
 standards, 469
 static seals, 97–102
 U-rings, lips, 291–2
De-waxing filters, 554–5
Diaphragms:
 applications, 352–3
 clamps, 350
 elastomer seals, 167
 flat/dished, 348–50
 forces, 349
 materials, 351–2
 reciprocating seals, 347–57
 resistance curves, 348–9
 rolling, 350–1
 strokes, 351
 see also Bellows seals

Diesel engines, 323
Dieseling effect, 451
DIN 2505 calculation method, 68–9
DIN 118511 hygienic fittings, 38, 39
DIN flanges, 66–7
DIN... : *see also* Standards
Directives:
    ATEX 95, 582–98
    IPPC, 501
Disadvantages:
    bellows seals, 172
    centrifugal seals, 266
Dished diaphragms, 348–50
Dispersion: vortex flow, 185–6
Dissolved air damage, 449–50
Distortion: lip seals, 140–1, 433
Distributor seal geometry, 285–6
Domestic double glazing systems, 40–1
Door window seals, 41–3
Double acting piston seals, 318–19
Double acting wipers, 311, 312
Double balance line features, 200
Double glazing systems, 40–1
Double seals:
    gas barriers, 204–6
    mechanical, 203–6
    mixer seals, 209–10
Double... : *see also* Dual...
Dovetail grooves, 15
Drawn tube scoring, 298
Drinking water pump emissions, 554–5
Dry compressors, 332–40
    piston rings, 333, 334
    piston/rod arrangements, 334, 335
    rider rings, 333, 334
    ring materials, 337–40
Dryer emissions, 554–5
Dry gas seals, 179–81, 215–19
    back-to-back seals, 218–19
    deposits, 443–4
    failure guide, 443–4
    gas films, 180
    groove patterns, 179–80
    mechanical damage, 444
    secondary seal problems, 444
    single, 215–16
    tandem seals, 216–18
Dual gas barrier seals, 209–10
Dual pressurised mechanical seals, 547, 551

Dual seals, 198–9, 204, 542, 546–7, 551, 595–6
Dual unpressurised mechanical seals, 542, 546–7, 551, 595–6
Dust lips, 129–30
Dust removers, 310, 311
    see also Excluders
Dykes (L-shaped) piston rings, 326–7
Dynamic mechanisms: lip seals, 109–20
Dynamic modulus, 377

Earthmoving equipment, 176–9
Edges:
    groove damage, 430
    profile comparisons, 144–5
    rectangular section seals, 289–90
Efficiency:
    abradable labyrinth seals, 248
    pumps, 438
EHSR *see* Essential health and safety requirements
Elastomer bellows seals:
    alternatives, 187–8
    mechanical, 160–1, 165–8
    ship stern shafts, 211–12
Elastomer O-rings:
    arrangements, 8
    backup rings, 18
    compression variations, 11
    cross-section tolerances, 14
    dovetail grooves, 15
    extrusion, 16–18
    flange assemblies, 13
    gap limits, 17
    groove design, 12–15
    permitted extrusion gaps, 16
    radial sealing, 14
    sealing action, 10
    selection, 12–15
    standard sizes, 12
    surface texture, 16
    tandem grooves, 16
    thermal expansion, 11
    triangular grooves, 15
    workings, 9–11
Elastomers, 358–93
    abbreviations, 382
    alternative sections, 25–43
    applications, 368, 381–93
    basics, 362–7

bellows, 355
bonded seals, 31–4
 chemical attack, 433–4
 compounds, 365–6
 compression set, 369–72
 diaphragm seals, 167
 dynamic sealing mechanisms, 109–20
 elevated temperature, 373–5
 energizers, 283–4, 313–14
 fabric seals, 149–50
 fillers, 363, 364
 fluid compatibility, 369, 370
 fluid resistance, 367–9, 389–90
 gaskets, 78–9
 integral backup flange seals, 30–1
 ISO 1629 abbreviations, 382
 lip seals, 108
 low temperature properties, 375–8
 L section seals, 30
 mechanical properties, 378
 metal scrapers, 310–11
 permeation, 378–81
 properties, 359–62
 PTFE bonded, 120–1
 Quad rings, 27
 rectangular section rings, 26–7
 reinforced, 355
 resilience, 381
 scrapers, 310–11
 selection factors, 367–81
 spring seals, 34–5
 stress relaxation, 372–3
 stress strain properties, 359, 360, 361
 swell testing, 369, 370
 temperature properties, 373–5
 tensile strength, 374, 375
 thermal expansion, 11
 T-seals/rings, 28–9
 U-rings, 30
 uses, 359–62
 X section rings, 27–8
 see also Polymers
Elastomer seals:
 blisters, 422
 compression set failures, 420–2
 energized plastic, 150–2
 extrusion failure, 418–19
 failure guide, 417–23, 428–36
 flash, 418–19
 fractures, 422
 gas damage, 422
 glossary, 454–8
 hardening failure, 420–1
 hydraulic applications, 282–314
 leakage from new, 420
 O-ring failure, 420
 PTFE lined, 124–7
 reciprocating seals, 294–6
 reciprocating type, 282–314
 rotary type, 148–59
 shrinkage, 423
 softening failures, 421
 surface cracking, 417–18
 swell failure, 419
 see also Elastomer bellows seals
Emerging techniques:
 materials, 539–40
 rotodynamic equipment, 556
 valves, 569–72
Emissions:
 BAT guidance notes, 504–7
 compression packing, 237
 compressors, 552
 current levels, 523–4, 564
 fugitive, 504–7
 pumps, 541–2, 544
 sources, 505–7
 standards, 467
 static seals, 101
 valves, 564–5
EN see Standards
End stop buffers, 319–20
Enhanced lubrication seals, 157
EPDM/EPM see Ethylene propylene
Equalizer package, 178
Equipment: repairs, 587
E-rings, 47–8
Erosion: toroidal vortices, 184–5
ESA see European Sealing Association
Essential health and safety requirements (EHSR), 588
Ethylene acrylic (AEM), 387
Ethylene propylene (EPM, EPDM), 387
EU see Polyurethane (polyether)
European ATEX legislation, 582–98
European Sealing Association (ESA):
 documents, 578
 testing methodologies, 569–70
 see also BAT guidance notes

Excluders:
  agricultural equipment, 131
  axial, 129, 130–1
  cassettes, 133–4
  construction equipment, 131
  dust lips, 129–30
  hydraulic applications, 310–14
  leakages, 310, 311
  lip seals, 127–33
  package, 178
  pneumatic cylinder seals, 318
  pump bearings, 239–41
  radial dust lips, 128
  steel rolling mill bearings, 132–3
  see also Wipers
Exfoliated graphite material, 229, 234–6
Exfoliated materials, 82–4
Expanded graphite material, 229
Expansion joint terminology, 458–63
Expeller faces, 239
Externally pressurized metal O-rings, 23
External pusher seals, 174–5
Extreme conditions: metal O-rings, 22–3
Extrusion: elastomer O-rings, 16–18
Extrusion damage:
  elastomer seals, 418–19
  plastic seals, 424
  reciprocating seals, 449, 450–1
Extrusion gaps, 16

Fabric reinforced elastomers, 355
Fabric seals, 149–50
Faces:
  cone geometry, 213
  expeller faces, 239
  laser machining, 181
  lubrication enhancement, 181–3
  mechanical seals, 176–8
  radial expellers, 239
Failure guide, 417–53
  clearance seals, 446–7
  compression packing, 445–6
  dry gas seals, 443–4
  elastomer seals, 417–23
  fasteners, 535
  flanges, 536
  gaskets, 426–8, 535–6
  joints, 534–7
  mechanical seals, 437–43
  metal seals, 425–6

plastic seals, 423–5
reciprocating seals, 447–52
rotary seals, 428–47
static seals, 417–28
Fasteners:
  failure, 535
  load monitoring type, 76–8
  standards, 466
  see also Bolts/Bolting
FEA analysis: mechanical seals, 164
FEPM (TFE/P) *see* Tetrafluorethylene propylene copolymer
Ferrofluid magnetic seals, 268, 270, 271–2
  see also Magnetic fluid seals
FFKM *see* Perfluoroelastomer
Fibre gaskets *see* Compressed fibre ...
Fillers, 363, 364, 399–400
FIP *see* Formed in place ...
Fitting compression packing, 220, 225–8
FKM *see* Fluorocarbon
Flange gaskets *see* Gaskets
Flanges:
  anaerobic sealants, 52–6
  arrangements, 522
  assembly procedures, 527–34
  BAT guidance notes, 520–40
  compact, 92–6
  configurations, 66
  connectors, 92–6
  contact forces, 93
  corrosion, 536–7
  cost of BAT, 538–9
  elastomer O-rings, 13
  faces, 63, 66
  failure, 536
  hinge points, 132–3
  integral backup seals, 30–1
  leakage sources, 514
  machining faces, 63
  metal jacketed gaskets, 89
  polyurethane seals, 31
  re-tightening, 534
  rigidity, 54
  safety, 534–7
  standards, 65–7, 470
  surface finish, 63–4
  tagging, 533–4
  tightening patterns, 531–3

torque calculation, 530–1
T-seals, 29
types, 65–7
see also Flexible flanges
Flash: elastomer seals, 418–19
Flat diaphragms, 348–50
Flax packing, 230, 342
Flexible flanges:
  application techniques, 59–60
  FIP sealing, 56–60
  RTV silicone dispensing, 58–9
  seal comparison testing, 58
  service/repair, 60
  see also Flanges
Flexible graphite products, 560–2
Fluid film transfer: lip seals, 111
Fluid power: standards, 467, 472
Fluid resistance: elastomers, 367–9, 389–90
Fluids:
  compatibility problems, 433–4
  mechanical seal effects, 440–1
  sealed, static seals, 100
Fluorocarbon (FKM) polymers, 362–3, 388–90
  dynamic modulus, 377
  fluid resistance, 389–90
  low temperature, 376–7, 391–2
  stiffness, 376–7
  tan δ plots, 377
Fluorosilicone (FVMQ/FMQ), 376–7, 390
Flushing: pumps, 224
FMQ *see* Fluorosilicone
Food industry fittings, 36–8
Formed metal bellows seals, 172–3
Formed in place (FIP) seals/gaskets, 51–60
  application techniques, 55–6
  flange design, 57–60
  flange rigidity, 54
  flexible flanges, 56–60
  sealants available, 56
  surface requirements, 54–5
Fractures: elastomer seals, 422
Fretting damage: metal seals, 425
Friction:
  elastomer reciprocating seals, 294–6
  PTFE lined elastomer seals, 125
  rectangular section seals, 295, 297

Fuel permeation: plastic backup rings, 31
Fugitive emissions, 504–7
  current levels, 523–4
  measurement, 569–70
  sources, 505–7
FVMQ *see* Fluorosilicone

Gaps: piston rings, 336
Gas:
  compressors, 251, 264, 265
  decompression, 422
  films, 180
  permeation, 379–80
  specified, 570–1
Gas barrier seals:
  double, 204–6
  dry gas, 179–81
  dual, 209–10, 551
  pressurised, 551
Gas damage: elastomer seals, 422
Gas seals:
  ferrofluid magnetic type, 271, 272
  reactive, 271
  rotary seals, 275–6
Gas supply system, 205
Gas turbines, 250, 257–8
Gaskets:
  applications, 78–92, 522
  Aramid fibre, 80
  BAT, 520–3
  behaviour, 62–5
  best practice, 72
  calculation methods, 67–71
  carbon fibre type, 81
  categories, 61
  centralisation, 529
  compressed fibre, 80
  cork, 79
  corrugated metal, 87–8
  cost index, 539
  definition, 61
  degradation indicators, 427–8
  elastomers, 78–9
  exfoliated materials, 82–4
  failure guide, 426–8, 535–6
  fibres, 64–5
  flange face configurations, 66
  flange face machining, 63
  formed in place, 51–60
  glossary, 458–63

Gaskets: (continued)
  handling, 526–7
  installation, 72–3, 529, 579
  Kammprofile types, 86–7
  leakage, 426–7, 524
  leak rate vs operating stress, 64
  load tightness, 64–5
  materials, 526–7
    standards, 470–1
    tensile strength, 63
  metal jacketed, 88–90
  metal-to-metal, 90–2
  mica, 83
  multilayer type, 81
  non-metallic, 78–84
  premature failure, 427
  premium mixed fibre, 80
  PTFE, 81–2
  selection, 524–6
  semi-metallic, 84–90, 525
  soft, 525
  spiral wound, 84–7
  standards, 67, 469, 470–1
  storage, 526–7
  stress relaxation, 62–3
  types, 78–92, 539
  undefined compressed fibre, 80
  see also Bolted joints/gaskets
Generic BAT, 509–19
  costs, 519
  good operating practices, 509
  maintenance, 511
  management systems, 509–10
  process design, 510–11
  training, 510
Glands:
  arrangements, 222–5
  cartridge assemblies, 190
  cross-section, 221
  isometric view, 223
  plates, 190
  see also Compression packing
Glass:
  fillers, 399–400
  glazing seals, 40
  packing, 232, 344
  walls, 40
Glazing see Window seals
Glossary, 454–89
  compression packing, 454–8
  elastomer seals, 454–8

  expansion joints, 458–63
  gaskets, 458–63
  mechanical seals, 463–5
  plastic seals, 454–8
Good operating practices, 509
  see also BAT guidance notes
Graphite... : see also Carbon...
Graphite braided ribbon, 236
Graphite cores, 89
Graphite filament packing, 232–3, 344–5
Graphite loaded sintered silicon carbide, 408
Graphite products, 560–2
Graphite ribbon, 236
Graphite rings, 234–5
Graphite tape, 561
Grayloc connectors, 95
Grooves:
  automotive cylinder heads, 25
  dry gas seals, 179–80
  elastomer O-rings, 12–15, 24
  grooving failures, 436–7
  lip seal failure, 434–5
  metal O-rings, 24, 25
  piston rings, 324–5
  reciprocating seal damage, 449–50
  sharp edges, 430
  spiralling failure, 420
  spiral patterns, 179, 180
  wavy profile, 151

Hallite seals, 293
Handling gaskets, 526–7
Handling RTV silicone, 59
Hardening failure, 420–1
Hard/hard mechanical seal face combinations, 410
Hazard assessment: ignition, 592–4
Head flow curves: pumps, 438
Heat transfer: mechanical seals, 165, 166
Heels, 436, 447–8
High precision spindles, 271
High pressure plunger pumps, 341–2
High pressure rotary seals, 149–51
High reliability hygienic fittings, 38–9
Hinges: flanges, 132–3
Hipres bellows accumulators, 356
History, 482, 484

Hole slot seals, 249–52
  industrial gas compressors, 251
  leakage, 252
Hollow shaft feed-throughs, 271, 272
Honeycomb seals, 249–52
Hornet packing, 232
Housings:
  design, 183–6, 190–1
  eccentricity, 126
  ISO 7425, 309–10
  mechanical seals, 183–6
  plastic wipers, 312–14
  shaft eccentricity, 126
  slurry pumps, 185–6
  snap fitting, 316, 317
  standards, 309–10
  tapered seals, 185
  vapour, 183–4
  wipers, 312–14
Hydraulic accumulators, 356
Hydraulic applications:
  coatings, 299, 301
  cylinders, 309
  elastomer seals, 282–314
  excluders, 310–14
  piston seals, 304, 307–8
  plastic seals, 282–314
  powder materials, 299, 300
  reciprocating seals, 282–314
  rod seals, 304, 305–6
  standards, 309
  tandem seals, 299–304
Hydraulic cylinders: piston rings, 340–1
Hydraulic forces: mechanical seals, 161–2
Hydraulic nut/bolt tensioning, 75
Hydraulic rotary unions, 149–50
Hydrodynamic waves, 156
Hydroform bellows, 354–5
Hydrogen: compressors, 340
Hydrogenated nitrile (HNBR), 386
Hydrostatically balanced piston rings, 336–7
Hydrostatic lubrication, 159
Hygienic fittings, 36–9
Hygienic seals:
  bellows seals, 172–3
  metal bellows seals, 172–3
  mixer seals, 209–10
  standards, 467, 468

Ignition hazard assessment, 592–4
IIR *see* Butyl
Impellers: circulation, 190–1
Inch series: elastomer O-rings, 12
Industrial elastomer bellows seals, 167–8
Industrial gas compressors, 251
Industry design codes, 99–100
Inert gases, 340
Information sources, 442–3, 502
In-line feed-throughs, 271
Inpro seals, 239–40
Inspection: assembly procedures, 528
Installation:
  compression packing, 562–4
  cost index, 539
  gaskets, 72–3, 529, 579
  joints, 578–9
  packing, 579
Integral backup flange seals, 30–1
Integral bearing seals, 138–9
Intermediate piston rings, 328–9
Internal cartridge seals, 174–5
Internal combustion engines, 321–32
  automotives piston rings, 323
  losses, 322
  marine diesels, 323
  piston ring construction, 324–5
  piston ring design, 325–31
  see also Piston rings
Internally pressurized metal O-rings, 21, 23
International Dairy Federation (IDF) hygienic fittings, 36
Interseal pressure:
  buildup, 450–1
  piston seals, 302
  reciprocating seals, 450–1
Invar 36 (nickel alloy), 414
IPPC Directive, 501
ISO:
  1629, 382
  5597:1987, 309
  6195, 312, 313, 314
  6547:1981, 309
  7425, 309
  address, 475
  see also Standards

Joints:
  dry piston rings, 335–6
  failure, 534–7

Joints: (*continued*)
  glossary, 458–63
  installation, 578–9
  piston rings, 332, 333
  segmented ring seals, 263–4

Kammprofile gaskets, 86–7
Keystone piston rings, 326
Kiln furnace emissions, 554–5
Kunstfeld seal pumping, 112–13

Labyrinth seals, 238–49
  balance piston, 246, 248
  bearings, 138, 238
  brush seals, 255
  machine tool spindles, 241–2
  materials, 248–9
  porous inserts, 241
  radial expeller faces, 239
  rotor dynamics, 249
  swirl reducers, 249, 250
  turbo machinery, 242–9
Lantern rings, 223
Large diameter mechanical seals, 209–15
  ship stern shafts, 210–12
  water turbines, 213–14
Large volatile organic compounds (LVOCs), 506
Laser machining: faces, 181
LCC *see* Life-cycle cost guide
Leaf seals, 256–8
  configurations, 257–8
  manufacture, 257
Leakage:
  axial scoring, 298
  blowby, 286, 287
  clearance seals, 345–6
  compressors, 514
  determination, 511–13
  drawn tubes, 298
  excluders, 310, 311
  fibre gaskets, 64
  flanges, 514
  gaskets, 426–7, 524
  hole slot seals, 252
  honeycomb seals, 252
  key sources, 514–16
  mechanical seals, 197–8
  metal seals, 425–6
  monitoring, 511–13

  open-ended lines, 514
  O-ring energized plastic seals, 286, 287
  pipe work, 514
  plastic seals, 286, 287, 423
  prevention, 516–17
  pumps, 514–15
  rate vs operating stress, 64
  reciprocating seals, 286, 287, 310, 311
  safety valves, 515
  sampling point, 515
  sources, 514–16
  static seals, 101
  tank lids, 515–16
  turbo machinery, 245–6
  valves, 515, 516, 559
Legal obligations, 501, 582–98
Lens rings, 91–2
Life-cycle cost guide (LCC), 549–52
  see also Costs
Life expectancy:
  lip seals, 118–19
  static seals, 100–1
Lift promotion: lubrication, 182
Limits and fits: standards, 466
Lips:
  angles, 144–5
  forces, 125
  optimized static, 157
  pressure distortion, 142
  PTFE lined elastomer seals, 125
  supports, 141, 142
  U-rings, 291–2
  wipers, 316
Lip seals, 105–48
  axial cracking, 432–3
  basic design, 106–9
  bearing seals, 137–40
  blind assembly, 430, 431
  blisters, 433–4
  carbon deposits, 430–3
  casings, 108
  cassettes, 133–4
  with centrifugal seals, 267, 268
  chemical attack, 433–4
  circumferential splits, 435–6
  contact band distortion, 110–11
  cracking, 432–3
  cross-sections, 107–8
  debris, 434
  distortion, 110–11, 140–1, 433

dynamic sealing mechanism, 109–20
edge profile comparisons, 144–5
elastomer casings, 108
elastomer chemical attack, 433–4
elastomer seals, 124–7
excluders, 127–33
failure guide, 428–36
fluid compatibility problems, 433–4
fluid film transfer, 111
grooving failure, 434–5
heavy wear, 434–5
life expectancy, 118–19
lip supports, 141
marine, 145–7
material performance limits, 116, 118
mechanical damage, 430
metal casings, 108
multi barrier, 147
no apparent damage/no debris failure, 429
performance limits, 116–20
power consumption effect, 119–20
pressure
　applications, 140–8
　distortion, 140–1
　lip angles, 144–5
PTFE, 120–3
PTFE lined, 124–7
pumping aids, 114–16
radial dust lips, 128
rust, 434, 435
shafts
　misalignment, 123
　runout, 118–19, 123, 126
　speed limits, 116–17
　surface damage, 430, 431
speed limits, 116–17, 142–3
standards, 471
stress profile, 110
temperature, 116, 117, 142–3
texture, 112–13
viscosity effect, 119–20
V-ring seals, 134–6
Liquid contaminants, 310–11
Liquid ring seals, 264–7
Liquid sealing, 273–5
Live loading:
　compression packing, 226–8
　standard plus enclosed, 227
　valves, 564, 565–6

Load discs: bolting, 75–6
Load monitoring fasteners, 76–8
Load tightness: gaskets, 64–5
Longitudinal grooves, 449–50
Losses: internal combustion engines, 322
Low emission valve sets, 234–5
Low load C-rings, 46
Low pressure seals: wipers, 311, 312
Low temperature stiffness, 376–7
L section seals, 30
L-shaped (Dykes) piston rings, 326–7
Lubrication:
　assembly procedures, 528–9
　face enhancement, 181–3
　hydrostatic, 159
　lift promotion, 182
　retention, 316
　see also Positive lubrication...
LVOCs *see* Large volatile organic compounds

Machine tool spindles, 241–2, 243
Machining flange faces, 63
Magnetically energised mechanical seals, 176
Magnetic fluid seals, 267–72
　arrangements, 267, 269
　attributes, 268–71
　temperature, 270
　torque, 270
　vacuum, 267, 270
　wear, 269
　see also Ferrofluid magnetic seals
Magnets fitted to pistons, 320
Maintenance, 102, 511
Management systems, 509–10
Manufacture:
　brush seals, 253, 254
　carbon, 403–4
　leaf seals, 257
　lip seals, 120–1
　PTFE seals, 120–1, 124–7
　static seals, 100, 101–2
Marine:
　diesel engine piston rings, 323
　stern shaft seals, 210–12
　stern tube seals, 145–7
Marking ATEX components, 588–90
Material ratio curves, 484–5

Materials, 358–416
  bolting/bolts, 73, 74
  carbon, 403–5
  composite seals, 157–8
  cured, 363, 364
  diaphragms, 351–2
  dry gas compressor rings, 337–40
  elastomer thermal expansion, 11
  emerging techniques, 539–40
  gaskets, 526–7
    standards, 470–1
    tensile strength, 63
  labyrinth seals, 248–9
  lip seals, 112–13, 116, 118
  metal O-rings, 22–3, 24
  properties, standards, 473–4
  soft metal overlay, 414–15
  standards, gaskets, 470–1
  static seals, 97–102
  tensile strength, gaskets, 63
  texture, lip seals, 112–13
  see also Elastomers; *individual materials*
Measurement:
  arithmetic mean, 478–9
  compression set, 371
  fugitive emissions, 569–70
  material ratio curves, 484–5
  root mean square, 479–80
  Rsk skewness, 482–4
  Rt parameter, 480–1
  Rz parameter, 481–2
  surface texture, 477–86
Mechanical bearing seals *see* Bearing seals
Mechanical damage, 436
  dry gas seals, 444
  lip seals, 430
Mechanical properties:
  elastomers, 378, 392
  plastics, 395–7
Mechanical seals, 159–219
  abrasive solids, 441–2
  applications, 186–219
  arrangements, 186–219
  assemblies, 188–90
  ATEX legislation, 582–98
  auxiliary systems, 585–6
  basic design, 160–1
  basic types, 165–83
  bearing seals, 175–9
  bellows seals, 160–1, 165–6, 170–2, 174
  cartridge assemblies, 188–90
  chemical attacks, 174–5, 440–1
  costs, 549–52
  dead end single seals, 186–8
  double seals, 203–6
  dual unpressurised, 542
  elastomer bellows seals, 165–6
  European legislation, 582–98
  face combinations, 410
  faces, 176–8
  failure guide, 437–43
  FEA analysis, 164
  fluid effects, 440–1
  glossary, 463–5
  heat transfer, 165, 166
  high
    pressure, 206
    speed, 207
    temperature, 206
  housing design, 183–6
  hydraulic forces, 161–2
  information sources, 442–3
  large diameter, 209–15
  leakage containment, 197–8
  magnetically energised, 176
  maximum temperature, 595–6
  mixer seals, 208–11
  multiple seal systems, 198–203
  operation method, 161–5
  pressure, 163
  problem types, 438–9
  pumps, 542, 543–8
  pusher seals, 160–1, 168–70
  random failures, 440–1
  relative costs, 551
  single seal backup arrangements, 196–8
  slurry seals, 208
  standards, 471
  sudden failure, 439–40
  temperature, 163
  types, 160–1
  wearout, 442
Metal bellows, 353–7
  high speed seals, 207
  hygienic seals, 172–3
  valve stems, 353–5
  welded, 356

Index 611

Metal casings: lip seals, 108
Metal jacketed gaskets, 88–90
Metallic foil packing, 233, 345
Metallic gaskets, 61, 526
  see also individual gasket types
Metal O-rings, 20–5
  applications, 21–3
  automotive cylinder heads, 25
  design comparison, 24, 25
  materials, 22–3, 24
  spring energized, 46–7
  temperature, 22–3, 24
  types, 21
  see also C-rings; Metal seals
Metal reinforcement, 316–17
Metals, 410–14
  nickel alloys, 412–14
  soft metal overlay, 414–15
  stainless steel, 411–12
  standards, 474
  stress strain range, 359, 360
Metal scrapers, 310–11
Metal seals:
  alternative designs, 45–8
  corrosion, 425–6
  C-rings, 45–6
  E-rings, 47–8
  failure guide, 425–6
  fretting damage, 425
  leakage, 425–6
  low load C-rings, 46
  sigma seals, 48
  spring energized, 46–7
  transient leakage, 426
  see also Metal O-rings
Metal-to-metal gaskets, 90–2
Metric series: elastomer O-rings, 12
Mica gaskets, 83
Mixer emissions, 554–5
Mixer seals, 208–11
  debris wells, 209, 211
  double seals, 209–10
  hygienic seals, 209–10
Modified Quad rings, 27
Monitoring leakage, 511–13
'Moulded in place' seals, 48–50
MPQ see Silicone
MPVQ see Silicone
MTU Aero Engines, 256
Multi-axial feed-through, 271, 272

Multi barrier lip seals, 147
Multilayer gaskets, 81
Multi part steel rail piston rings, 331
Multiple seal systems, 198–203
Multi-spring pusher seals, 168–9
MVQ see Silicone

NADUOP SIMRIT pneumatic piston
    seals, 320
Napier stepped piston rings, 328
Narrow bore seal housing, 183
NBR (nitrile), 366, 383–4
Nickel alloys, 412–14
Nitrile (NBR), 366, 383–4
Nomenclature:
    pusher seals, 161
    viscoseals, 260
    see also Glossary
Non-contamination seals, 269
Non-metallic gaskets, 61, 78–84
    see also individual gasket types
Non-metallic pumps: PTFE sleeves, 174–5
Non-pressurized metal O-rings, 21, 22
Non-SI units: conversion factors, 573–4
Nut/bolt stresses, 71
    see also Bolt...

Octagonal gaskets, 90–1
Oil consumption: piston rings, 332, 333
Oil control:
    multi part steel rail piston rings, 331
    piston rings, 324, 329–31
    slotted cast iron piston rings, 329–30
Oilfield rock bits, 148–9
Oil films, 286, 288, 289–90
Oil lubricated stern tubes, 145–7
Open-ended line leakage, 514
Optimized static lips, 157
Organizations, 486–9
    research, 488–9
    standards, 475–6
    trade and industry, 486–8
O-rings, 8–25
    damage, 420, 452
    elastomers, 8–19
    energized plastic seals, 286, 287
    installation, 563
    labyrinth seals, 239–40
    leakage, plastic seals, 286, 287
    metal, 20–5

612   Index

O-rings (continued)
  perfluoroelastomer type, 421
  plastic, 19–20
  reciprocating seals, 285
  rotary seals, 148–9
  spiral failure, 285, 420, 452
  standards, 467–8
OT/EOT see Polysulphide
Oval metal gaskets, 90–1
Overheating damage, 436
Oxidizing gases: compressors, 340

PA see Polyamides
Packing:
  aromatic polymer fibre, 343–4
  cotton, 342
  flax, 342
  glass, 344
  graphite filament, 344–5
  installation, 579
  metallic foil, 345
  PTFE, 343
  ramie, 343
  reciprocating seal types, 342–5, 557
  see also Compression packing
PAI see Polyamide-imide
Parallel bores, 184–6
Peak parameter history, 482
PEEK see Polyetheretherketone
PEK: mechanical properties, 396
Perfluoroelastomer (FFKM), 392–3, 421
Perfluoroether, 391–2
Performance:
  fibre gaskets, 64–5
  lip seals, 116–20
  viscoseals, 260–1
Permeation: elastomers, 378–81
PI see Polyimide
Piezo-electric actuators, 42
Pins: segmented ring seals, 264
Pipe work leakage, 514
Piston rings, 321–41
  applications, 339
  automotives, 323
  barrel faced, 325–6
  butt joints, 324–5
  chamfer, 329
  coatings, 324–5, 331–2
  compression, 325–8
  compressors, 332–40

  construction, 324–5
  design, 325–31
  dry compressors, 333–6
  equalizing passages, 327–8
  gaps, 336
  grooves, 324–5
  hydraulic cylinders, 340–1
  hydrostatically balanced, 336–7
  inserts, 324
  intermediate, 328–9
  internal combustion engines, 321–32
  joints, 332, 333, 335–6
  Keystone, 326
  marine diesel engines, 323
  materials, 337–40
  Napier stepped, 328
  numbers of, 333, 334
  oil consumption, 332, 333
  oil control, 324, 329–31
  pressure, 327–8, 333, 334–5
  reciprocating compressors, 557–8
  two part, 336
  wear resistant coatings, 324–5
Piston seals:
  applications, 304, 307–8
  arrangements, 304, 307–10
  back-to-back type, 302, 303
  clip over, 320
  interseal pressure, 302
  pneumatic cylinders, 318–20
  venting, 302, 303
Pistons fitted with magnets, 320
Plan 11, 193
Plan 12, 192–3
Plan 13, 192
Plan 14, 192–3
Plan 21, 193
Plan 23, 193–4
Plan 31, 194–5
Plan 32, 194–5
Plan 41, 195–6
Plan 52, 199–200
Plan 62, 196–7
Plan 72, 201, 204
Plan 74, 219
Plan 75, 202
Plan 76, 202
Plastics, 393–403
  alternative sections, 43–4
  applications, 397–403

backup rings, 31
benefits, 394
disadvantages, 395
elastomer energized seals, 150–2
elastomer energizers, 283–4
fuel permeation, 31
hydraulic applications, 282–314
lip seals, 147–8
mechanical properties, 395–7
O-rings, 19–20
properties, 395–7
rotary seals, 148–59
stress strain range, 359, 360
types, 397–403
U-rings, 153–4
U-springs, 283–4
wipers, 312, 313–14
see also individual polymers; Polymers; PTFE; Reciprocating seals

Plastic seals:
cold start leakage, 425
extrusion flash, 424
failure guide, 423–5, 428–36
glossary, 454–8
leakage, 423, 425
static leakage, 425
wear damage, 424–5

PL seals *see* PTFE lined elastomer seals
Plungers: powder materials, 299, 300
Pneumatic applications, 318–19, 321
Pneumatic cylinder pistons:
clip over piston seals, 320
end stop buffers, 319–20
U-seal design, 318, 319
Pneumatic cylinder seals, 315–21
examples, 315–16
excluders, 318
piston seals, 318–20
press fit, 316–17
snap fitting, 316, 317
Pneumatic rod seals, 316–18
press fit, 316–17
retaining beads, 317–18
snap fitting, 316, 317
Polishing effect on surfaces, 296, 298
Polyacrylic (ACM), 386–7
Polyamide-imide (PAI), 401–2
Polyamides (PA), 396, 397
Polybenzimidazoles, 396
Polyester properties, 396, 398

Polyetheretherketone (PEEK), 256, 396, 401
Polyethylene (ultrahigh molecular weight) (UHMWPE), 398
Polyimide (PI), 396, 403
Polymers:
bellows, 353
fluorocarbon, 362–3
reciprocating seal failure guide, 447–52
representation, 363, 364
water absorption, 402
see also Elastomers
Polymer seals:
cellular, 253
rub-tolerant seals, 246
see also Plastic... ; PTFE
Polyphenylene sulphide (PPS), 396, 398–9
Polysulphide (OT/EOT), 386
Polytetrafluoroethylene *see* PTFE
Polyurethane (AU or EU), 31, 384–5
POM *see* Acetal
Porous inserts, 241
Ports: high pressure, 149–50
Positive lubrication rotary seals, 155–9
applications, 155–8
geometry, 156
hydrodynamic waves, 156
water turbines, 158–9
Powder materials, 299, 300
Power consumption: lip seals, 119–20
Power losses: V-ring seals, 135
PPS *see* Polyphenylene sulphide
Premature failure: gaskets, 427
Premium mixed fibre gaskets, 80
Press fit seals:
plastic wipers, 313, 314
pneumatic cylinder rods, 316–17
Pressure:
centrifugal seals, 265, 267
conversion factors, 576
distortion, lip seals, 140–1, 142
elastomer reciprocating seals, 294–5
equalizing passages, 327–8
interseal, 302
lip angles, 144–5
lip seals, 126, 140–8
lip supports, 142
mechanical seals, 163, 206
piston rings, 327–8, 333, 334

Pressure: (*continued*)
  plastic lip seals, 126, 147–8
  plastic U-rings, 153–4
  PTFE lined seals, 126
  PTFE seals, 44
  reciprocating seals, 294
  rectangular section seals, 290
  rod seal damage, 449
  rolling diaphragms, 351, 352
  rotary seals, 273–4, 276
  seals in series, 202–3
  static seals, 98–9
  see also High pressure...
Pressure ports: high pressure, 149–50
Pressure vessel standards, 466
Pressurized gas barrier double seals, 204–5
Pressurized metal O-rings, 21, 22, 23
Pressurized water reactor, 202–3
Prevention: leakage/spills, 516–17
Problem types, 5–6, 438–9
  see also Failure guide
Process design, 510–11
Process plant bearings, 239
Process valves, 355–6
Properties of materials:
  carbon materials, 405, 406
  plastics, 395–7
  silicon carbide, 405, 407
  standards, 473–4
PTFE:
  applications, 399–401
  bellows seals, 174
  bonded to elastomers, 120–1
  fillers, 399–400
  gaskets, 81–2
  lip seals, 120–3
  mechanical properties, 396
  non-metallic pump sleeves, 174–5
  packing, 230–1, 343
  pneumatic applications, 321
  sheathed O-rings, 19
  spring energized seals, 43–4, 152–5, 321
  types, 120–1
  U-rings, 152
  washers, 120–1, 123
  see also Plastic
PTFE lined (PL) elastomer seals, 124–7
  advantages, 125–7
  automobiles, 125–6
  cost, 126
  design/manufacture, 124–7
  flexibility, 125
  friction power loss, 125
  lip forces, 125
  pressure, 126
  shafts, 125, 126
  speed advantages, 125–6
  storage, 126
  temperature, 126
Pumps:
  aids, 114–16
  BAT guidance notes, 541–52
  BAT recommendations, 548–9
  bearing excluders, 239–41
  bearing seals, 175–6
  centrifugal seals, 265
  dual pressurised seals, 547
  dual unpressurised seals, 546–7
  efficiency points, 438
  emission management, 541–2, 544
  with flushing, 224
  head flow curves, 438
  high pressure plunger type, 341–2
  lantern rings, 223
  leakage sources, 514–15
  life-cycle cost guide, 549–52
  lip seals, 114–16
  mechanical seals, 542, 543–8
  packing, 223
  powder materials, 299, 300
  relative life-cycle cost guide, 549–52
  reliability, 542–3
  sealless drive systems, 547–8
  single seals, 545–7
  see also Slurry pumps
Pusher seals, 160–1, 168–70
  corrosive applications, 174–5
  large single springs, 169–70
  multi-spring, 168–9
  nomenclature, 161
  pairs, 198–9, 204
  stern shafts, 212
  unbalanced, 187–8
PVRC calculation method, 69

Quad rings, 27
Quench provision: steam, 196

Ra *see* Arithmetic mean
Radial dust lips, 128
Radial expeller faces, 239
Radial sealing, 14, 144–5
Ramie fibre packing, 230, 343
Reaction bonds: silicon carbide, 405, 408
Reactive gas seals, 271
Reactors:
  emissions, 554–5
  pressurized water, 202–3
Recesses: bonded seals, 32
Reciprocating compressors, 557–8
Reciprocating seals, 282–357
  background, 283–5
  bellows, 347–57
  burnt areas, 451–2
  clearance seals, 345–7
  compression packing, 341–5
  diaphragms, 347–57
  distributor seal geometry, 285–6
  elastomer seals, 282–314
  extrusion damage, 449, 450–1
  failure guide, 447–52
  Hallite seals, 293
  heel wear, 447–8
  hydraulic applications, 282–314
  interseal pressure buildup, 450–1
  leakage, 310, 311
  longitudinal grooves, 449–50
  packing types, 342–5
  piston rings, 321–41
  plastic seals, 282–314
  pneumatic cylinder seals, 315–21
  polymer failure guide, 447–52
  pressure, 294
  scoring damage, 450
  sealing mechanism, 286–99
  spiral failure, 285
  squeeze seals, 283
  stability, 285–6
  U-seals configurations, 283–4
  see also Piston seals; Rod seals
Reciprocating shafts: BAT guidance notes, 557–8
Rectangular section rings, 26–7
Rectangular section seals:
  edges, 289–90
  friction, 295, 297
  interseal pressure, 302

oil films, 286, 288, 289–90
pistons, 302
pressure distortion, 290
viscosity effect, 295, 297
Reducing gases: compressors, 340
Reinforced elastomers: bellows, 355
Relative costs: mechanical seals, 551
Reliability:
  pumps, 542–3
  standards, 467
Repairs:
  equipment, 587
  flexible flanges, 60
Research organizations, 488–9
Resilience: elastomers, 381
Resistance curves: diaphragms, 348–9
Retaining beads, 317–18
Re-tightening flanges, 534
Retraction, 376
Retrofitted brush seals, 255
Rider rings, 333, 334
Rigidity: flanges, 52–3, 54
Ring joint type (RJT) hygienic fittings, 36
Risk assessment, 586–7
Rock bits: oilfields, 148–9
Rod seals:
  applications, 291, 304, 305–6
  arrangements, 304
  double acting wipers, 311, 312
  examples, 293
  pneumatic, 316–18
  pressure damage, 449
  reciprocating seals, 291
  in tandem, 302–4
Rolling diaphragms, 350–1, 352
Rolling element bearings, 137
Root mean square (Rq), 478–80
RotaBolt, 76–7
Rotary gas unions, 271
Rotary-linear feed-throughs, 271
Rotary machine standards, 466–7
Rotary seals, 105–281
  abrasives, 274
  alternative elastomer seals, 148–59
  alternative plastic seals, 148–59
  chemical applications, 274
  clearance seals, 237–67
  compression packing, 219–37
  elastomer seals, alternative, 148–59
  failure guide, 428–47

Rotary seals (*continued*)
  gas sealing, 275–6
  high pressure, 150–1
  lip failure, 428–36
  lip seals, 105–48
  liquid sealing, 273–5
  magnetic fluid seals, 267–72
  mechanical seals, 159–219
  O-rings, 148–9
  plastic, 148–59
  positive lubrication, 155–9
  pressure, 273–4, 276
  selection, 273–81
  slurries, 274
  speed, 275
  standards, 471
  vacuum, 273, 275
  vapour pressure, 274, 275
Rotodynamic equipment:
  BAT guidance notes, 541–56
  emerging techniques, 556
  emissions, 554–5
  large, 555
  pump BAT guidance notes, 541–52
Rotor dynamics, 249
Rq *see* Root mean square
Rsk skewness, 482–4
Rt parameter, 480–1
RTV silicone, 58–9
Rub-tolerant polymer seals, 246
Rust: lip seals, 434, 435
RX joints, 91
Rz (ISO) parameter, 481–2

S3xxxx stainless steels, 411–12
Sacrificial excluders, 131
Safety: flanges, 534–7
Safety valves, 515
Sampling points: leakage, 515
SCC *see* Stress corrosion cracking
Scrapers: elastomers, 310–11
Screw threads: viscoseals, 258–9
Sealants: FIP seals/gaskets, 56
Sealed bearings, 138–9
Sealed fluids: static seals, 100
Sealing ability *see* Load tightness
Sealing beads: cured in place seals, 50
Sealing coefficients: viscoseals, 260–1
Sealing mechanism: reciprocating seals, 286–99

Sealing plates: integral, 33–4
Sealing technology BAT guidance notes, 490–581
Sealless pump drive systems, 547–8, 551
Sealol metal bellows, 171
Seating loads: metal O-rings, 22–3
Secondary seals, 444
Segmented ring seals, 262–4
  joint designs, 263–4
  pins, 264
  springs, 263, 264
Selection:
  elastomer O-rings, 12–15
  elastomers, 367–81
  gaskets, 524–6
  rotary seals, 273–81
  static seals, 7–8
Self-aligning bearings, 241, 242
Self-centring bonded seals, 32–3
Self-sintered silicon carbide, 408
SEM bearing package, 178
Semi-metallic gaskets, 61, 84–90, 525
  see also individual gasket types
Service: flexible flanges, 60
Sewn fabric bellows, 355
Shafts:
  BAT guidance notes, 557–8
  housing eccentricity, 126
  lip seals
    runout, 118–19, 123, 126
    speed, 116–17
    surface damage, 430, 431
  misalignment, lip seals, 123
  packing types, 229–33
  PTFE lined elastomer seals, 126
  reciprocating, 557–8
  runout, 118–19, 123, 126
  seal standards, 471
  self-aligning bearings, 241, 242
  speeds, lip seals, 116–17
  stationery mode, 239–40
  surface damage, lip seals, 430, 431
  texture, 430
  wear, 125, 436–7
  see also Rotary seals
Shielded bearings, 137–8, 238
Ships *see* Marine
Shoed brush seals, 254–5
Shrinkage: elastomer seals, 423
Sigma seals, 48

Silicon carbide, 405–8
  graphite loaded sintered, 408
  properties, 405, 407
  reaction bonds, 405, 408
  self-sintered, 408
Silicon nitrile, 409
Silicone (MVQ, MPQ, MPVQ), 387–8
SIMRIT pneumatic piston seals, 320
Single seal cartridge assemblies, 188–9
  plus auxiliaries to API plans, 192–6
  plus circulation, 190–6
Single seals:
  advanced technology type, 545–6
  backup arrangements, 196–8
  dry gas seals, 215–16
  maximum temperature, 595–6
  pumps, 545–7
  unpressurised mechanical, 595–6
Single spring pusher seals, 169–70
SI units: conversion factors, 573, 575–7
Slotted cast iron oil control rings, 329–30
Slurry pumps: packing, 224
Slurry seals, 208
  pump housings, 185–6
  rotary seals, 274
SM Coatings Service, 301
Snap fitting housings/seals, 316, 317
Softening failures: elastomers, 421
Soft gaskets, 525
Soft metal coatings, 48
Soft metal overlay, 414–15
Soft packing *see* Compression packing
Solid metal O-rings, 21, 23
Sources: leakage, 514–16
Spare seals, 587
Specified gas, 570–1
Speed:
  elastomer reciprocating seals, 295, 296
  lip seals, 142–3
  magnetic fluid seals, 269
  mechanical seals, 207
  metal bellows seals, 207
  PTFE lined elastomer seals, 125–6
  rotary seals, 275
  see also Velocity
Spill prevention, 516–17
Spiral packing, 562–3

Spirals:
  failure, 285
  groove patterns, 179, 180
  O-ring damage, 420, 452
Spiral wound gaskets, 84–7
Springback: metal O-rings, 22–3
Spring energized:
  metal seals, 46–7
  PTFE seals, 43–4, 152–5, 321
  U sections, 43
Spring loaded stuffing boxes, 341–2
Springs: segmented ring seals, 263, 264
Spring seals, 34–5
Squealing, 437
Squeeze seals, 283
Stability: reciprocating seals, 285–6
Stainless steel, 411–12
Standards, 466–76
  aerospace materials, 474
  cylinders, 309
  elastomer O-ring sizes, 12
  flanges, 65–7
  gaskets, 469–71
  general, 466–7
  housings, 309–10
  hydraulic applications, 309
  hygienic seals, 468
  materials, 470–1
  metals, 474
  organizations, 474–6
  O-rings, 467–8
  rotary machines, 466–7
Standing Committee: ATEX, 585
Static seals, 7–104
  accessibility, 102
  alternative
    elastomer sections, 25–43
    metal seals, 45–8
    plastic sections, 43–4
  bolted joints/gaskets, 61–96
  design, 97–102
  emissions, 101
  failure guide, 417–28
  gaskets, 61–96
  industry design codes, 99–100
  leakage, 101, 425
  life expectancy, 100–1
  maintenance, 102
  manufacture, 100, 101–2
  materials selection, 97–102

Static seals (*continued*)
  O-rings, 8–25
  plastic, 425
  pressure, 98–9
  sealed fluids, 100
  selection, 7–8
  temperature, 98
Stationery mode: shafts, 239–40
Steam quench provision, 196
Steel rail oil control piston rings, 331
Steel rolling mill bearings, 132–3
Stern shaft seals, 211–12
Stern tube seals, 145–7
Stick slip, 437
Storage:
  gaskets, 526–7
  PTFE lined elastomer seals, 126
Strain measurement: bolts, 77–8
Stress corrosion cracking (SCC), 537
Stresses:
  lip seals, 110
  nuts/bolts, 71
Stress relaxation: elastomers, 372–3
Stress strain range:
  elastomers, 359, 360, 361
  metals, 359, 360
  plastics, 359, 360
Strokes: diaphragms, 351
Stud tightening patterns, 531–3
Stuffing boxes, 341–2
Sudden failure: mechanical seals, 439–40
Surface cracking: elastomer seals, 417–18
Surfaces:
  flange finish, 63–4
  formed in place seals/gaskets, 54–5
  maximum temperature, 595–6
  polishing effect, 296, 298
  see also Texture
Swell, 369, 370, 419
Swirl reducers, 249, 250
Symmetry, 482–4

Tagging: flanges, 533–4
Tandem grooves, 16
Tandem seals:
  dry gas seals, 216–18
  hydraulic applications, 299–304
  rods, 302–4
Tan δ plots, 377, 379–80
Tank lid leakage, 515–16

Tapered seals, 185
Taper-lok connectors, 94–6
Temperature:
  abradable materials, 249
  conversion factors, 577
  elastomers, 373–8
  fluorocarbon, 391–2
  lip seals, 116, 117, 142–3
  magnetic fluid seals, 270
  mechanical seals, 163, 206
  metal O-rings, 22–3, 24
  PTFE lined elastomer seals, 126
  static seals, 98
  surface maximum, 595–6
Tensile strength: elastomers, 374, 375
Terminology *see* Glossary
Testing:
  flange seals, 58
  methodologies, 569–70
  swell, 369, 370
Tetrafluoroethylene propylene copolymer (FEPM) (TFE/P), 390–1
Texture:
  arithmetic mean measurement, 478–9
  elastomer O-rings, 16
  lip seals, 112–13
  material ratio curves, 484–5
  measurement, 477–86
  root mean square, 479–80
  Rsk skewness, 482–4
  Rt parameter, 480–1
  Rz parameter, 481–2
  shafts, 430
  standards, 474–5
TFE/P *see* Tetrafluoroethylene propylene copolymer
Thermal expansion: elastomers, 11
Thermal fluid chambers, 209–10
Thermoplastic polyester elastomer (TPC-ET), 385
Three seals in series, 202–3
Tightening:
  bolts/nuts, 72–6, 531–3
  compression packing, 226
  patterns, 531–3
Tolerances:
  elastomer O-ring cross-sections, 14
  stack-up assessment, 447, 448
Tools: assembly procedures, 527
TORLON polymers, 402

Toroidal vortices, 184–5
Torque:
    flanges, 530–1
    magnetic fluid seals, 269
Torque discs, 76
Torque vs tension: bolting, 73–4
TPC-ET *see* Thermoplastic polyester elastomer
Trade and industry organizations, 486–8
Training, 510
Transient leakage, 426
Trax rollers, 53
Triangular grooves, 15
Tribology: carbon, 404–5
Truload, 77–8
T-seals/rings, 28–9
Tube problems, 298–9
Tungsten carbide, 408–9
Turbines: *see also* Gas turbines
Turbo machinery, 242–9
    abradable seals, 246–8
    labyrinth arrangements, 244–5
    leakage calculation, 245–6
    rub-tolerant polymer seals, 246
Two part piston rings, 336
TX marks, 596

UHMWPE *see* Polyethylene...
Ultrasound: extension measurement, 75
Unbalanced pusher seals, 187–8
Undefined compressed fibre gaskets, 80
U-rings:
    back-to-back seals, 302, 303
    early designs, 291–2
    elastomers, 30
    lips, 291–2
    plastic, 153–4
    PTFE, 152
U-seals:
    configurations, 283–4
    pneumatic cylinder pistons, 318, 319
    reciprocating seals, 283–4
U sections: spring energized, 43
U-springs, 283–4

Vacuum, 126
    automobiles, 126
    magnetic fluid seals, 267, 270
    metal O-rings, 22–3
    rotary seals, 273, 275
    sealing, 22–3

Valves:
    BAT guidance notes, 559–72
    BAT recommendations, 566–7
    braided packing, 560
    compression packing, 559–64
    control type, 569
    control valve packing, 237
    current emission levels, 564
    emerging technologies, 569–72
    emission levels, 564–5
    exfoliated graphite material, 234–5
    leakage, 515, 516, 559
    live loading, 564, 565–6
    low emission sets, 234–5
    metal bellows, 353–5
    on-off type, 568
    packing types, 233–7
    process valves, 355–6
    relative costs, 567–9
    stems, 354–5
Vane height effects, 265, 267
Vapour: housing design, 183–4
Vapour pressure: rotary seals, 274, 275
Vehicle door window seals, 42–3
Velocity:
    plastic U-rings, 153–4
    see also Speed
Venting
    metal O-rings, 21, 22
    piston seals, 302, 303
Vermiculite gaskets, 83–4
Viscoseals, 258–62, 446–7
    clearance ratios, 260–1
    nomenclature, 260
    performance, 260–1
    screw threads, 258–9
    sealing coefficients, 260–1
Viscosity, 119–20, 295, 297
Viton fluoroelastomer, 389
Volatile organic compounds (VOCs), 506, 507–8, 517–18
Vortex flow, 185
V-ring seals, 134–6
    air purge labyrinths, 242, 244
    clamping bands, 136
    power loss, 135

Washers, 31–2, 120–1, 123
Wastewater pumps, 265
Water absorption: polymers, 402

Water cooling: ferrofluid magnetic seals, 270
Water pumps:
  automotives, 165, 166, 189
  dead end single seals, 186–7
  see also Pumps
Water turbines, 158–9, 213–14
Wavy profile grooves, 151
Wavy spring seals, 170
Wear:
  heels, 447–8
  lip seal failure, 434–5
  magnetic fluid seals, 269
  shafts, 436–7
Wearout: mechanical seals, 442
Wear resistant coatings, 324–5
Wear resistant inserts, 324

Welded metal bellows, 170–2, 356
Wells: debris, 209, 211
Wide footprint seals, 157
Wide parallel bores, 184–5
Windback seals *see* Viscoseals
Window seals, 39–43
Wipers, 310–11
  double acting, 311, 312
  housings, 312, 313
  lips, 316

XNBR *see* Carboxylated nitrile
X section rings, 27–8

Yarn braided packing, 560

Zones: ATEX, 582–3